Lecture Notes in Artificial Intelligence   9246

Subseries of Lecture Notes in Computer Science

LNAI Series Editors

Randy Goebel
University of Alberta, Edmonton, Canada
Yuzuru Tanaka
Hokkaido University, Sapporo, Japan
Wolfgang Wahlster
DFKI and Saarland University, Saarbrücken, Germany

LNAI Founding Series Editor

Joerg Siekmann
DFKI and Saarland University, Saarbrücken, Germany

More information about this series at http://www.springer.com/series/1244

Honghai Liu · Naoyuki Kubota
Xiangyang Zhu · Rüdiger Dillmann
Dalin Zhou (Eds.)

Intelligent Robotics and Applications

8th International Conference, ICIRA 2015
Portsmouth, UK, August 24–27, 2015
Proceedings, Part III

Springer

Editors
Honghai Liu
University of Portsmouth
Portsmouth
UK

Naoyuki Kubota
Tokyo Metropolitan University
Tokyo
Japan

Xiangyang Zhu
Shanghai Jiao Tong University
Shanghai
China

Rüdiger Dillmann
Karlsruhe Institute of Technology
Karlsruhe
Germany

Dalin Zhou
University of Portsmouth
Portsmouth
UK

ISSN 0302-9743   ISSN 1611-3349   (electronic)
Lecture Notes in Artificial Intelligence
ISBN 978-3-319-22872-3   ISBN 978-3-319-22873-0   (eBook)
DOI 10.1007/978-3-319-22873-0

Library of Congress Control Number: 2015946083

LNCS Sublibrary: SL7 – Artificial Intelligence

Springer International Publishing AG Switzerland is part of Springer Science+Business Media
(www.springer.com)

Preface

The Organizing Committee of the 8th International Conference on Intelligent Robotics and Applications aimed to facilitate interactions among active participants in the field of intelligent robotics, automation, and mechatronics. Through this conference, the committee intended to enhance the sharing of individual experiences and expertise in intelligent robotics with particular emphasis on technical challenges associated with varied applications such as biomedical application, industrial automations, surveillance, and sustainable mobility.

The 8th International Conference on Intelligent Robotics and Applications was most successful in attracting 228 submissions by researchers from 20 countries addressing the state-of-the art developments in robotics, automation, and mechatronics. Owing to the large number of valuable submissions, the committee was faced with the difficult challenge of selecting the most deserving papers for inclusion in these lecture notes and presentation at the conference. For this purpose, the committee undertook a rigorous review process. Despite the high quality of most of the submissions, a total of 172 papers were selected for publication in three volumes of Springer's *Lecture Notes in Artificial Intelligence* as subseries of *Lecture Notes in Computer Science*, with an acceptance rate of 75.4 %. The selected papers were presented at the 8th International Conference on Intelligent Robotics and Applications held during August 24–27, 2015, in Portsmouth, UK.

The contribution of the Technical Program Committee and the reviewers is deeply appreciated. Most of all, we would like to express our sincere thanks to the authors for submitting their most recent work and to the Organizing Committee for their enormous efforts to turn this event into a smoothly run meeting. Special thanks go to the University of Portsmouth for their generosity and direct support. Our particular thanks are due to Alfred Hofmann and Anna Kramer of Springer for enthusiastically supporting the project.

We sincerely hope that these volumes will prove to be an important resource for the scientific community.

June 2015

Honghai Liu
Naoyuki Kubota
Xiangyang Zhu
Rüdiger Dillmann
Dalin Zhou

Organization

International Advisory Committee

Jorge Angeles	McGill University, Canada
Tamio Arai	University of Tokyo, Japan
Hegao Cai	Harbin Institute of Technology, China
Tianyou Chai	Northeastern University, China
Jiansheng Dai	King's College London, UK
Han Ding	Huazhong University of Science and Technology, China
Toshio Fukuda	Nagoya University, Japan
Huosheng Hu	University of Essex, UK
Oussama Khatib	Stanford University, USA
Yinan Lai	National Natural Science Foundation of China, China
Zhongqin Lin	Shanghai Jiao Tong University, China
Guobiao Wang	National Natural Science Foundation of China, China
Kevin Warwick	University of Reading, UK
Bogdan M. Wilamowski	Auburn University, USA
Ming Xie	Nanyang Technological University, Singapore
Youlun Xiong	Huazhong University of Science and Technology, China
Huayong Yang	Zhejiang University, China

General Chair

Honghai Liu	University of Portsmouth, UK

General Co-chairs

Rüdiger Dillmann	Karlsruhe Institute of Technology, Germany
Jangmyung Lee	Pusan National University, Korea
Xiangyang Zhu	Shanghai Jiao Tong University, China

Program Chair

Naoyuki Kubota	Tokyo Metropolitan University, Japan

Program Co-chairs

Patrick Keogh	University of Bath, UK
Jeremy Wyatt	University of Birmingham, UK
Shengquan Xie	University of Auckland, New Zealand

Publicity Chairs

Darwin G. Caldwell   Istituto Italiano di Tecnologia, Italy
Yutaka Hata   University of Hyogo, Japan
Ning Xi   Michigan State University, USA

Award Chairs

Jianda Han   Chinese Academy of Sciences, China
Sabina Jesehke   RWTH Aachen University, Germany
Angelika Peer   University of the West of England, UK

Publication Chairs

William A. Gruver   Simon Fraser University, Canada
Jie Zhao   Harbin Institute of Technology, China
Tom Ziemke   University of Skövde, Sweden

Organized Session Chairs

Frank Guerin   University of Aberdeen, UK
Paolo Remagnino   Kingston University, UK
Bram Vanderborght   Vrije Universiteit Brussel, Belgium
Zhouping Yin   Huazhong University of Science and Technology,
                            China

Organizing Committee Chairs

Kaspar Althoefer   King's College London, UK
Angelo Cangelosi   University of Plymouth, UK
Shengyong Chen   Zhejiang University of Technology, China
Feng Gao   Shanghai Jiao Tong University, China
Ping Li   Liaoning Shihua University, China
Robert Riener   ETH Zurich, Switzerland
Chunyi Su   Concordia University, Canada
Caihua Xiong   Huazhong University of Science and Technology,
                            China

Technical Theme Committee

Chris Burbridge   University of Birmingham, UK
Jiangtao Cao   Liaoning Shihua University, China
Charles Fox   University of Sheffield, UK
Lars Kunze   University of Birmingham, UK
Takenori Obo   University of Malaya, Malaysia

Takahiro Takeda   Tokyo Metropolitan University, Japan
Xiaolong Zhou   Zhejiang University of Technology, China
Yahya Zweiri   Kingston University, UK

Local Arrangements Chairs

Zhaojie Ju   University of Portsmouth, UK
Hui Yu   University of Portsmouth, UK

Secretariat

Dalin Zhou   University of Portsmouth, UK

Contents – Part III

Mobile Robots and Intelligent Autonomous Systems

Intelligent System and Cybernetics

Stiffening Mechanisms of Soft Robots

Robot Mechanism and Design

Robotic Vision, Recognition and Reconstruction

Active Control in Tunneling Boring Machine

Industrial Robot and Its Applications

Mobile Robots and Intelligent Autonomous Systems

Autonomous Frontier Based Exploration
for Mobile Robots

K. Verbiest[✉], S.A. Berrabah, and E. Colon

Department of Mechanical Engineering, Royal Military Academy,
Av. de la Renaissance 30, 1000 Brussels, Belgium
{kristel.verbiest,sidahmed.berrabah,eric.colon}@rma.ac.be

Abstract. Autonomous exploration of an unknown environment by a mobile robot can be beneficial as robots can navigate in unknown environments without maps being supplied. Also it provides the possibility to produce maps without human interaction. Frontier based exploration is used here as the method for exploration. Some simulation and real world results with a Pioneer 3-AT are presented and discussed. Both simulation and real world experiments use ROS.

1 Introduction

Most mobile robots require some sort of map when navigating and performing tasks in unstructured environments. Usually the territory is mapped in advance by some form of human interaction. As a result, most navigating robots become useless when placed in unknown environments. While many robots can navigate using maps, and some can map what they can see, few can explore autonomously beyond their immediate surroundings. Exploration has the potential to free robots from this limitation. Exploration is defined as the act of moving through an unknown environment and building a map that can be used for subsequent navigation. A good exploration strategy is one that generates a complete or nearly complete map in a reasonable amount of time.

Using ROS and some of its packages, exploration with a Pioneer 3-AT robot is performed in an unknown environment; an overview is given in section 2. The approach used here is based on the detection of frontiers, regions on the border between open space and unexplored space. The methods of the ROS exploration packages employed here are discussed in more depth in section 3. In section 4 some results in simulation and real world experiments are shown.

2 ROS

There are several frameworks available to develop robot software. ROS [1] (Robot Operating System) is run here on the robot and a remote base station overseeing activities. It offers support for various robot hardware platforms and sensors. ROS provides standard operating system services such as hardware abstraction, low-

© Springer International Publishing Switzerland 2015
H. Liu et al. (Eds.): ICIRA 2015, Part III, LNAI 9246, pp. 3–13, 2015.
DOI: 10.1007/978-3-319-22873-0_1

level device control, implementation of commonly-used functionality, message-passing between processes, and package management. It is based on a graph architecture where processing takes place in nodes that may receive, post and multiplex sensor, control, state, planning, actuator and other messages. The library is geared toward a Unix-like system. ros-pkg consist of user contributed packages that implement functionality such as simultaneous localization and mapping, planning, perception, simulation etc.

For mapping the environment the *gmapping* package is used. The *gmapping* package [2] contains a ROS wrapper for GMapping [3]. GMapping is a highly efficient Rao-Blackwellized particle filer to learn grid maps from laser range data. The *gmapping* package provides laser-based SLAM (Simultaneous Localization and Mapping). Using *gmapping*, we can create a 2-D occupancy grid map from laser and pose data collected by a mobile robot. The map is created incrementally and in real-time during the robot motion.

For navigation purposes the *move_base* package [4] provides an implementation to move the robot to a goal point. The *move_base* node links together a global and local planner to accomplish its global navigation task. The global and local planner can be substituted by other planner implementations; they just have to adhere to interfaces laid out in the *move_base* package. The global planner provides a path and the local planner will send appropriate velocity and angular commands to move the robot over the current segment of the global path taking into account incoming sensor data.

For exploration two available ROS packages are used. The *explore* package [5] and the *hector_exploration_planner* [6] package. The *explore* package is a frontier-based exploration library; it generates goals for the *move_base* package. It actively builds a map of its environment by visiting unknown areas. The *hector_exploration_planner* is based on the exploration transform approach presented in [7]. *hector_exploration_planner* provides a planner that generates both goals and associated paths for the exploration of unknown environments.

3   Frontier Exploration

3.1   Method

The goal of exploration is to produce a map of an unknown environment. The method consists out of finding frontiers. Frontiers are the boundary between explored and unexplored space. During exploration, robots navigate to frontiers [8] to extend the known area until the complete (reachable) environment is explored. Which frontier to go to next can be controlled based on the cost of a frontier. The path is planned to the next goal frontier. The robot moves towards this goal autonomously while avoiding obstacles on its way. As the robot continues to move, new information about the environments comes in, leading to the formation of new frontiers. If there are no frontiers left, exploration is complete and the area is mapped.

In both the *explore* package and the *hector_exploration_planner* package occupancy grids are used as an input. In order to get new information, one must go to a frontier that separates known from unknown regions, see figure 1. Such a frontier is a cell in the occupancy grid that is marked as *free* but has a neighboring cell that is marked *unknown*. A segment of adjacent frontier cells is considered as a potential target if it is large enough so that the robot could pass it. If more than one potential target is detected in the occupancy grid, then the frontier which has the lowest cost associated with it is selected.

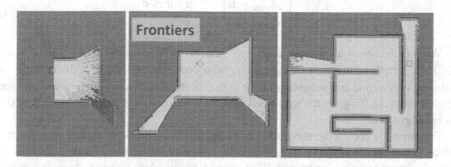

Fig. 1. Screenshots of an exploration missions in the ROS Rviz simulator. The current frontier cells are marked in red.

3.2 Explore package

The explore package returns a list of frontiers, sorted by the planners estimated cost to visit each frontier. The frontiers are weighted by a simple cost function (1), which will prefer frontiers that are large and fast and easy to travel to. The cost of a goal frontier depends on several factors: the distance between the robot and the frontiers, change in orientation to face the frontier and expected information gain of the frontier.

$$cost = weight_1.distance$$

$$+ weight_2.orientationChange$$

$$- weight_3.frontierSize \qquad (1)$$

3.3 Hector_exploration_planner Package

The exploration transform is used in the *hector_exploration_planner* package. Based on the knowledge of the environment that the robot has already acquired, the algorithm calculates a path to the next interesting frontier. This approach takes into account the distance to the next frontier and the difficulty of the path for the robot. Those difficulties can result from narrow passages but also from wide, open spaces where the sensors cannot detect any landmark.

The *Exploration Transform* Ψ of a cell c to reach a target cell c_g is defined as follows:

$$\Psi(c) = \min_{c_g \in F} \left(\min_{c \in \chi_c^{c_g}} \left(l(C) + \alpha \sum_{c_i \in C} c_{danger}(c_i) \right) \right) \qquad (2)$$

$$c_{danger}(c_i) = \begin{cases} (d_{min} - d)^2, & if\ d \le d_{min}) \\ 0 &, \quad else \end{cases} \qquad (3)$$

or

$$c_{danger}(c_i) = \begin{cases} (d_{min} - d)^2, & if\ d \le d_{min} \\ (d_{opt} - d)^2, & else \end{cases} \qquad (4)$$

With F the set of all frontier cells, $\chi_c^{c_g}$ the set of all possible paths from c to c_g, $l(C)$ the length of the path C, $c_{danger}(c_i)$ the cost function for the 'discomfort' of entering cell c_i, and α a weighting factor ≥ 0. d is the distance between cell c_i and the closest obstacle. To go to the target cell from any free cell, it is sufficient to follow the steepest gradient.

Equation (3) forces the robot to stay away from obstacles, no matter how far the robot gets away from them. This bears the risk that the sensors (with their limited range) cannot see any landmarks. This should be avoided, because the robot needs landmarks for localizing itself and for extending the map. Therefore, the cost function is extended in a way that the robot is encouraged to stay away from obstacles at a distance of d_{opt} and never gets closer than d_{min}, see equation (4).

Among the parameters that will influence how the environment is explored are the following:

- **inflate obstacles** (true / false)

  Inflate the detected obstacles with dimension of the robot in the path planning space to avoid collisions.

- **plan in the unknown** (true / false)

  Planned paths can go through unknown space.

- **wall following** (true / false)

  With wall following option set, the path planned for the robot will be less costly when at an optimal distance from obstacles.

- **security constant α**

  α is a weighing factor ≥ 0 that determines how safe the path is. Close proximity to obstacles is punished with higher value of α. This gives the possibility to avoid narrow passages.

- **goal angle penalty**

The robot is encouraged to stay on its current path, to avoid unnecessary turning. A cost is added to frontiers which require the robot to turn; this cost is proportional to the angle difference. The closer the frontier to the robot, the harder it will be refrained from turning.

- **minimum obstacle distance**

The minimum distance needed between the robot and obstacles to have a safe path. Paths planned for the robot will be less costly if they exceed this specified distance from the obstacles.

4   Results

4.1   Simulation Worlds

Simulation experiments are performed with the help of the 2D mobile robot simulator Stage [9]. Stage simulates a world as defined in a .*world* file. This file tells stage everything about the world, from obstacles, to robots and other objects. The simulation worlds used here are shown in figure 2.

4.2   Real Worlds

The real world experiments are performed in office corridors and labs with the Pioneer 3-AT robot, shown in figure 3. The Pioneer 3-AT is a 4 wheel drive robot, equipped with an embedded PC and 2D SICK LMS 200 Laser. The scanner measures distances in a plane up to 80m (cm accuracy) with an angular resolution of 0.5–1°. The field of view is 180° resulting in 181–361 range measurements.

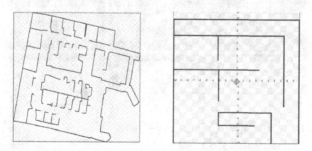

Fig. 2. Screenshots showing the simulation worlds in stage.

Fig. 3. Pioneer3-AT robot, equipped with 2D SICK LMS 200 laser.

4.3   Explore Package Results

In Figures 4 and 5 respectively the construction of the map during exploration with the *explore* package are illustrated by the image sequences obtained in simulation and the real world. The robot gradually explores the environment by planning paths to the generated goals and moving towards them while avoiding obstacles until exploration is finished.

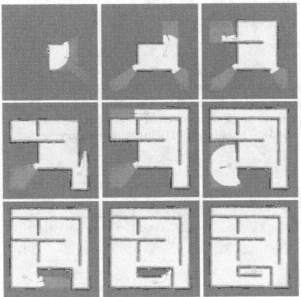

Fig. 4. Consecutive stages in frontier exploration mission in simulation.

The default local planner in ROS has been substituted by our own local planner which adheres to the *move_base* interfaces; as the currently available ROS planners are error prone. The *gmapping* package is used to produce the maps.

A tradeoff can be made between possible information gain of the frontier and the location and orientation of the frontier with respect to the robot's current position.

By varying the weights (1) of the different costs the robot can display different behavior while exploring the scene.

When two frontiers have more or less the same cost, it is possible that the robot will alternate between the two of them. By penalizing changes to the robot's current orientation, alternating between two frontiers can be reduced. Code has been added to take the distance between the robot and the frontier into account when adding extra costs due to orientation change. For frontiers in close proximity to the robot, the cost for turning is higher.

If the focus is on information gain, the robot explores the frontiers that gain the most information first. This is suited for an exploration mission that needs as much information as possible in a limited time frame. If the mission is to explore the whole environment, it is not needed to visit high information yielding frontiers first, as this will increase total exploration time and travel cost. Also, frontiers with high gain factor have no guarantee to produce the most information gain.

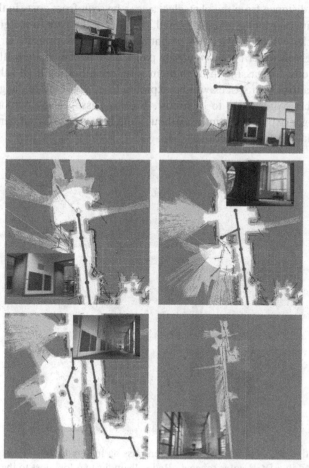

Fig. 5. Consecutive stages in frontier exploration mission in real world.

4.4   Hector_exploration_planner Package Results

The consecutive screen shots in figure 6 and 7 respectively show an exploration mission with the exploration transform in simulation and in real world. The exploration transform is visualized in a RGB color scheme. Red cells indicate a low cost distance to the next safe frontier; violet/blue cells indicate a high cost. The robot gradually explores the environment by planning paths to the generated goals and moving towards them while avoiding obstacles.

Instead of using the global planner in *move_base*, the trajectory is obtained from the steepest descent of the exploration transform, allowing several cost factors to come into play. The default local planner in ROS has been substituted by our own local planner which adheres to the *move_base* interfaces; as the currently available ROS planners are error prone. The *gmapping* package is used to produce the maps.

Figure 8 shows a comparison of an exploration mission where the wall following option is set and not set. If the wall following is enabled, the robot is encouraged to stay at optimal distance from the obstacles, as can be seen from costs in the figure.

When security constant α is set too small, the goal or the path to the goal can be too close to obstacles which can lead to potential collisions. Setting lower values for α will generally give shorter but more unsafe paths. Also narrow paths will no longer be avoided as the cost for passing through them is reduced. Setting high values for α will produce safe paths. Open spaces will be explored first, as the more dangerous narrow passages will be avoided due to high cost. Narrow passages are explored last. This can lead to longer exploration times. Although the behavior to avoid narrow passages can be beneficial in some scenarios.

Fig. 6. Consecutive stages in frontier exploration mission in simulation. The exploration transform is visualized in a RGB color scheme. Red cells indicate a low cost to the next safe frontier, violet/blue cells indicate a high cost.

Setting the security constant α high will prevent the robot going through narrow spaces. But with high enough goal angle penalty, if the hallway is not too narrow it will finish the hallway before switching to other frontiers.

Figure 9 shows the traversed path of the robot with and without wall following for the particular simulation world in figure 6. The effect of the wall following option and the security distance on the exploration time is shown in figure 10. Exploration time increases with security distance. When the wall following option is set the exploration time is shorter for smaller security distances and longer for larger security distance. Of course a lot of other factors influence the results.

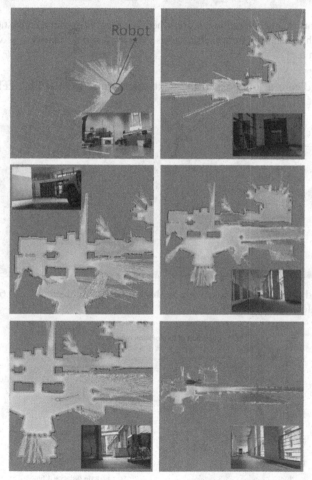

Fig. 7. Exploration of ground level building H of RMA. The exploration transform is visualized in a RGB color scheme. Red cells indicate a low cost to the next safe frontier, violet/blue cells indicate a high cost.

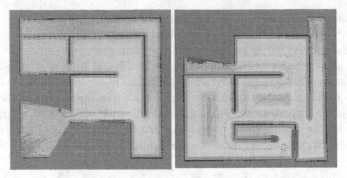

Fig. 8. Screenshots of an exploration mission where the robot is encouraged to stay at safe (left) and optimal (right) distances from the obstacles. The exploration transform is visualized in a RGB color scheme.

Red cells indicate a low cost to the next safe frontier, violet/blue cells indicate a high cost.

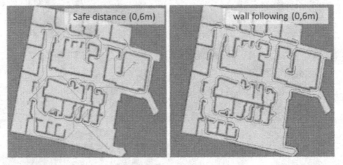

Fig. 9. Traversed path in simulation, with on the left for a safe distance of 0.6m, and on the right for wall following option with 0.6m.

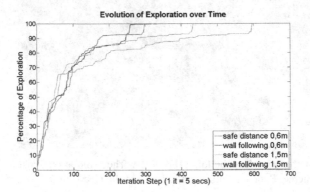

Fig. 10. Evolution of Exploration over time for different values of the security distance and wall following option for the specific case shown in figure 8.

5   Conclusion

In comparison with the *explore* package, the exploration transform method provides more options to manipulate the path the robot will follow by introducing extra costs with certain behavior. The used algorithms offer a reliable solution for autonomous exploration with a single mobile robot. They can cope with moving obstacles and replan when necessary. The next step is to let a multi-robot team explore and map an unknown environment. Each robot will produce partial maps; these maps of the individual robots will be merged in one global map. Based on this global map and the current positions of the robots in the team, it will be decided where each robot will go next to continue exploration.

References

1. Quigley, M., Gerkey, B., Conley, K., Faust, J., Foote, T., Leibs, J., Berger, E., Wheeler, R., Ng, A.: ROS: an open-source robot operating system. In: Proc. IEEE Int. Conf. on Robotics and Automation (ICRA) - Workshop on Open Source Robotics, ROS (Robot Operating System), May 2009. http://www.ros.org/
2. http://wiki.ros.org/gmapping
3. Grisetti, G., Stachniss, C., Burgard, W.: Improving grid-based slam with rao-blackwellized particle filters by adaptive proposals and selective resampling. In: Proc. of the IEEE International Conference on Robotics and Automation (ICRA) (2005)
4. http://wiki.ros.org/move_base, http://wiki.ros.org/navigation
5. http://wiki.ros.org/explore
6. Kohlbrecher, S., Meyer, J., Graber, T., Petersen, K., Klingauf, U., von Stryk, O.: Hector open source modules for autonomous mapping and navigation with rescue robots. In: Behnke, S., Veloso, M., Visser, A., Xiong, R. (eds.) RoboCup 2013. LNCS, vol. 8371, pp. 624–631. Springer, Heidelberg (2014)
7. Wirth, S., Pellenz, J.: Exploration transform: a stable exploring algorithm for robots in rescue environments In: IEEE International Workshop on Safety, Security and Rescue Robotics (SSRR), pp. 1–5 (2007)
8. Yamauchi, B.: A frontier-based approach for autonomous exploration. In: Proc. IEEE Int. Symp. Computational Intelligence in Robotics and Automation (CIRA), July 1997
9. Gerkey, B., Vaughan, R.T., Howard, A.: The player/stage project: tools for multi-robot and distributed sensor systems. In: Proceedings of the 11th International Conference on Advanced Robotics (ICAR 2003), Coimbra, Portugal, pp. 317-323, June 2003. http://playerstage.sourceforge.net/

Gene Regulatory Networks with Asymmetric Information for Swarm Robot Pattern Formation

Shuai Zhang, Xingguang Peng$^{(\boxtimes)}$, Yunke Huang, and Panpan Yang

School of Marine Science and Technology, Northwestern Polytechnical University,
Xi'an 710072, Shaanxi, The People's Republic of China
pxg@nwpu.edu.cn

Abstract. Gene Regulatory Networks (GRNs) play a central role in understanding natural evolution and development of biological organisms from cells. In this paper, inspired by limited neighbors' information in the real environment, we propose a GRN-based algorithm with asymmetric information for swarm-robot pattern formation. Through this algorithm, the neighbors' information will be only used once, swarm robots can collect limited neighbors' information to self-organize autonomously to different predefine shapes. Furthermore, a discrete dynamic evolvement model of cellular automaton of pattern formation is provided to demonstrate the efficiency and convergence of the proposed method. Various cases have been conducted in the simulation, and the results illustrate the effectiveness of the method.

Keywords: Gene regulatory networks · Swarm robots · Pattern formation · Asymmetric information · Cellular automaton

1 Introduction

A very simple rule has been revealed again and again by the nature that extremely complicated phenomenons can emerge from simple agents with limited interactions. In order to reveal the mechanism of complexity emerging from interactions of simple agents, some concepts have been proposed inspired by the behaviors of ant colony, school of fish, flock, etc, just like Swarm Intelligence, Synergetics, Artificial Intelligence, Self-Organized Network, Evolutionary Learning, Complexity [1], [2], [3], [4]. The solution to this problem, in a sense, may be a candidate to explain the origin of life.

Pattern formation is a challenging part of this area. Various shapes can be self-organized generated with no central controller under natural conditions. In order to reveal the mechanism of pattern formation in the nature and to apply to swarm-robot pattern formation, many methods have been explored, such as: the L-systems and iterated function systems in fractal theory [5], cellular automaton modeling of biological pattern formation [6], using morphogen gradient [7], leader following algorithms [8], [9], potential field algorithms [10], [4], gene regulatory networks for swarm-robot pattern formation [11], [12], [13], [14], [15]. Challenges and classifications of pattern formation in existing literature are reviewed in [16].

© Springer International Publishing Switzerland 2015
H. Liu et al. (Eds.): ICIRA 2015, Part III, LNAI 9246, pp. 14–24, 2015.
DOI: 10.1007/978-3-319-22873-0_2

On the other hand, the existence of asymmetric information is a common phenomenon in the nature. Taking an example of visual, almost all animals have asymmetric eye positions, in other words, animals can not get neighbors' information all around, visual information is obtained asymmetrically or limitedly. Based on biological evolutionary theory, we can figure out that animals obtain asymmetric information on purpose and have asymmetric methods to deal with these asymmetric information. This strategy using asymmetric information means agent needs collecting less information and avoids information redundancy problem. From the point of view of whole population, strategy using asymmetric information have contributed to the development and evolution of the population.

Taking an example of robots, when a robot deals with its neighbors' information, if the information have been used by its neighbors, these information will not be used again by the robot, the robot uses the rest of neighbors' information to compute its next time position. In other words, the information between two neighbors is only used once.

In this paper, we use gene regulatory networks with asymmetric information to study the influence of asymmetric information on swarm-robot pattern formation. A discrete dynamic evolvement model of cellular automaton of pattern formation is proposed to demonstrate the converge efficiency and numerical simulations confirm the effectiveness of the proposed model.

The rest of this paper is organized as follows. Section 2 presents a problem statement including the definitions of symmetric information and asymmetric information. Section 3 presents two cellular automaton models of pattern formation in swarm-robot system, consisting of a symmetric information model and an asymmetric information model. Section 4 introduces a gene regulatory network algorithm with asymmetric information for pattern formation. Section 5 presents numerical simulations containing pattern formation with symmetric information and asymmetric information in different initial condition. Conclusions and future work are given in Section 6.

2   Problem Statement

As our starting point, we propose the following definitions for the strategy using neighbors' information.

Definition 1. *Symmetric information is a kind of information that is public to all of its participators, each of the participators will use this information to make decisions.*

Definition 2. *Asymmetric information is a kind of information that is not public to all its participators, only part of the participators will use this information to make decisions.*

This paper considers the problem of how swarm robots self-organize to different predefine shapes driven by gene regulatory networks with asymmetric neighbors' information. It is assumed that global 2D position is available for the robots, and robots can only detect single directional neighbors. In other words, the robots have vision blindness.

3 Cellular Automaton Models of Pattern Formation in Interacting Cell System

Cellular automaton is a discrete dynamic system. It has no central controller and is rule-based evolvement model, usually used to simulate the natural phenomenons. It has become paradigms of self-organized complex systems in which collective behaviors arise from simple interaction rules.

The following two CA models give a strong confirmation that symmetric neighbors' information and asymmetric neighbors' information can both guide swarm robots to evenly target shapes.

The game of one-dimensional pattern formation is assumed . There are seven robots which are too close for each other in a line. They need using their neighbors' information to self-organize to a evenly line. Table. 1 presents initial position and final position of these seven robots.

Table 1. One-dimensional pattern formation game

Initial position	1 2 3 4 5 6 7
Final position	1 2 3 4 5 6 7

3.1 Cellular Automaton Model of Pattern Formation with Symmetric Neighbors' Information

Neighbors' position information in both sides is collected by each robot. That is to say the information between two neighbors is symmetric information and will be used duplicated twice. Evolution rules are as follows:

(1) The robot can detect neighbors in both sides.

(2) The robot just moves one grid or keeps still during one time step.

(3) If there are two neighbors in both sides, the robot will keep still. If there is one neighbor in one of the sides, the robot will move to the opposite side for one grid. If two robots occupy one grid, both two will leave away this grid at next time step.

Table. 2 presents the whole evolution process of seven robots with symmetric information from the initial position to the final position. It is easy to see that robot 2 and 3 occupy the same grid at time step 3, the same thing happens to robot 4 and 5. This situation should be avoided because of severe collision.

3.2 Cellular Automaton Model of Pattern Formation with Asymmetric Neighbors' Information

Neighbors' position information is detected single-directly for some environment reasons or hardware limitations. That means the information between two neighbors is asymmetric information and should be only used once. In the following

Table 2. CA model with Symmetric information

```
Initial position |    1 2 3 4 5 6 7
       1         |  1   2 3 4 5 6    7
       2         |  1 2    3 4 5    6 7
       3         |  1    23  4  56    7
       4         |  1 2    3 4 5   6   7
       5         |  1 2 3   4   5 6 7
       6         |  1 2      3 4 5     6 7
       7         | 1   2 3   4   5 6    7
       8         | 1 2      3 4 5    6   7
  Final position | 1  2   3   4   5   6   7
```

asymmetric model, each robot can only detect right side neighbors. Evolution rules are as follows:

(1) The robot can only detect right side neighbors.

(2) The robot just moves one grid or keeps still during one time step.

(3) If there is a right neighbor, the robot will move one left grid next time step. If there is no right neighbor, the robot will keep still next time step.

Table. 3 presents the whole evolution process of seven robots with asymmetric information from the initial position to the final position. we can draw the conclusion that asymmetric information can also guide swarm robots to a evenly target shape. This method needs fewer time steps and has no severe collision, that means asymmetric information has better converge efficiency.

Table 3. CA model with asymmetric information

```
Initial position |        1 2 3 4 5 6 7
       1         |        1 2 3 4 5 6    7
       2         |        1 2 3 4 5   6   7
       3         |        1 2 3 4  5   6 7
       4         |        1 2 3  4   5   6 7
       5         |       1 2   3   4   5 6 7
  Final position | 1   2   3   4   5   6   7
```

4 GRN Model with Asymmetric Information

The dynamics of the GRN for multi-robot construction can be described by the following equations[14]:

$$\frac{dg_{i,x}}{dt} = -a \cdot z_{i,x} + m \cdot p_{i,x} \tag{1}$$

$$\frac{dg_{i,y}}{dt} = -a \cdot z_{i,y} + m \cdot p_{i,y} \tag{2}$$

$$\frac{dp_{i,x}}{dt} = -c \cdot p_{i,x} + k \cdot f(z_{i,x}) + b \cdot D_{i,x} \tag{3}$$

$$\frac{dp_{i,y}}{dt} = -c \cdot p_{i,y} + k \cdot f(z_{i,y}) + b \cdot D_{i,y} \tag{4}$$

where $g_{i,x}$ and $g_{i,y}$ denote the x-axis position and y-axis position of robot i respectively. $p_{i,x}$ and $p_{i,y}$ denote the velocity-like property of robot i along the x-axis and y-axis respectively. $z_{i,x}$ and $z_{i,y}$ are the gradients which carry the information of target shape. $f(z_{i,x})$ and $f(z_{i,y})$ are sigmoid functions.

where $D_{i,x}$ and $D_{i,y}$ are the sum of neighbors' information, neighbors' distance information is collected to avoid collision in this paper. We have two strategies using neighbors' information as proposed in section 3. We have strategy using symmetric information :

$$D_{i,x} = \sum_{j=1}^{N_i} D_{i,x}^j \quad D_{i,y} = \sum_{j=1}^{N_i} D_{i,y}^j \tag{5}$$

and strategy using asymmetric information:

$$D_{i,x} = \sum_{j=1}^{i-1} D_{i,x}^j \quad D_{i,y} = \sum_{j=1}^{i-1} D_{i,y}^j \quad 1 \leq j \leq N_i \tag{6}$$

where N_i denotes the number of neighbors of robot i, and $D_{i,x}^j$ and $D_{i,y}^j$ are the distance function between robot i and robot j, which is defined as

$$D_{i,x}^j = \frac{(g_{i,x} - g_{j,x})}{\sqrt{(g_{i,x} - g_{j,x})^2 + (g_{i,y} - g_{j,y})^2}} \tag{7}$$

$$D_{i,y}^j = \frac{(g_{i,y} - g_{j,y})}{\sqrt{(g_{i,x} - g_{j,x})^2 + (g_{i,y} - g_{j,y})^2}} \tag{8}$$

Under the strategy using asymmetric information, robots are numbered clockwise and the robot on boundary $(-1, 0)$ is numbered 1 as show in Fig. 1. Robots can only detect low-number direction neighbors, so neighbors' information is only used once. For example, robot 6 can only select its neighbors from robot 1, 2, 3, 4 and 5.

Since the unit circle is a closed curve, the 1-st ($i = 1$) robot is treated in a special way to satisfy the boundary condition. Specifically 1-st robot uses all its neighbors' information.

Mathematically, two distance function matrixes with symmetric information and asymmetric information present the differences between two strategies clearly. Symmetric matrix:

$$\begin{pmatrix} 0 & D_{12} & D_{13} & \cdots & D_{1n} \\ D_{21} & 0 & D_{23} & \cdots & D_{2n} \\ D_{31} & D_{32} & 0 & \cdots & D_{3n} \\ \vdots & \vdots & \vdots & 0 & \vdots \\ D_{n1} & D_{n2} & D_{n3} & \cdots & 0 \end{pmatrix} \quad D_{ij} = -D_{ji} \quad D_{ii} = 0 \tag{9}$$

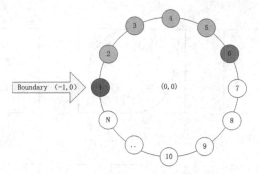

Fig. 1. The robots are numbered clockwise.

Asymmetric matrix or lower triangular matrix:

$$\begin{pmatrix} 0 & D_{12} & D_{13} & \cdots & D_{1n} \\ D_{21} & 0 & 0 & \cdots & 0 \\ D_{31} & D_{32} & 0 & \cdots & 0 \\ \vdots & \vdots & \vdots & 0 & \vdots \\ D_{n1} & D_{n2} & D_{n3} & \cdots & 0 \end{pmatrix} \qquad D_{ij} = -D_{ji} \quad D_{ii} = 0 \qquad (10)$$

The system's convergence to the target shape is proved according to the lyapunov theory [12]. The five parameters in the main system dynamics are optimized in [17].

5   Numerical Simulation

To evaluate the reliability and efficiency of the proposed methods, we perform a set of simulations using MATLAB.

In order to evaluate the evenness of the target shapes, we define the variance:

$$s^2 = \frac{\sum\limits_{i=1}^{n}(d_i - d_0)^2}{n} \qquad (11)$$

where d_i denotes distance between robot i and robot $i+1$ and d_0 denotes the expected distance value. The parameter s^2 should be as small as possible. In the following two cases, we assume that $s^2 = 0.001$ means that the uniform target shape is accomplished.

In order to ultimate uniform distribution of the robots, we define robots' neighbor range to be $d = \frac{L_{edge}}{N}$, where L_{edge} refers to the length of the target shape and N refers to the number of robots [18].

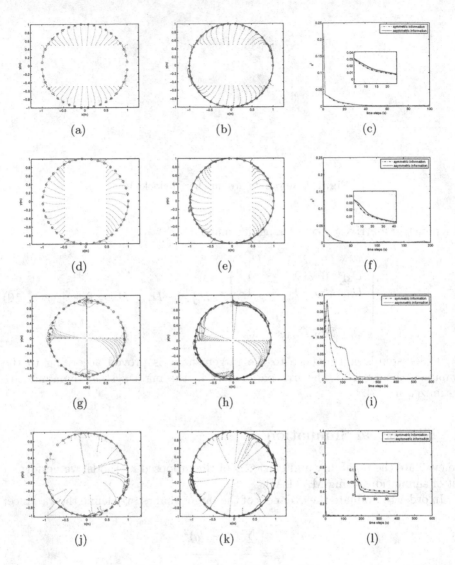

Fig. 2. The trajectories of 40 robots using GRN method to construct to a unit circle under different initial conditions. Initial position is denoted as a dot where final position is denoted as a small circle. (a) robots use symmetric information from two horizontal lines; (b) robots use asymmetric information from two horizontal lines; (c) s^2-time comparison curves from two horizontal lines; (d) robots use symmetric information from two upright lines; (e) robots use asymmetric information from two upright lines; (f) s^2-time comparison curves from two upright lines; (g) robots use symmetric information from two vertical lines; (h) robots use asymmetric information from two vertical lines; (i) s^2-time comparison curves from two vertical lines; (j) robots use symmetric information from a random square region; (k) robots use asymmetric information from a random square region; (l) s^2-time comparison curves from a random square region;

5.1   Case Study 1: Converge to a Unit Circle from Initial Symmetric Position

In this case, we deploy 40 robots to a unit circle from four different symmetric initial positions. Both strategies using neighbors' information are performed.

Fig. 2 shows that both strategies using symmetric and asymmetric information can guide the swarm robots to predefine target shapes. From the point of view of the converge time and efficiency, the four s^2-time comparison curves with different initial symmetric conditions show that there is no significant difference between two strategies under the same initial symmetric condition, but different initial conditions will lead to different converge time and efficiency no matter which strategy the robots uses.

5.2   Case Study 2: Converge to a Unit Circle from Initial Asymmetric Position

In this case, we deploy 40 robots to a unit circle from four different asymmetric initial positions. Both strategies using neighbors' information are performed.

Fig. 3 shows that if the robots are deployed to asymmetric shapes initially, the converge time is extended. Besides, we can figure it out that under the same initial asymmetric condition, the converge time of asymmetric strategy is largely shorter than that of symmetric strategy, that means the symmetric strategy needs less information but has better converge efficiency. Since the robots are always asymmetrically deployed in the real environment, the conclusion have a valuable meaning in application.

5.3   Results Analysis

Why the converge time of asymmetric strategy is largely shorter than that of symmetric strategy? In fact, robot collecting symmetric information moves very little or even don't move at each time step because of neutralization of distance information from its symmetric neighbors, while robot collecting asymmetric information doesn't face this problem, so it moves longer at each time step. Fig. 4(a) shows that under two horizontal lines initial symmetric position, total distance at each time step is not largely different, so the converge time is approximate. Fig. 4(b) shows that under one horizontal line asymmetric initial position, before 200 time step, total distance with asymmetric strategy is larger than that of symmetric strategy, so the converge time of asymmetric strategy is largely shorter than that of symmetric strategy.

5.4   Problem and Shortcoming

There are still many problems and shortcomings. As we can see from Fig. 2 and Fig. 3, the trajectories of second column of figures are more cluttered than that of first column. Fig. 3(e) and Fig. 3(k) show that there is a black regiment near the boundary position. In fact, when most robots have converged to the

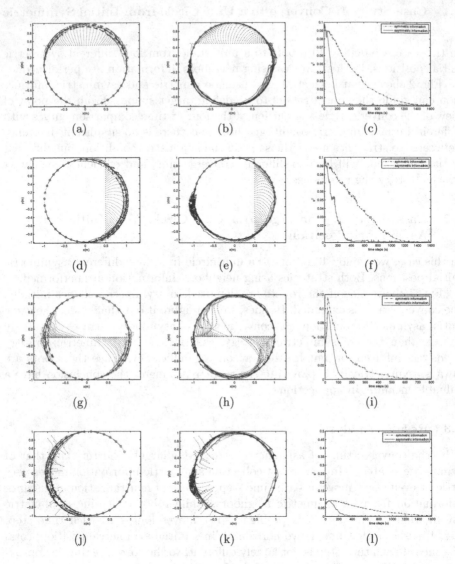

Fig. 3. The trajectories of 40 robots using GRN method to construct to a unit circle under different initial conditions. Initial position is denoted as a dot where final position is denoted as a small circle. (a) robots use symmetric information from one horizontal line; (b) robots use asymmetric information from one horizontal line; (c) s^2-time comparison curves from one horizontal line; (d) robots use symmetric information from one upright line; (e) robots use asymmetric information from one upright line; (f) s^2-time comparison curves from one upright line; (g) robots use symmetric information from x-axis line; (h) robots use asymmetric information from x-axis line; (i) s^2-time comparison curves from x-axis line; (j) robots use symmetric information from a random rectangle region; (k) robots use asymmetric information from a random rectangle region; (l) s^2-time comparison curves from a random rectangle region;

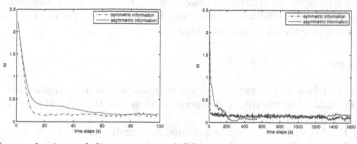

(a) two horizontal lines as initial (b) one horizontal line as initial
position   position

Fig. 4. Total distance at each time step

circle, there are still several unstable robots nearby the boundary position. These robots with their neighbors remain volatile for a long time that cause a black regiment and make trajectories more cluttered.

6   Conclusion and Future Work

In this paper, we have presented an asymmetric information-based gene regulatory network distributed control approach to multi-robot construction. Cellular automaton models of pattern formation and numerical simulation results show the effectiveness and advantages of the proposed method. The major conclusions are as follows:

(1) Both strategies using symmetric and asymmetric information can guide the robots to a predefine target shape, but the strategy using asymmetric information needs less neighbors' information.

(2) If the initial position is symmetric, there is no significant differences between two strategies, but if the initial position is asymmetric, the converge time of strategy using asymmetric information is largely shorter than that of symmetric information.

Numerical simulation results also show some problems, the trajectories under the strategy using asymmetric information are more cluttered and have longer total distance than that of symmetric information.

In the future, we will continue our research on asymmetric information and GRN-inspired multi-robot controllers. We will further investigate the proposed problems, especially boundary condition problem and use real robots to verify the effectiveness of the proposed method. We will also investigate the universality of asymmetric information in the natural world and compare the advantages and disadvantages of both strategies in detail.

Acknowledgments. This work has been supported by Graduate Starting Seed Fund of Northwestern Polytechnical University, National Nature Science Foundation of China under Grant 61105068, 61473233, Natural Science Foundation of Shaanxi Province, China under Grant 2014JQ8330, Fundamental Research Funds for the Central Universities of China under Grant 3102014ZD0042.

References

1. Merkle, D., Blum, C.: Swarm Intelligence: Introduction and Application (2008)
2. Waldrop, M.M.: Complexity: The emerging science at the edge of order and chaos. Simon and Schuster (1993)
3. Haken, H.: Erfolgsgeheimnisse der natur: synergetik, die lehre vom zusammenwirken. Deutsche Verlags-Anstalt (1981)
4. Olfati-Saber, R.: Flocking for multi-agent dynamic systems: Algorithms and theory. IEEE Transactions on Automatic Control **51**(3), 401–420 (2006)
5. Falconer, K.: Fractal geometry: mathematical foundations and applications. John Wiley and Sons (2004)
6. Deutsch, A., Dormann, S.: Cellular automaton modeling of biological pattern formation (2005)
7. Mamei, M., Vasirani, M., Zambonelli, F.: Experiments of morphogenesis in swarms of simple mobile robots. Applied Artificial Intelligence **18**(9–10), 903–919 (2004)
8. Alur, R., et al.: A framework and architecture for multirobot coordination. In: Rus, D., Singh, S. (eds.) Experimental Robotics VII. LNCIS, vol. 271, pp. 303–312. Springer, Heidelberg (2001)
9. Consolini, L., Morbidi, F., Prattichizzo, D., et al.: LeaderCfollower formation control of nonholonomic mobile robots with input constraints. Automatica **44**(5), 1343–1349 (2008)
10. Hsieh, M.A., Kumar, V., Chaimowicz, L.: Decentralized controllers for shape generation with robotic swarms. Robotica **26**(05), 691–701 (2008)
11. Jin, Y., Meng, Y.: Morphogenetic robotics: An emerging new field in developmental robotics. IEEE Transactions on Systems, Man, and Cybernetics, Part C: Applications and Reviews **41**(2), 145–160 (2011)
12. Guo, H., Meng, Y., Jin, Y.: A cellular mechanism for multi-robot construction via evolutionary multi-objective optimization of a gene regulatory network. BioSystems **98**(3), 193–203 (2009)
13. Guo, H., Meng, Y., Jin, Y.: Swarm robot pattern formation using a morphogenetic multi-cellular based self-organizing algorithm. In: 2011 IEEE International Conference on Robotics and Automation (ICRA), pp. 3205–3210. IEEE (2011)
14. Jin, Y., Guo, H., Meng, Y.: A hierarchical gene regulatory network for adaptive multirobot pattern formation. IEEE Transactions on Systems, Man, and Cybernetics, Part B: Cybernetics **42**(3), 805–816 (2012)
15. Oh, H., Jin, Y.: Evolving hierarchical gene regulatory networks for morphogenetic pattern formation of swarm robots. In: 2014 IEEE Congress on Evolutionary Computation (CEC), pp. 776–783 (2014)
16. Varghese, B., McKee, G.: A review and implementation of swarm pattern formation and transformation models. International Journal of Intelligent Computing and Cybernetics **2**(4), 786–817 (2009)
17. Guo, H.: Morphogenetic computing and reinforcement learning for multi-agent systems. Stevens Institute of Technology (2011)
18. Jin, Y., Guo, H., Meng, Y.: Robustness analysis and failure recovery of a bio-inspired self-organizing multi-robot system. In: Third IEEE International Conference on Self-Adaptive and Self-Organizing Systems, SASO 2009, pp. 154–164. IEEE (2009)

Universal Usage of a Video Projector on a Mobile Guide Robot

Ronny Stricker[✉], Steffen Müller, and Horst-Michael Gross

Neuroinformatics and Cognitive Robotics Lab,
Technische Universität Ilmenau, 98693 Ilmenau, Germany
ronny.stricker@tu-ilmenau.de
http://www.tu-ilmenau.de/neurob

Abstract. In this paper, we present a holistic approach to enable mobile robots using video projection in a situation aware and dynamic way. Therefore, we show how to autonomously detect wall segments that are suitable to be used as projection target in a dynamic environment. We derive several quality measures to score the wall segments found in the local environment and show how these scores can be used by a particle swarm optimization to find the best local projection position for the mobile robot.

Keywords: Mobile robot · Video projection · Position optimization

1 Introduction

User interaction plays a very important role for mobile service robots. It should be possible to easily use them and to get a real benefit from the offered service. However, current generations of these robots do have some shortcomings in presenting information in an appropriate and easily understandable manner. The idea of using video presenters on a mobile robot states back to the first Star Wars films and can help to improve the intelligibility of the information provided by the robot [1]. This is especially true for our tour robots Konrad and Suse that are used to guide and to tour people around in our multi-story faculty building [2]. In order to increase the acceptance of the tour guide scenario, the robots should not only talk to the user and give information on the on-board display, but also make use of walls next to the exhibit to display information. Since the exhibits can change over time, it will be beneficial if the robot can optimize its position and the wall used for projection depending on the local surroundings and the location of the current audience.

In addition to the problem of finding an appropriate position and wall segment used for projection, we also need the address the problem of projector calibration. As the robot is not usually oriented perpendicular to the wall, the projection will get distorted. This problem increases if the user itself is not oriented perpendicular to the projection wall. Therefore, we show how to implement

© Springer International Publishing Switzerland 2015
H. Liu et al. (Eds.): ICIRA 2015, Part III, LNAI 9246, pp. 25–36, 2015.
DOI: 10.1007/978-3-319-22873-0_3

two different projection modes that do rectify the image without the need of calibration images. In the first mode, the image is rectified according to the wall used for projection. In the second mode, the image is rectified according to the user position and therefore seams to be floating on the wall.

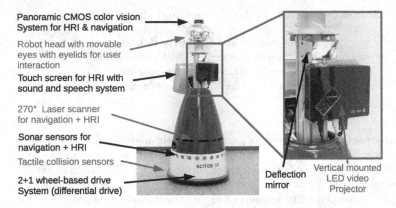

Fig. 1. Robot platform with mounted LED video projector and deflection mirror.

Fig. 1 shows the experimental platform used for our research. It is equipped with interaction devices, mainly a touch-display, as well as a couple of additional sensors enabling autonomous navigation and perception of people and obstacles in the robots environment. In addition to this regular setup, the test platform has been equipped with a LED video projector (ViewSonic PLED-W500). The projector is mounted together with a deflection mirror in a vertical position below the robot head to guarantee minimum space requirements. It does deliver 500 lumens of brightness at a maximum power consumption of 120 W, which is of course questionable for the desired field of application. However, current generations of small LED based projectors have already doubled or tripled brightness and can easily replace our projector used for demonstration purpose.

To deal with the different problems arising from the dynamic projector position optimization the paper is structured as follows: After a brief overview of the related work and the presentation of the prerequisites of our work we are giving a detailed overview of our proposed method and the involved score functions in Sec. 4. After the distortion correction has been explained in Sec. 5, we are showing some experiments and finish with a conclusion in Sec. 7.

2   Related Work

Several methods can be found, that are dealing with aspects of the problem we have described above. The largest group has been emerged during the last few years and tries to integrate a video projector onto a guide robot. One example is given in [3]. The authors are using a pan and tiltable video projector to display addition information of exhibits onto the wall. Furthermore, the projector is used to project buttons on the ground, that can be activated by means of the users feet. In [4] a guide robot is used to augment a guiding tour by projecting directly onto different exhibits in order to highlight the parts explained but also to simulate ancient computer models by projecting directly onto a switched off monitor. The authors of [5] are using the humanoid robot NAO to project information on walls in a home environment. However, the robot position as well as the projection surface are predefined in this approach. Two examples of methods combining video projection with gesture detection are given in [6,7].

Since the surfaces used for projection are predefined in all the methods stated above, none of these tries to find an optimal projection surface in the local environment dynamically. Most of the methods do even rely of fixed and predefined robot positions that are aligned perpendicular to the wall.

Current methods for camera based projection calibration do rely on predefined patterns that are displayed during initialization phase [8,9]. Since the robot changes its pose relative to the projection surface once it is moving, a closed loop algorithm, as used by other methods for projection, correction cannot be applied.

3   Prerequisites

In order to build and to explain the projector position optimization, we rely on different components that are not in the scope of this paper.

First, we are using MIRA [10] as software framework in order to combine all the different modules in an easy and efficient way. The integrated transformation framework of MIRA enables us to step back and forth between the different coordinate frames (robot frame, map frame, person frame) easily.

Second, we also need to take the user position, view direction and walk direction into account. To reliably track persons in the local environment we are using the probabilistic multi-hypotheses people detection and tracking system developed in our lab over the last eight years [11]. It is based on a 7D Kalman filter that tracks the position, velocity, and upper body orientation of the respective persons assuming an uncertain random acceleration. The tracker processes the detections of different, asynchronous observation modules – namely a 2D laser-based leg detector, a face detector, a motion detector, and an upper-body shape detector. The leg detector in its initial version is based on the boosted classifier approach of [12]. The face detection system utilizes the well-known face detector of Viola & Jones [13]. Finally, we apply an upper body shape detector based on Histograms of Oriented Gradients (HOG) [14]. A detailed description of the person detector and tracker and the tracking results of comparing evaluation studies on different data sets with increasing difficulty is given in [11].

Third, we are relying on mapping and localization algorithms proven to work robust during several years [15]. These algorithms include the generation of a local map (8x8 meters in our application) and covers the local surroundings seen by the robot so far.

4 Finding Optimal Projection Surface and Position

Finding an optimal projection position in our eyes falls into two tasks. First, the detection of walls in the local surrounding that are candidates for a projection target. Second, we need to take the user and the robot position into account to score the wall candidates in order to obtain the best projection surface. Three aspects are of importance during scoring. The wall needs to be visible to the user and should show an appropriate distance and view angle. Furthermore, it should be possible to project onto the wall. Therefore, it should show a suitable brightness and color and should not have any dominant structure (no signs or posters should cover the wall). Moreover, the wall should be in range of the projector, which again sets requirements on the distance and angle of the wall. The person visibility and wall structure related demands do not depend on the current robot position, therefore, we do relate to these requirements as *robot position independent score functions* in the remainder of this paper. In turn the other requirements are *robot position dependent*.

Walls in the local surrounding can be very long, the suitability is likely to fluctuate heavily at different positions. Therefore, we break the walls apart by dividing them into different wall segments, that are scored independently. It would be possible to use a real 3d segmentation for this step. However, it would increase the computational complexity a lot and cause only a slight benefit in our scenario since we are not able to pan or tilt the projector. Therefore, we are using wall segment with fixed width (30 cm in our application as a trade of between computational complexity and spacial resolution) and a fixed hight of 1.2 meter (1 meter above the ground) during the segmentation phase.

Having a closer look at the requirements and at the application scenario reveals that the overall problem is twofold. The first task is to find the best suited wall segment(s) for the current position of the robot. The second task is more general and comprises finding of the best robot position, so that the best wall segment(s) in the local surrounding can be used for projection (Fig. 2). The second task involves the first one, since we need to evaluate the maximum score for different robot locations and orientations in the local surrounding. Since evaluating all possible robot poses in the local surrounding is way to expensive, we apply a particle swarm optimization (PSO) [16] to find the optimal projection position. Therefore, every particle returns the score of the best segment(s) that can be used for projection from this single position.

The two tasks can be applied directly to our application scenario where we can use the PSO approach to find the best location for augmenting the exhibit presentation.

In the remainder of this section we explain the wall extraction and the different score functions in more detail.

4.1 Wall Segment Extraction

For wall extraction we do not rely on a pre-build map, since the public environment is likely to change so that walls can be covered by trolleys, posters or

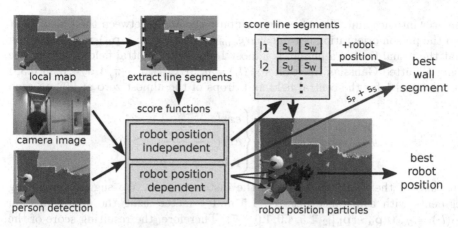

Fig. 2. Overview of process to find the best wall segment (red) and robot position (blue). The robot position independent scores need to be computed only once and can be stored with the segment. The robot position dependent scores need to be recomputed every time the robot pose changes. The image at the bottom right shows the result of a PSO with 200 particles with the best particle colored in red.

other obstacles. Therefore, we are using the local map that does contain the local surroundings as seen by the robot so far.

The wall segmentation is performed using a *Random Sample Consensus* (RANSAC) [17] based algorithm for line fitting on all the points marked as obstacle in the local map. The algorithm searches and returns the line hypothesis with the most points supporting the hypothesis. Removing the supporting points (that also determine the start and endpoint of the line) and repeating the process as long as lines with a specified support can be found, results in the extraction of the best n lines for the local surrounding which correspond to the walls. These lines are split into smaller line segments with a fixed length (30 cm in our application).

4.2 Robot Position Independent Cost Functions

The first cost functions to be discussed are independent of the robot's position and need to be computed only once for every run of the PSO. Please note that we try to use Gaussian score function whenever possible to help the PSO particles to find a gradient if they are far away from the optimum.

User Dependent Segment Visibility. To check if a segment is visible to the user. The line of sight between the user (defined by its position p_u and its orientation normal n_p) and the segment i (defined by its position p_{l_i} and orientation normal n_{l_i}) is free of obstacles. Therefore, we trace the local map between the points p_u and p_{l_i} and set the obstacle score s_O to zero if we found an obstacle and to 1 otherwise. Furthermore, we need to take the visual field of the user into account. Therefore, we compute the angle between the line of sight and the person orientation normal as $\alpha_{Segment} = acos((\mathbf{p_u} - \mathbf{p_{l_i}}) \cdot \mathbf{n_p})$. Assuming that the visual field of a human is almost 180° with a central field of 90° we are using a spitted Gaussian function $s_{VF}(l_i) = g_{sp}(\alpha_{Segment}, -\pi/4, \pi/4, 0.35, 0.35)$ that returns 1 for the central field and drops of the almost zero at 90° on each side.

$$g_{sp}(x, lo, up, s_1, s_2) = \begin{cases} exp(-\frac{(x-lo)^2}{2s_1^2}), x < lo \\ exp(-\frac{(x-up)^2}{2s_2^2}), x > up \\ 1, lo \leq x \leq up \end{cases} \tag{1}$$

The last of the tree sub scores, is the distance score. We suggest preferring segments with a distance between 1.5 and 4 meter using the split Gaussian $s_D(l_i) = g_{sp}(||\mathbf{p_u} - \mathbf{p_{l_i}}||_2, 1.5, 4.0, 1.0, 1.7)$. Therefore, the resulting score of the user dependent segment visibility s_U becomes: $s_U(l_i) = s_O(l_i) \cdot s_{VF}(l_i) \cdot s_D(l_i)$.

Projection Suitability. It is important that the wall segments used for projection purpose are of a bright and uniform color and are clear of obstacles.

Using a calibrated camera we can extract the image regions associated with the single segments and analyze them in terms of color and structure. We project the 3D-world position of the segment edges into the image space and extract the gray-scale image region (Fig. 3). Afterwards, the average image value is derived. This value should not be too close to white, as ceiling light might outshine the projected image, nor should it be to dark or have any extreme color cast. Therefore, we are using a Gaussian function aiming at an average gray value of 2/3 of the maximum possible gray value to compute brightness subscore $s_B(l_i) = Gauss(Avg(l_i), \mu = 170, \sigma^2 = 50)$. The structure of the segment is analyzed by computing the magnitude of the Sobel-filtered image region in x and y direction. Since the segments can become unsuitable for projection even if the magnitude is far away from the maximum value, we use a low threshold for the score function that allows a maximum average gradient magnitude of 40: $s_G(l_i) = max(0, 1.0 - AvgMag(l_i)/40)$. Thus, the resulting wall projection suitability score gets $s_W(l_i) = s_B(l_i) \cdot s_G(l_i)$.

The drawback of this approach is that not all the segments are visible to the camera and that they can be shadowed by persons and thus get wrong score values. To avoid this problem, the module stores a local segment history (same 8x8 meter environment as the local map). If segments are shadowed by persons (details on how to detect shadowing are given in the Sec. 4.3), the suitability score is not updated. The same applies if a segment is currently used for projection, since the projector changes the score of the segment. If the score cannot be obtained, we take the score from a similar segment in the history buffer.

4.3   Robot Position Dependent Cost Functions

The second group of cost functions depends on the robot pose and therefore needs to be computed for every single robot pose hypothesis.

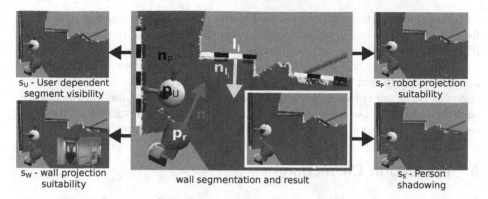

Fig. 3. Results of the different score functions for an exemplary situation. The line segments are color coded according to the result of the score functions with blue being th lowest score. Please note that the line segments do vary slightly since the images were recorded consecutively and the RANSAC algorithm varies due to its random nature.

Robot Projection Suitability. This cost function is mostly related to the limitations of the video projector and combines three different sub scores. First, similar to the person visibility function, the line of sight between $\mathbf{p_r}$ and $\mathbf{p_{l_i}}$ needs to be free of obstacles (s_O). Second, the projector offers only a limited aperture angle, therefore segments that cannot be covered by the projector need to get a very low score. We compute the angle between segment normal and robot with $\alpha_{Robot} = acos((\mathbf{p_r} - \mathbf{p_{s_i}}) \cdot \mathbf{n_r})$ and combine it with the split Gaussian to match the projector opening angle of $60°$: $s_{PA}(l_i) = g_{sp}(\alpha_{Segment}, -\pi/6, \pi/6, 0.05, 0.05)$. The last subscore regards the distance between projector and wall. Since the projector has only limited brightness and the focus is fixed, the projector needs to stay within a certain distance range to offer acceptable projections. Therefore, we select the optimum distance to be within 1.4 and 2 meter and let the score drop on both ends: $s_D(l_i) = g_{sp}(||\mathbf{p_r} - \mathbf{p_{l_i}}||, 1.4, 2, 0.5, 1.5)$. Again, the combined score s_P becomes $s_P(l_i) = s_O(l_i) \cdot s_{PA}(l_i) \cdot s_D(l_i)$.

Person Shadowing. Segments that, according to the robot position, are shadowed by or next to a person should not be used for projection for two reasons. First, it makes a proper projection impossible if the person is blocking the projected image. Second, a person can be dazzled. Therefore, we penalize segments, when the angle between the line of sights of the segment and all person hypothesis p_i is too low: $alpha_{min} = \min_{\forall p_i}(acos((\mathbf{p_r} - \mathbf{p_i}) \cdot (\mathbf{p_r} - \mathbf{l_i})))$ $s_S(l_i) = 1.0 - Gauss(\alpha_{min}, \mu = 0, \sigma^2 = 0.1)$.

4.4 Selecting the Best Wall Segment for Projection

To find the best wall segment for a given robot position $\mathbf{p_r}$ and orientation $\mathbf{n_r}$ the robot position dependent (s_P, s_S) and independent score functions (s_U, s_W) have to be computed. Afterwards, the scores of the different functions are multiplied for every segment independently. We use multiplication for this step since the different constraints cannot compensate each other. Segments that are below a certain threshold will be rejected and removed from the set of segments. In a final step, adjacent wall segments get merged, whereby the score of the new segment is the sum of its sub-segments. This guarantees that large segments are preferred. Finally, the segment with the highest score is returned as the best segment for projection for the current robot and user positions.

5   Projection Correction

As already stated above, the projected image gets distorted if the projector is not aligned perpendicular to the projection wall. Fortunately, the distortion correction is relatively easy for unbowed, straight walls if the angle between the projector and the wall is known.

Since the position and orientation of the wall segment used for projection is known from the position optimization, we can compute coordinates of the rectified image in the image space by means of vector geometry (Fig. 4).

Once we have obtained the image coordinates of the distorted and the rectified image, we can compute a point correspondence matrix (homography matrix) to distort the input image for the projector so that it is displayed correctly on the wall segment.

5.1   Wall Rectified Projection

Knowing the video projector pose $\mathbf{p}$ and its aperture angle, we can compute the intersection points of the four image corners (lu, ru, ll, rl) with the plane H defined by the target wall segment S. Afterwards we transform these 3D coordinates $(lu_{3D}, ru_{3D}, ll_{3D}, rl_{3D})$ on the plane to obtain 2D coordinates (lu_H, ru_H, ll_H, rl_H) of the beamer edge points. After the x coordinates of these points have been adapted to stay within the region defined by the desired wall segment l_i, we compute the maximum rectangle with the aspect ratio defined by the projector $(lu_{max'}, ru_{max'}, ll_{max'}, rl_{max'})$. These points are transformed back to 3D coordinates and projected back into the beamer frame to get the points $lu_{max}, ru_{max}, ll_{max}$ and rl_{max}. Together with the points lu, ru, ll and rl they are defining point correspondences, which are used to compute the planar homography matrix that defines the image distortion. The resulting projection rectifies the image according to a perpendicular observer position.

5.2   User Adaptive Projection

An alternative rectification takes the position of the observer into account and rectifies the image in order to make it appear perpendicular to the axis of view. Therefore, we define a plane V_o that is 10 cm in front of the observer in direction of the segment center c. This plane is oriented perpendicular to the observer view

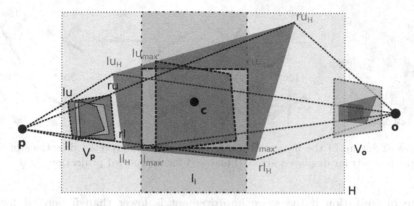

Fig. 4. Illustration of the image rectification process. The projector edge coordinates are projected into the wall or observer coordinate system. Afterwards, they are rectified and projected back into the projector frame.

direction and has the normal vector $c-o$. In contrast to the observer independent approach, the 3D coordinates $lu_{3D}, ru_{3D}, ll_{3D}, rl_{3D}$ are projected onto this plane V_o using the vectors from the corner coordinates to the observer. Therefore, the maximum rectangle is computed on the observer plane V_o before the coordinates are projected back onto the wall plane and from there to the projector space in order to yield the homography matrix.

6 Experiments

Experiments have been conducted with the actual robot platform in our faculty building and using a robot simulator.

We have tested the extraction of suitable wall segments and the projection rectification for different exhibits and situations in our lab. Results for two situations are given in Fig. 6. For every location the local surrounding (8x8 m) of the robot and the user position was taken into account. The algorithm was able to generate a rectified projection on wall segments that are of good visibility to the user. However, the position optimization tends to generate positions with acute projection angle as these positions are good in terms of user visibility and maximum projection distance. If maximum projector brightness is an issue, we recommend integrating an additional module for rating the robot to wall angle.

An analysis of the runtime of the different algorithm steps on the robot platform running an Intel Core i7-3612QM was conducted. As a result, segment detection (18.6 ms) and computation of the robot position independent modules (6.1 ms) is required only once for every new person detection. It has to be noted, that the quality of the optimization highly depends on a proper line extraction of the RANSAC algorithm. Although, it fluctuates a bit, it has proven to work well enough for wall segments with a minimum length of 60 cm. The robot position dependent modules need to be computed for every new particle hypotheses. During our tests, we configured the PSO to use 500 particles and

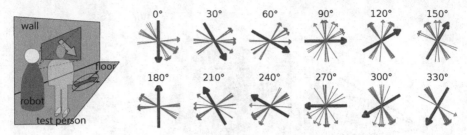

Fig. 5. *left*: Set up of the projection experiment. *right*: interpreted directions of the test persons (red) for displayed arrow (blue) and for unrectified projection.

to stop optimization if the score improvement is lower than 5 percent for 5 generations. This leads to an average of 7 iterations in order to obtain a stable optimum position and requires 35.1 ms of run time on average.

The update rate of the goal position optimization is limited to 10 Hz, which is the update rate of the laser range finder and thus the frequency of the person detection. Furthermore, the person detection accuracy does fluctuate over time requiring smoothing of the person pose as small changes in the person pose do have a high impact on the person specific projection. Therefore, the person dependent projection is always a bit behind if the person is moving fast. However, people are unlikely to move very fast during the presentation of an exhibit.

In order to test the benefit of the projection rectification, we have set up the experiment given in Fig. 5. The robot was located 50 cm in front of a wall, facing the wall at an acute angle causing strong image distortions. We have asked 11 test persons to stand 1 meter next to the robot and to view the images projected by the robot with three different methods: (a) no correction, (b) wall rectified correction, (c) user adaptive projection. First of all, we showed them a video with the different projection methods and ask which one was most comfortable to view. Since only one person prefers the uncorrected projection, image rectification can improve the presentation quality. However, the type of preferred correction is person dependent (4 votes for (b) and 6 votes for (c)) and should be inquired at the beginning of a tour. In the second stage we examined, how the different projections affect the interpretation of direction arrows that might point to the next exhibit. Therefore, we consecutively presented an arrow rotated randomly between 0° and 360° (with 30° increment) on the projection area and asked the test persons to interpret the shown arrow in the floor coordinate system (Fig. 5). Although, the average deviation of subjective angle perception decreases from 62.7° for (a) to 46.81° (b) and 49.09° (b) it hast to be noted, that the person specific preference is more important than the chosen rectification method. Therefore, the diagrams for (b) and (c) are almost identical to the one given in Fig. 5. It seems that the viewing angle of the test person to the arrow position on the wall has more influence than the actual type of projection and therefore should be considered while generating instruction arrows.

Fig. 6. Example image for the wall segment extraction and projection rectification. left: unmodified and distorted projection. middle: projection after segment extraction and rectification. right: direction information displayed on the wall.

7   Conclusion and Future Work

This paper describes a new method that takes the user position into account in order to dynamically extract wall segments that are suitable for video projection. Furthermore, we show how to embed the scoring of the wall segment extraction into an PSO framework in order to obtain the best projection position for the robot in the local surrounding.

Continuing our work, we are currently working on the integration of the robot position scoring into the local motion planning algorithm of the robot. We are using a dynamic window approach (DWA) [18] for motion planning, that samples the possible velocity commands of the robot within a certain prediction time window and scores them by different navigation objectives (e.g. Distance to obstacles, follow a path, follow a person with a given distance,..) [19,20]. We already extended this approach to also score the motion trajectories in terms of projection suitability. First experiments already show promising results as the average projection score during driving can be increased significantly. This allows the robot to improve the experience during a guided tour by projection information (e.g. directions) onto walls during driving.

Furthermore, the work presented so far is designed to work with one user only. Therefore, in future we want to find out if and how the method can be extended to work with multiple users at a time.

References

1. Kwon, E., Kim, G.J.: Humanoid robot vs. projector robot: exploring an indirect approach to human robot interaction. In: Proc of 5th ACM/IEEE International Conference on Human-Robot Interaction (HRI), pp. 157–158, March 2010
2. Stricker, R., Müller, S., Einhorn, E., Schrter, C., Volkhardt, M., Debes, K., Gross, H.-M.: Interactive mobile robots guiding visitors in a university building. In: Proc. 21st IEEE Int. Symposium on Robot and Human Interactive Communication (RoMan), Paris, France, pp. 695–700, September 2012
3. Sasai, T., Takahashi, Y., Kotani, M., Nakamura, A.: Development of a guide robot interacting with the user using information projection basic system. In: International Conference on Mechatronics and Automation, pp. 1297–1302, August 2011
4. Donner, M., Himstedt, M., Hellbach, S., Boehme, H.-J.: Awakening history: preparing a museum tour guide robot for augmenting exhibits. In: Proc. European Conference on Mobile Robots (ECMR), pp. 337–342, September 2013

5. Panek, P., Edelmayer, G., Mayer, P., Beck, C., Rauhala, M.: User acceptance of a mobile LED projector on a socially assistive robot. In: Wichert, R., Eberhardt, B. (eds.) Ambient Assisted Living. ATSC, vol. 2, pp. 77–92. Springer, Heidelberg (2012)
6. Choi, S.-W., Kim, W.-J., Lee, C.H.: Interactive display robot: projector robot with natural user interface. In: Proceedings of the 8th ACM/IEEE International Conference on Human-Robot Interaction (HRI 2013), pp. 109–110. IEEE Press, Piscataway (2013)
7. Ishii, K., Zhao, S., Inami, M., Igarashi, T., Imai, M.: Designing laser gesture interface for robot control. In: Gross, T., Gulliksen, J., Kotzé, P., Oestreicher, L., Palanque, P., Prates, R.O., Winckler, M. (eds.) INTERACT 2009. LNCS, vol. 5727, pp. 479–492. Springer, Heidelberg (2009)
8. Moreno, D., Taubin, G.: Simple, accurate, and robust projector-camera calibration. In: Second International Conference on 3D Imaging, Modeling, Processing, Visualization and Transmission (3DIMPVT), pp. 464–471, October 2012
9. Audet, S., Okutomi, M.: A user-friendly method to geometrically calibrate projector-camera systems. In: Conference on Computer Vision and Pattern Recognition Workshops. IEEE Computer Society, pp. 47–54, June 2009
10. Einhorn, E., Langner, T., Stricker, R., Martin, Ch., Gross, H.-M.: MIRA - middleware for robotic applications. In: Proc. IEEE/RSJ Int. Conf. on Intelligent Robots and Systems (IROS 2012), Vilamoura, Portugal, pp. 2591–2598, October 2012
11. Volkhardt, M., Weinrich, Ch., Gross, H.-M.: Multi-modal people tracking on a mobile companion robot. In: Proc. 6th European Conference on Mobile Robots (ECMR 2013), Barcelona, Spain, pp. 288–293, September 2013
12. Arras, K.O., Mozos, O.M., Burgard, W.: Using boosted features for the detection of people in 2d range data. In: IEEE International Conference on Robotics and Automation 2007, pp.3402–3407 (2007)
13. Viola, P., Jones, M.: Rapid object detection using a boosted cascade of simple features. In: Proc. of the IEEE Computer Society Conference on Computer Vision and Pattern Recognition (CVPR), vol. 1, pp. 511–518 (2001)
14. Weinrich, Ch., Vollmer, Ch., Gross, H.-M.: Estimation of human upper body orientation for mobile robotics using an SVM decision tree on monocular images. In: Proc. IEEE/RSJ Int. Conf. on Intelligent Robots and Systems (IROS 2012), Vilamoura, Portugal, pp. 2147–2152, October 2012
15. ROREAS project. www.roreas.org
16. Kennedy, J., Eberhart, R.: Particle swarm optimization. In: IEEE International Conference on Neural Networks, vol. 4, pp. 1942–1948, November/December 1995
17. Fischler, M.A., Bolles, R.C.: Random sample consensus: a paradigm for model fitting with applications to image analysis and automated cartography. Commun. ACM **24**(6), 381–395 (1981)
18. Fox, D., Burgard, W., Thrun, S.: The dynamic window approach to collision avoidance. IEEE Robotics & Automation Magazine **4**(1), 23–33 (1997)
19. Gross, H.-M., Debes, K., Einhorn, E., Müller, St., Scheidig, A., Weinrich, Ch., Bley, A., Martin, Ch.: Mobile robotic rehabilitation assistant for walking and orientation training of stroke patients: a report on work in progress. In: Proc. IEEE Int. Conf. on Systems, Man, and Cybernetics (SMC 2014), San Diego, USA, pp. 1880–1887, October 2014
20. Einhorn, E., Langner, T.: Pilot - modular robot navigation for real-world-applications. In: Proc. 55th Int. Scientic Colloquium, Ilmenau, Germany, pp. 382–393. Verlag ISLE 2010

Effects of Mirrors in Mobile Robot Navigation Based on Omnidirectional Vision

Mohammad Hossein Bamorovat Abadi[1]([envelope])
and Mohammadreza Asghari Oskoei[2]

[1] Department of Electrical, Biomedical and Mechatronics Engineering,
Qazvin Branch, Islamic Azad University Qazvin, Qazvin, Iran
M.bamorovvat@ymail.com
[2] Faculty of Mathematics and Computer Science,
University of Allameh Tabataba'i, Tehran, Iran
oskoei@atu.ac.ir

Abstract. In this paper, we present a omnidirectional vision-based navigation system that includes three approaches: obstacle avoidance based on sonar vision, direction estimation based on sonar vision and confection method of obstacle avoidance and direction estimation. This paper peruses effects of the mirror in omnidirectional vision applied to mobile robot navigation, as well. We design and establish four mirrors: small non-uniform pixel-density hyperbolic mirror, small uniform pixel density hyperbolic mirror, large non-uniform pixel density hyperbolic mirror and spherical mirror. This paper provides autonomous navigation for a mobile robot in an unknown environments. We use omnidirectional images without any prior calibration and detects static and dynamic obstacles. Our experiments operates in indoor environment with our particular sonar vision. The result show that small uniform pixel density hyperbolic mirror have best performance and big non-uniform pixel density hyperbolic mirror have weak performance in vision base mobile robot navigation. Also, the experimental results show acceptable performance considering computation costs in our sonar vision algorithm.

Keywords: Omnidirectional vision · Mobile robot · Vision navigation · Sonar vision

1 Introduction

The purpose of mobile robot navigation is moving in a structured or unstructured environment, and transferring to the target.Vision based mobile robot navigation, in a structured environment, without having any prior knowledge of the environment, is a very powerful capability for robots. A major advantage of image based navigation systems, no need to have other sensors, and thus reduce the cost.

Omnidirectional vision sensors for mobile robots are valuable because they are provide full visibility of the surrounding environment of the robot, just in a frame [4]. This is important because knowing the positions of objects in the

© Springer International Publishing Switzerland 2015
H. Liu et al. (Eds.): ICIRA 2015, Part III, LNAI 9246, pp. 37–48, 2015.
DOI: 10.1007/978-3-319-22873-0_4

surroundings of the robot, will help robots to perform tasks such as matching or intuitive navigation map. 360 degrees field of view reduce visual perceptual bad view. In addition, use of omnidirectional vision system in a very dynamic environment and a burst of consecutive images of the environment around the robot, provides easy to use tracing techniques to follow the desired objects. Since access to the systems of all popular way for us not possible, therefore, in this work, we design and build our own special omnidirectional system.

In this work, we combine two main behaviors, which allow the robot to navigate in different environments. First, estimate the path and the second, identify barriers and lack of dealing with them which takes place during the pursuit. Both actions are performed using sonar vision algorithms [1] directly on omnidirectional images [2]. Both techniques are implemented simultaneously. The important point is that, omnidirectional image does not require any calibration and we do not change the appearance of the image all the way, as well. Also, all the steps can be performed directly on the original omnidirectional images.

The rest of the paper is as follows. In the section 2, we've talked about special omnidirectional system which created. In the section 3, we have explained the sonar vision algorithms. Also in this section, we describe the algorithm for estimating the path, avoiding collisions with obstacles and also remove the light reflection from the surface. In the section 4, we present how the robot navigation. Then in the section 5, presents the experimental results and discuss them. Finally, Section 6 presents the conclusion and future plans we have.

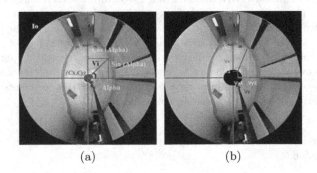

(a)   (b)

Fig. 1. Examples of the image captured by our catadioptric system

2   Omnidirectional Vision

The sense of sight is one of the ways to collect information from the environment. Usually, normal camera used for machine vision as the visual system. In contrast, we believe that as many visual nature is designed for different tasks, we also need to design different visual sensors for robots and use them for different tasks. Due to problems such as cost and technical limitations, it is not always possible. But omnidirectional vision and especially catadioptric vision are more flexible than the photography systems. Omnidirectional vision sensor for mobile robots are valuable because they are provide full visibility of the environment of the robot, just in a frame [1]. In this work, we used catadioptric vision[1]. We used Spherical mirror, small uniform pixel density hyperbolic mirror, small non-uniform pixel density hyperbolic mirror and big non-uniform pixel density hyperbolic mirror. We use [4,5,10] to build our mirrors. For this reason, we have implemented our design, build and optimize omnidirectional vision system with financial support, the Science and Technology Park of Gilan[2].

3   Sonar Vision

In this work, we developed a method based on sonar vision algorithm for omnidirectional images that known as the OmniVisual sonars (OV-sonars) introduced in [2]. The omnidirectional sonar including virtual beams [1,2] which is a circle from the center and all the way up to the point where a gradient of the image would continue. The important thing is that, we have not done any calibration on the camera image and the images are processed directly in the mirror and get to navigation. In this method, we directly use the images, all the way, to the navigation and images does not become to the panoramic image or the birds eye image. Therefore, processing speed has gone up as a result of the reaction of the robot path goes up against obstacles.

According to [2] k-th sonar $V_k = \{v_1, ..., v_r\}$ shows apeak of r pixel of the omnidirectional image. any pixel v_i tally with:

$$v_i = I_o\{i \sin \alpha + C_x, i \cos \alpha + C_y\} \tag{1}$$

$$\alpha = k.2\pi/N_s$$

where Io is the main 2D omnidirectional image with semidiameter r and center in (Cx, Cy) and N_s is the number of sonars proceed for each image. Figure 1a determine method of producing a visual omnidirectional sonar.

To do this, first we detect of the image edge with Sobel edge detection algorithm. Then the gray image with an appropriate threshold are converted to black and white image. Then with the help of equation 1, sonar vision production and then we estimated path and the obstacles identified. Since our image is black and white, each pixel has a value of 255 was the first visual sonar, as it will prevent detection. To calculate the final motion vector that represents the path is correct and unstructed path we took advantage of the unit vectors. As such, we've created a vector of motion using visual sonar vectors. Unit vectors added together and all the sonar target point representing the path is open and unstructed to obtain. Then we point the ultimate goal and direction from the point of origin to

---

[1] A kind of omnidirectional vision system which uses of mirrors.
[2] This research was conducted by Grant No.2306 and Grant No.1591, generously funded by the Science and Technology Park of Gilan.

the point of final motion vector drawing that represents us. Calculating a path through the following equations:

$$V_{xi} = V_{x2} - V_{x1} \tag{2}$$
$$V_{yj} = V_{y2} - V_{y1} \tag{3}$$
$$f_k(x, y) = V_{xi} + V_{yj} \tag{4}$$

In this work, we predict the path for navigation, and along the way we will use the method of obstacle avoidance. An important point is that, in all phases of navigation, just the way, we use visual sonar. For obstacle avoidance we use the mothod in which Boyan et al [2] have proposed to have developed and innovative approach to predict the path will produce the visual sonar. Figure 1b define that how we obtain the unit vectors of a visual sonars.

3.1   Obstacle Avoidance Based Sonar Vision

To identify obstacles and deal with them in accordance with [2] we use 17 visual sonar. But changes in the scattering sonar created (Figure 2a).

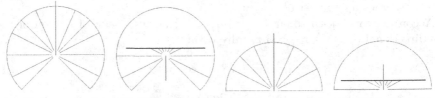

(a) Without obstacle (b) With obstacle (c) Without obstacle (d) With obstacle

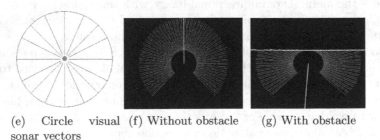

(e)   Circle   visual   (f) Without obstacle   (g) With obstacle
sonar vectors

Fig. 2. Visual sonar vectors

We have to create a semi-spherical sonar, were inspired by [7]. But in the experiments, we found that there are problems in navigation. So we created our own special sonar (figure 2). We have formed the sonar because: 1-Generally, no obstacles behind valuable for navigation, thus the behind barriers would not affect to navigation. 2-The vector that represents the outcome of the move when there is no obstacle nowhere the final vector selects straight forward if there is no obstacle in the way when Boyan scattering sonar not choosing the right track. 3-When forward path is completely blocked the resultant vector indicates the reverse direction, in case if sonar in semicircular form, do not choose a path. 4-In this case, do not need our robots have a circular shape. Figure 2a and 2b display our sonars. Figure 2c and 2d display semicircle sonars. As you see despite the length of the final vector is small but just right direction is detected. In the section 3.2 you will see how we will solve this problem.

According figure 2e, you can see that in the case of Sonar are fully when there is no obstacle in the final vector is zero. In the figure 2c and 2d, you can see that when the Sonar semicircle despite a correct diagnosis in the absence of an obstacle course but when there is an obstacle in the path can not be detected correctly. In the figure 2, you can see that when the sonar are as follows: in both cases, unobstructed and obstructed, the resultant vector, correctly detects the correct path.

3.2 Direction Estimation Based Sonar Vision

We used an innovative method for estimating path. To do this, we use a lot of visual sonar to estimate the path. Unlike [2] that a large number of visual sonar to detect small obstacles sees fit, we hypothesized that the large number of sonar due to the resultant vector recognizes the right track and will determine the path, and we will show the right path to take. The reason for this claim is that due to the large number of sonars, despite dealing with small obstacles, but the end result Sonar on the right track and will be open. So, we created a visual sonar for every 2.5 degree. To complete the picture 360 degrees, we took the 144 sonar use. The tests found that the estimated path of the robot do not need back sonars, as well. Therefore, to estimate the direction of 108 sonar, we use the form in figure 3a and 3b. As you see in 3a the best direction estimated when there is no obstacle and correct path estimated when obstacle is in the robot forward.

3.3 Remove the Light Reflected from the Surface

Usually this happens on the surface gloss with light sources and the issue of domestic environments are very common. Landscape reflected light is like an obstacle. To solve this problem, we decided to detect reflected light and forget them. In [2] boyan said that a pixel belongs to a reflected light when the HSV color space has a low saturation value (S), and a high value (V) is. Gradient of pixels with large difference $V - S$ do not need be considered as an obstacle. So unlike [2] only the value $w(i)$ with a certain amount of more than 100 tests were compared. If the number is above 100, it is not a barrier and if the difference is below 100, the obstacle is detected.

$$w(i) = V - S \tag{5}$$

$$\begin{cases} \text{Not Obstacle, if } w(i) \geq 100 \\ \text{Obstacle , if } w(i) < 100 \end{cases} \tag{6}$$

Where $v(i)$ and $s(i)$, are value, and saturation of pixel i in the HSV color space.

In all experiments and implementations in both the path and the minimal estimate of the barrier, the algorithm used to remove the light reflected from the surface. The following figure shows an example of the implementation of this algorithm. As you can see, all three pictures show the moment of testing. In figure 3a picture is related to the obstacle avoidance, and image 3b, the estimated path is. In both cases, the reflected light is detected correctly. In figure 3c, you can see a complete form of the reflection light on the floor.

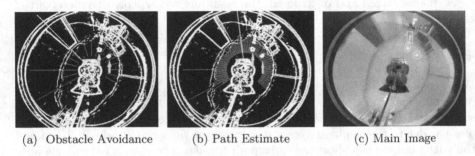

(a) Obstacle Avoidance   (b) Path Estimate   (c) Main Image

Fig. 3. Remove the light reflected from the surface

4   Navigation

We use a very simple method for robot navigation. To obtain the direction of the robot, we average the final angles of estimate an obstacle avoidance vectors to obtain the angle of the robot motion. Equation 7, use to obtain the final vector of estimation vector and obstacle vector, as well. Algorithm 1, also, show how robot navigation in our project.

$$f^* = \sum_{k=1}^{N_s} f_k(x, y) \tag{7}$$

In this algorithm, when the final vector angle is between -2 to 2, robot is moving forward. Otherwise, the robot rotates to the right or to the left. As you can see, our motion algorithm is very simple. However, despite this simplicity very accurate and in almost all cases the robot has detected the correct path. Also nowhere with not hit any obstacle. Neither fixed nor prevent movable barrier.

Algorithm 1. Navigation

INPUT: $\alpha \leftarrow$ Angle where path estimate is achieve
INPUT: $\beta \leftarrow$ Angle where obstacle avoidance is achieve
OUTPUT: $SA \leftarrow$ Turn speed
OUTPUT: $SF \leftarrow$ Forward speed

$\gamma = (\alpha + \beta)/2$
Begin
   if $(2 \geq \gamma$ and $\gamma \geq -2)$ **then**
      SA= $\gamma/10$, SF = 0.2
   end if
   if $((\gamma > 2$ and $\gamma \leq 10)$ or $(\gamma < -2$ and $\gamma \geq -10))$ **then**
      SA= $\gamma/100$, SF = 0.15
   end if
   if $((\gamma > 10$ and $\gamma \leq 25)$ or $(\gamma < -10$ and $\gamma \geq -25))$ **then**
      SA= $\gamma/100$, SF = 0.1
   end if
   if $((\gamma > 25$ and $\gamma \leq 40)$ or $(\gamma < -25$ and $\gamma \geq -40))$ **then**
      SA= $\gamma/100$, SF = 0.06
   end if
   if $\gamma > 40$ or $\gamma < -40$ **then**
      SA= $\gamma/100$, SF = 0
   end if
End

---

5 Experiments and Results

In this work, we put the robot in an empty and straight corridor without any obstacles with the help of three sonar vision algorithms [3], we testd each mirrors, separately.

The purpose of these tests, in general, are to evaluate the performance of the navigation algorithms rather than the mirror. No barrier of this stage of the experiments was to study the algorithms and mirror, in terms of the path. In these experiments, we examined the end vector results of all mirrors of the three algorithms and compared them together. Compared parameters are: 1. distortion and vibration data to the main line, 2. The distribution of the data to the path and 3. The number of frames to reach the goal.

We examine the number of frames, because this agent is representing the whole time of the robot moves. Whatever the number of frames is more, the robot has spent more time to reach the destination.

Figure 5 shows the performance of the mirror, for the three algorithms. These figures show that big non- uniform pixel density hyperbolic mirror and spherical mirror, got through with a lot of distortion and vibration in the route, Whereas, small non-uniform pixel density hyperbolic mirror and small uniform

---

[3] Obstacle avoidance, path estimate and ultimate algorithm (Average of the obstacle avoidance and path estimate together.)

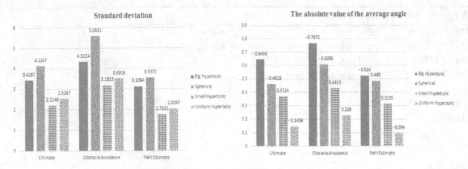

(a) Standard Deviation   (b) The Absolute Value of Average of Angle

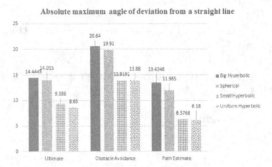

(c)   The Absolute Angle of Deviation from a
Sright Line

Fig. 4. Statistical analysis

pixel density hyperbolic mirror, completed the route, smoother and with much less volatility.

From the perspective of the number of frames, as it is known, the small non-uniform pixel density hyperbolic mirror, have minimum number of frames which is indicative of the fact that the robot has arrived faster than other states. This is due to short variations in the path of the robot. Hereafter, we call mirrors such this: SNU[4], SU [5], SM [6], LNU [7].

Table 1. Compare the distribution of the data in the algorithm

1	Path estimate algorithm
2	Ultimate algorithm
3	Obstacle avoidance algorithm

---

[4] Small non-uniform pixel-density hyperbolic mirror.
[5] Small uniform pixel density hyperbolic mirror.
[6] Spherical mirror.
[7] Large non-uniform pixel density hyperbolic mirror.

Table 2. Compare the performance of the mirror in various surveys.

P	Frames	Vibration	Standard deviation	Average angle	Absolute angle of deviation
1	SNU	SNU	SNU	SU	SU
2	SU	SU	SU	SNU	SNU
3	SM	SM	LNU	SM	SM
4	LNU	LNU	SM	LNU	LNU

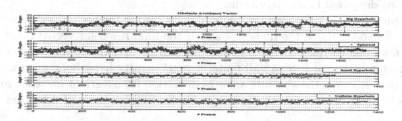

(a) Obstacle Avoidance Algorithm Vectors

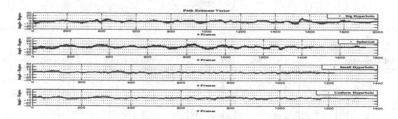

(b) Path Estimate Algorithm Vectors

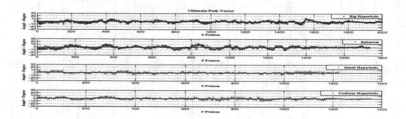

(c) Ultimate Algorithm Vectors

Fig. 5. Performance of the mirrors for three algorithms

If we compare three algorithms, we find that, in all the mirrors, the data has not uniform dispersion. As can be seen, the path estimation algorithm is having the lowest dispersion of data, whereas, the obstacle avoidance algorithm is having the greatest dispersion of data.

Compare the number of the mirror in various surveys is in table 2 and compare the distribution of the data in the algorithm is in table 1. As you can see, the best performance is for the first one and the worst performance is for the last one.

Figure 4a shows the standard deviation of each mirror for each algorithm, separately. Here, we evaluated the performance of the mirrors for each algorithm. As you can see, in this case, the spherical mirror has the highest standard deviation, and small non-uniform pixel density hyperbolic mirror has the lowest standard deviation in all algorithms.

Figure 4b shows the average angle of each mirror for each algorithm, separately. As you can see, the highest angle of the final resultant vector Related to large non-uniform pixel density hyperbolic mirror and the lowest angle of the final resultant vector is for small uniform pixel density hyperbolic mirror.

Figure 4c shows the absolute angle of deviation from a sright line each mirror for each algorithm, separately. In this case, the highest angle of deviation from

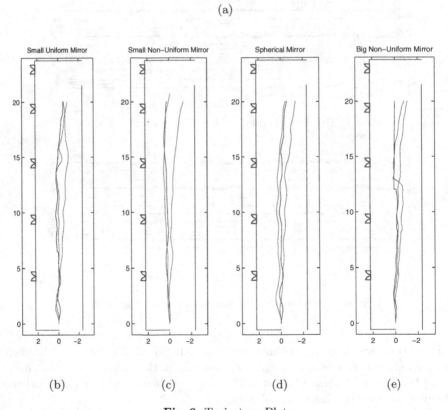

Fig. 6. Trajectory Plots

a sright line Related to large non-uniform pixel density hyperbolic mirror and the lowest angle of deviation from a sright line is for small uniform pixel density hyperbolic mirror.

Figure 6 shows how the robot moves in the corridor. In this section we've tested the performance of the robot for each mirror and each algorithm, separately. As is known, the robot is able to correctly discern the correct path, and continue to move along the corridor in all cases. In all the mirrors the best performance in terms of movement is for path estimate algorithm (blue lines) and the worst performance in terms of movement is for obstacle avoidance algorithm (red lines). In all cases, the best performance in terms of movement is for small non-uniform pixel density hyperbolic mirror and the worst performance in terms of movement is for spherical mirror.

Compare the best performance in terms of movement for mirrors, as follow: 1. SNU (figure 6c), 2. SU (figure 6b), 3. LNU (figure 6e) and 4.SM (figure 6d).

6 Conclusions and Future Work

In this work, we have developed a method, which is capable of robot navigation in different environments. To do this, we used the omnidirectional vision system. The type of omnidirectional vision system was catadioptric. For experiments we used four mirrors: Spherical mirror, small uniform pixel density hyperbolic mirror, small non-uniform pixel density hyperbolic mirror and big non-uniform pixel density hyperbolic mirror. We also use the sonar vision algorithm for navigation. Sonar vision algorithm consists of three parts: 1. path estimate, 2. obstacle avoidance and 3. ultimate.

This method can be used in different environments, with no previous knowledge of the environment. This can be done by any of the omnidirectional vision system and without any calibration and changing in the picture mode. Also, this method has low computational cost. By combining this method and omnidirectional vision system, and due to the low computational cost, almost robots in all cases, be able to identify the correct way and the route, fixed and moving obstacles easily detected and avoid dealing with them.

Path estimate algorithm would greatly improve previous forms of this method [2], the uncertainty as it was. Especially in indoor, light reflected from the surface is detected by a particular method, and it was not chosen as a barrier and to ease the path continues. It is still uncertain, and this method can not be trusted 100%.

In general, compare the best performance of mirrors, as follow: 1. small non-uniform pixel-density hyperbolic mirror, 2. small uniform pixel density hyperbolic mirror, 3. large non-uniform pixel density hyperbolic mirror and 4. spherical mirror, and compare the best performance for algorithms, as follow: 1. path estimate algorithm, 2. ultimate algorithm and 3. obstacle avoidance algorithm.

Drawback of this method is dependent on the ambient lighting. A major disadvantage of this method is dependent on edge detection. In environments that can not be properly edge detection, this method will be in trouble. In continuation of our research, we are looking at ways that can reduce dependence on edge detection and improve the problem [8]. Also working on other control methods for this kind of robot motion, becomes softer [9].

Acknowledgments. We would like to thank Dr.A.M.Shahri, Mohammadreza Beyad and Yasan Majidi for many useful suggestions and helps related to this work.

References

1. Lenser, S., Veloso, M.: Visual sonar: fast obstacle avoidance using monocular vision. In: Proceedings of the 2003 IEEE/RSJ International Conference on IEEE Intelligent Robots and Systems, (IROS 2003), vol. 1, pp. 886–891 (2003)
2. Bonev, B., Cazorla Quevedo, M.Á., Escolano Ruiz, F., et al.: Robot navigation behaviors based on omnidirectional vision and information theory. Red de Agentes Físicos (2007)
3. Baker, S., Nayar, S.K.: A theory of single-viewpoint catadioptric image formation. International Journal of Computer Vision **35**(2), 175–196 (1999). Springer
4. Nayar, S.K.: Catadioptric omnidirectional camera. In: Proceedings of the 1997 IEEE Computer Society Conference on IEEE Computer Vision and Pattern Recognition, pp. 482–488 (1997)
5. Ishiguro, H.: Development of low-cost compact omnidirectional vision sensors and their applications. In: Proc. Int. Conf. Information Systems, Analysis and Synthesis, pp. 433–439 (1998)
6. Kochan, A.: HelpMate to ease hospital delivery and collection tasks, and assist with security. Industrial Robot: An International Journal **24**(3), 226–228 (1997). MCB UP Ltd
7. Chung, Y.-C., Wang, C.H., Wang, J.M., Lin, S.C., Chen, S.W.: Integration of omnidirectional and movable cameras for indoor surveillance. In: IPPR Conference on Computer Vision, Graphics and Image Processing (CVGIP) (2004)
8. Chang, C.-K., Siagian, C., Itti, L.: Mobile robot monocular vision navigation based on road region and boundary estimation. In: 2012 IEEE/RSJ International Conference on Intelligent Robots and Systems (IROS), pp. 1043–1050. IEEE (2012)
9. Krüsi, P., Pivtoraiko, M., Kelly, A., Howard, T.M., Siegwart, R.Y.: Path set relaxation for mobile robot navigation. Eidgenössische Technische Hochschule Zürich, Autonomous System Lab (2010)
10. Gächter, S., Pajdla, T.: Mirror design for an omnidirectional camera with a uniform cylindrical projection when using the SVAVISCA sensor. Research Reports of CMP, OMNIVIEWS Project, Czech Technical University in Prague, 3 (2001)

Autonomous Science Target Touchability Evaluation: A Fuzzy Logic-Based Approach

Chen Gui and Changjing Shang[✉]

Department of Computer Science, Institute of Mathematics,
Physics and Computer Science, Aberystwyth University,
Aberystwyth SY23 3DB, UK
{chg12,cns}@aber.ac.uk

Abstract. Currently, for Mars science target selection, the task of determining whether or not it is possible for a robot arm to touch a target is accomplished by human operators and scientists on Earth. The development of useful on-board autonomous touchability techniques would greatly reduce human intervention. It would be advantageous if the rover could evaluate autonomously whether the robot arm would be able to place an instrument against an identified science target. In this paper we propose a new approach to the problem of autonomous science target touchability evaluation. We assess the touchability of a potential science target in terms of its size (the number of pixels of the science target in the image), SV (the science value of the science target), distance (the reachable distance of a robot arm), and orientation (the angular regions of the arm's shoulder azimuth). In particular, the plane in front of the arm is divided into a number of partitions, which are ranked with the different touchability levels by the use of a fuzzy rule-based system. Simulations on the rank of science object touchability are carried out, via software and hardware implementation. Based on the real data gathered from the cameras and the Schunk arm experimental results successfully verify the validity of the proposed approach.

Keywords: Fuzzy logic · Target touchability · Autonomous evaluation

1 Introduction

Owing to the high cost and risk of manned space exploration missions, space agencies primarily concentrate on unmanned planetary exploration. The most advanced exploration robots that have been deployed for planetary exploration are the Mars Exploration Rover (MER, "Spirit" and "Opportunity") and the Mars Science Laboratory (MSL, "Curiosity"). They both have an identical manipulation scenario, consisting of four main stages for science target exploration: (a) The rover on Mars transmits the images captured from the navigation camera to the operators/scientists on Earth, with an interesting target being manually selected by the ground scientists in a stereo range map. (b) A target tracker enables the rover to autonomously drive to its goal position while

© Springer International Publishing Switzerland 2015
H. Liu et al. (Eds.): ICIRA 2015, Part III, LNAI 9246, pp. 49–61, 2015.
DOI: 10.1007/978-3-319-22873-0_5

avoiding obstacles, and to reach the goal position with the precision of a few centimetres. (c) The scientists artificially designate the sampling point on the scientific goal from the downlinked images. (d) A variety of instruments on-board are utilised in order to sample and analyse the science target.

In order to determine whether a science goal can be acquired, both MER and MSL adopt a rigid approach using the robot arm workspace. For instance, in terms of the Curiosity rover the workspace volume is an upright cylinder 80cm diameter, 100cm high, positioned 105cm in front of the rover when it is on a smooth flat terrain [1]. Therefore, the current mechanical strategy employed by MER and MSL is that the science target is deemed to be able to be acquired just when it is within the robot arm workspace. This strategy does not consider the cases where an interesting target is surrounded by other rocks that can not be traversed in the 'rock garden', which is not able to get into the robot arm workspace but is just on the edge of the workspace. This problem can be addressed if the robot arm workspace is variable based on the distinct characteristic of a given target. We propose an autonomous flexible approach to adjusting automatically the robot arm workspace in relation to the science value score (SV) regarding the scientific goal. This strategy is based on the use of fuzzy logic techniques, capable of increasing appropriately the magnitude of the robot arm workspace when the science value score is high.

Our work is inspired by the observation that the application of fuzzy logic in planetary exploration has recently gained significant recognition. Traditionally, the difficulty level a rover may encounter when attempting to traverse a region of terrain in a no priori knowledge environment, is classified through the use of traversability index [2]. Howard et al. [3] extended this approach by combining it with a fuzzy map representation that reflects the traversal difficulty of the terrain. It concentrates on planning over an optimally safe path of minimum traversal cost. Mahmound [4] utilised a fuzzy adaptation technique that examines the population of paths throughout the execution of the underlying algorithm while adjusting operator probabilities to attain better solutions for path planning. Fuzzy logic has also seen applied to planetary landing and the tier-scalable robotic planetary reconnaissance. For example, in addressing the issue of landing site selection [5], the score of each potential candidate landing site is obtained from sensor measurements that are fed into a fuzzy system which decides on spatial and temporal dependence. Furfaro et al. [6] built a fuzzy system where the appropriate past/present water/energy indicators can be acquired when the tier-scalable mission framework is deployed, and used to estimate the habitability on Mars. Also, in dealing with the first-stage of the aforementioned operational scenario, Barnes et al. [7] and Pugh et al. [8], proposed that a fuzzy rule based expert system (KSTIS 1.0) could be used. Such a system adopts knowledge elicitation from a planetary geologist to obtain the primary clues regarding the geological background (*Structure*, *Texture* and *Composition*) of Martian rocks, and generates a useful science value score (SV) with respect to each rock in an image.

The rest of this paper is organised as follows. A brief overview of the proposed touchability system framework is presented in Section 2, with a focus on the construction of the linguistic fuzzy sets and the fuzzy rules associated with the underlying attributes that are utilised by the system. Results of computer simulation are reported in Section 3, as part of the verification of the proposed approach. Further experimental results though hardware implementation are described in Section 4, comparing the proposed approach against the performance attainable by a domain expert. The paper is concluded in Section 5.

2 Proposed Touchability System

The structure of the proposed fuzzy logic-based touchability system is shown in Figure 1. It consists of six main components: a vision system, the Knowledge based Science Target Identification System (KSTIS 1.0), a fuzzification module, a fuzzy inference engine, a rule base and a defuzzification module. In this system the input data i_d, i_o, i_s are: the distance between the robot arm base and the centroid of a possible science target, the orientation of the arm's shoulder azimuth, and the size of the target in the image, respectively. The output o_t from the proposed fuzzy system is the touchability probability for each identified scientific target. KSTIS 1.0 assists in ground-based interpretation of scientific targets [7,8], from the description of the *Structure, Texture* and *Composition* of a scientific target whose values are provided by scientists/experts on Earth, giving the score of *Science Value* i_{SV}.

The proposed approach is easy to comprehend and is simple to implement. It basically adopts the general structure of a conventional fuzzy logic controller, with the functionalities of its key components described below. Note that although simplistic in implementation, the underlying techniques adopted are well formed with solid mathematical foundations that have been developed in the field of fuzzy logic control. This forms a sharp contrast with typical existing mechanisms to assess the touchability that are largely based on use of ad hoc methods.

Fuzzification. This module maps the crisp input values onto their corresponding linguistic fuzzy terms. This involves the four physical properties indicated previously: Size, Distance, Orientation and SV of a possible target.

Size (i_s): The bounding area of a given object within an image is charactered as the size of the object. One way to identify the surroundings of an object such as that used by MER is to form a detailed DEM (Digital Elevation Model) by accomplishing stereo matching over the full images. However, in order to obtain just the essential information on size efficiently, only 5 points per object are herein applied for stereo matching as shown in Figure 2. The minimum rectangle (A, B, C and D) for each edge inscribes the leftmost, rightmost, uppermost and bottommost points (P3, P4, P1 and P2) of the object, respectively, while E, F, G and H are the middle points of the line segments 'AB', 'BC', 'CD' and 'AD', respectively. The point C0 is the cross point of the line segments 'EG' and 'HF' and is the centroid of the object. P1, P2, P3 and P4 represent the stereo

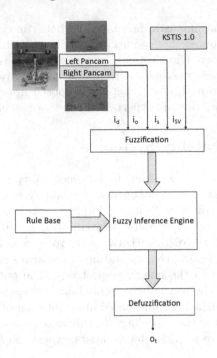

Fig. 1. Fuzzy logic touchability evaluation system.

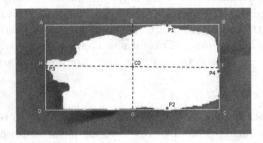

Fig. 2. Stereo matching points selection.

matching points, whose three dimensional frame values are then derived by the external and internal parameters of the cameras. The membership functions of these fuzzy sets are empirically defined as given in Figure 3.

Distance (i_d)*:* This is a significant variable in this study, whose span is provided by the length of the robot arm. Figure 4 illustrates the distance between the original point O in the mobile robot arm base and the centroid (C) of an object. The membership functions of the relevant fuzzy sets are given in Figure 5.

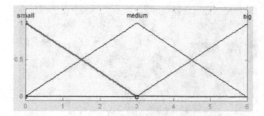

Fig. 3. Membership functions for Size (i_s).

Fig. 4. Distance between arm and object.

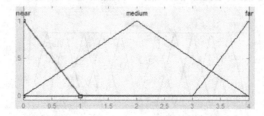

Fig. 5. Membership functions for Distance (i_d).

Orientation (i_o): This is the angle formed by the straight line defined relative to the heading of the rover, and the straight line that connects the projection of the centroid of the object with the reference arm, as shown in Figure 6. As highlighted in the figure, C' is the projection of C on the plane that is constituted by the X and Y axes. θ is an angle between the straight line OC' and Y axis, and is the orientation. In Figure 7, the orientation in front of the rover is divided into six regions that are represented by six linguistic fuzzy sets $\{very-bad(VB), bad(B), very-soso(VS), soso(S), good(G), very-good(VG)\}$. The "very-good", "good", "soso", "very-soso", "bad" and "very-bad" are sectors at $\pm15°$ (Red), between $\pm15°$ and $\pm30°$ (Turquoise), between $\pm30°$ and $\pm45°$ (Yellow), between $\pm45°$ and $\pm60°$ (Green), between $\pm60°$ and $\pm75°$ (Orange), and between $\pm75°$ and $\pm90°$ (Pink) relative to the heading of the

Fig. 6. Orientation between arm and object.

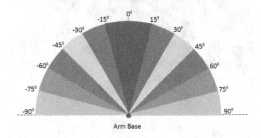

Fig. 7. Decomposition of orientation regions.

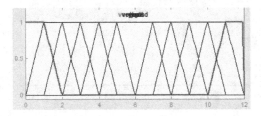

Fig. 8. Membership functions for Orientation (i_o).

rover, respectively. The membership functions of these sets are shown in Figure 8 in which "0", "2", "4", "6", "8", "10" and "12" are corresponding to $-90°$, $-60°$, $-30°$, $0°$, $30°$, $60°$ and $90°$, respectively.

Science Value (SV) (i_{SV}): This is a score between 0 and 9999, computed by KSTIS 1.0. It is represented by one of the three linguistic fuzzy sets {LOW, MEDIUM, HIGH}, with the corresponding membership functions defined as given in Figure 9.

Note that as illustrated above, all fuzzy sets used in this system are implemented with triangular membership functions. This is mainly due to the relative simpler computation this type of fuzzy set entails as compared to the use of typical alternatives such as trapezoidal or Gaussian functions. The employment of

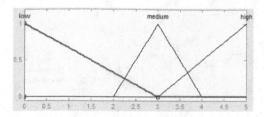

Fig. 9. Membership functions for SV (i_{SV}).

triangular membership functions is also partially because of the relative ease in communicating the underlying mathematical concepts between the knowledge engineers and domain experts.

Inference Mechanism. This is responsible for decision-making in the fuzzy system through fuzzy reasoning. It achieves two tasks: (1) to determine the extent to which each rule in the rule base is associated with the current situation as characterised by the inputs; and (2) to derive a conclusion by firing the best matching rule. Seventy-four rules are included in the rule base, including the 72 general rules shown in Figure 10 and the following two specific ones (that human experts believe to be of significance for the present investigation):

- IF Size is SMALL THEN TIndex is VERYLOW
- IF SV is LOW THEN TIndex is VERYLOW

Defuzzification. The output of the fuzzy inference mechanism is mapped onto a crisp value, called *Touchability Index* by this module. There are a number of methods that can be used to implement this inverse operation of fuzzification. The "COG defuzzification" is herein used to combine the outputs represented by the implied fuzzy sets from all rules that at least partially match the inputs to form a single overall output. The *Touchability Index* is represented by seven fuzzy sets {VERYLOW, LOW, MEDIUMLOW, MEDIUM, MEDIUMHIGH, HIGH, VERYHIGH}, whose membership functions are shown in Figure 11.

3   Software Simulation

This section presents experimental results of computer-based simulation, comparing the resulting *Touchability Index* of mock objects in terms of their ranks with that given by a human expert. The system is implemented using MATLAB Fuzzy ToolBox simulator and involves 9 artificially created rock objects, of three different types: small (10 × 15), medium (20 × 15) and big (30 × 20). Here, the length of the Curiosity rover arm has been employed for simulation experiments, which is 2.3 meters from the front of the rover body. The three science value scores used are 35, 65 and 105. In Table 1, *Length× Width* is the size of an object.

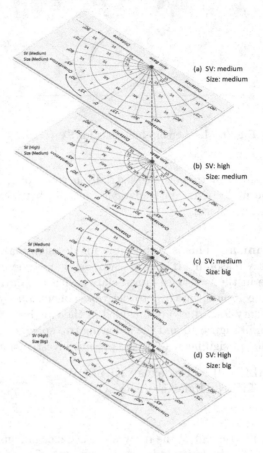

Fig. 10. Rule base for touchability (VL-VeryLow, L-Low, ML-MediumLow, M-Medium, MH-MediumHigh, H-High, VH-VeryHigh).

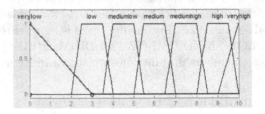

Fig. 11. Membership functions for Touchability Index (o_t).

Figure 12 shows the resulting rock ranking. The centre of the frame is the arm base, the SV is represented by colour, with the relevant colour coding defined in Table 2, and the size of the rock is depicted by the diameter of the colour circle. It summarises that the touchability sequence of these rocks can be intu-

Table 1. Experimental data for simulation.

Rock No.	Length	Width	SV	Orientation	Distance
hline 1	10	15	105	12	132
2	10	15	65	-65	166
3	10	15	35	50	111
4	20	15	105	-17	161
5	20	15	65	-33	126
6	20	15	35	72	151
7	30	20	105	5	148
8	30	20	65	32	167
9	30	20	35	-46	112

itively ranked as shown in Table 3, where the rank is sorted with respect to
the magnitude order of the *Touchability Index*, *TIndex*. These results compare
perfectly with those given by the human expert, demonstrating the validity of
the proposed approach.

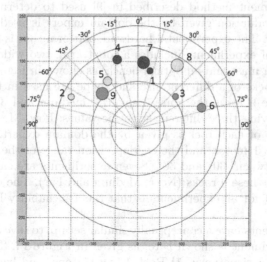

Fig. 12. Simulated experiment environment.

Table 2. Correspondence between SV and colour.

Corresponding Colour							
SV Score	<20	20-39	40-59	60-79	80-99	100-119	>120

Table 3. Simulation-based experimental result.

RockNO.	Human Rank	TIndex(%)	TRank
1	5	35.2	5
2	7	23.8	7
3	8	22.3	8
4	2	88.4	2
5	4	54.1	4
6	9	18.4	9
7	1	96.6	1
8	3	67.5	3
9	6	34.9	6

4 Hardware Implementation

This section reports results on the use of real rocks and data to further verify the validity of the proposed approach for fuzzy logic-based touchability evaluation through hardware implementation. All images taken are segmented manually, with the measurement method described in [9] used to determine the size of each rock. The evaluation given by the domain expert is used as the ground truth in this study. The experimental hardware platform that is implemented to perform this set of experiments includes a robot arm, two wide-angle cameras (WACs), a camera mast and an optical bench, as shown in Figure 13.

A total of 9 rocks of a different size and shape are used in this experimentation. Within the 9 rocks there are 3 small, 3 medium and 3 big ones, each type of rock involving three different scientific values: Low, Medium and High. The science value of each rock is given by the domain expert. The length of the Schunk arm is 1 meter, and the overlap vision range of the two cameras is approximately between -30° and 30°. Seven experiments are carried out based on the location of these 9 rocks (Near, Medium and Far). The evaluation from the domain expert for all experiments is that the Touchability Index should be at least 80%.

As all experiments have a conceptually similar set up, to save space, only that for the first experiment is shown here, as given in Figure 14. This experiment involves four target situations: (1) Rock 1 is a big one, and has a high science value; its distance to the robot arm is medium. (2) There is a low scientific value for the small Rock 2 with a medium distance. (3) A high science value and a medium distance are assumed for small Rock 3. (4) Small rock 4 possesses a medium scientific value and is of a near distance to be robot arm. The evaluation of the touchability given by the domain expert for these four rocks is that only Rock 1 is touchable. Figure 14 shows the result produced by the Schunk arm. The Touchability Index and the relevant measurement computed are presented in Table 4.

As specified by the domain expert, a positive result is achieved if the computed Touchability Index over a certain rock is greater than 80%. In this first experiment only the Touchability Index over Rock 1 (92.5%) is greater than

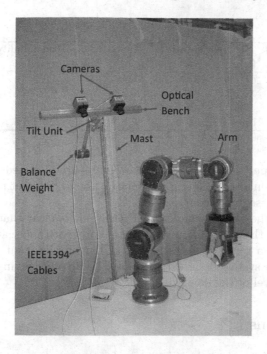

Fig. 13. Experimental platform.

Fig. 14. Top left image: captured by the left camera. Top right image: captured by the right camera. Bottom image: result of touchability computation.

80%. This means that the robot arm can reach out for this rock but not the rest. This result matches well with the evaluation of the domain expert. Similar results are obtained for experiments 4 and 6.

In experiment two the Touchability Index over all rocks is less than 80%, so the robot arm cannot reach any rocks. The evaluation of the touchability given

Table 4. Results from the first experiment.

Rock No.	Size (cm^2)	Orientation	Distance (cm)	Science Value	Touchability Index
1	737.86	18°	76.4	120	92.5%
2	26.33	-6°	75.3	40	9.76%
3	24.4	-25°	52.7	100	9.75%
4	109.46	20°	37.6	90	21.6%

by the domain expert is the same. Similar results are obtained for experiments three, five and seven also, although different rocks and different numbers of rocks are involved in those experiments.

Summarising the above experimental results, it is clear that the evaluation outcomes of using the present approach match fully with those attainable by the domain expert in a range of target settings. This implies that the fuzzy logic-based touchability system designed herein is capable of achieving the experience and knowledge level of the domain expert.

5 Conclusions

In this paper, a fuzzy logic-based system for autonomous touchability evaluation of space science targets has been presented. The membership functions and fuzzy rules have been devised and the defuzzification mechanism identified. The approach has been implemented in both software and hardware.Simulation-based experimentation has shown the validity of the proposed system, which has been further confirmed by the results of seven independent experiments on real settings, over different rock locations, sizes and science values. The system has proven to be able to achieve the performance of a domain expert. Whilst successful, further research remains. This includes an investigation of whether the approach is sensitive to the use of different fuzzification and defuzzification methods, and a study of how fuzzy rules may be learned from historical missions.

References

1. Anderson, R.C., Jandura, L., Okon, A.B.: Collecting Samples in Gale Crater, Mars; an Overview of the Mars Science Laboratory Sample Acquisition, Sample Processing and Handling System. Space Science Reviews **170**, 57–75 (2012)
2. Seraji, H.: Traversability index: a new concept for planetary rovers. In: Proceedings of IEEE International Conference on Robotic and Automation (1999)
3. Howard, A., Seraji, H., Werger, B.: Fuzzy terrain-based path planning for planetary rovers. In: Proceedings of EEE International Conference on on Fuzzy Systems, pp. 316–320 (2002)
4. Mahmound, T.: Hybrid Intelligent Path Planning for Articulated Rovers in Rough Terrain. Fuzzy Sets and Systems **159**, 2927–2937 (2008)

5. Navid, S., Homayoun, S.: Landing site selection using fuzzy rule-based reasoning. In: Proceedings of International Conference on Robotics and Automation, pp. 4899–4904 (2007)
6. Furfaro, R., et al.: The Serach for Life Beyond Earth through Fuzzy Expert Systems. Planetary and Space Science **56**, 448–472 (2008)
7. Barnes, D., Pugh, S., Tyler, L.: Autonomous science target identification and acquisition (ASTIA) for planetary exploration. In: Proceedings of International Conference on Intelligent Robots and Systems, pp. 3329–3335 (2009)
8. Pugh, S., Barnes, D., Tyler, L.: AUPE-A PanCam emulator for the ExoMars 2018 mission. In: Proceedings of International Symposium on Artificial Intelligence, Robotics and Automation in Space (2012)
9. Gui, C., Barnes, D., Pan, L.: A Method for matching desired non-feature points to size martian rocks based upon SIFT. In: Proceedings of Towards Autonomous Robotic System Conference (2014)

Simultaneous Localization and Mapping Based on (μ+1)-Evolution Strategy for Mobile Robots

Yuichiro Toda[✉] and Naoyuki Kubota

Graduate School of System Design, Tokyo Metropolitan University,
6-6 Asahigaoka, Hino, Tokyo, Japan
toda-yuuichirou@ed.tmu.ac.jp, kubota@tmu.ac.jp

Abstract. Simultaneous Localization and Mapping (SLAM) is one of the most important capabilities for autonomous mobile robots, and many researches have been proposed demonstrating the effective SLAM methods. However, these SLAM methods sometimes require assumptions such as the sensor model, which is difficult to implement and use the SLAM methods. In our previous work, a SLAM method based on Evolution Strategy (ES) was proposed and the on-line SLAM in indoor environments was realized. However, the definition of the map building method was not clear. Therefore, we propose a SLAM method based on a simple map building and search method. In this paper, we explain our autonomous mobile robot system and propose our SLAM method based on (μ+1)-ES. The experimental results show the effectiveness of the proposed method.

Keywords: SLAM · Occupancy grid map · Evolution strategy · Intelligent robotics

1 Introduction

Simultaneous Localization and Mapping (SLAM) is a fundamental problem for an autonomous mobile robot because the robot searches an unknown environment and perform a decision making according to a situation in the environment [1]. Various types of methods for SLAM have been proposed such as Extended Kalman Filter (EKF) SLAM, Graph SLAM, visual SLAM. The EKF SLAM algorithm applies the EKF to online SLAM using maximum likelihood data associations. In the EKF SLAM, feature-based maps are used with point landmarks [2]. Graph SLAM solves a full SLAM problem in offline using all data obtained until the current time, e.g., all poses and all features in the map. Therefore, Graph SLAM has access to the full data when building the map [3,4]. Furthermore, cooperative SLAM (C-SLAM) has been also discussed in the study of multi-robot systems [5, 6].

In our previous work, we proposed a SLAM and initial self-localization method for multi-robot system based on the map sharing approach [7]. In the proposed method, one leader robot performs SLAM based on occupancy grid mapping. Both of localization and map building are performed by (μ+1)-Evolution Strategy (ES) [8]. However,

© Springer International Publishing Switzerland 2015
H. Liu et al. (Eds.): ICIRA 2015, Part III, LNAI 9246, pp. 62–69, 2015.
DOI: 10.1007/978-3-319-22873-0_6

our map building method is not stable because the map value uses an empty, occupied, partially occupied and unknown, and the definition of the partially occupied is not clear. Therefore, we apply a simple occupancy grid map method to our SLAM method for defining the map building method more clearly. Next, we propose the localization method based on (μ+1)-ES whose fitness function is composed of the summation of the map value and one penalty function for realizing on-line SLAM. In our SLAM, it is easy to implement the method, and our SLAM method can build the map and localize the robot position accurately.

This paper is organized as follows. Section 2 explains our robot system for an autonomous mobile robot. Section 3 proposes an SLAM method based on (μ+1)-ES. Finally, we show several experimental results of our SLAM method by using SLAM benchmark datasets.

2   SLAM

2.1   Mobile Robots

At first, we explain the hardware specification of a mobile robot (Table 1). We use an omni-directional robot with four omni-wheels and DC motors (Fig. 1). The robot can move to different omni-direction by changing the combination of output levels to motors. Basically, the action outputs of the robot are direct forward movement and rotation at the same position to avoid the slip appeared as noise in SLAM. Furthermore, the robot changes the moving direction only when the robot conducts obstacle avoidance. In addition, we use a laser range finder (LRF, URG04-LN) for SLAM.

Table 1. Specification of Omni-directional mobile robot

Diameter	300 mm
Height	177 mm
Weight	8 kg (approximately)
Maximal Speed	1.5 km/h
Operating Time (Battery)	1 hour
Maximal Payload Weight	15 kg
Communication Method	Wi-Fi (2.4 GHz)

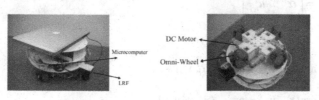

Fig. 1. Omni-directinal mobile robot

2.2   Procedure for SLAM

In our previous work, we use the occupancy grid map based on the following map value [9].

$$map(x,y)=\begin{cases} 1 & \text{(occupied)} \\ 0.5 & \text{(partially occupied)} \\ 0 & \text{(unknown)} \\ -1 & \text{(empty)} \end{cases}. \tag{1}$$

Here the value of all cells is initialized at 0. Figure 2 shows the concept of the occupancy grid map. However, almost all cells excluding the empty and unknown cells are partially occupied cells because the cell including an object is defined as the partially occupied cell unless the object size fits the cell size correctly. Therefore, we use the simple definition of the occupancy grid map as follows;

$$map_t(x,y)=\frac{\text{hit}_t(x,y)}{\text{hit}_t(x,y)+\text{err}_t(x,y)}. \tag{2}$$

where $\text{hit}_t(x,y)$ and $\text{err}_t(x,y)$ are the number of measurement and through points of LRF until the tth step, respectively. The measurement data is represented by (d_i, θ_i), $i=1,2, ..., M$, $j=1,2, ..., L$, where d_i is measurement distance from LRF; θ_i is the angle of the measurement direction; M is the number of total measurement directions; L_i $(=\left[\alpha^{Res}\cdot d_i\right])$ is the number of resolution for the map building by the occupancy grid model. Therefore, the map is updated by following procedure

---

Algorithm 1. Map-update

for $i=1$ to M do

    for $j=1$ to L_i do

$$u_{i,j} = \frac{j}{L_i}\left(d_i\cos(\theta_i + r_p)\right) + x_p \tag{3}$$

$$v_{i,j} = \frac{j}{L_i}\left(d_i\sin(\theta_i + r_p)\right) + y_p \tag{4}$$

$$x_{i,j} = \left[\alpha^{Map}\cdot u_{i,j}\right] \tag{5}$$

$$y_{i,j} = \left[\alpha^{Map}\cdot v_{i,j}\right] \tag{6}$$

      if $j = L_i$ then

        $\text{hit}_t(x_{i,j}, y_{i,j}) = \text{hit}_{(t-1)}(x_{i,j}, y_{i,j}) + 1 \tag{7}$

      else

        $\text{err}_t(x_{i,j}, y_{i,j}) = \text{err}_{(t-1)}(x_{i,j}, y_{i,j}) + 1 \tag{8}$

      endif

    endfor

endfor

where (x_p, y_p) is the position of the mobile robot; r_p is the posture; d_i is measurement distance from LRF in the ith direction; θ_i is the angle of the measurement direction; α^{MAP} is the scale factor mapping from the real world to the grid map.

Next, we explain out localization method. Basic localization algorithm is the almost same as out previous work. We apply (μ+λ)-ES for estimating the correct robot pose where μ and λ indicate the number of parent and offspring population generated in a single generation, respectively. ES can be used easily to discuss the formulation of strategy of artificial evolution in evolutionary computing [8]. We use (μ+1)-ES to enhance the local hill-climbing search as a continuous model of generations, which eliminates and generates one individual in a generation. A candidate solution is composed of numerical parameters of revised values to the current position ($g_{k,1}, g_{k,2}$) and rotation angle ($g_{k,3}$). In (μ+1)-ES, only an existing solution is replaced with the candidate solution generated by the crossover and mutation. We use the elitist crossover and adaptive mutation. Elitist crossover randomly selects one individual, and generates an individual by combining genetic information between the selected individual and best individual in order to obtain feasible solutions from the previous estimation result rapidly. Next, the following adaptive mutation is performed to the generated individual,

$$g_{k,h} \rightarrow g_{k,h} + \left(\alpha_h \cdot \frac{f_{max} - f_k}{f_{max} - f_{min}} + \beta_h \right) \cdot N(0,1) \tag{9}$$

where f_k is the fitness value of the kth individual, f_{max} and f_{min} are the maximum and minimum of fitness values in the population; $N(0,1)$ indicates a normal random value; α_h and β_h are the coefficient and offset, respectively. A Fitness value of the kth candidate solution is calculated by the following equation,

$$fit_k = p_t^{occ} \left(x_{i,L}, y_{i,L} \right) \cdot \sum_{i=1}^{M} map_t \left(x_{i,L}, y_{i,L} \right)$$

$$p_t^{occ} \left(x_{i,L}, y_{i,L} \right) = \frac{\sum_{i=1}^{M} hit_t' \left(x_{i,L}, y_{i,L} \right)}{\sum_{i=1}^{M} hit_t' \left(x_{i,L}, y_{i,L} \right) + \sum_{i=1}^{M} err_t' \left(x_{i,L}, y_{i,L} \right)} \tag{10}$$

$$hit_t' \left(x_{i,L}, y_{i,L} \right) = \begin{cases} 1 & \text{if } hit_t \left(x_{i,L}, y_{i,L} \right) > 0 \\ 0 & \text{else if } err_t \left(x_{i,L}, y_{i,L} \right) > 0 \end{cases}$$

$$err_t' \left(x_{i,L}, y_{i,L} \right) = \begin{cases} 1 & \text{if } err_t \left(x_{i,L}, y_{i,L} \right) > 0 \\ 0 & \text{else if } hit_t \left(x_{i,L}, y_{i,L} \right) > 0 \end{cases}$$

where the summation of the map values is basic fitness value in (μ+1)-ES and p_t^{occ} indicates a penalty function. The summation of the map values is high if the estimation result is high. Furthermore, the penalty function has low value if many measurement points exist on empty cells. Therefore, this problem is defined as a maximization problem. Actually, we can estimate the robot pose by using only the summation of the map values. However, the estimation method sometimes gets stuck in local optima according to the environment if we use only the summation of the map value. Therefore, we use the penalty function p_t^{occ} for avoiding the situation. The localization based on $(\mu+1)$-ES is finished when the number of iteration reaches the maximum number of iteration T. Algorithm 2 shows the total procedure of our SLAM method. Our SLAM procedure is very simple algorithm and it is easy to implement.

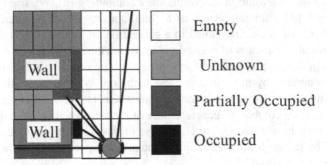

Fig. 2. Concept image of the occupancy grid map in our previous work [7]

Algorithm 2. Total Procedure of SLAM
1. $t = 0$, $hit_t(x,y) = 0$ and $err_t(x,y) = 0$
2. Input the LRF data
3. if $t \neq 0$ then
4.   Estimate the robot pose using $(\mu+1)$-ES
5. end if
6. Perform Map-update
7. $t = t + 1$
8. return to step 1

3   Experimental Results

We conducted an experiment of the proposed method by using two SLAM benchmark datasets [10,11]. We used only measurement data of LRF form these datasets. Figure 3 shows the ground truth of each dataset. In this experiment, we used two conditions. Condition 1 used the fitness function with p_t^{occ}. Condition 2 used the fitness function without p_t^{occ}. Our proposed algorithm was run on 3.5GHz 6-Core Intel Xeon E5 processor. Table xx shows the parameters using these experiments.

Figure 4 and 5 show an example of the experimental result of Condition 1 in each dataset. In these results, our proposed method can correctly localize and build the map compared to Fig. 4 and 5. Figure 6 shows the transition of variance of the best fitness value in each time step. The variance values are stable between until about the 1400th step in both results. On the other hand, Condition 1 is more stable than Condition 2 from about the 1400th step. Fig. 7 shows a failed example of Condition 2. In Fig. 7, red circles mean that the localization is often failed in these areas because (μ+1)-ES gets stuck in local optima in these areas if the fitness function does not include the penalty function p_t^{occ}. Table 3 shows experimental results of computational time in each dataset. In both results, the average of the computational time is about 18 [ms], and we consider that this computational time is enough for on-line SLAM. In this way, SLAM based on (μ+1)-ES can build the map and localize the robot position by designing the suitable fitness function according to the map building method.

Table 2. Setting parameters for the experiments

Number of trial in each experiment	10
Number of parents μ	100
Maximum number of iterations T	1000
Coefficients for adaptive mutation α_1, α_2	10.0
Coefficients for adaptive mutation α_3	1.0
Offset for adaptive mutation β	0.01
Cell size α^{Map}	100

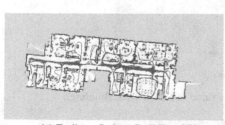
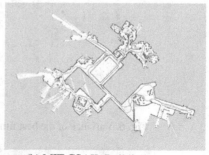

(a) Freiburg Indoor Building 079   (b) MIT CSAIL Building

Fig. 3. Correct map building result using the benchmark datasets [11].

(a) Map building   (b) Localization

Fig. 4. Experimental results of map building and localization in Freiburg Indoor Building 079. In (b), red line indicates localization result.

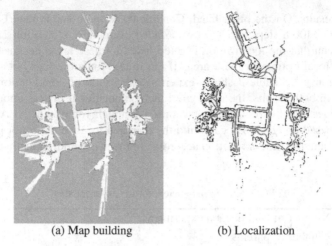

(a) Map building   (b) Localization

Fig. 5. Experimental results of map building and localization in MIT CSAIL Building. In (b), red line indicates localization result.

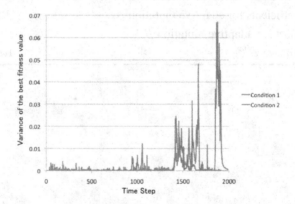

Fig. 6. Variance of the best fitness value in MIT CSAIL Building.

Fig. 7. A failed example of Condition2.

Table 3. Experimental results of computational time [ms]

	Freiburg Indoor Building 079	MIT CSAIL Building
Average	18.8	18.6
Variance	3.71	3.75
Max	28	26
Min	12	11

4   Summary

In this paper, we proposed our SLAM based on Evolution Strategy (ES). Our SLAM method changed the definition of the occupancy and grid map. Next, we proposed our localization method based on (μ+1)-ES. The fitness function of the ES used the summation of the map value and one penalty function based on occupied and empty cells. The experimental results showed that our SLAM method realized on-line SLAM by using the SLAM benchmark datasets. However, our SLAM method sometimes gets stuck in local optima because the search strategy of (μ+1)-ES is based on hill climbing search. Therefore, we will change the search method for combining the local and global search based on ES.

References

1. Thrun, S., Burgard, W., Fox, D.: Probabilistic Robotics. The MIT Press (2005)
2. Huang, G.P., Trawny, N., Mourikis, A.I., Roumeliotis, S.I.: Observability-based consistent EKF estimators for multi-robot cooperative localization. Journal Autonomous Robots **30**(1), January 2011
3. Folkesson, J., Christensen, H.I.: Closing the Loop With Graphical SLAM. IEEE Transactions on Robotics **23**(4), 731–741 (2007)
4. Kaess, M., Ranganathan, A., Dellaert, F.: iSAM: Incremental Smoothing and Mapping. IEEE Transactions on Robotics **24**(5), 1365–1378 (2008)
5. Pinheiro, P., Wainer, J.: Planning for multi-robot localization. In: da Rocha Costa, A.C., Vicari, R.M., Tonidandel, F. (eds.) SBIA 2010. LNCS, vol. 6404, pp. 183–192. Springer, Heidelberg (2010)
6. Choi, J., Choi, M., Nam, S.Y., Chung, W.K.: Autonomous topological modeling of a home environment and topological localization using a sonar grid map. Journal Autonomous Robots **30**(4), May 2011
7. Toda, Y., Kubota, N.: Self-localization Based on Multi-resolution Map for Remote Control of Multiple Mobile Robots. IEEE Transactions on Industrial Informatics **9**(3), 1772–1781 (2013)
8. Fogel, D.B.: Evolutionary Computation. IEEE Press (1995)
9. Thrun, S.: Learning Occupancy Grid Maps With Forward Sensor Models. Autonomous Robots, Springer **25**(2), 111–127 (2003)
10. Burgard, W., Stachniss, C., Grisetti, G., Steder, B., Kummerle, R., Dornhege, C., Ruhnke, M., Kleiner, A., Tardos, J.D.: A comparison of SLAM algorithms based on a graph of relations. In: Intelligent IEEE/RSJ International Conference on Robots and Systems, IROS 2009, pp. 2089–2095, October 10-15, 2009
11. slam benchmarking. http://kaspar.informatik.uni-freiburg.de/~slamEvaluation/index.php

Novel Model for SLAM in UAS

Alexander Harman, Olga Duran, and Yahya Zweiri[✉]

Kingston University London, Roehampton Vale Campus, London SW15 3DW, UK
{a.harman,o.duran,y.zweiri}@kingston.ac.uk
http://www.kingston.ac.uk

Abstract. The Simultaneous Localization and Mapping (SLAM) problem is of great significance within the modern field of unmanned systems. However, many current methodologies have high cost implications, utilising expensive Light Detection and Ranging (LIDAR) or Charged Coupled Devices (CCD) sensors to obtain information pertaining to the local topology of the device. The objective of this paper is to reduce the inherent cost of SLAM by generating a motion model which is suitable for use with the low cost Microsoft Kinect sensor system.

A novel filtering methodology is developed which can separate the static and dynamic accelerations in order to compute a full 6 DOF pose estimate from a 3 axis–accelerometer, suitable for application as a SLAM motion model. The filter is seen to operate in constant time, at a frequency sufficient for on–line implementation to a suitable level of accuracy for use with SLAM.

Keywords: Filtering · Kinect sensor · SLAM problem · Unmanned Aerial Systems (UAS)

1 Introduction

Many current automated applications such as Unmanned Aerial Systems (UAS) equipped with way-point following capabilities are capable of traversing a pre-defined flight path without human interaction. To enable such a functionality, it is necessary for the UAS to possess accurate information with respect to its current position; within external environments this information is obtained using a GPS receiver. If the pose is known, the flight path should provide a transformation from the initial to the final position whilst obeying the constraints imposed by the vehicle dynamics and any interim obstacles. Such a transform is achievable through the application of a path planning algorithm such as *Dubins Curves* or *Pythagorean Hodographs*. However, within urban environments, or areas in which the GPS signal is intentionally or otherwise disrupted, the localisation method is unsuitable. Therefore an alternate method of localisation must be developed.

High accuracy Inertial Navigation System (INS) can be used to provide a complete estimate of the current pose [5]. A number of researchers have utilised GPS and INS navigation systems for this purpose; typically through the application of a *Kalman Filter* (KF) for data fusion [6]. Such techniques are shown

© Springer International Publishing Switzerland 2015
H. Liu et al. (Eds.): ICIRA 2015, Part III, LNAI 9246, pp. 70–79, 2015.
DOI: 10.1007/978-3-319-22873-0_7

to provide both increased accuracy when GPS and INS data are available and adequate accuracy during short term GPS denial [6].

The GPS and INS data fusion techniques are seen to perform well for short term GPS isolation; however, this is not the case if the UAS should operate indoors or during GPS denial. If *a priori* knowledge of the environment is available, the system can be provided with a map of it's environment and the problem is reduced to simple localisation within this framework. Within this environment, the distance measurements to recognisable features can be obtained with Light Detection and Ranging (LIDAR) or through stereo vision realised by CCD sensors; the distance measurements can then be used to localise the system within the map. If however, no *a priori* knowledge of the environment is available, the system is faced with a dilemma. It is not possible to establish the localisation of the UAS within an environment without knowledge of the local topology. This led to the formulation of the Simultaneous Localisation and Mapping (SLAM) problem. The SLAM posterior in this format possesses no closed form solution and therefore an Extended Kalman Filter (EKF) can be used.

The suitability of the *Kinect* for application to SLAM problems is currently the subject of much debate [1][7]. This paper presents further evidence towards this discussion and investigates the practicabilities of SLAM with small monitoring angle and high error sensor systems.

The motion model is a prerequisite for the implementation of all versions of SLAM. When considering the EKF it is seen that the algorithm naturally subdivides into the estimation and update phases. During the estimation phase, an approximation of the system pose is obtained as a function of the control input. This is represented as the state transition probability [2]. The motion model can be obtained through the application of inertial dead–reckoning, which allows the motion model to be applied to any system which is equipped with the required sensor payload. In addition to this it extends the motion model to systems for which a dynamic model is difficult, if not impossible to derive for example a hand held *Kinect*. This paper presents the derivation of an inertial dead–reckoning system which utilises only a three–axis accelerometer. This enables feature based SLAM at a fraction of the normal cost. The motion model uses a novel filter for separating the static and dynamic acceleration signals which is shown to be more effective than existing solutions.

2   Motion Model

The motion model can be represented by some probabilistic function of the system dynamics or via interpretation of some form of robot odometry. If the UAS is considered as a whole it is a relatively simple matter to derive a dynamic model for a quad–rotor system [3]. However, a more generalised motion model can be obtained through the application of inertial dead–reckoning, which allows the motion model to be applied to any system which is equipped with the required sensor payload. In addition, it extends the motion model which is difficult to derive for a hand held Kinect.

2.1   Static Acceleration Filtering

The raw data acquired from the three–axis accelerometer is in fact the sum of two distinct accelerations; static and dynamic acceleration. If the static acceleration component can be isolated, it provides information with respect to the devices rotational motion with respect to earth [9]. It is seen that a 6 DOF pose estimate is calculated from a three–axis accelerometer provided that the static and dynamic accelerations can be reliably separated by some form of post processing.

Usually, it is assumed that the gravity vector can be removed via the application of a high pass filter. However discrete FIR filters must be of an impractically high order which is unsuitable for this application. In this paper, a unique filtering methodology is presented which overcomes many of the obstructions to achieving inertial dead–reckoning with a three–axis accelerometer.

2.2   Frequency Domain Methodology

To test the influence of gravity within the overall acceleration measurement which is expected to appear at a constant frequency across each accelerometer axis, accelerometer data was gathered from a *Kinect* at $100Hz$ which was rotating at approximately $1Hz$ about the y–axis (pitch rotations). The results of the experiment were transformed into the frequency domain by means of a **F**ast **F**ourier **T**ransform (FFT).

From figure 1, it is clear that the gravity vector can be identified by the amplitude corresponding to $9.81ms^{-2}$ within the frequency domain of the magnitude of the measured acceleration values. This observation permits the frequency of the gravity vector to be obtained and therefore isolated from each of accelerometer axes within the frequency domain. The resulting filtered signal can then be transferred into the time domain by the application of an Inverse Fast Fourier Transform (IFFT).

This algorithm was applied to accelerometer readings provided by the Freiburg dataset [8], the results of which can be seen in figure 2. It can be seen that this filter has removed the influence of gravity much more precisely than the FIR HPF. In this case the gravity vector was near DC, and therefore it can be seen that the resulting dynamic acceleration values closely mimic the raw data. Furthermore it should be noted that, for a signal length of 1000 samples, the algorithm completed in $7.0493 \times 10^{-4}s$ which permits a data acquisition frequency of $1.418kHz$, approximately 15 times faster than the FIR HPF.

2.3   Mathematical Motion Model

The acceleration vector a obtained directly from a three axis accelerometer is in–fact the sum of measured static, g and dynamic acceleration, α values such that:

$$\begin{Bmatrix} a_{x,b} \\ a_{y,b} \\ a_{z,b} \end{Bmatrix} = \begin{Bmatrix} g_{x,b} \\ g_{y,b} \\ g_{z,b} \end{Bmatrix} + \begin{Bmatrix} \alpha_{x,b} \\ \alpha_{y,b} \\ \alpha_{z,b} \end{Bmatrix} \tag{1}$$

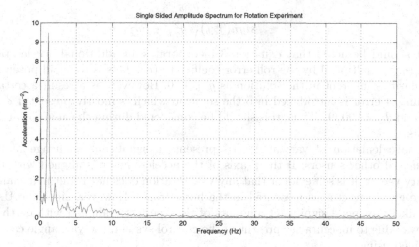

Fig. 1. The Single Sided Amplitude Spectrum for a *Kinect* Sensor Subjected to Pitching at $f \approx 1Hz$

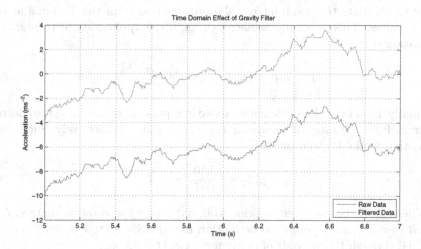

Fig. 2. The Effect of Applying the Designed FIR HPF Filter Kernel on Raw Accelerometer Data

Once the dynamic acceleration is obtained in the body reference frame, the static acceleration can be calculated using equation 1. From these values the rotational and translational positions can be obtained.

If the static acceleration is considered, pitch (θ) and roll (ϕ) can be obtained as follows.

$$tan(\theta) = \frac{-g_{x,b}}{\sqrt{g_{y,b}^2 + g_{z,b}^2}} \qquad (2)$$

$$tan(\phi) = \frac{g_{y,b}}{sign(g_{x,b})\sqrt{g_{z,b}^2 + \beta g_{x,b}^2}} \tag{3}$$

It should be noted that equation 3 incorporates an additional acceleration term $(g_{x,b})$ multiplied by the roll error coefficient (β). This is to avoid potential singularities inherent in the solution as $g_{z,b} \to 0$. However, as a result a certain amount of error is introduced into the solution which is mostly present if $\phi \to \pm90$ and $\theta \to \pm90$ [9]. Thus, this should be considered during the analysis of the results.

The calculation of yaw angle (ψ) presents a significant challenge in the absence of other sensors. If the z–axis of the accelerometer is aligned with the gravity vector, it is self evident that any yaw rotation does not result in a change of the gravity vector as measured in the body frame. However, since the UAS is unlikely to spend significant amounts of time perfectly aligned with earth in this way due to inaccuracies present in the control system, the yaw angle can be obtained using:

$$tan(\psi) = \frac{g_{y,b}}{g_{x,b}} \tag{4}$$

Once the roll, pitch and yaw estimates have been obtained, it is then necessary to calculate the resultant translational motion from the dynamic acceleration values. This is readily achieved through the application of trapezoidal integration such as [4]:

$$\int_{t_{t-1}}^{t_t} \alpha dt = (t_t - t_{t-1})\left[\frac{\alpha^t + \alpha^{t-1}}{2}\right] \tag{5}$$

Finally, from the values calculated above it is possible to assemble the control vector u_t from the rates of change in x, y, z, ϕ, θ and ψ respectively. Such that the motion model is given by [2]:

$$\bar{\mu}_t = \mu_{t-1} + \left[\begin{array}{c|c} R & 0 \\ \hline 0 & I \end{array}\right] u_t + \Gamma \tag{6}$$

Where $\bar{\mu}_t$ is the state vector prior to update, R is the rotation matrix, I is a 3×3 identity matrix and Γ is the motion model noise matrix which has been obtained through the analysis of stationary *Kinect* data as:

$$\Gamma = \begin{bmatrix} 1.4801 & 0 & 0 & 0 & 0 & 0 \\ 0 & 0.6319 & 0 & 0 & 0 & 0 \\ 0 & 0 & 1.3042 & 0 & 0 & 0 \\ 0 & 0 & 0 & 0.0064 & 0 & 0 \\ 0 & 0 & 0 & 0 & 0.002 & 0 \\ 0 & 0 & 0 & 0 & 0 & 0.002 \end{bmatrix} \tag{7}$$

3   Results

Utilising the motion model, it is possible to assemble an algorithm which is capable of estimating the 6 DOF pose estimate from raw accelerometer data.

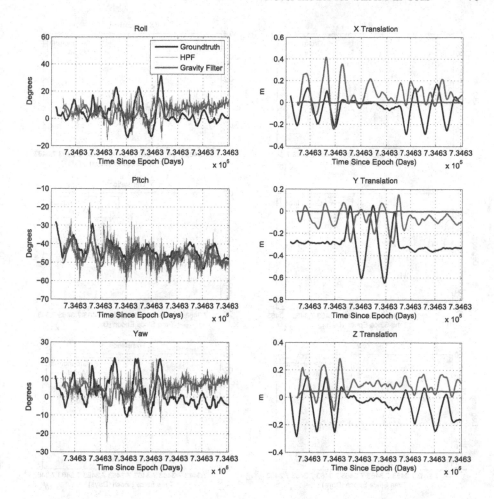

Fig. 3. A Comparison of the Pose Estimate Obtained by the FIR HPF and Frequency Based Gravity Filter

The results obtained in this paper using the motion model to compute pose estimates from accelerometer data provided by the Freiburg datasets are presented next. In addition a version of the Relative Pose Error (RPE) technique [8] has been utilised to calculate the percentage error between the corresponding ground truth and pose estimate rates of change.

A comparison between the FIR HPF and the frequency domain gravity filter has been presented in figure 3 to detail the difference in pose estimates obtainable by current methodologies and the technique proposed in this paper. Qualitative analysis of the plots clearly show that the frequency domain gravity filter has been significantly more successful in obtaining the true pose. Two distinct observations are made with respect to the FIR HPF technique; the rotational estimates exhibit high signal to noise ratio and no information has been produced

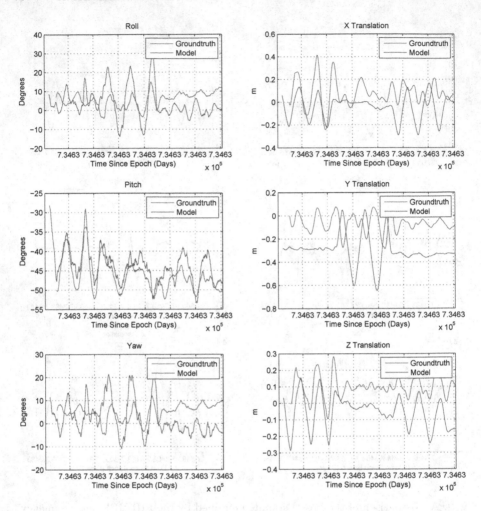

Fig. 4. The Results Obtained by the Motion Model for the Freiburg1 XYZ Dataset

with respect to translational motion. Since the width of this region is a direct function of the order of the filter, it is possible to affect a sharper transition between the pass and the stop bands, however this has an associated increase in computational expense. At this stage, it is prudent to note the difference in execution times between the two methodologies as compared to the duration of the dataset. The Frieburg dataset utilised for this analysis consists of 14156 accelerometer samples gathered at a frequency of $498.784Hz$ which results in a total real–time length of $28.38s$. When obtaining pose estimates utilising the FIR HPF, the total processing time was observed to be $152.85s$, as compared to the frequency domain gravity filter's $7.05s$. From this, it can clearly be seen that

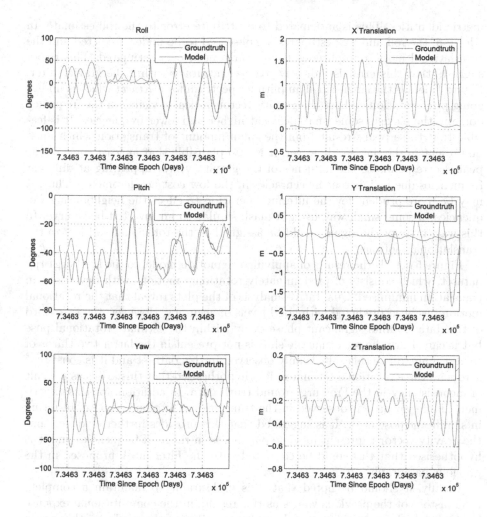

Fig. 5. The Results Obtained by the Motion Model for the Freiburg1 RPY Dataset

the proposed gravity filter offers significant computational savings and permits the data processing to occur in real–time for this dataset.

It has been demonstrated that the proposed methodology provides an improvement over current methodologies, it is then necessary to assess the filters performance across a range of different motions. Figure 4 details the results for the Freiburg1 xyz dataset which consists of predominately translational motions, with deliberate rotations kept to a minimum. It can be seen that the motion model has provided a good estimate of the pitch angle as well as x and z–axis translational motion; however, the roll and yaw angles as well as y–axis translation have not been estimated as reliably. It can be seen that the pitch angle of the Freiburg1 xyz dataset is operating close to the error region of the trigonometric identities. This is anticipated to contribute error to the roll estimate. In addition, the roll and yaw estimates are observed to be under predicted and the translational motion amplitudes have clearly been over estimated. This paper suggests that this error is indicative of over accuracy within the gravity filter. Figure 1 details the single sided amplitude spectrum for a *Kinect* which is undergoing pitch rotation. Whilst the gravity vector is clearly visible in the frequency domain, there exists some small peaks at higher and lower frequencies. It is feasible that these peaks result from the small amount of translation which must be present in the dataset; however, it is also plausible that at least some of the peaks corresponds to the influence of the gravity vector appearing at different frequencies due to inherent inaccuracies in the low cost accelerometer. Thus, it may be more prudent for the filtering operation to effect the neighbouring frequencies in some small way using a mask similar to an inverted bell–curve. In this way, more of the signal would be assigned to the rotational and less to the translational motions.

Figure 5 details the results of applying the motion model to the Freiburg1 rpy dataset, which consists of predominately rotational motions with deviations in translation minimised. Qualitative analysis of the plots reveal that the rotational motion has been predicted with good levels of accuracy. However, the first third of the data exhibits significant phase delay within the estimate rotational pose but it can clearly be seen that the shift is not present in the latter two thirds of the estimate. Since this shift is only observed in this dataset and it is consistent across all three rotational estimates, it is hypothesised that this arises as a result of inconsistency in the Freiburg ground truth data. In addition, for this experiment, large variance is observed in the translational motion estimate, resulting in significant pose error. It is suggested that this provides further evidence that the gravity vector is present in some part across a range of frequencies and it is hypothesised that this could be corrected with the filter mask proposed in the previous paragraph.

Finally, it should be noted that it is undesirable to maintain a complete time history of the previous values as this results in the computational expense increasing with every time–step. In order to mitigate these effects, this paper suggests a methodology similar to that utilised to perform a sliding FFT. This means that a buffer of previous acceleration values should be maintained and rotated as each new data sample becomes available, discarding the oldest sample. The length of this buffer has direct implications on the computational complexity of the filter and its accuracy. A longer time history of values increases the resolution of the FFT and thus the accuracy with which the effects of gravity can be removed.

4   Conclusions

This paper presented a motion model for SLAM which utilises the low–cost Microsoft Kinect sensor. The motion model is capable of obtaining a 6 DOF pose estimate with only the data available from the three–axis accelerometer.

Such model permits SLAM to be performed without the requirement of complex data fusion techniques. However, it should be noted that such a technique could not be used outside the SLAM framework as it is not possible to achieve the required level of accuracy for pure inertial dead–reckoning. The limitations of the proposed filter have been highlighted and suggestions towards minimising the effects of these shortcomings have been proposed. All the requirements of the motion model has been addressed. Furthermore, the proposed model provides a significant hardware cost reduction over conventional sensors utilised for this type of localisation problem; which should in turn contribute to the financial viability of disposable military or consumer grade unmanned systems.

References

1. Kamarulzaman, K., Syed, M., Ali, S., Ammar, Z.: Performance Analysis of the Microsoft Kinect Sensor for 2D Simultaneous Localization and Mapping (SLAM) Techniques. Sensors **14**, 23365–23387 (2014). ISSN 1424-8220
2. Thrun, S., Burgard, W., Fox, D.: Probabilistic Robotics. The MIT Press, Cambridge (2006)
3. Fahimi, F.: Autonomous Robots: Modelling, Path Planning and Control. Springer, New York (2009)
4. Singh, K.: Engineering Mathematics Through Applications. Palgrave Macmillan, London (2003)
5. Wang, J., Garratt, M., Lambert, A., Wang, J.: Integration of GPS/INS/vision sensors to navigate unmanned aerial vehicles. In: International Archives of the Photogrammetry, Remote Sensing and Spatial Information Sciences, ISPRS, vol. 37, part. B1, pp. 963–369 (2008)
6. Sasiadek, J., Wang, Q., Zeremba, M.: Fuzzy adaptive kalman filtering for INS/GPS data fusion. In: IEEE International Symposium on Intelligent, Control, pp. 181–189 (2000)
7. Yin-Tien, W., Chin-An, S., Jr-Syu, Y.: Calibrated kinect sensors for robot simultaneous localization and mapping. In: Proceedings of the 8th International Conference on Sensing Technology, pp. 104–109, Liverpool, UK, September 2-4, 2014
8. Sturm, J., Engelhard, N., Endres, F., Burgard, W., Cremers, D.: A Benchmark for the Evaluation of RGB-D SLAM Systems. In: Proc. of the International Conference on Intelligent Robot Systems (IROS) (2012)
9. Pedley, M.: Tilt Sensing Using a Three-Axis Accelerometer. Recent Advances in Computer Science and Information Engineering, Free-scale Semiconductor, no. AN3461, version 6, (2013). http://cache.freescale.com/files/sensors/doc/app_note/AN3461.pdf?fpsp=1

Intelligent System and Cybernetics

A New Sequential Approximate Optimization Approach Using Radial Basis Functions for Engineering Optimization

Pengcheng Ye, Guang Pan$^{(\boxtimes)}$, Qiaogao Huang, and Yao Shi

School of Marine Science and Technology, Northwestern Polytechnical University,
Xi'an, People's Republic of China
{ypc2008300718,panguang601,huangqiaogao,sy880408}@163.com

Abstract. For most engineering optimization problems, it is difficult to find the global optimum due to the unaffordable computational cost. To overcome this difficulty, a new sequential approximate optimization approach using radial basis functions is proposed to find the global optimum for engineering optimization. In the approach, the metamodel is constructed repeatedly to replace the expensive simulation analysis through the addition of sampling points, namely, extrema points of response surface and minimum point of density function. Optimization algorithms simulated annealing and sequential quadratic programming are employed to obtain the final optimal solution. The validity and efficiency of the proposed approach are tested by studying several mathematic examples and one engineering optimization problem.

Keywords: Sequential approximate optimization · Engineering optimization · Radial basis functions · Metamodel

1 Introduction

Engineering optimization for large-scale systems is generally sophisticated and time-consuming. A large number of function evaluations will be needed to find the global optimum via traditional optimization methods. It is reported that it takes Ford Motor Company about 36-160h to run one crash simulation [1], which is unacceptable in practice. If these computationally intensive models are directly used for engineering optimization, the computational burden would be unaffordable. For such engineering optimization, only a limited number of function evaluations are feasible [2]. To alleviate this computational burden, Metamodel which is often called surrogate model or response surface, as a widely used approximate model replacing the expensive simulation analysis has been proposed. Therefore, a preferable strategy is to utilize the metamodels instead of the expensive high fidelity models in the optimization process, termed as approximate optimization or surrogate-based optimization [3,4,5].

In this paper, a new sequential approximate optimization (SAO) approach using radial basis functions (RBF) [6,7,8] for engineering optimization is presented. Firstly, a certain number of initial sampling points are generated in the design space

© Springer International Publishing Switzerland 2015
H. Liu et al. (Eds.): ICIRA 2015, Part III, LNAI 9246, pp. 83–93, 2015.
DOI: 10.1007/978-3-319-22873-0_8

for constructing the initial metamodel with radial basis functions. Then, in order to utilize the geometrical feature of metamodels, extrema points of the response surface, as new optimal sampling points, are added to the sample. Moreover, an effective function [9,10] called the density function constructed by using the RBF for determining the sparse region in the design space is considered. The metamodel is constructed repeatedly through the addition of new sampling points, namely, extrema points of response surface and minimum point of density function found by the optimization algorithm simulated annealing (SA). Afterwards, the more accurate metamodel would be constructed by repeating the procedure of adding points to the sample adaptively and sequentially until the terminal or convergence criterion is satisfied. Finally, the global optimal solution of metamodel is found by SA. However, it may only find the near-optimal solution mainly due to the stochastic search feature of SA that cannot guarantee convergence to the exact optimal solution. Therefore, the near-optimal solution is taken as the starting point for the gradient-based optimization algorithm sequential quadratic programming (SQP) to find the exact global solution. Certainly, several mathematic examples and one engineering problem are tested to demonstrate the validity and efficiency of the proposed sequential approximate optimization approach.

The remainder of the paper is organized as follows. Approximate method radial basis functions and the sequential optimization sampling method are introduced in Section 2, and then the new sequential approximate optimization approach is proposed in Section 3. In section 4, some examples are tested to show the validity of the proposed SAO approach. Eventually, conclusions are drawn in Section 5.

2   Background

2.1   Radial Basis Functions

Radial basis functions was originally developed by Hardy [11] in 1971 to fit the irregular topographic contours of geographical data. The output of RBF network $\hat{f}(x)$, which corresponds to the response surface, has the general form of

$$\hat{y} = \hat{f}(x) = \sum_{i=1}^{N} \lambda_i \phi(\|x - x_i\|) = \Phi \cdot \lambda \tag{1}$$

where N is the number of sampling points, x is a vector of the design variables, x_i is a vector of design variables at the ith sampling point, $\|x - x_i\|$ is the Euclidean norm, $\Phi = [\phi_1, \phi_2, \ldots \phi_N] (\phi_i = \phi(\|x - x_i\|))$, $\lambda = [\lambda_1, \lambda_2, \ldots \lambda_N]^T$, ϕ is a basis function, and λ_i is the coefficient for the ith basis function. The approximate function $\hat{y}$ is actually a linear combination of some basis functions with weight coefficients λ_i. The basis functions multiquadric and Gaussian are best known and most often applied. The multiquadric is used in this paper because it's nonsingular and simple to use [12].

Although the approximate method RBF is good for nonlinear responses, it has been verified to be unsuitable for linear responses [13]. To make the RBF appropriate for linear responses, an RBF model can be augmented with a linear polynomial given as

$$\hat{f}(x) = \sum_{i=1}^{N} \lambda_i \phi(\|x - x_i\|) + \sum_{j=1}^{m} b_j p_j(x) \tag{2}$$

where $p_j(x)=[1, x_1,..., x_d]$ is a linear polynomial function, $m=d+1$ is the total number of terms in the polynomial, d is the number of design variables, and b_j is the weight coefficient. Equation (2) consists of N equations with $N+m$ unknowns. The additional m equations can be obtained from the following m constraints:

$$\sum_{i=1}^{N} \lambda_i p_j(x_i) = 0, \quad for \ j = 1,2,...,m \tag{3}$$

2.2   The Sequential Optimization Sampling Method

The accuracy of metamodels owes to the approximate methods and sampling methods primarily [5]. The sequential optimization sampling method proposed in this paper includes the procedure of adding extrema points of response surfaces and adding sampling points in sparse regions [9,10] by using density function.

Extrema Points of Response Surface. General knowledge tells us that adding sampling points at the site of valley and peak of response surfaces to the samples would improve the accuracy of metamodels at the greatest extent. The valley and peak of the response surfaces are extrema points of functions. In mathematics, the extrema are the largest or smallest values that the function takes a point within a given neighborhood.

McDonald [28] found that the RBF models created with Eq (1) are twice continuously differentiable when employing multiquadric as basis function for all $c \neq 0$. Considering the model with n dimensions in Eq (2), the gradients of the equation is

$$\frac{\partial \hat{f}}{\partial x} = \sum_{i=1}^{N} \frac{\lambda_i \phi'(r_i)}{r_i(x)}(x - x_i)^T + b_j \tag{4}$$

where $\phi'(r_i) = \partial \phi / \partial r_i$, $r_i(x) = \|x - x_i\|$. For multiquadric RBF model, $\phi(r_i) = (r_i^2 + c^2)^{1/2}$, $\phi'(r_i) = r_i(r_i^2 + c^2)^{-1/2}$.

The Hessian matrix can be calculated from Eq (4) as

$$H(x) = \frac{\partial^2 \hat{f}}{\partial x^2} = \sum_{i=1}^{N} \frac{\lambda_i}{r_i(x)}\left[\phi'(r_i) \cdot I + \left(\phi''(r_i) - \frac{\phi'(r_i)}{r_i(x)}\right)\right](x - x_i)^T \tag{5}$$

Similarly, $\phi''(r_i) = \partial^2 \phi / \partial r_i^2$. For multiquadric RBF model, $\phi''(r_i) = c^2(r_i^2 + c^2)^{-3/2}$.

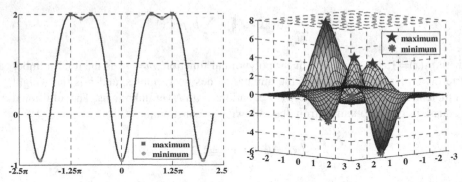

Fig. 1. Extrema points in 2-D   **Fig. 2.** Extrema points in 3-D

Let Equation (4) be equal to zero and solved

$$\frac{\partial \hat{f}}{\partial x} = 0, x = \left(x_1, x_2, \cdots x_N \right) \tag{6}$$

Then, the extrema points can be obtained. Fig. 1 and Fig. 2 separately show the extrema points in the case of 2-D, 3-D. The red square ('□') and blue pentagon ('★') indicate maxima points, meanwhile, the green dot ('.') and red asterisk ('*') indicate minima points. Once an approximate model has been created, we can obtain coordinates of extrema points.

Minimum Point of Density Function. Adding the extrema points of the response surfaces to the samples will lead to a local approximation with high accuracy. However, only the successive additions of the extrema points may result in finding the local minimum. It is necessary to add new sampling points in the sparse region. The addition of sampling points in the sparse region will lead to the global approximation. By this way, it is possible to avoid falling into the local minimum. Thus, global and local approximation will be achieved simultaneously through the above sequential sampling method. In this sequential sampling method, it is important to find the sparse region in the design space. To achieve this, the density function which is proposed by Kitayama et al. [9,10] is constructed using the RBF network in this paper. This density function generates local minimum in the sparse region, so that the minimum of this function can be taken as a new sampling point. The detail procedure of constructing the density function can refer to the literature [9,10].

3   The New Sequential Approximate Optimization Approach

Following the background knowledge introduced in the previous sections, the sequential approximate optimization approach using radial basis functions is proposed in this research to find the optimal solution for engineering optimization problems. Fig. 3 shows the flowchart for the proposed SAO approach.

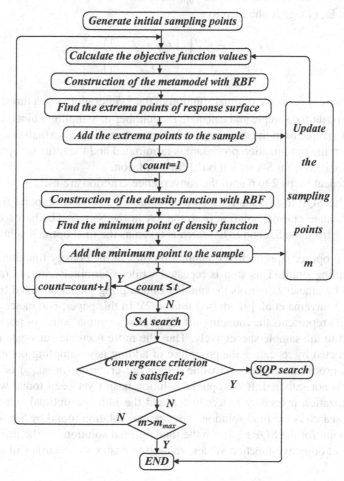

Fig. 3. Flowchart of the sequential approximate optimization approach

The approach can be roughly divided into three phases: (1) construct the initial metamodel in the design space. (2) Add new sampling points to the sample to update the metamodels successively. (3) SA and SQP search. The following are the steps of the sequential approximate optimization approach.

Step 1: Generate a certain number of sampling points in the whole design space using the Latin hypercube design (LHD).

Step 2: Analyse and calculate the exact objective function values of sampling points through simulation analysis programs or experiments.

Step 3: Construct the metamodel to replace the practical problem by using the RBF.

Step 4: Obtain the new sampling points i.e., extrema points of response surface and the minimum point of the density function and add them to the sample.

Step 5: Use the optimization algorithm SA to find the optimal solution of the metamodel. If the optimal solution f_{opt}^k satisfies the convergence criterion formated as Eq (7), go to the Step 8.

$$\frac{1}{2}\left(\left|\frac{f_{opt}^k - f_{opt}^{k-1}}{f_{opt}^{k-1}}\right| + \left|\frac{f_{opt}^{k-1} - f_{opt}^{k-2}}{f_{opt}^{k-2}}\right|\right) \le \varepsilon \tag{7}$$

where f_{opt}^k is the optimal solution in the kth iteration. ε is a threshold value.

Step 6: Update the sample and calculate the number of sampling points. If the number of sample m in the design space is larger than the m_{max} which is a certain number, the optimization procedure is terminated and the ultimate optimal solution obtained from SA search is the final solution.

Step 7: Repeat Steps 2 to 6 until the convergence criterion are met.

Step 8: Use SQP to find the final optimal solution. The starting point for SQP is the ultimate optimal solution from the SA in the Step 5. The best solution resulting from the SA and SQP search is taken as the final optimal solution.

In the proposed approach, the minimum point of the density function is taken as a new sampling point. This step is repeated while a terminal criterion (*count* $\le t$) is satisfied. Parameter t controls the number of sampling points obtained by the density function. Kitayama et al. [9] advised $t=int(n/2)$. In this paper, parameter $t=int(n/2)+1$, where $int()$ represents the rounding-off. Similarly, extrema points of response surfaces are added to the sample successively. Thus the more accurate surrogate model would be constructed by repeating the procedure of adding new sampling points to the sample. This procedure is repeated while a terminal criterion ($m>m_{max}$) or convergence criterion is not satisfied. If the optimal solution hasn't yet been found when $m>m_{max}$, the optimization procedure is terminated and the ultimate optimal solution obtained from SA search is the final solution. The optimal solution found by SA is taken as the starting point for the SQP to search the final optimal solution. At the final SQP search stage, exact objective function values are used to ensure the accuracy of solution.

4   Results and Discussions

In this section, the validity of the proposed SAO approach will be examined through some well-known mathematic examples and one engineering design optimization problem. In all of the numerical examples, the parameter ε in section 3 is set to 10^{-3}. The experimental results are calculated on a PC with an Inter Core i3 3.3GHz CPU.

4.1 Tests Using Several Benchmark Optimization Problems

The validity of proposed SAO approach is tested by five typical mathematic examples. These examples and their global minimum $\hat{f}_{min}$ are listed in Table 1 (see Appendix). For each test example, ten trials are carried out to avoid unrepresentative numerical results. In each of these examples, the initial number of sampling points generated by the LHD is set to twenty and the maximum number of sampling points m_{max} is set to 50. The results are shown in Table 2. It is clear from Table 2 that the proposed SAO approach is valid for the benchmark problems considered here.

Table 2. Results of benchmark problems

	Test 1	Test 2	Test 3	Test 4	Test 5
Minimum of global optima	-1.9863	0.3986	0.0121	-6.5483	-9.6276
Maximum of global optima	-1.9733	0.3998	0.0335	-6.5327	-9.6265
Average of global optima	-1.9812	0.3991	0.0227	-6.5446	-9.6273
Standard deviation of global optima	0.0236	0.0026	0.0368	0.0146	0.0004

Table 4. Comparisons results with other SAO approaches

No.	MPS [15]		HAM [14]		Proposed approach	
	NFE	$\hat{f}_{min}$	NFE	$\hat{f}_{min}$	NFE	$\hat{f}_{min}$
1	41.2	-1.0308	49.6	-1.0310	43.5	-1.0308
2	304.6	1.1721	72.8	0.0537	68.6	0.0328
3	44.7	0.7012	88.4	0.5367	58.6	0.5363
4	139.5	3.4341	160.6	0.0030	112.0	0.0018
5	57.0	0.0388	81.2	0.0053	51.7	0.0069

4.2 Comparison with Other Sequential Approximate Optimization Approaches

In this section, the proposed SAO approach has been tested using five mathematic examples listed in Table 3 (see Appendix), and compared with two other SAO approaches including Hybrid and adaptive meta-modelling (HAM) [14] and Mode-pursuing sampling (MPS) [15]. To avoid unrepresentative numerical results, ten runs are carried out for each example. In addition, the number of initial sampling points for constructing the metamodels is separately 25, 48, 25, 64, 36. The largest number of sampling points m_{max} is set to twice as much as the initial number of sampling points.

For engineering optimization problems, computational time is largely proportional to the number of black-box function evaluations. Therefore, reducing the function evaluations is an important item in the SAO approaches. From this viewpoint, the number of function evaluations (NFE) is thus used to measure computation efficiency. It is obvious that the global minimum is also shown to illustrate the validity of the approaches. The proposed SAO algorithm is compared with other SAO approaches through benchmark problems. The average values of the number of function evaluations (NFE) and global minimum $\hat{f}_{min}$ are shown in Table 4.

As shown by the results in Table 4, the proposed SAO approach perform well in terms of both NFE and global minimum for all test examples. MPS and HAM approaches require more expensive function evaluations, making them unsuitable to solve the engineering optimization problems. Meanwhile, the global minimum obtained by the proposed SAO approach is the best relatively.

4.3   Application to an Engineering Optimization Problem

The validity of the sequential approximate optimization approach is tested by a typical mechanical design optimization problem involving three design variables i.e., tension/compression spring design. Coello [16] and Ray [17] have used this as a benchmark problem in the structural optimization. The schematic of the tension/compression spring is shown in Fig. 4. It consists of minimizing the weight of a tension/compression spring subject to constraints on minimum deflection, shear stress, surge frequency, limits on outside diameter and on design variables. Three variables are identified: diameter d, mean coil diameter D, and number of active coils N. In this case, the variable vectors are given by

$$X = (d, D, N) = (x_1, x_2, x_3) \tag{8}$$

The mathematical model of the optimization problem is expressed as

$$\begin{aligned}
\min \quad & f(X) = (2 + x_3) x_1^2 x_2 \\
s.t. \quad & g_1(X) = 1 - \frac{x_2^3 x_3}{71785 x_1^4} \le 0 \\
& g_2(X) = \frac{4 x_2^2 - x_1 x_2}{12566 (x_2 x_1^3 - x_1^4)} + \frac{1}{5108 x_1^2} - 1 \le 0 \\
& g_3(X) = 1 - \frac{140.45 x_1}{x_2^2 x_3} \le 0 \\
& g_4(X) = \frac{x_1 + x_2}{1.5} - 1 \le 0
\end{aligned} \tag{9}$$

The ranges of the design variables $x_1 \in [0.05, 2]$, $x_2 \in [0.25, 1.3]$, $x_3 \in [2, 15]$ are used.

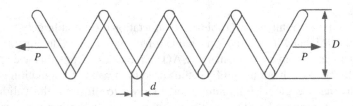

Fig. 4. Diagram of tension/compression spring design

The problem formulated above is a nonlinear constraint problem. Now assuming the objective and constraint functions defined by Eq (9) are computation-intensive functions and thus the reduction of the number of function evaluations is considered. Hence, metamodels of objective and constraint functions are constructed by RBF. 100 initial sampling points are generated by LHD for constructing the initial metamodel. Besides, the largest number of sampling points m_{max} is set to 200.

For a fair comparison, 11 trials are performed referring to the past researches [16,17]. The average values obtained by applying the proposed SAO approach are shown in Table 5. It is clear from Table 5 that the function evaluations are drastically reduced in comparison with those in the past researches. Meanwhile, the global minimum obtained by the proposed SAO approach is the best.

Table 5. Comparison of results on the optimum design of tension/compression spring

	x_1	x_2	x_3	g_1	g_2	g_3	g_4	NFE	$\hat{f}_{min}$
Coello [16]	0.0515	0.3517	11.632	-0.0021	-0.0001	-4.026	-0.731	900000	0.0127
Ray [17]	0.0504	0.3215	13.980	-0.0019	-0.0129	-3.899	-0.752	1291	0.0134
proposed approach	0.0516	0.3536	11.512	-0.0035	0	-4.030	-0.730	178.6	0.0127

5   Conclusion

In this paper, a sequential approximate optimization approach using radial basis functions is proposed to find the global optimum for engineering optimization. In the approach, the extrema points of response surfaces are taken as the new sampling points in order to improve the local accuracy. In addition, new sampling points in the sparse region are obtained for better global approximation. The density function constructed by using radial basis functions has been applied to determine the sparse region. Thus, global and local approximation will be achieved simultaneously through the above sequential sampling method. In the proposed approach, the metamodel is constructed successively until the terminal criterion or convergence criterion is satisfied. In this way, the more accurate metamodel would be constructed by repeating the procedure of adding new points to the sample adaptively and sequentially. The final optimal solution will be found by SA and SQP search.

In order to examine the validity of the proposed SAO approach, five benchmark problems and one engineering design optimization problem have been tested. In addition, the proposed SAO approach is compared with other SAO approaches through five numerical examples. The ultimate results demonstrate the validity and effectiveness of the proposed sequential approximate optimization approach.

The new sequential approximate optimization approach has been proposed, which provides another effective way for solving engineering optimization problems. However, the performance in high dimension is not good sometimes. The limitations mentioned above need to be resolved in future work.

Acknowledgements. Authors would like to thank everybody for their encouragement and support. The grant support from National Natural Science Foundation of China (51375389 and 51279165) are greatly acknowledged.

Appendix

Table 1. Benchmark problems and Global minimum

No.	Function	Design space	$\hat{f}_{min}$
1	$f(x) = x_1^2 + x_2^2 - \cos(18x_1) - \cos(18x_2)$	$-1 \le x \le 1$	-2
2	$f(x) = \left(x_2 - 5.1/(4\pi^2)x_1^2 + 5/\pi x_1 - 6\right)^2 + 10(1 - 1/(8\pi))\cos(x_1) + 10$	$-5 \le x_1 \le 10$ $0 \le x_2 \le 15$	0.398
3	$f(x) = -10\dfrac{\sin\sqrt{x_1^2 + x_2^2 + eps}}{\sqrt{x_1^2 + x_2^2 + eps}}, eps = 10^{-15}$	$-3\pi \le x \le 3\pi$	0
4	$f(x) = 3(1 - x_1)^2 \cdot e^{-x_1^2 - (x_2+1)^2} - 10\left(\dfrac{x_1}{5} - x_1^3 - x_2^5\right)$ $\cdot e^{-x_1^2 - x_2^2} - \dfrac{1}{3}e^{-(x_1+1)^2 - x_2^2}$	$-3 \le x \le 3$	-6.55
5	$f(x) = x_1 \sin(x_2) + x_2 \sin(x_1)$	$-2\pi \le x \le 2\pi$	-9.629

Table 3. Test examples and Global minimum

No.	Function	Design space	$\hat{f}_{min}$
1	$f(x) = 4x_1^2 - 2.1x_1^4 + x_1^6/3 + x_1x_2 - 4x_2^2 + 4x_2^4$	$-2 \le x \le 2$	-1.0316
2	$f(x) = \sum_{i=1}^{n} x_i^2/200 - \prod_{i=1}^{n} \cos(x_i/\sqrt{i}) + 1$	$-100 \le x \le 100$	0
3	$f(x) = \left(1.5 - x_1(1 - x_2)\right)^2 + \left(2.25 - x_1(1 - x_2^2)\right)^2$ $+ \left(2.625 - x_1(1 - x_2^3)\right)^2$	$-2 \le x \le 2$	0.5233
4	$f(x) = 100\left(x_2 - x_1^3\right)^2 + (x_1 - 1)^2$	$-10 \le x \le 10$	0
5	$f(x) = \left(x_1^2 + x_2 - 11\right)^2 + \left(x_1 + x_2^2 - 7\right)^2$	$-6 \le x \le 6$	0

References

1. Gu, L.: A comparison of polynomial based regression models in vehicle safety analysis. In: ASME Design Engineering Technical Conferences-Design Automation Conference, pp. 9–12. ASME Press, Pennsylvania (2001)

2. Regis, R.G., Shoemaker, C.A.: A stochastic radial basis function method for the global optimization of expensive functions. Informs Journal on Computing **19**, 497–509 (2007)
3. Jin, R., Chen, W., Simpson, T.W.: Comparative studies of metamodeling techniques under multiple modeling criteria. Struct. Multidisc. Optim. **23**, 1–13 (2001)
4. Park, J., Sandberg, I.W.: Universal approximation using radial basis function networks. Neural Computing **3**, 246–257 (1991)
5. Wang, G.G., Shan, S.: Review of metamodeling techniques in support of engineering design optimization. J. Mech. Des. **129**, 370–380 (2007)
6. Babu, G.S., Suresh, S.: Sequential projection-based metacognitive learning in a radial basis function network for classification problems. IEEE Transactions on Neural Networks and Learning Systems **24**, 194–206 (2013)
7. Rashid, K., Ambani, S., Cetinkaya, E.: An adaptive multiquadric radial basis function method for expensive black-box mixed-integer nonlinear constrained optimization. Engineering Optimization **45**, 185–206 (2013)
8. Vuković, N., Miljković, Z.: A growing and pruning sequential learning algorithm of hyper basis function neural network for function approximation. Neural Networks **46**, 210–226 (2013)
9. Kitayama, S., Arakawa, M., Yamazaki, K.: Sequential approximate optimization using radial basis function network for engineering optimization. Optimization and Engineering **12**, 535–557 (2011)
10. Kitayama, S., Srirat, J., Arakawa, M.: Sequential approximate multi-objective optimization using radial basis function network. Structural and Multidisciplinary Optimization **48**, 501–515 (2013)
11. Hardy, R.L.: Multiquadric equations of topography and other irregular surfaces. Journal of Geophysical Research **76**, 1905–1915 (1971)
12. Micchelli, C.A.: Interpolation of scattered data: distance matrices and conditionally positive definite functions. Constructive Approximation **2**, 11–22 (1984)
13. Krishnamurthy, T.: Response surface approximation with augmented and compactly supported radial basis functions. In: 44th AIAA/ASME/ASCE/AHS/ASC Structures, Structural Dynamics, and Materials Conference, pp. 3210–3224. AIAA Press, Virginia (2003)
14. Gu, J., Li, G.Y., Dong, Z.: Hybrid and adaptive meta-model-based global optimization. Engineering Optimization **44**, 87–104 (2012)
15. Wang, L.Q., Shan, S., Wang, G.G.: Mode-pursuing sampling method for global optimization on expensive black-box functions. Engineering Optimization **36**, 419–438 (2004)
16. Coello, C.C.A.: Use of a self-adaptive penalty approach for engineering optimization problems. Computers in Industry **41**, 113–127 (2000)
17. Ray, T., Saini, P.: Engineering design optimization using swarm with an intelligent information sharing among individuals. Engineering Optimization **33**, 735–748 (2001)

Impact of Traffic Information Feedback Strategies on Signalized Two-Route Traffic Systems

Yang Wang, Yanyan Chen[(⊠)], and Xin Li

Beijing Key Lab of Traffic Engineering, College of Metropolitan Transportation,
Beijing University of Technology, Beijing 100124, China
{wang_yang,cdyan}@bjut.edu.cn, 583635349@qq.com

Abstract. The existing traffic information feedback strategies were originally proposed and evaluated for the two-route traffic systems without signal lights. However, traffic flow in urban traffic systems is rather complicated largely due to the presence of traffic signals. By introducing the signalized two-route traffic systems, this paper has investigated the eleven traffic information feedback strategies from system and user aspects for the first time, based on a set of two-route scenarios which simulate various combinations of traffic signal timing and location. The experimental results reveal the following findings. None of the test strategies can effectively improve road capacity for the scenarios where traffic lights are installed on both routes. The traffic information feedback strategies which use global information can reduce travel time to varying degrees, when dynamic vehicles become dominant on the roads. Furthermore, the eleven information feedback strategies have also been examined from the user equilibrium aspect and the corresponding results imply three congestion-based strategies can better approximate the user equilibrium as compared to the others. Finally, experiments demonstrate that the most inefficient strategy to achieve system optimality is the travel time feedback strategy.

Keywords: Intelligent transportation systems · Traffic information feedback strategy · Nagel-schreckenberg model · Signalized traffic systems

1   Introduction

To understand the traffic dynamics and solve the related congestion issue, researchers have proposed a number of theories [1][2] and realized that the informed travel can significantly increase the probability to avoid congestion [3]. Over the last decade, many researches have been dedicated to the traffic information feedback strategy for the situation where a driver can use the information provided at entrance to make a route choice. The work presenting the travel time feedback strategy (TTFS) by Wahle et al. [3], has triggered great interest of researchers to deliver more advanced information feedback strategies. In 2001, Lee et al. [4] developed a new information feedback strategy, named as mean velocity feedback strategy (MVFS). It has been proved that MVFS is more efficient than TTFS as MVFS diminishes the lag effect of TTFS [4][5].

© Springer International Publishing Switzerland 2015
H. Liu et al. (Eds.): ICIRA 2015, Part III, LNAI 9246, pp. 94–105, 2015.
DOI: 10.1007/978-3-319-22873-0_9

However, the information provided in MVFS can not reflect the real situation as MVFS incorporates the fragile velocity caused by the random mechanism of the Nagel-Schreckenberg (NS) model [6]. In 2005, Wang et al. [7] introduced a new traffic information feedback strategy, called the congestion coefficient feedback strategy (CCFS). It has been reported that CCFS outperforms TTFS and MVFS in terms of road capacity as CCFS provides congestion information [8]. However, CCFS is unable to reflect the influence of the distance of congestion to the vehicle newly entering the route. Aiming to enhance CCFS, a number of improved information strategies have been successively proposed in recent years. Dong et al. [8] developed a weighted congestion coefficient feedback strategy (WCCFS) by incorporating a linear weighting function to CCFS. Another notable improvement to CCFS is the exponential function feedback strategy (EFFS) which uses an exponential function to account for the distance effect of each congestion cluster to the vehicle newly arrived [9]. Dong et al. [10] proposed alternative way to add weight for each congestion cluster, yielding the corresponding angle feedback strategy (CAFS). CAFS calculates the angle between the two edges of each congestion cluster in reference to a site located vertically above the entrance instead of the cluster length as in CCFS, WCCFS, EFFS, in order to embed the cluster location information. To explicitly include the cluster length information, the improved congestion coefficient feedback strategy (ICCFS) incorporates CAFS into CCFS by using the corresponding angles as the weights for each congestion coefficient [11]. Nonetheless, CAFS and ICCFS have limited applications in that they are impossible to accurately calculate the angles of congestion clusters for the non-straight roads (e.g. the S-shape roads), as reported in Ref. [9]. In 2010, Dong et al. [12] developed a new information strategy, called vehicle number feedback strategy (VNFS), by examining the number of en-route vehicles. In 2012, Chen et al. [13] introduced the vacancy length feedback strategy (VLFS), by which drivers will be guided to the route with the larger number of empty sites from the entrance to the vehicle nearest to the entrance. In the same year, the authors from the same group (as in Ref. [13]) proposed two new information feedback strategies, namely the time flux feedback strategy (TFFS) and the space flux feedback strategy (SFFS), and found SFFS is the best one but TFFS generally performed same as that without information guidance based on the simulated two-route scenarios with two exits [14]. In 2013, Xiang and Xiong [15] proposed a weighted mean velocity feedback strategy (WMVFS) by applying a linear weighting function for each site in order to eliminate the influence of positioning errors (e.g., GPS error).

So far, all existing information feedback strategies are examined on either symmetrical or asymmetrical two-route systems with appropriate modifications [3]. However, travel delay in the urban road traffic system is largely caused by traffic signals, which is not considered in the previous researches. Therefore, this paper firstly introduces a signalized two-route traffic system and then evaluates eleven information feedback strategies based on a set of signalized two-route scenarios.

The rest of the paper is organized as follows. The next section is dedicated to an introduction of NS model and signalized two-route traffic system. The results obtained from a series of simulations are presented and discussed in Section 3, and this paper is concluded in Section 4.

2   NS Model and Signalized Two-Route Traffic System

2.1   NS Mechanism

The NS mechanism is a cellular automation model frequently adopted in analyzing traffic flow, as it is a simple model capable of mimicking the fundamental features of the real traffic follow, like stop-and-go wave, phantom jams, and the phase transition. This section provides a brief review on the NS mechanism for single lane traffic (and an excellent overview can be found in Ref. [1]).

In the NS model, the road is subdivided into a number of cells with equal length and each cell corresponding to 7.5m can be either empty or occupied by only one vehicle with a velocity ranging from 0 to v_{max} with integer interval (v_{max} is the maximum velocity permitted and set to be 3 cells/time step, corresponding to 81 km/h). The density ρ of a road of length L that carries N vehicles is defined as $\rho = N/L$. At each discrete time step (corresponding to 1 second), all en-route vehicles are subject to a motion update according to the following rules (parallel dynamics):

Rule 1. Acceleration: $v_i(t + 1/3) \rightarrow \min\{v_i(t) + 1, v_{max}\}$;
Rule 2. Deceleration: $v_i(t + 2/3) \rightarrow \min\{v_i(t + 1/3) + 1, g_i(t)\}$;
Rule 3. Randomization: $v_i(t + 1) \rightarrow \max\{v_i(t + 2/3) - 1, 0\}$ with probability p;
Rule 4. Movement: $x_i(t + 1) \rightarrow x_i(t) + v_i(t + 1)$.

Here, $x_i(t)$ and $v_i(t)$ are defined to be the location and the velocity of vehicle i at time t, respectively, and $g_i(t)$ is the number of empty cells in front of the vehicle i at time t.

2.2   A Signalized Two-Route Traffic System

It is assumed that a two-route traffic system includes traffic lights, which are operated in a fixed cycle manner. At each time step, a vehicle arrives at the entrance of the two-route system (i.e., the arrival rate r_a is 1) and will choose one route to enter if the first cell of the chosen route is empty. This means the vehicle will be deleted as the entrance of the desired route is occupied by the other vehicle arrived earlier. After entering the route, the vehicle moves through the route by following the NS rules as described in previous subsection. An en-route vehicle will check the gaps from its current position to the preceding vehicle and the stop-line of a signal intersection when the signal is red and stop at the stop-line if the gap from the stop-line is smaller. Note that the scenarios presented in this paper assume no vehicles at signal intersections will be diverted out and no vehicles will be added at any intersections (e.g., such intersections are often constructed for pedestrian crossing street in real world). When an en-route vehicle reaches the route end, it will be removed.

Furthermore, two different types of drivers are introduced in the system: static and dynamic drivers. The static driver will randomly choose a route to enter regardless of the information provided, while the dynamic driver will take advantage of the information to make a route choice. The rates of dynamic and static drivers are set to be S_d and $1-S_d$, respectively.

The two-route system with one entrance and two exits is frequently adopted to evaluate the different information strategies due to its simplicity [7][12][14]. Such traffic system simulates the situations where the two facilities with the same functions are located on two routes (e.g., two banks on different locations).

2.3   The Related Definitions

The flux of the routes is frequently adopted to describe road capacity and is defined as follows:

$$F = V_{mean}\rho = V_{mean}\frac{N}{L} \tag{1}$$

where V_{mean} represents the mean velocity of N vehicles on the route of length L, ρ is the traffic density of the route.

Furthermore, the traffic system operating over T time period can be evaluated by calculating the so-called average flux as:

$$F_{avg} = \frac{\sum_{i=1}^{n}\sum_{t=1}^{T}f_{ij}}{T \times n} \tag{2}$$

Here, f_{ij} is the flux of the ith route of n routes at time t.

3   Simulation Results and Discussion

The eleven information feedback strategies (TTFS, MVFS, CCFS, WCCFS, EFFS, CAFS, ICCFS, VNFS, VLFS, SFFS, WMVFS) have been examined in terms of individual cost and system optimality by performing a series of simulations on the two-route traffic systems with two exits and one exit. The individual cost is measured by the time required by individual vehicle to traverse the chosen route, while the average flux for the two routes is used to indicate the system optimality. All simulation results presented here were performed over 25000 iterations.

The eleven information feedback strategies can be classified into two categories: one is called global information strategy as they use the information reported by all vehicles and the other called local information strategy as only the information from a proportion of en-route vehicles is used. The strategies which apparently fall into the global information strategy are TTFS, MVFS, CCFS, WCCFS, CAFS, ICCFS, and WMVFS, while VNFS, VLFS, and SFFS belong to the second category. Although EFFS computes congestion coefficient using the information from all en-route vehicles, the exponential function employed considerably weights the vehicles near to the entrance, behaving similarly to the local information strategy. Consequently, EFFS is classified into the local information strategy in this paper. CCFS, WCCFS, EFFS, CAFS, and ICCFS compute congestion coefficients and therefore are called congestion-based strategies here. While TTFS is a time-based strategy, MVFS and WMVFS use the velocity information of en-route vehicles. VNFS provides the information on the number of a part of en-route vehicles. In contrast, only the distance from entrance to the vehicle closest to the entrance is provided by VLFS.

The evaluation on the eleven strategies were conducted based on 6 symmetric two-route scenarios, 2000 cells in length with one entrance and two exits, constructed by combining different signal timings and locations, as shown in Table 1.

Table 1. Traffic signal configurations for the two-route scenarios

Scenario	Route	Red light on	Green light on	Signal light location
1	A	50	80	1000
	B	100	60	1000
2	A	100	80	1000
	B	40	70	1000
3	A	60	80	1000
	B	60	80	1000
4	A	50	50	200
	B	50	50	200
5	A	50	50	1800
	B	50	50	1800
6	A	50	50	200
	B	50	50	1800

In reality, the rate of dynamic vehicles (S_d) may vary in a wide range and thus it is interesting to see the performance of different strategies in response to the changes in S_d. Fig. 1 shows the average fluxes obtained by taking the mean values over 10 independent runs. In general, TTFS has performed worst from the road capacity aspect, even though its performance is similar to the others when the proportion of dynamic vehicle is typically below 0.6 around. This finding is consistent to those previously reported for the symmetric two-route systems without signal lights due to the lag effect [13][15]. The road capacities realized by TTFS, MVFS, CCFS, WCCFS, ICCFS, and WMVFS are shrinking quickly while the others are apparently insensitive, as dynamic vehicles increasingly dominate on the roads for the scenarios except 4. The majority (i.e., EFFS, VNFS, VLFS, and SFFS) of the insensitive strategies belong to the category of local information strategy and consequently the lack of globe information can largely account for the insensitiveness. Furthermore, the results in Fig. 1 imply that all test strategies are unable to improve the system performance as the average fluxes have not been increased with information provision. This can be explained by the fact that all test strategies do not incorporate any information on the delay caused by traffic signals. Although scenarios 4 and 5 (in Fig. 1(d) and (e)) have same traffic signal configurations for the both routes, much less variations in average flux resulted from the strategies except TTFS can be observed for scenario 4. Notice that the locations of signal lights in scenario 4 are close to the entrance and therefore congestion near to the entrance is likely to be severe due to the traffic signal. Such important information can also be captured by the local information strategies which primarily focus on the local area near to the entrance. However, the local information strategies are unable to incorporate the congestion information when the traffic lights are located far from the entrance as the case in scenario 5. The first three scenarios have the signal lights located on the same sites but different timings and the different average fluxes presented in Fig. 1(a), (b), and (c) imply the signal timings have impact on road capacity. The average fluxes resulted from the last three scenarios suggest the averaged flux is influenced by the location of traffic light.

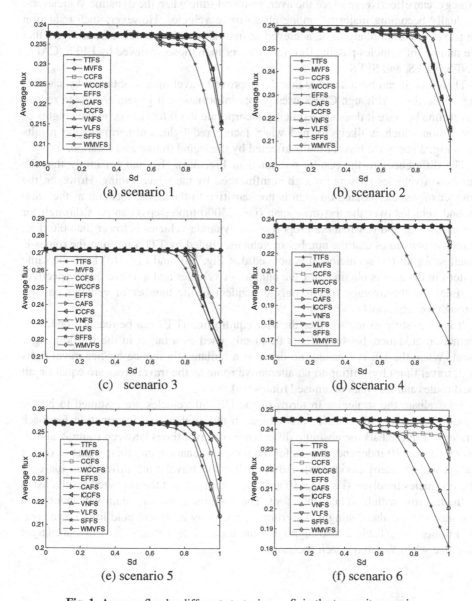

Fig. 1. Average flux by different strategies *vs.* S_d in the two-exit scenarios.

As road users always attempt to minimize their travel costs when choosing a route, the eleven strategies have been evaluated for the six scenarios in terms of travel cost measured by the time taken to pass through the chosen route. Fig. 2 shows the travel time averaged over 10 independent runs for the different proportions of dynamic vehicles ranging from 0 to 1 with the interval of 0.05. In general, TTFS, MVFS, CCFS, WCCFS, ICCFS, and WMVFS, which fall into the category of the global information strategy, can effectively reduce the average travel time when the dynamic vehicles are gradually becoming majority among all en-route vehicles. However, such reduction on travel cost is unlikely to be achieved at free of cost. It is evident from Fig. 3 that the number of vehicles passing through is lower than those achieved by EFFS, CAFS, VNFS, VLFS, and SFFS.

Therefore, it can be concluded that the improved travel time is obtained by satisficing traffic flow. Although CAFS uses global information, it generally fails to reduce travel time because it does not explicitly incorporate the information on the lengths of congestion which is likely formed when facing red light. Furthermore, the results from Fig. 2 imply the travel time is affected by the signal timing and location.

The differences in the results presented in Fig. 3(a), (b), and (c) imply that the number of vehicles passing through is influenced by the signal timing. However, the number of vehicle passing through is not sensitive to the signal location as the 5800 around vehicles over the test time span (i.e., 25000 time steps) can pass through for scenarios 4, 5, and 6, when the proportion of dynamic vehicles is lower than 0.6. Fig. 3 also demonstrates that the number of vehicles guided by TTFS through the routes is smallest for all test scenarios. It is noticed that Fig. 3(d) and Fig. 1(d) have the same pattern in the results obtained for the number of vehicles and average flux respectively, implying the average flux is only dependent to the number of vehicles passing through for scenario 4.

It is interesting to see whether the user equilibrium (UE) can be reached under information guidance, because UE is frequently used as a target in the traffic assignment. When the UE is reached, no driver can unilaterally reduce his/her travel costs (i.e., travel time) by shifting to an alternative route as the travel costs are equal on all used routes and less than on unused routes [16].

To evaluate the strategies in terms of the UE, all vehicles are assumed to be the dynamic type, i.e., all vehicles are subject to the guidance of information feedback strategy. Table 2 lists the absolute differences of travel times between route A and B, averaged over 10 independent runs for the six test scenarios, and their statistics (mean values and standard deviations). Additionally, the travel time differences for only static vehicles involved (i.e., $S_d = 0$) are also computed and the averaged results listed in the last row with bold and italic font. The maximum and minimum values for each test scenario are also distinguished from the others by italic and bold font, respectively. Furthermore, Table 2 also highlights the travel time differences which are larger than that when S_d is 0 for each scenario.

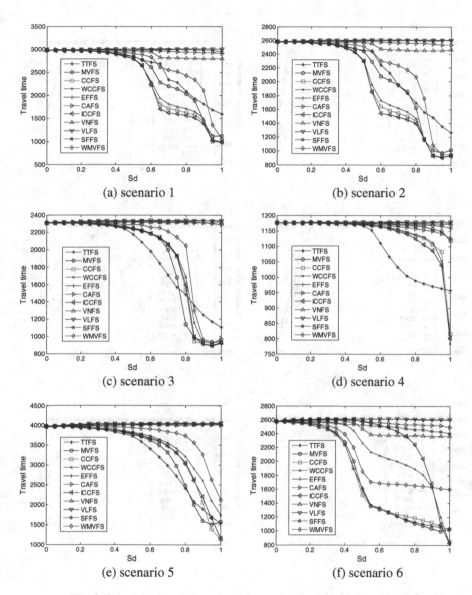

Fig. 2. Travel time by different strategies *vs.* S_d in the two-exit scenarios.

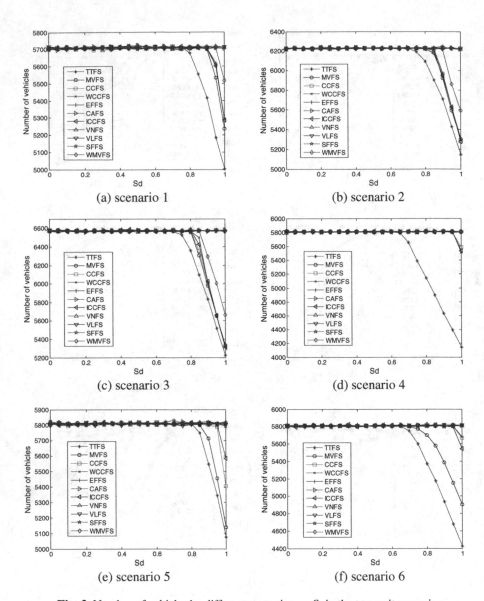

Fig. 3. Number of vehicles by different strategies *vs.* S_d in the two-exit scenarios.

Table 2. Travel time differences ($\times 10^3$) between route A and B by different strategies when S_d is 1 in the two-exit scenarios.

Scenario	1	2	3	4	5	6
TTFS	0.5188	0.2557	*0.0221*	*0.0097*	0.0277	0.0772
MVFS	0.2291	0.1406	0.0072	0.0009	0.0264	0.0445
CCFS	0.1317	0.0954	**0.0040**	0.0024	**0.0081**	**0.0035**
WCCFS	0.1313	0.0929	0.0043	**0.0006**	0.0141	0.0741
EFFS	1.8034	1.2234	0.0111	0.0048	0.0268	2.8431
CAFS	1.6597	1.1215	0.0064	0.0035	*0.0362*	2.9302
ICCFS	**0.1208**	**0.0883**	0.0043	0.0007	0.0187	0.0642
VNFS	1.4096	0.9443	0.0204	0.0033	0.0175	2.4388
VLFS	1.8137	1.2212	0.0139	0.0041	0.0167	2.8479
SFFS	*1.9206*	*1.2683*	0.0151	0.0053	0.0245	2.4583
WMVFS	0.2498	0.1105	0.0043	0.0025	0.0085	0.9390
Mean value	0.9080	0.5965	0.0103	0.0034	0.0205	1.3383
Standard deviation	0.7959	0.5435	0.0067	0.0026	0.0087	1.3405
$S_d = 0$	*1.8126*	*1.2302*	*0.0172*	*0.0040*	*0.0387*	*2.8018*

It can be found that the travel times between the two routes can be mostly equalized by CCFS, WCCFS, and ICCFS which are all congestion-based strategies. This can be understood by the fact that these three strategies directly incorporate the congestion caused by the traffic signal. In contrast, the majority of the strategies which enlarges the gap of travel times between the two routes are local information strategies. The worst one in terms of the UE for scenario 1 and 2 is SFFS, while TTFS is the worst for scenario 3 and 4. For scenario 5, all strategies can reduce the gap of travel times between the two routes. As the same traffic signal configurations are arranged for the two routes in scenario 3, 4, and 5, the travel time differences between the two routes are small (which is reflected by the mean values in Table 2). The mean value and standard deviation for scenario 4 are smallest among all test scenarios. One possible explanation is that the traffic lights located near to the entrance in scenario 4 enable the local information strategies to incorporate the congestion resulted from the traffic signals.

4   Conclusions

In the urban traffic system, traffic signals contribute a large proportion of travel delay. This paper has investigated the eleven information feedback strategies have been examined from system and user aspects for the first time, based on the developed two-route traffic systems controlled by traffic signals.

A number of conclusions can be drawn from the experimental results obtained for a set of simulated scenarios. None of the test strategies is able to improve the road capacity achieved without guidance for the two-route systems with traffic lights installed on both roads. This inefficiency is likely rooted in the fact that none of them explicitly contains the information on the congestion caused by traffic signal. To improve the road capacity for the signalized roads, further work is required with a special focus on the incorporation of traffic signals. Secondly, the average flux and the number of vehicles obtained from all experiments indicate that TTFS is the worst one in terms of system optimality when the dynamic type of vehicles become majority among all en-route vehicles. Thirdly, TTFS, MVFS, CCFS, WCCFS, ICCFS, and WMVFS, which belong to the category of global information strategy, can effectively reduce the average travel time when the proportion of dynamic vehicles is large. Finally, the experimental results for travel time indicate that the congestion-based strategies of CCFS, WCCFS, and ICCFS can better approximate the UE, implying the congestion information is more important in the signalized traffic system. Nevertheless, the examination on the eleven strategies is limited to the traffic system with signalized interactions where the flow rates do not change.

Therefore, it would be interesting to further investigate the information strategies on more complex systems with traffic lights. Also, the further research will focus on a new information strategy which can accommodate the travel delay rooted from the traffic signal.

Acknowledgement. The project was funded by the National Basic Research Program of China (Grand No. 2012CB723303) and the Scientific Research Foundation for the Returned Overseas Chinese Scholars, State Education Ministry (32004011201201).

References

1. Chowdhury, D., Santen, L., Schadschneider, A.: Statistical Physics of Vehicular Traffic and Some Related Systems. Phy. Rep. **329**, 199–329 (2000)
2. Helbing, D.: Traffic and Related Self-driven Many-particle Systems. Rev. Mod. Phys. **73**, 1067–1141 (2001)
3. Wahle, J., Bazzan, A.L.C., Klugl, F., Schreckenberg, M.: Decision Dynamics in a Traffic Scenario. Physica A **287**, 669–681 (2000)
4. Lee, K., Hui, P.M., Wang, B.H., Johnson, N.F.: Effects of announcing global information in a two-route traffic flow model. J. Phys. Soc. Jpn. **70**(35), 07–3510 (2001)
5. Yokoya, Y.: Dynamics of Traffic Flow with Real-time Traffic Information. Phys. Rev. E **69**, 016121 (2004)
6. Nagel, K., Schreckenberg, M.: A cellular automaton model for freeway traffic. Journal de Physique I **2**(2), 2221–2229 (1992)
7. Wang, W.X., Wang, B.H., Zheng, W.C., Yin, C.Y., Zhou, T.: Advanced Information Feedback in Intelligent Traffic Systems. Phys. Rev. E **72**, 066702 (2005)
8. Dong, C.F., Ma, X., Wang, B.H.: Weighted Congestion Coefficient Feedback in Intelligent Transportation Systems. Phys. Lett. A **374**, 1326–1331 (2010)

9. Chen, B.K., Dong, C.F., Liu, Y.K., Tong, W., Zhang, W.Y., Liu, J., Wang, B.H.: Real-time Information Feedback based on a Sharp Decay Weighted Function. Computer Physics Communications **183**, 2081–2088 (2012)
10. Dong, C.F., Ma, X.: Corresponding Angle Feedback in an Innovative Weighted Transportation System. Physics Letters A **374**, 2417–2423 (2010)
11. Dong, C.F., Paty, C.S.: 2011 Application of adaptive weights to intelligent information systems: An intelligent transportation system as a case study. Information Sciences **181**, 5042–5052 (2011)
12. Dong, C.F., Ma, X., Wang, B.H.: Advanced Information Feedback Strategy in Intelligent Two-route Traffic Flow Systems. Science China, Information Sciences **53**(11), 2265–2271 (2010)
13. Chen, B.K., Xie, Y.B., Tong, W., Dong, C.F., Shi, D.M., Wang, B.H.: A Comprehensive Study of Advanced Information Feedbacks in Real-time Intelligent Traffic Systems. Physica A **391**, 2730–2739 (2012)
14. Chen, B.K., Tong, W., Zhang, W.Y., Sun, X.Y., Wang, B.H.: Flux Information Feedback Strategy in Intelligent Traffic Systems. European Physics Letter **97**, 14001-7
15. Xiang, Z.T., Xiong, L.: A Weighted Mean Velocity Feedback Strategy in Intelligent Two-route Traffic Systems. Chin. Phys. B **22**(2), 028901-10 (2013)
16. Wardrop, J.G.: Some theoretical aspects of road traffic research. In: Proc. Inst. Civ. Eng., Part II, vol. 1(2), pp. 325–3783 (1952)

A Reservation Multichannel MAC Protocol Utilize Blind Source Separation

Yang Yu[✉], Jie Shi, Ke He, and Peng Han

School of Marine Science and Technology, Northwestern Polytechnical University,
Xi'an 710072, China
nwpuyuy@nwpu.edu.cn

Abstract. Due to the long propagation delay of acoustic waves and limited available bandwidth of the underwater acoustic channels, it takes a long delay for the handshaking between the transmitting/receiving nodes to avoid the data/control packets collision. This will decrease the channel utilization ratio, and increase the power consumption of the nodes. In this paper, a reversal multichannel media access control protocol using blind source separation for underwater acoustic sensor networks is presented. Compared with existing multichannel protocols, the proposed MAC protocol adds channel reservation information in handshaking packet, and uses the blind source separation to deal with the control packets collision. It makes the channel reservation will be finished before the current data transmission ending. At the end of this paper, the simulation results show that the network throughput and energy consumption of the proposed MAC protocol are much better than that of multichannel MACA on the heavy network traffic condition.

Keywords: Multichannel MAC protocols · Underwater acoustic networks · Blind source separation · Deep networks

1 Introduction

The major issue of media access control protocol is to transmit data effectively in shared channel without collisions. As for multiple nodes sharing the same channel, a receiving-transmitting collision usually occurs, which means, the data receiving will be affected by data sending of neighboring nodes. In the terrestrial wireless network, the wave speed of radio is so fast that the propagation delay between nodes can be neglected, and the receiving- transmitting collision was prevented by channel sensing and guard time. However, for underwater acoustic networks, the wave speed of underwater acoustic is so slow (only 1500m/s) that the propagation delay between nodes cannot be neglected. For the limited available channel bandwidth of the underwater acoustic channel, the channel sensing and guard time that are used for preventing channel collision will significantly decrease the channel utilization ratio. These challenges make it difficult to design a efficient MAC protocol for underwater acoustic networks. To deal with these

© Springer International Publishing Switzerland 2015
H. Liu et al. (Eds.): ICIRA 2015, Part III, LNAI 9246, pp. 106–115, 2015.
DOI: 10.1007/978-3-319-22873-0_10

challenges, multi-channel MAC protocol which is introduced by terrestrial sensor networks has been one of the most active research problems in recent years. Reference [1–3] indicates that multi-channel MAC protocol can greatly improve the network throughput, decrease channel access delay and the energy consumption. Reference [4] proposes the hidden node problem of multi-channel RTS / CTS and gives the solution. In [5], it proposes a CDMA-based multichannel ALOHA scheme which has gained excellent network performance. Multi-channel MAC protocol separates control channel and data channel, allowing multiple nodes transmitting data in parallel in different channels, and the network throughput and energy consumption of multi-channel protocol perform are better than that of single-channel protocols [6].

In this paper, we present a new multichannel MAC protocol: BSS-RM-MAC, which utilize the blind source separation (BSS) and adopts key point of multi-channel protocol. Based on the MACA protocol, it uses RTS / CTS packet which contain the reservation information of data channel to reserve the channel before ending of the current data transmitting, and use the BSS to handle the collision of the control packets. The channel utilization ratio will be improved by using these mechanisms. In this paper, the underwater acoustic channel bandwidth for data transmission is discussed in section 1. In section 2, we briefly discusses the BSSRM-MAC protocol, gives the packet format, and the mechanism to avoid collision of the control packets. Section 3 shows the simulation results of network throughput and energy consumption of BSSRM-MAC protocol in simulator, and a conclusion is made in section 4.

2   Bandwidth of Data Transmission

Due to the high attenuation and the noise of the marine environment, the available bandwidth for data transmission is highly restricted. In [7] it provides the relationship between signal to noise ratio at the receiver - $SNR(l, f)$ and the marine environmental noise - $N(f)$ and the propagation attenuation $A(l, f)$.

$$SNR(l, f) = 10 \log P - 10 \log A(l, f)N(f) - 10 \log \Delta f \qquad (1)$$

Where P represents the transmit power of source node and represents the narrowband bandwidth with f as the center frequency. It can be seen from (1) that the signal to noise ratio at the receiver is related to product of the attenuation and noise (AN product - $A(l, f) \times N(l, f)$). As shown in Figure 1, the available bandwidth is only about $20KHz$ when the distance between transmitting and receiving is about $1000m$ or less.

3   BSS-RM-MAC Protocol

In the multiple-channel MAC protocols, the data packets and control packets are transmitted independently in the data channels and control channels, and the probability of collisions between packets are decreased. Researchers in [6]

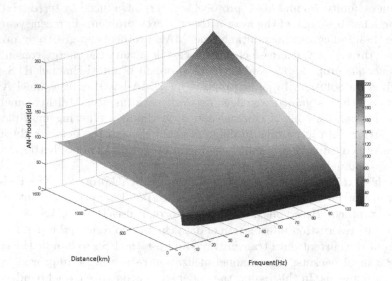

Fig. 1. The relationship of the AN product and the distance l and frequency f

found that the throughput and energy consumption performance of the underwater acoustic networks will increase with the number of the data channels. However, the frequency gap between adjacent data channels decrease the channel utilization ratio, and the collision between control packets can't be avoided. To deal with this problem, we use reservation mechanism to improve the channel utilization ratio, and use BSS to prevent the collision between control packets.

3.1   Overview of the Protocol

The BSS-RM-MAC can be used in multi-hop network. It adds the reservation time slot information to RTS/CTS control packet. The reservation information includes beginning time and ending time of source node/destination node that using the data channel. All nodes in the network establish the time slots information of neighbors by the channel sensing. When transmitting data, the source node search for an idle time slot in the time slots list, and embeds the idle time information within the broadcasting RTS packet, to initiate a data transfer reservation. Then the destination node judges the potential collision with neighbors time slot information which is established form RTS/CTS packet. If no collision occurs, it broadcasts a CTS packet, to inform the source node of the successful reservation. Since BSS-RM-MAC protocol reserves the future transmitting time slot, the reservation process can be carried out when the control channel is idle, regardless of the data channel status. Hence, it can decrease the time interval between two data transmissions, improving the data channel utilization ratio. The pseudo code of the BSS-RM-MAC as follows.

Pseudo Code of The BSS-RM-MAC

```
if you receive a contention tone (CTD) while idle;
    Set Reservation blocking state to true;
    Clear at end of current reservation period (RP);
    Start handshaking sensing;
    if you receive a RTS/CTS/RTSW/CTSW;
        Decode future Transmitting /Receiving start and end time;
        Map the decoded Time Slot to my selfs DC schedule
        check the corresponding state;
        if Self RE state is true;
            Send RTSW;
        if RTSs address is myself
          & RE blocking state is false
          & Self RE state is false;
            Send CTS;
            Set the Self RE and TR blocking state to true ;
        if RTSs address is others or CTSW;
            Set the RE blocking state to true ;
        if CTS & Self TR state is true;
            Send CTSW;
        if CTS or RTSW;
            Set the TR blocking state to true ;
        Wait for this current reservation period end;
    End handshaking sensing;
else stay at idle state;

When application invokes MAC send;
Find a TR unblocking slot for TR and insert it to RTS;
if reservation blocking state is true;
    wait for the end of current RP and attempt in next RP;
else transmit contention tone;
    if (contender count (CTC) > 1)
        Compute w uniformly from [0,CTC];
        Backoff w RP(s)
        if CTD; while in backoff
            Set Reservation blocking state to true;
            Clear at end of current RP;
            Wait for end of current RP;
        else backoff ends;
            Go to line 2 and repeat contention
    else send RTS including reservation time slot;
        Set the RE blocking;
        Set Self TR stat to true in DC schedule;
        if receive CTS;
            data reservation successful;
```

```
     if receive RTSW;
          data reservation failed;
          Clear the RE Blocking in DC schedule;
          go to line 2;
     if receive CTSW;
          data reservation failed;
          Clear the Self TR state in DC schedule;
          go to line 2;
     if nothing;
          Clear RE blocking;
          Clear self TR state in DC schedule;
          Go to line 2;
Wait for end of current contention period;
```

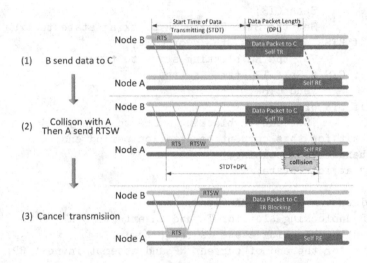

Fig. 2. In the first phase, B sent RTS to A. Next, A receive the RTS, and check the schedule of himself. It calculate the data packet arriving time, and sent a RTSW because of the collision. The third phase, B received the RTSW, clear the Self TR state and set the TR blocking to true in the reservation time slot.

The source node (transmitting) and the destination node (receiving) have obtained enough prior knowledge of idle period before the actual data transmission starting by the channel sensing, and the success probability of the data channel reservation will be improved by using this approach. However, the inevitable reservation collisions will lead the mismatch of neighbors idle time slot, resulting in the collisions of the data channel reservation. To resolve this problem, BSS-RM-MAC adds the cancel mechanism which is explained as: neighbors can judge the collisions existing based on the reservation information broadcast by source node /destination node. If a collision occurs, neighboring node broadcasts a warning packet immediately to inform the collision period, as shown in the Figure 2.

3.2   The Control Packet Format

The control channel of BSS-RM-MACA protocol use 2 control packet formats with different length: 32bits Reservation & Warning Packet and 11bits Reservation Cancel Packet, which can been distinguished by the first three bit.

Fig. 3. Reservation & Warning Packet format

Figure 3 is the 32bits reservation & warning packet format. The highest three bit TCP of first byte represents the type of control packet while the lowest five bit represents the reservation time length. The highest three bit TCP is explained as: when TCP=001b, the packet represents RTS packet broadcasted by source node; when TCP=111b, it represents the CTS packet broadcasted by destination node; when TCP=000b, it represents RTS warning packet broadcasted by neighbors of source node, which means a collision between this reservation and receiving time schedule of source node neighbors; when TCP=100b, it represents CTS warning packet broadcasted neighbors of destination node, which means a collision between this reservation and sending time slot of destination node neighbors. DPL represents data packet length.

3.3   Deal with the Collision Between Control Packets

Due to the BSS-RM-MAC dose not require time synchronization between nodes, the collision probability between control packets will be very high when the load of the network is heavy. To deal with this problem, the BSS method which utilize deep neuron networks has been used to separate the mixing control packets according to the different angle of arriving of the different packets. This blind source separation system consists of the following four stages: (1) extraction of the low-level features ; (2) training of the deep networks ; (3) estimation of the probabilities that each T-F unit belongs to different sources and generation of the soft mask ; and (4) reconstruction of the target signal from the soft mask and mixture signal.

As shown in Figure 4, the inputs to the system are the two channel underwater acoustic mixtures. We perform short-time Fourier transform (STFT) for each channel, and obtain the T-F representation of the input signals, $X_L(m, f)$ and

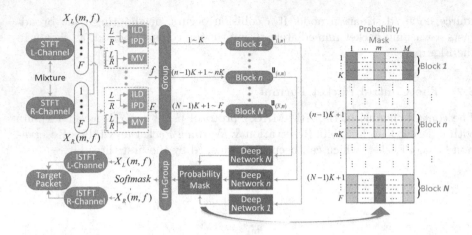

Fig. 4. The architecture of the proposed system using deep neural network based time-frequency masking for blind source separation.

$X_R(m, f)$ where $m = 1, \cdots, M$ and $f = 1, \cdots, F$ are the time frame and frequency bin indices respectively. The low-level features, i.e. mixing vector (MV), interaural level and phase difference (IPD/ILD), are then estimated at each T-F unit. Next, we group the low-level features into N blocks (only along the frequency bins f). The block n includes K frequency bins $((n-1)K+1, \cdots, nK)$, where $K = \frac{F}{N}$. We build N deep networks with each corresponding to one block and use them to estimate the direction of arrivals (DOAs) of the sources. Through unsupervised learning and the sparse autoencoder [8] in deep networks, high-level features (coded positional information of the sources) are extracted and used as inputs for the output layer (i.e. the softmax regression) of the networks. The output of softmax regression is a source occupation probability (i.e. the soft mask) of each block (through the un-group operation, T-F units in the same block are assigned with the same source occupation probability) of the mixtures. Then the sources can be recovered applying the softmask to the mixtures followed by the inverse STFT (ISTFT).The deep networks are pre-trained by using a greedy layer-wise [9] training method.

4 Simulation Results

We use single channel RTS/CTS and Multichannle RTS/CTS as baseline. Consider the node space is 400 to 600 meters in the simulation. In [10], it provides the optimal frequency band of energy transmission of 20kHz to 53kHz on this condition. Hence, the total available bandwidth is 33kHz and the maximum network throughput is 32kbps while the underwater acoustic propagation seed is 1500 m/s. The parameters of single-channel MACA can be set as: the RTS/CTS packet length is 3 byte; the maximum data packet length is 4096 byte and the transmitting time is 1024ms; the minimum data packet length is 128 byte and the transmitting time is 32ms; the maximum propagation delay is 400ms. The parameters of multiple-channel protocol can be set as: the reservation & warning packet length is 4 byte and other parameters of data packet are the same as those of single-channel protocol. The number of data channel is 2, using the maximum and minimum data packet length respectively. Considering the 1 kHz band gap between data channel and control channel, the final bandwidth of data channel, which is divided into 2 sub-channels, is 23 kHz. Meanwhile, provided that the control channel bandwidth is 8 kHz and the 48 nodes uniformly distributes in a rectangular area of 34km, forming a multi-hop network.

The data transmission process of the node is a Poisson process with parameter λ. Parameter λ determines the network traffic. Next we will analyze the impact of the network traffic to network throughput and energy consumption with different λ in the simulation, provided that the length of data packet is 256 byte, and λ increases from 0.05 to 0.5, equal to the corresponding network traffic becoming heavier.

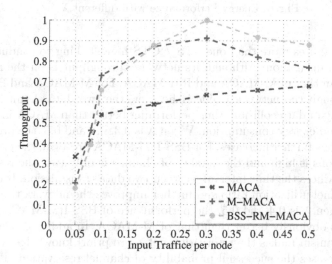

Fig. 5. Throughput performance with different λ

The simulation result shows that the probability of data transmission collisions is low when the network traffic is light, and the time interval between two data transmission is longer as well. Hence, the network throughput and energy consumption of MACA is optimal. Owing to the low collision probability of data transmission, BSS-RM-MACA protocol just transmits RTS/CTS control packet similar to M-MACA. Thus the network throughput and energy consumption of these two protocols are almost the same. However BSS-RM-MACA uses a longer control packet in theory which makes its performance slightly worse than M-MACA. When λ is 0.1, 0.2, the probability of data collisions begins to increase

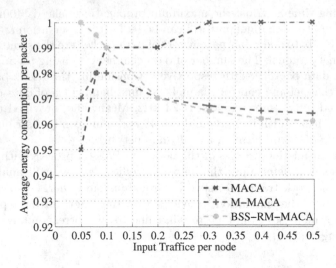

Fig. 6. Energy Performance with different λ

and MACA requires more than one RTS / CTS handshaking to conduct a successful data transmission. Although its network throughput is on the rise, it is still worse than that of M-MACA and RM-MACA. For M-MACA and BSS-RM-MACA, although the network traffic is heavier, the two data channels reduce the probability of data collision, thus performing better than MACA in network throughput and energy consumption. When λ is 0.3, 0.4 and 0.5, the probability of data collisions further increases. For BSS-RM-MACA protocol, the probability of data collisions is high and the number of channel is only two, the reservation mechanism reduces the time interval between two data transmission. It enhances the data channel utilization ratio and further improves the network throughput. Especially when λ is 0.4 and 0.5, the performance of BSS-RM-MACA network throughput is obviously better than that of M-MACA. In addition, the channel sensing mechanism makes the reservation get more priori knowledge which significantly increases the successful probability of channel reservation. Therefore, with the increasing network traffic, the energy consumption of BSS-RM-MACA does not significantly increases and it shows the best performance.

5   Conclusion

In this article, we describe the reservation process of BSS-RM-MACA protocol and the estimation method of data transmission time slot, and compare the performance of MACA, multi-channel MACA and BSS-RM-MACA protocol. The results shown that, on the light network traffic conditions, MACA performs the best, and the network throughput / energy consumption performance of BSS-RM-MACA and multi-channel MACA is almost equivalent. As the increasing of network traffic, the network throughput of BSS-RM-MACA is better than that of multi-channel MACA, but the energy performance is inferior to multi-channel MACA. On the heavy network traffic, BSS-RM-MACA is optimal. It explains that BSS-RM-MACA protocol improves the network throughput at the expense of energy consumption (caused by channel sensing and adding the reservation time to the control packet).

Acknowledgments. This research was supported partially by the Natural Science Basis Research Plan in Shaanxi Province of China (Program No.2014JQ8355).

References

1. Tzamaloukas, A., Garcia-Luna-Aceves, J.: A receiver-initiated collision-avoidance protocol for multi-channel networks. In: Proceedings ot the IEEE Twentieth Annual Joint Conference of the IEEE Computer and Communications Societies, INFOCOM 2001. IEEE (2001)
2. So, J., Vaidya, N.H.: Multi-channel mac for ad hoc networks: handling multi-channel hidden terminals using a single transceiver. In: Proceedings of the 5th ACM international symposium on Mobile ad hoc networking and computing. ACM (2004)
3. Han, Y.S., Deng, J., Haas, Z.J.: Analyzing Multi-Channel Medium Access Control Schemes with Aloha Reservation. IEEE Transactions on Wireless Communications 5(8), 2143–2152 (2006)
4. Mo, J., So, H.-S., Walrand, J.: Comparison of Multichannel Mac Protocols. IEEE Transactions on Mobile Computing 7(1), 50–65 (2008)
5. Zhou, Z., Peng, Z., Cui, J.-H., Jiang, Z.: Handling Triple Hidden Terminal Problems for Multichannel Mac in Long-Delay Underwater Sensor Networks. IEEE Transactions on Mobile Computing 11(1), 139–154 (2012)
6. Pompili, D., Melodia, T., Akyildiz, I.F.: A Cdma-Based Medium Access Control for Underwater Acoustic Sensor Networks. IEEE Transactions on Wireless Communications 8(4), 1899–1909 (2009)
7. Zhou, Z., Peng, Z., Xie, P., Cui, J.-H., Jiang, Z.: Exploring Random Access and Handshaking Techniques in Underwater Wireless Acoustic Networks. EURASIP Journal on Wireless Communications and Networking 2013(1), 1–15 (2013)
8. Bengio, Y., Courville, A., Vincent, P.: Representation learning: A review and new perspectives. IEEE Transactions on Pattern Analysis and Machine Intelligence 35(8), 1798–1828 (2013)
9. Bengio, Y., Lamblin, P., Popovici, D., Larochelle, H.: Greedy layerwise training of deep networks. Advances In Neural Information Processing Systems 19, 153 (2007)
10. Stojanovic, M.: On the Relationship between Capacity and Distance in an Underwater Acoustic Communication Channel. ACM SIGMOBILE Mobile Computing and Communications Review 11(4), 34–43 (2007)

Design and Implementation of a Robot System Architecture Based on Agent Technology and Delegate Mechanism

Jialin Yu, Yonghua Yan$^{(\boxtimes)}$, and Jianrong Zhang

State Key Laboratory of Mechanical System and Vibration,
School of Mechanical Engineering,
Shanghai Jiao Tong University, Shanghai 200240, China
{yjl_sjtu,yhyan,jrzhang}@sjtu.edu.cn

Abstract. The purpose of this paper is to build a robot system architecture based on agent technology and delegate mechanism. Agent technology has penetrated in various fields and delegate mechanism has been widely used to make software architectures scalable. This paper focuses on three aspects: (i) introduction and principle of agent and delegate; (ii) design of a robot system architecture based on agent technology and delegate mechanism; (iii) implementation and verification of the robot system architecture by corresponding experiments. Based on agent technology and delegate mechanism, the robot system architecture can be developed in different developers and period.

Keywords: Agent technology · Delegate mechanism · Robot system architecture

1 Introduction

Etzioni proposed that agent technology was ninety-nine percent computer science and one percent artificial intelligence (AI), so agent technology needs to rely on a lot of basic computer science, such as communication technology and programming language [1]. Over the past few decades, robot systems have been widely used in various fields, while object oriented programming(OOP) has becoming the mainstream programming language. However, the robot systems based on OOP are complicated because of invoking functions from different classes frequently, and the whole architectures is unclear because classes of OOP cannot match the independent robot modules completely. In addition, the most criticized issue is that the objects of OOP are not appropriate to human thinking and unable to keep up with the pace of artificial intelligence because they are static and have no capable of reactivity and pro-activeness. Considering practical applications, such as inserting a simple robot module, the whole program must be analyzed and handled, it costs a lot of time and effort, affects the efficiency of development greatly and increases difficulty of debugging. Therefore, agent technology is used in the design of a robot system architecture in this paper.

© Springer International Publishing Switzerland 2015
H. Liu et al. (Eds.): ICIRA 2015, Part III, LNAI 9246, pp. 116–125, 2015.
DOI: 10.1007/978-3-319-22873-0_11

Agent technology originated from artificial intelligence, the concept of agent had been put out in 1960s and developed in 1990s. Nowadays, agent oriented programming (AOP) has been known and recognized by the majority of programming enthusiasts, it is inheritance and development of object oriented programming. While agent technology is penetrating in various fields and becoming a promising way to develop many complex applications, ranging from electronic commerce to industrial process control[2]. However, there is not a uniform definition about agent[3], Wooldridge defined it as "an autonomous decision making system, which senses and acts in some environment"[4]. In this paper, agent technology is more inclined to be defined as "an autonomous body which runs on the dynamic environment is relatively independent of other bodies and responses by perception of dynamic environment." Although the theoretical study of agent has made great progress, practical applications still exist considerable hysteresis[5]. At present, the agent language is still developing, most of the agent applications are still using Java, C++ and C#, while C++ and platform of Visual Studio 2008 is used to build the robot system architecture in this paper.

With agent technology, there are still some problems. For one thing, the function and size of traditional architectures are set in stone and there is less flexibility in the architectures because they are cannot be trimmed. For another, the traditional architectures are not scalable and are difficult to be configured because the program is protected and encapsulated. If other developers want to develop the architecture, it will cost too much, this is "pull one hair and the whole body is affected". Therefore, delegate mechanism is used here in the design of the robot system architecture. With delegate mechanism, other developers only need to choose configuration or provide new functions to delegate mechanism, the architecture can be personalized and the new functions can be feasible. Delegate mechanism which focuses abstract structure type has been used widely by C# or JAVA language, while delegate mechanism which focuses specific applications is built by C++ language here.

In this paper, the concept of agent technology and delegate mechanism is firstly introduced. And then the robot system architecture based on agent technology and delegate mechanism is designed. Finally, the feasibility of robot system architecture is verified by corresponding experiments.

2   Principle of Agent Technology and Delegate Mechanism

2.1   Principle of Agent Technology

Principle of agent technology is mainly reflected in agent characteristics, agent structure and agent communication modes. There are a lot of agent characteristics, which associated with robot system are autonomy, social ability, reactivity and pro-activeness[6].

1) Autonomy

Agents can start spontaneously and control their own behavior and internal state without the direct intervention of humans or others;

2) Reactivity

Agents can be aware of their environment and respond with changes of environment;

3) Pro-activeness

Agents can actively meet the target behavior.

4) Social ability

Agents can communicate with other agents by certain language and cooperation with other agents to finish the task;

Agent structure shows the operation way inside agent. There are a variety of forms, which is used in the robot system should contain a message queue, a message processor and a lot of agent methods at least[7]. The message queue is used for storing message, which is usually programmed by an array or a list. The essence of the message processor is a thread, which is used to deal with the messages from message queues and invoke agent methods. After agents start spontaneously, the agent processor is detecting messages and environment circularly and constantly. The essence of agent methods are the member function of OOP classes, which are important to finish specific tasks. When the agent processor receives the messages or is aware of changes of environment, it responds and invokes certain agent methods to finish movement[8]. Agent structure diagram is shown in Fig. 1.

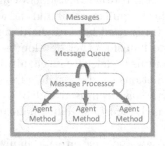

Fig. 1. Agent structure diagram

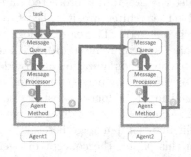

Fig. 2. Diagram of agent communication by message passing model

Agent communication mode decides the way of messages flow between different agents. There are five basic communication modes between agents: no communication model, message passing model, scheme passing mode, the blackboard mode and the agent language model[9]. Message passing model is used in this paper because it is easy and effective to satisfy the needs of robot system architecture design.

Message passing model means that messages are the media between agents, and messages should be formulated and ruled in advance. Only the messages that conform to these rules and formats are useful. When a particular state (for example, the agent get messages from other agents, or the agent is aware of some changes in the environment) happens, agent can make corresponding actions.

The process of agent communication is shown in Fig. 2. Firstly, Agent1 receives the messages from the task, then the messages are stored in message queue, message processor is detecting the messages in message queue cyclically and constantly, when detecting the useful message, message processor invokes corresponding agent method, then the agent method makes operation and sends messages to Agent2, while Agent2's message processor is detecting the messages of message queue, after Agent'2 detects the messages, it invokes corresponding agent method, then the task is finished. After the task has been finished, the agent method of Agent2 sends messages to Agent1, and then Agent1 cannot detect the useful messages any more. Agent1 and Agent2 work together to finish a task, this is the communication process between different agents by message passing model[10].

2.2 Principle of Delegate Mechanism

There is less flexibility in traditional architectures and they are difficult to be developed as they cannot be trimmed, while delegate mechanism in java or C# language focuses on abstract data type. Therefore, delegate mechanism which focuses on specific application about robot system is built in C++ language.

There are a data queue, a serial of registration and initialization functions and a serial of library files in delegate mechanism of this paper. The data queue is in charge of storing the first address of registered functions. Registration functions are charge of providing a way to configure functions to other developers, while initialization functions are in charge of invoking registered functions by invoking the pointers in date queue. The C++ function points are used to finish the process of registration and the process of initialization which are known collectively as process of configuration. The library files are in charge of storing functions that could be registered. In the process of configuration, developers only register the functions needed, and they can register the new functions that are not in library. With delegate mechanism, the architecture is scalable and can be developed.

3 Design of System Architecture

3.1 Design of Agent-Based System

As the agent concept cannot appear in a vacuum, a robot system architecture based on agent technology is designed[11]. There are several functions in traditional robot system architecture, such as configuring parameters, running in different coordinates and running in different model. With the needs of traditional robot system, five agents are designed: Agent Interface, Agent Management, Agent Motion-mode, Agent Interpolation and Agent Controller. In this design, every agent module of the new system has the ability of finishing a simple task[12], while a serial of agent modules can finish a complex task, introduction of every agent function is as follows:

1) *Agent Interface* is the media between the user and machine, which is in charge of providing display and operation for the user.

2) *Agent Management* is in charge of internal data management and external serial management. Internal data consist of reduction ratio, encoder resolution and joint size. External serial management is in charge of reading the value of absolute encoder.

3) *Agent Motion-mode* is in charge of provide several different mode of motion for the user, which includes jog mode, job mode etc. Job motion mode means moving according to job instructions that written in advance. After analysing these instruction, Agent Motion-mode decomposes them into a serial of object points.

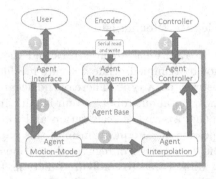

Fig. 3. Structure and information flow of multi-agent

4) *Agent Interpolation* is in charge of receiving the object points from Agent Motion-mode and interpolating into a lot of refined points with professional interpolation method.

5) *Agent Controller* is in charge of managing functions from the controller providers (Controller from GOOGOLTECH company in this paper is used), such as "GT_Open()" and "GT_SetEncPos()".

In this paper, a basic agent module named BaseAgent is designed in advance. BaseAgent has the simplest agent structure: a simple message queue, a message processor that starts spontaneously and cycles continuously and many agent methods that are virtual functions in nature and have nothing meaning. The five agents are derived by the basic module. Fig. 3 is the structure and information flow of multi-agent.

3.2   Improvement with Delegate Mechanism

There are many advantages in delegate mechanism of this paper, the most representative one is that it is feasible to support open and scalable architecture. With the delegate mechanism, both a simplified architecture and a powerful architecture can be built, in other words, an open architecture can be built, which is a bit like the embedded system. In addition, the robot system architecture is scalable, other developers can replace or add functions according to their needs.

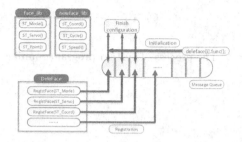

Fig. 4. Process of configuration

In the design, a serial of library files about robot system is built, such as "face_lib", "key_lib" and "Ipolation_lib". There are a lot of functions in every library file, other developers can configure them and get their own architecture. While the developers can write new functions in new library files, such as "new-face_lib" in Fig. 4.

A data queue which is programmed by arrays and structures is designed to store the first address of functions. The developers can register functions in library files with the registration functions of the delegate mechanism, and then the first address of functions is written in data queue. There are many registration functions, such as "RegistFace()", "RegistKey()" and "RegistPara()".

After process of registration, these registration functions is configured, which is called process of initialization. There are many initialization functions like "InitFace()", "InitKey()" and "InitPara()".

With process of registration and initialization, the functions registered in data queue are configured one by one. Fig. 4 is the process of interface registration and initialization with the delegate mechanism.

4   Implementation of System Architecture

The process of system program running is shown in Fig. 5. The program begins from the entrance function "OnInitDialog()", and then is the process of registration and initialization, after configuration finished, the five agents start to run, and the message processors are the state of detection, at that time, the user can control the machine to finish movement.

Fig. 6 reflects the information flow of Jog-mode movement. Firstly, the user choose jog mode from the interface and the message "P_Motionmode" is turned to 1(1 means the message is valid), when message processor of Agent Interface detects the changes, it invokes "Change_Mode()" which changes the state to Jog-mode.

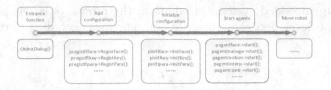

Fig. 5. The process of system program running

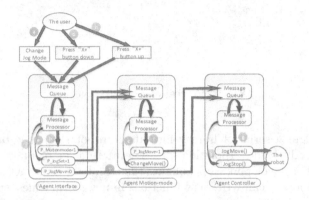

Fig. 6. The information flow diagram of jog movement

When the user pressed the "Y+" button, the message "P_JogSet" is turned to 1, in the same way, the message processor detects the changes and invokes corresponding function to set jog parameter and turns message "P_JogMove" to 1, at last, "JogMove()" is invoked and the robot starts to run in jog mode. When the user press the button up, the message "P_JogMove" is turned to 0, subsequently, "JogStop()" is invoked and the robot is stop.

5   Experimental Studies

5.1   Experimental Setup

The system architecture is used in a six-axis industrial robot based on Windows CE 5.0 , which is shown in Fig. 7. The control cabinet of the robot is shown in in Fig. 8. In order to verify the feasibility of the system architecture, corresponding experiments are carried out.

The experiments collect the working time of every agent when Y-axis is turn from the position in Fig. 9 to the position in Fig. 10 in jog mode. Working time is the time of detecting messages and invoking corresponding functions, which is different from running time. In other words, agents are always in running time after activated, only the agents detect effective messages are in working time.

Fig. 7. Programs of registration

Fig. 8. Programs of initialization

Fig. 9. Programs of registration

Fig. 10. Programs of initialization

5.2   Feasibility Analysis of System Architecture

As shown in Fig. 11, only three agents (Agent Interface, Agent Motion-mode and
Agent Controller) receive valid messages and are in working time. The "Power
On" button is put down at point a, the span "ab" is the time that Agent Interface detects the message "P_Power" and makes it changed. Subsequently, Agent
Controller detects the changed message "P_Power" and makes the power open in
span "cd". The time between b and c is 15 milliseconds, which is the time Agent
Controller detects the message "P_Power". The jog mode is chose at point e,
and Agent Motion-mode detects the messages at point f, and the point g and h
is similar in the operation of putting "Y+"button down. These spans reflect the
time of detecting corresponding messages, they are usually 5 to 16 milliseconds.
When "Y+" button is put up at point m, the movement is stop slowly in span
"np". As shown in Fig. 11, the working time of Agent Controller is far more
than the others because the most time-consuming operation is finished in Agent
Controller.

Fig. 12 is made by collecting the message state of agents and it reflects
message state diagram of agents in jog mode. After Agent Controller detects
"P_Power" turn to 1(1 means power on message is valid) at point b, the corresponding function makes the power open and "P_Power" is turn to 0 quickly

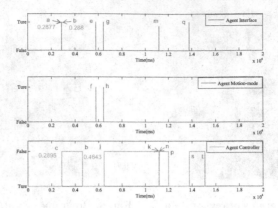

Fig. 11. Sequence diagrams of agents working time in jog mode

in order the operation is done once, similarly, when Agent Controller detects "P_Power" turn to 2(2 means power off message is valid) at point d, the corresponding function makes the power close and "P_Power" is turn to 0.

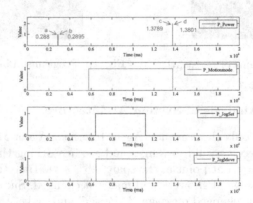

Fig. 12. Message state diagram of agents in jog mode

Although the experiments are simple, they can reflect working time and state of every agent in the process of the system architecture running.

6   Conclusions

Agent technology is a promising development which can provide a solution to many important software problems. In this paper, the robot system architecture combines agent technology with the robot systems creatively, while it is a simplified version of agent technology and cannot finish many things that theoretical agent technology promises to do. Although the robot system architecture is a small step towards intelligent agent technology, the big step is not far with efforts of other developers. With delegate mechanism, more personalized configuration can be easily completed because the robot system architecture is scalable and can be developed. In the future, the robot system architecture can be widely used in the field of robot systems.

Acknowledgments. This research was supported in part by National Key Basic Research Program of China under Grant 2013CB035804, National Natural Science Foundation of China under Grant 51120155001 and U1201244 and was sponsored by Shanghai Economic and Information Technology Commission (No.CXY-2013-21).

References

1. Etzioni, O.: Moving up the information food chain: deploying softbots on the world wide web. In: Proceedings of the National Conference on Artificial Intelligence, pp. 1322–1326 (1996)
2. Jennings, N.R., Wooldridge, M.: Applying agent technology. Applied Artificial Intelligence an International Journal 9(4), 357–369 (1995)
3. Bellifemine, F.L., Poggi, A., Rimassa, G.: Developing multi-agent systems with jade. In: Castelfranchi, C., Lespérance, Y. (eds.) ATAL 2000. LNCS (LNAI), vol. 1986, pp. 89–103. Springer, Heidelberg (2001)
4. Wooldridge, M., Jennings, N.R.: Intelligent agents: Theory and practice. The Knowledge Engineering Review 10(02), 115–152 (1995)
5. Wooldridge, M.: An introduction to multiagent systems. Wiley & Sons 4(2), 125–128 (2011)
6. Wooldridge, M.: Agent-based software engineering. IEE Proceedings-Software 144(1), 26–37 (1997)
7. Lee, J., Barley, M., Systems, M.A., Web, S.: Intelligent agents and multi-agent systems. Studies in Fuzziness & Soft Computing 2(12), 67–96 (2009)
8. Rao, A.S., Georgeff, M.P., et al.: BDI agents: From theory to practice. ICMAS 95, 312–319 (1995)
9. Luck, M., d'Inverno, M., et al.: A formal framework for agency and autonomy. ICMAS 95, 254–260 (1995)
10. Olfati-Saber, R., Fax, J.A., Murray, R.M.: Consensus and cooperation in networked multi-agent systems. Proceedings of IEEE 95(1), 215–233 (2007)
11. Corchado, J.M., Bajo, J., Paz, Y.D., Tapia, D.I.: Intelligent environment for monitoring alzheimer patients, agent technology for health care. Decision Support Systems 44(2), 382–396 (2008)
12. Maes, P., et al.: Agents that reduce work and information overload. Communications of the ACM 37(7), 30–40 (1994)

A Programming by Demonstration with Least Square Support Vector Machine for Manipulators

Jingdong Zhao, Chongyang Li, Zainan Jiang[(✉)], and Hong Liu

State Key Laboratory of Robotics and System,
Harbin Institute of Technology, Harbin 150001, China
jiangzainan@hit.edu.cn

Abstract. This paper presents a method of programming by demonstration, aiming at instructing the manipulator to accomplish tasks with obstacles in the way or with strict motion paths. Least square support vector machine (LS-SVM), based on the principle of structure risk minimization, is employed to achieve better generalization and reproduced trajectories with higher accuracy. Furthermore, the velocity field method is applied to maintain the convergence of reproduced trajectories and smooth the motion. Finally, a series of obstacle avoidance experiments with a 7-DOF manipulator are conducted to verify the feasibility of the proposed method.

Keywords: Programming by demonstration · LS-SVM · Velocity field method · Obstacle avoidance

1 Introduction

In recent years, with the advance of robotics and their widespread applications, manipulating tasks have become more and more complex. Manual programming, which demands the expertise of the programmer and could hardly adjust to the environment, is tedious, and more and more difficult to meet the requirements from changeable occasions. Thus, there arises the demand of the manipulator trajectory planning method, which is able to operate conveniently and adapt to the environment.

Programming by demonstration (PbD) is an effective solution for the last issue [1-3]. This method endows the manipulator with the ability to learn demonstrations from human operators. In PbD, some most commonly used methods for trajectory planning are as follows: methods based on statistics [4-6], methods based on dynamical system (DS)[7-8], methods based on spline[9].

Many researchers modeled and reproduced trajectories employing the statistics-based methods. For example, Tso [4] modeled and reproduced the motion of the human by Hidden Markov Models (HMM); Calinon [5-6] encoded a series of trajectories by Gaussian Mixture Models (GMM), and reproduced the trajectories in different environments using Gaussian Mixture Regression (GMR). Ijspeert [7] firstly proposed encoding the motion by DS in the form of nonlinear differential equations. Learning algorithms can be integrated into the framework of DS, such as GMM [10] and locally weight regression (LWR) [11].

© Springer International Publishing Switzerland 2015
H. Liu et al. (Eds.): ICIRA 2015, Part III, LNAI 9246, pp. 126–137, 2015.
DOI: 10.1007/978-3-319-22873-0_12

However, the above methods mainly focused on point-to-point motions, rather than the obstacle avoidance during tasks. In this paper, we present a new method of PbD. With the DS as its framework, this method uses least square support vector machine (LS-SVM) to encode the motion and adds the velocity field method nearby the target position. This paper completes the theoretical derivation and analysis of this method, and validates the feasibility by experiments.

2   Overview

Fig. 1 shows the architecture of the manipulator system, using PbD as its foundation.

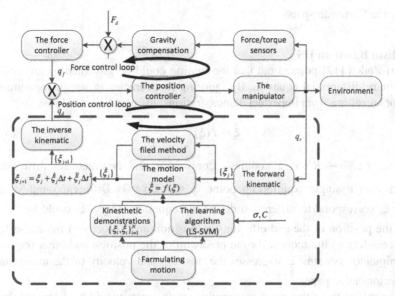

Fig. 1. The manipulator system

The manipulator system was composed of two closed loop: the position control loop and the force control loop. For protecting the manipulator, the force control loop corrected the desired joint position of the manipulator when it received the supernumerary force from the environment. The LS-SVM-based PbD was applied in the position control loop, which reproduced the trajectory of the task by regression. The whole work was done in two phases.

Phase 1: a global motion model, formulated by recursive positions or DS principle respectively, with more details illustrated in Section 2.1, was established based on the principle of LS-SVM with the information from kinesthetic demonstrations.

Phase 2: the trajectory was reproduced from its initial. By recalling the global motion model iteratively, the trajectory would converge to the end position of the task.

2.1   Formulism

Formulism Based on Position
This formulism for encoding the motion could be summarized as follows: the position of the next cycle was obtained in accordance with the current position by a certain conversion. In other words, it encoded the motion with the position-based recursive expression:

$$\xi_{i+1} = f(\xi_i) \tag{1}$$

$\xi_i \in R^n$ indicates the current position, while $\xi_{i+1} \in R^n$ indicates the next position of the trajectory. Each entry of the vector ξ_i represents the coordinate in the joint space or the Cartesian space.

Formalism Based on DS
Khansari-Zadeh [12] pointed out that the motion could be governed by an autonomous ordinary differential equation. By ignoring inaccuracies in sensor measurements and errors resulting from imperfect demonstrations, the equation would be:

$$\dot{\xi} = f(\xi) \tag{2}$$

Where $f : R^n \to R^n$ is a nonlinear continuous and continuously differentiable function with a single equilibrium point $\dot{\xi}^* = f(\xi^*) = 0$ .Different input state variables ξ correspond to different order of the equations (e.g., ξ could be the joint angles, the position of the end-effector, the position and velocity of the end-effector, etc.). According to the request for the precision of the position and other requests of the manipulator system, ξ expresses the position and velocity of the manipulator's end-effector in this paper.

The impact of these two ways to reproduce trajectories would be discussed through experiments in a later section.

2.2   Kinesthetic Demonstrations and the Learning Algorithm

In this paper, the kinesthetic demonstrations were the position and velocity of N given points on L trajectories. There were many methods to acquire the kinesthetic demonstrations, such as visual detecting and location tracking. In our research, human operator demonstrated the skill to the manipulator by dragging it on the model of the zero force control.

To acquire the formulated relationship by kinesthetic demonstrations, LS-SVM was introduced in this paper. This algorithm is based on the principle of structure risk minimization [13]. Compared with previous algorithms used in PbD (GMM, HMM and so on), LS-SVM has better generalization. And LS-SVM could be trained faster than the traditional SVM [14].

For the set of N given demonstrations on L trajectories $\{\xi_i,\dot{\xi}_i\}_{i=1}^{N}$, $\xi_i \in R^n$ represents the input variable, and $\dot{\xi}_i \in R^n$ indicates the desired output of ξ_i. The regression function of ξ is $f(\xi)=\mathbf{w}^T\varphi(\xi)+b$, and its LS-SVM-based regression problem can be transformed into an optimization problem with constraints:

$$\begin{cases} \min_{w,b,e} J_{LS} = \min_{w,b,e}(\dfrac{1}{2}\mathbf{w}^T\mathbf{w}+\dfrac{C}{2}\sum_{i=1}^{N}e_i^2) \\ s.t. \quad y_i-[\mathbf{w}^T\varphi(\xi_i)+b]=e_i, i=1,2,...,N \end{cases} \tag{3}$$

Where b defines the bias of the regression function, e_i is the training error of demonstration $\{\xi_i,\dot{\xi}_i\}$, and $\varphi(\bullet)$ indicates the mapping from demonstrations to kernel feature space. $C>0$ is the penalty function, which plays an important role on balancing the generalization of algorithm and the tolerance error. The influence of C to the reproduced trajectories would be discussed in a later section. The output equation of LS-SVM is:

$$f(\xi) = \sum_{i=1}^{N}\alpha_i k(\xi_i,\xi)+b \tag{4}$$

$$\begin{cases} b=\dfrac{\mathbf{E}_v^T(\mathbf{K}+C^{-1}\mathbf{I})^{-1}\mathbf{Y}}{\mathbf{E}_v^T(\mathbf{K}+C^{-1}\mathbf{I})^{-1}\mathbf{E}_v} \\ \alpha = (\mathbf{K}+C^{-1}\mathbf{I})^{-1}(\mathbf{Y}-b\mathbf{E}_v) \end{cases} \tag{5}$$

$\mathbf{E}_v=[1,1,...,1]^T$, $\alpha=[\alpha_1,\alpha_2,...,\alpha_N]^T$, $\mathbf{Y}=[y_1,y_2,...,y_N]\in R^{N\times1}$, $\mathbf{K},\mathbf{I}\in R^{N\times N}$, $\mathbf{I}$ is the unit matrix, and $\mathbf{K}$ is the kernel matrix. Our method used Gaussian kernel (6) to map the feature space to the infinite dimension.

$$K_{i,j}=k(\xi_i,\xi_j) = \exp(-\dfrac{\|\xi_i-\xi_j\|^2}{\sigma^2}) \tag{6}$$

σ is the coverage of Gaussian kernel. And the effect of σ to the reproduced trajectories would be discussed in a later section.

Without the loss of generality, we take the constructing of DS model as an example to demonstrate the establishment of global motion models. Within the DS model, the kinesthetic demonstrations are $\{\xi_i,\dot{\xi}_i\}_{i=1}^{N}$. Wherein, the input variable $\xi_i = [x,y,z,\dot{x},\dot{y},\dot{z}]$ is detected by the sensors of the manipulator, and $x\in R, y\in R, z\in R$ are the coordinates of the manipulator's end-effector in the Cartesian space, and $\dot{x}\in R, \dot{y}\in R, \dot{z}\in R$ are the velocity components. The output variable $\dot{\xi}_i =[\dot{x},\dot{y},\dot{z},\ddot{x},\ddot{y},\ddot{z}]$ is calculated by $\dot{\xi}_i = (\xi_i-\xi_{i-1})f_s$, and f_s indicates the sampling frequency of the sensors. In order to obtain the mapping from ξ to $\dot{\xi}$, kinesthetic demonstrations were used for training the DS using LS-SVM. As the system had multiple-input and multiple-output (MIMO), the six output variables needed to be trained respectively. Then, the motion was represented as six output equations, such as:

$$\dot{x} = f_x(\xi) = \sum_{i=1}^{N} \alpha_{\dot{x},i} k(\xi_i, \xi) + b_{\dot{x}} \tag{7}$$

In the light of the output equations, the position of the next time would be calculated using the current position and velocity of the manipulator, through the equation of motion along three directions, such as:

$$x_{i+1} = x_i + \dot{x}_i \Delta t + \frac{1}{2} \ddot{x}_i \Delta t^2 \tag{8}$$

Where Δt indicates the step time, which is the control period of the manipulator. However, if the running time of the manipulator bot needed to be changed, Δt could be increase or decrease appropriately. Each position on this trajectory was calculated according to the last position by this method.

2.3   Velocity Field Method

When demonstrating the manipulator to approach the target position precisely, the human operator needed to drag the manipulator repetitively to aim at the target point, which would result in the tremble of the reproduced trajectory at the target point.

To avoid the tremble, this paper presents a method based on the velocity field. The method was applied nearby the target position, to ensure the reproduced trajectory could converge smoothly and quickly. Fig. 2 shows the velocity field:

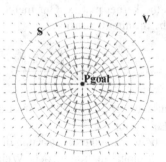

Fig. 2. The velocity field

Pgoal represents the target position, and S is a circle with the center Pgoal. The direction of the arrow indicates the normal direction of circle S, which represents the velocity direction of the current position. Arrow density is in inverse proportion to the velocity. Thus, the formula about the velocity in the velocity field could be obtained:

$$\mathbf{V} = \mathbf{k}(\mathbf{P} - \mathbf{P}_{goal}) \tag{9}$$

$\mathbf{V} \in R^3$ denotes the velocity on the position $\mathbf{P} \in R^3$. $\mathbf{k}$ is used to ensure the manipulator's velocity continuous when it enters the velocity field, which was obtained as follows:

$$\mathbf{k} = \frac{\mathbf{V}_{in}}{\mathbf{P}_{in} - \mathbf{P}_{goal}} \qquad (10)$$

$\mathbf{V}_{in}, \mathbf{P}_{in} \in R^3$ are the velocity and position when the manipulator enter the velocity field.

After obtaining the velocity $\mathbf{V}$, the next position could be calculated combined with the current position and Δt (same with Δt in eq. (8)). From eq. (9), it can be found that with $(\mathbf{P} - \mathbf{P}_{goal})$ decreasing, $\mathbf{V}$ is reduced continuously to zero at the target position. Fig. 3 – Fig. 4 shows the effect of applying the method base on the velocity field.

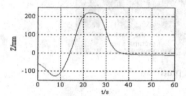

Fig. 3. The reproduced trajectory in Z-axis without the velocity filed method

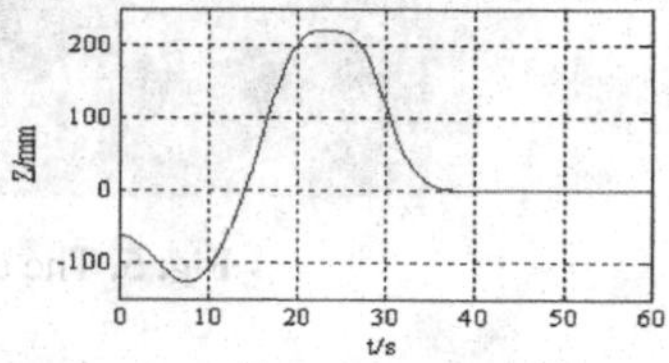

Fig. 4. The reproduced trajectory in Z-axis with the velocity filed method

Fig. 3 shows the result without the velocity field method, and it converges to -12mm finally. However, the result in Fig. 4 converges to 0 (the target position).

3   Experiments and Discussions

3.1   Experiments Setup

A series of experiments were conducted on the experimental platform, shown in Fig. 5, to verify above algorithms. The platform is mainly consisted of the 7-DOF manipulator, the central controller, the central control computer, the teleoperation computer, two typical obstacles and other subsystems such as the power system and the displayer.

The manipulator was taught to bypass the two obstacles starting from the left edge of the black rectangle area in Fig. 5, and terminate at the target position at the right. In this task, the diameter of the manipulator's end-effect was 120mm, and the center distance between the two obstacles was 350mm. One of the two obstacles was a cylinder with a diameter of 40mm, and the other one was a cuboid with the maximum upper surface of 80mm*40mm. Thus, in order to accomplish the task, the position of the manipulator had to be controlled precisely during the task (the orientation was locked). The purpose of this experiment was to verify whether the method above could reproduce the trajectories satisfying the strict requirement of the position.

Human operator guided the manipulator to demonstrate the task. To facilitate the calculation, the coordinate system $\{x', y', z'\}$ (Fig. 5) was established over the target position. And there was a fixed positional relationship between the target position and the global coordinate system $\{x, y, z\}$. The target position was acquired by the camera.

Fig. 5. The experiment platform

3.2   Result and Discussion

Aiming at reproduced trajectories, this paper presented three evaluation indexes: the convergence rate, the velocity characteristic, and the success rate.

The convergence rate *con* represented the rate of the reproduced trajectories which could converge to the target position. And it could be calculated as follows:

$$con = \frac{n_c}{N} \tag{11}$$

n_c expresses the number of the convergent trajectories. N indicates the total number of reproduced trajectories.

The velocity characteristic indicated the smoothness of velocity. And it limited the velocity was slower than 50mm/s and the acceleration was smaller than 20mm/s^2.

The success rate *suc* was the rate of completing the task. And it could be calculated as follows:

$$suc = \frac{n_s}{N} \tag{12}$$

N is the number of the reproduced trajectories which could complete the task.

As mentioned above, the three facts including the formulism methods of encoding the motions, the variables C in eq. (3) and σ in eq. (6) would affect the reproduced trajectories. In this section, taking advantage of the same demonstrations which is obtained by the experiments and changing the three facts referred above, we would compare and analyze the difference of results.

The Effect of the Formalism Methods
Different formalism methods meant different methods of encoding the motion, and would influence the information extracted from the demonstrations during training. Obviously, the formulism based on DS used the value of the sampling frequency of the sensors f_s and the single-step running time Δt while training and regressing. Regression results of the two formalism methods were shown in Fig. 6-Fig. 11:

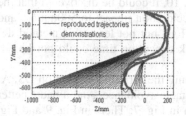

Fig. 6. The reproduced trajectories (based on positions)

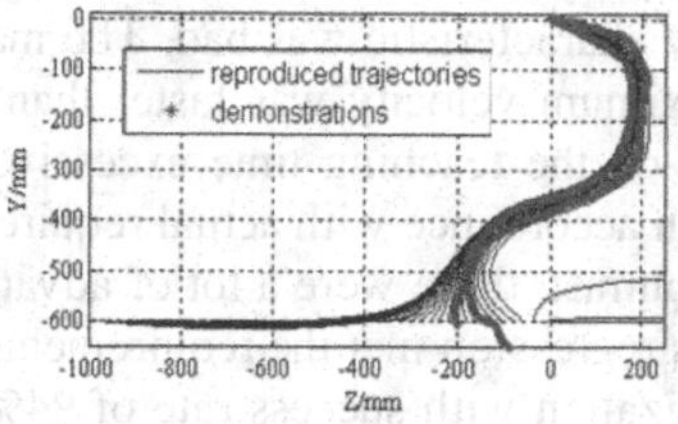

Fig. 7. The reproduced trajectories (based on DS)

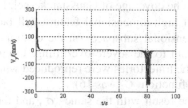
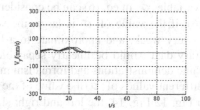

Fig. 8. The velocities of reproduced trajectories in Y-axis (based on positions)

Fig. 9. The velocities of reproduced trajectories in Y-axis (based on DS)

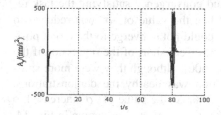
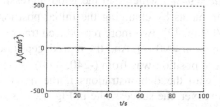

Fig. 10. The acceleration of reproduced trajectories in Y-axis (based on positions)

Fig. 11. The acceleration of reproduced trajectories in Y-axis (based on positions)

Because the left edge of the black rectangle area was at -600 in Y-axis of $\{x', y', z'\}$, the initial positions of the blue trajectories varied from (-1000, -600) to (0, -600), which took an interval of 20mm. During the execution of this task, the position changed very little (less than 10mm) in the X-axis. Therefore, the figures only showed the position in the Y-axis and Z-axis.

Comparing Fig. 6 with Fig. 7, it was found that the regression of the reproduced trajectories using the formalism based on positions (almost coincided with one of the demonstration trajectories) was better than those employing the formalism based on DS, if its initial position neared the demonstrations. But the reproduced trajectories using this formalism method had some drawbacks as follows: firstly, its generalization was bad ($con = 22\%$), and there would be local convergence, if the initial positions varied from (-1000, 600) to (-200, 60) as shown in Fig. 6. The success rate of that was only 20%. And then, from Fig. 8 and Fig. 10, it could be discovered that the velocity characteristic was bad. The maximum acceleration impact to 500mm/s^2, and the maximum velocity was faster than 250mm/s. Last but not least, this method depended on the teaching time excessively, and task execution time could not be adjusted in accordance with actual requirements.

In contrast, there were a lot of advantages of using the formulism based on DS. At first, its regression met the requirements of the obstacle avoidance. And it had a better generalization with success rate of 94%, as shown in Fig. 7. Then, in Fig. 9 and Fig. 11, the reproduced trajectories were smooth, and relied little on the teaching time.

The Influence of σ in Gaussian Kernel

The variable σ in eq. (6) can be considered to be the coverage of Gaussian kernel. In eq. (6), if the value of σ is large, the weight of high-dimensional feature will become very weak, and approximate to a low-dimensional subspace; on the contrary, arbitrary data would be linearly separable, and lead to the severe overfitting. Using the same demonstrations and formalism method, the trajectories were reproduced with the different values of σ (the value of the penalty function C was 16).

In Fig. 12-Fig. 14, the left and right showed the result with the same σ, and the differences between them were the restrictions such as the boundary, the obstacles (adding the compensation of the end-effector's diameter) and so on. If the value of σ was 300, as shown in Fig. 13, there were 47 reproduced trajectories of 50 ($suc = 94\%$) which had the good regression and convergence satisfying the task requirements by changing the initial position. When the value of σ was reduced to 210 (Fig. 12), two more reproduced trajectories could not converge to the target position ($con = 90\%$). And the reproduced trajectories were short of the regression if the initial position was from (-240, -600) to (-1000, -600), although they were more similar with the demonstrations if the initial position was nearby the demonstrations. Moreover, the success rate in Fig. 12 was only 26%. As the value of σ increased to 1166, all reproduced trajectories were convergent ($con = 100\%$) as shown in Fig. 14. But they had the bad regression and had the huge difference with demonstrations. Thus, 8% of them could meet the requirements ($suc = 8\%$).

As a result, the smaller the value of σ, and the worse success rate and convergence of the reproduced trajectories. And the larger the value of σ, the better the convergence

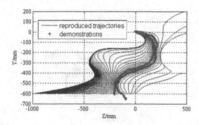
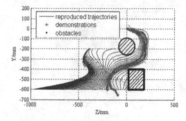

Fig. 12. The reproduced trajectories ($\sigma = 300$)

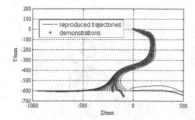
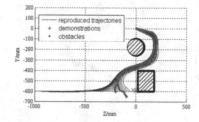

Fig. 13. The reproduced trajectories ($\sigma = 210$)

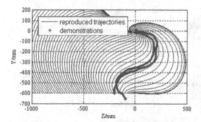
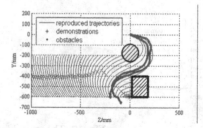

Fig. 14. The reproduced trajectories ($\sigma = 1166$)

and the worse success rate. So we should select the appropriate σ based on the actual system, the regression and convergence, to gain the best success rate.

The Effect of the Penalty Function C.

The main role of the penalty function C in eq. (3) is to balance the generalization of algorithm and the tolerance error. Fig. 15-Fig. 17 show the results with different value of C, using the same demonstrations and σ ($\sigma = 300$).

Comparing Fig. 15 and Fig. 16, the reproduced trajectories were dissimilar with the demonstrations if the value of C was smaller. At the same time, they were not good at the regression, and only 8 trajectories could ensure the task ($suc = 16\%$) in Fig. 15. And comparing Fig. 16 and Fig. 17, increasing value of C would lead to the regression deteriorate, though the results were more similar with the demonstrations as the initial position was near the demonstrations, and the success rate in Fig. 17 was 52%. However, it also

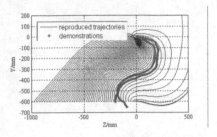
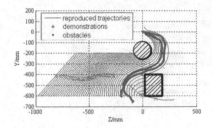

Fig. 15. The reproduced trajectories ($C = 0.2$)

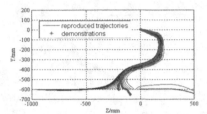
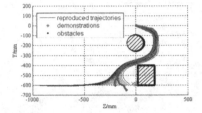

Fig. 16. The reproduced trajectories ($C = 16$)

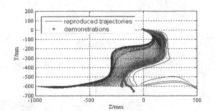
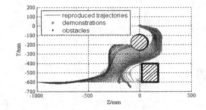

Fig. 17. The reproduced trajectories ($C = 50$)

could be concluded that the value of C had a little effect on the convergence of the trajectories ($con = 92\%$ in Fig. 15, $con = 94\%$ in Fig. 16 and Fig. 17).

4   Summary

In this paper, a new method of PbD was presented to meet the precise requirements of obstacle avoidance in tasks. Utilizing LS-SVM, this method has better generalization, and can obtain trajectories from wide-ranging initial positions with a few demonstration trajectories (only two in this paper). Furthermore, for obstacle avoidance and boundary constraints, the variables of σ and C can be adjusted according to the requirements of tasks. In addition, this method applies the velocity field method nearby the target position to maintain the convergence of the reproduce trajectories and smooth the velocity. In future work, we will make use of force sensors and visual sensors to monitor the real-time changes in the working environment of the manipulator, and exploit the online avoidance algorithm to improve the reproduced trajectories by PbD.

Acknowledgments. This work is in part supported by the National Natural Science Foundation of China (Grant No. 51305097), Natural Science Foundation of Heilongjiang Province(Grant No.E201428), the Fundamental Research Funds for the Central Universities(Grant No.HIT.KISTP.201410), and the Self-Planned Task(No.SKLR201404B) of State Key Laboratory of Robotics and System(HIT)

References

1. Billard, A., Calinon, S., Dillmann, R., Schaal, S.: Robot programming by demonstration. In: Siciliano, B., Khatib, O., (eds.) Handbook of Robotics, ch. 59. Springer, New York (2008)
2. Schaal, S.: Is imitation learning the route to humanoid robots? Trends in Cognitive Sciences **3**(6), 233–242 (1999)
3. Dong, S., Williams, B.: Learning and recognition of hybrid manipulation motions in variable environments using probabilistic flow tubes. International Journal of Social Robotics **4**(4), 357–368 (2012)
4. Tso, S.K., Liu, K.P.: Hidden Markov model for intelligent extraction of robot trajectory command from demonstrated trajectories. In: Proceedings of The IEEE International Conference on Industrial Technology (ICIT 1996), pp. 294–298. IEEE (1996)
5. Calinon, S., Guenter, F., Billard, A.: On learning, representing, and generalizing a task in a humanoid robot. IEEE Transactions on Systems, Man, and Cybernetics, Part B: Cybernetics **37**(2), 286–298 (2007)
6. Calinon, S., Billard, A.: A probabilistic programming by demonstration framework handling constraints in joint space and task space. In: IEEE/RSJ International Conference on Intelligent Robots and Systems, IROS 2008, pp. 367–372. IEEE (2008)
7. Ijspeert, A.J., Nakanishi, J., Schaal, S.: Learning attractor landscapes for learning motor primitives (2002)
8. Schaal, S.: Dynamic movement primitives-a framework for motor control in humans and humanoid robotics. In: Adaptive Motion of Animals and Machines, pp. 261–280. Springer, Tokyo (2006)
9. Ude, A., Atkeson, C.G., Riley, M.: Planning of joint trajectories for humanoid robots using B-spline wavelets. In: Proceedings of IEEE International Conference on Robotics and Automation, ICRA 2000, vol. 3, pp. 2223–2228. IEEE (2000)
10. Gribovskaya, E., Billard, A.: Learning nonlinear multi-variate motion dynamics for real-time position and orientation control of robotic manipulators. In: 9th IEEE-RAS International Conference on Humanoid Robots, Humanoids 2009, pp. 472–477. IEEE (2009)
11. Bentivegna, D.C., Atkeson, C.G.: Learning from observation using primitives. In: Proceedings of IEEE International Conference on Robotics and Automation, ICRA 2001, vol. 2, pp. 1988–1993. IEEE (2001)
12. Khansari-Zadeh, S.M., Billard, A.: Learning stable nonlinear dynamical systems with gaussian mixture models. IEEE Transactions on Robotics **27**(5), 943–957 (2011)
13. Vapnik, V.N.: The Nature of Statistical Learning Theory. Springer, New York (1995)
14. Suykens, J.A.K., Vandewalle, J.: Least Square Support Vector Machine Classifiers. Neural Processing Letters **9**(3), 293–300 (1999)

Stiffening Mechanisms of Soft Robots

Stiffness Control of Soft Robotic Manipulator for Minimally Invasive Surgery (MIS) Using Scale Jamming

S.M.Hadi Sadati[1(✉)], Yohan Noh[1], S. Elnaz Naghibi[2], Kaspar Althoefer[1], and Thrishantha Nanayakkara[1]

[1] Center for Robotics Research, Department of Informatics, King's College London, Strand, London WC2R 2LS, UK
{seyedmohammadhadi.sadati,yohan.noh,kaspar.althoefer, thrish.antha}@kcl.ac.uk
[2] School of Engineering and Materials Science, Queen Mary, University of London, Mile End Rd., London E1 4NS, UK
s.e.naghibi@qmul.ac.uk

Abstract. Continuum and soft robotics showed many applications in medicine from surgery to health care where their compliant nature is advantageous in minimal invasive interaction with organs. Stiffness control is necessary for challenges with soft robots such as minimalistic actuation, less invasive interaction, and precise control and sensing. This paper presents an idea of scale jamming inspired by fish and snake scales to control the stiffness of continuum manipulators by controlling the Coulomb friction force between rigid scales. A low stiffness spring is used as the backbone for a set of round curved scales to maintain an initial helix formation while two thin fishing steel wires are used to control the friction force by tensioning. The effectiveness of the design is showed for simple elongation and bending through mathematical modelling, experiments and in comparison to similar research. The model is tested to control the bending stiffness of a STIFF-FLOP continuum manipulator module designed for surgery.

Keywords: Stiffness control · Soft robot · Continuum manipulator · Layer jamming · Scale

1 Introduction

Fabrication of variable stiffness material [1] and also variable stiffness soft manipulators, mostly designed by inspiration from octopus arms [2], and wearable robots have been widely investigated recently. They have numerous applications specially in soft surgeries where their deformable structure is beneficial to improve manoeuvrability, control and sensing [3], with less invasive interactions with organs [4-6].

The idea of jamming has been used in stiffness control through increasing the friction. Granular jamming has been utilized in design of flexible manipulators [4], [5]

The original version of this chapter was revised: An incorrect version of an author's name was published. This has been corrected. The correction to this chapter is available at https://doi.org/10.1007/978-3-319-22873-0_52

and also variable stiffness joints [4], [6]. As an instance, Cheng and co-workers [4] have obtained a wider stiffness range than Jiang and co-workers [5] through granular jamming for a soft manipulator. J. Santiago and co-workers [7] presented a new scaly layer jamming design with a wire driven actuation method without any pneumatic actuation which shows some improvements in smaller design possibilities. However, granular jamming apparatus is too bulky and not appropriate for wearable applications. The hysteresis loss is also inevitable in this kind of jamming. On the other hand, layer jamming, although suitable for wearable purposes, has the problem of design, fabrication and modelling because of their multilayer structure which is necessary for the manipulator to resist buckling and the long flap length which is required to achieve large deformations [8], [9].

The layer jamming design proposed by Kim et. al. recalls the bird feathers and controls normal tension due to bending to maintain the stiffness [9]. In this research, a new simpler design inspired by the role of scales in flexural stiffness of a real fish and snake skin presented [10]. Scales with jagged contact surfaces presented here are capable of stiffness control through jamming that resists the backbone coil torsion shear force with very low hysteresis (Fig. 1). Placing the scales on a helix backbone guarantees their end-face-to-face contact even in large deformations despite their small size. Here, we have used cable tension as in catheter navigation systems [11] to engage the scales and increase the friction. However, to our best knowledge, a design combining these ideas was not previously investigated for stiffening purposes in robotics. To model the scaled helical spring, the theoretical background already existing for the helical springs under eccentric loadings [12], [13] is used.

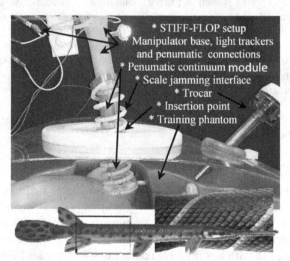

Fig. 1. STIFF-FLOP continuum manipulator with scale jamming for surgery (top), the helix like formation of scales in a fish [10] (down)

We showed that a continuum manipulator arm can be inserted in the designed scaled spring for stiffness control purposes (Fig. 1). Here, we have used a single module belonging to STIFF-FLOP manipulator, designed for surgical applications

[14], [15], inside the spring, to verify the stiffening property of the scales in bending. The module uses a set of Mckibben pneumatic actuators to achieve constant curvature bending in different directions. An alternative setup to test the proposed design in future is a tendon deriven shrinkable soft manipulator developed by F. Maghooa A. Stilli, et. al. where the manipuletor elongation is as important as its bending [16][17].

In this paper, a simple mathematical model is presented considering the elastic behaviour of the spring and the scales and LuGre friction model. The model is verified through experiments in simple elongation and bending tests. To test the bending stiffness a STIFF-FLOP module [14] is used to show the effectiveness of the idea. Finally, the advantageous of the proposed design is addressed in comparison with similar recent research.

2   Modelling

A simple piecewise static model is derived using the Castigliano's theorem to model the jammed and moving states of the scale jamming, once under external force pure tension (f_e) as in Fig. 1a-c and once under external momentum pure bending (τ_e) as in Fig. 1d, e. Spring has a rectangular cross section ($w \times h$) and the wire length (l_w), lead angle (γ) and the nominal diameter of the spring (d_s) considered to be fixed, and the lead angle changes are small and neglected. The stresses due to pure normal and shear forces are neglected since their effects are small compared to the torques effects [12]. Model parameters are presented in Table 1. The jammed scales behave similar to a rigid spring up to a breaking point related to their normal force (f_t) (elastic phase) and cause Coulomb friction damping afterwards as they move relatively (plastic phase). The scales jam again and behave like a spring if the external force reduces but the jammed state initial length (l_{c0}) should be updated during the plastic phase. The scales damping effect is modelled as a force (f_{sc}) opposing the external force and the spring deformation. The jagged surface slope angle (α) should be considered to calculate the breaking force and cancels the hysteresis at the end of a full return cycle ideally. However, a small hysteresis due to backlash between the slopes is observed and modelled by correcting the breaking force limit for the return half cycle equal to $2f_{Cb}$ where $f_{Cb} = r_{eff}.f_t.\mu_c/d_s$ is the base Coulomb friction force.

Pure Tension

For the pure tension, the spring coefficient (k_s) can be derived based on Castigliano's theorem as in [12][13] and the LuGre friction model is used to calculate the Coulomb friction torque (τ_C) due to scales' relative slip velocity (v) [18].

$$l_w = \pi d_s n_c, r_{eff} = r_i + \frac{j_{sc}}{a_{sc}}, v = \frac{2r_{eff}\, l_s}{n_c d_s}, \tag{1}$$

$$\tau = \frac{f_e d_s}{2}, u = \int_0^{l_w} \frac{\tau^2}{2\,G\,j} dl_w, \Delta l_s = \frac{\partial u}{\partial f_e}, \tag{2}$$

$$k_{(j,G)} = f_e/\Delta l_s = 4Gj/(n_c \pi d_s^3), \tag{3}$$

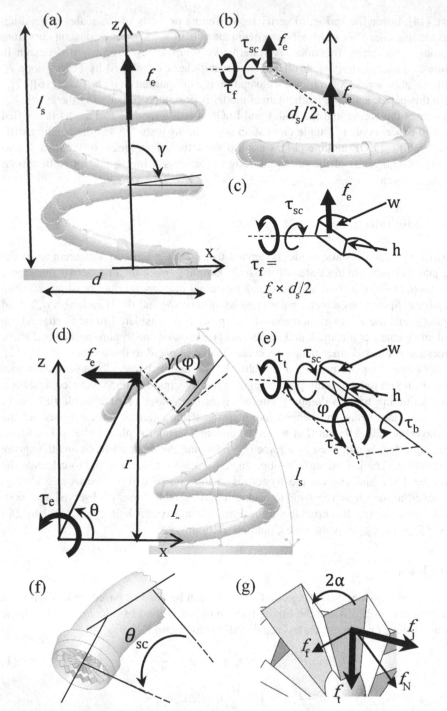

Fig. 2. Model parameters for pure tension (a-c), pure bending (d, e), and scale curvature (f) and forces on contact jagged surface (g).

$$i = f_j \sin(\alpha) - f_t \cos(\alpha) - \text{sgn}(\dot{l}_s)\mu_s \left(f_j \cos(\alpha) + f_t \sin(\alpha)\right), \tag{4}$$

$$l_{c0} = \begin{cases} l_{s0} & t = 0 \\ l_{c0} & \text{sgn}(i\dot{l}_s) \le 0 \lor |f_e| \le f_s - 2f_{cb}, \\ l_s - f_{sc}/k_{(j_{sc},G_{sc})} & \text{otherwise} \end{cases} \tag{5}$$

$$C_C = \text{sgn}(\dot{l}_s)(\mu_c + (\mu_s - \mu_c)e^{-\left(\frac{v}{v_0}\right)^2} + \sigma_s|v|), \tag{6}$$

$$f_j = f_t(\cos(\alpha) + C_C \sin(\alpha))/(\sin(\alpha) - C_C \cos(\alpha)), \tag{7}$$

$$f_{sc} = \begin{cases} 0 & t = 0 \\ k_{(j_{sc},G_{sc})}(l_s - l_{c0}) & \text{sgn}(i\dot{l}_s) \le 0 \lor |f_e| \le f_s - 2f_{cb}, \\ 2f_j r_{\text{eff}}/d_s & \text{otherwise} \end{cases} \tag{8}$$

$$f_s = k_{(j_s,G_s)}\Delta l_s \to f_e = f_{sc} + f_s, \tag{9}$$

where l_s is the spring length and l_s' is its deformation rate, r_{eff} is the effective radius and a_{sc} is the scale surface area, u is the strain energy due to torsional torque (τ), I is an indicator determining the slip condition, C_C is the Coulomb friction coefficient based on LuGre model, f_j and f_t are the forces acting on the jagged surface due to torsion and wire tension respectively, $f_s=2.r_{\text{eff}}.f_t.\mu_s/d_s$ is the limiting static friction, $j_s=wh(w^2+h^2)/3$ and $j_{sc}=\pi(r_o^4 - r_i^4)/4$ are the polar moment of inertia for the wire and scales.

Pure Bending

Assuming a constant curvature bending of the spring axis with constant ($l_s=l_{s0}$) under uniform external bending moment (τ_e) caused by an external force (f_e) at its end in which is similar to the case of having a constant curvature soft manipulator insides the spring. Here, θ is the spring bending angle and θ' is its time dependant rate, φ is the coil twist angle, τ_t and τ_b are the torsional and bending torques respectively, $\tau_{Cb} = r_{\text{eff}}.f_t.\mu_c$ is the base Coulomb friction torque, $\tau_S= r_{\text{eff}}.f_t.\mu_s$ is the limiting static friction torque $I_s=wh^3/12$ and $I_{sc}=\pi(r_o^4 - r_i^4)/2$ are the cross section area moment of inertia along the principal radial axis. Here the breaking occurs only in scales at $\varphi=n\pi$ positions. i and f_j are derived as in Eq.s 4 and 7.

$$r = l_{s0}/\theta, \dot{\theta} = \dot{x}_e/(r\sin(\theta)), v = -r_{\text{eff}}\dot{\theta}, \tag{10}$$

$$\tau_e = f_e r_e, \tau_t = \tau_e \cos(\phi), \tau_b = \tau_e \sin(\phi), \tag{11}$$

$$\Delta\theta = \frac{\partial u}{\partial f_e}, u_t = \int_0^{2\pi} \frac{\tau_t^2}{2Gj}\frac{d_s}{2}d\phi \to k_{t(J,G)} = \frac{2Gj}{n_c\pi d_s}, \tag{12}$$

$$u_b = \int_0^{2\pi} \frac{\tau_b^2}{2EI}\frac{d_s}{2}d\phi \to k_{b(I,E)} = \frac{2EI}{n_c\pi d_s}, \tag{13}$$

$$\theta_{c0} = \begin{cases} \theta_0 & t = 0 \\ \theta_{c0} & \mathrm{sgn}(i\dot\theta) \le 0 \ \lor \ |\tau_e| \le \tau_S - 2\tau_{Cb}, \\ (\theta - \tau_{sc}/k_{t(j_{sc},G_{sc})}) & \text{otherwise} \end{cases} \tag{14}$$

$$C_C = \mathrm{sgn}(\dot\theta)(\mu_c + (\mu_s - \mu_c)e^{-\left(\frac{v}{v_0}\right)^2} + \sigma_s|v|), \tag{15}$$

$$f_{sc} = \begin{cases} 0 & t = 0 \\ k_{t(j_{sc},G_{sc})}(\theta - \theta_{c0}) & \mathrm{sgn}(i\dot\theta) \le 0 \ \lor \ |\tau_e| \le \tau_S - 2\tau_{Cb}, \\ f_j r_{\text{eff}} & \text{otherwise} \end{cases} \tag{16}$$

$$\tau_S = \left(k_{t(j_s,G_s)} + k_{b(I_s,E_s)} + k_{b(I_{sc},E_{sc})}\right)\Delta\theta, \tag{17}$$

$$\rightarrow \tau_e = \tau_{sc} + \tau_s, \ f_e = \tau_e/r, \tag{18}$$

Table 1. Modelling Paramaters and Simulation and Experimental vaslues

E_s: Spring wire Young modulus, 60.7 [GPa]
G_s: Spring wire Shear modulus, 23.8 [GPa]
E_{sc}: Scales Young modulus for experiments, 1.28; for simulation, tension: 5.39, bending: 0.15×5.39 [GPa]
G_{sc}: Scales Shear Modulus for experiments, 0.32; for simulation, tension: 1.33, bending: 0.15×1.33 [GPa]
w: Cross section width, 1.2 [mm]
h: Cross section height, 0.2 [mm]
l_{s0}: Spring initial length in tension, 17.1 [mm]
l_{s0}: Spring initial length in torsion, 35 [mm]
d_{s0}: initial mean diameter, 38.2 [mm]
γ_0: Initial helix angle, tension: 87.3, bending: 94 [deg]
n_c: Coil turns, tension: 3, bending: 5
φ_{sc}: Scale curvature angle, 30 [deg]
n_{sc}: Number of scales, $2n_c\pi/\varphi_{sc}$
r_o: Scale end surface outer radius, 2.35 [mm]
r_i: Scale end surface inner radius, 0.8 [mm]
μ_s: Static friction coefficient, 0.03
μ_c: Coulomb friction coefficient, 0.01
v_0: Stribeck velocity, 1e-4
σ_s: Viscosity coefficient, -2e-3
r_0: Module bend curvature initial radius, for initially straight 1000, initially bent 30 [mm]
θ_0: Bend curvature initial angle, l_{s0}/r_0
α: Jagged surface slope angle, 50 [deg]
r_e: External force action radius in bending, 50 [mm]
σ_B: STIFF-FLOP module viscosity coefficient, -2e-2
k_B: STIFF-FLOP module stiffness constant, 33.8 [N/m]
t_f: Simulation and Experiment cycle time, 120 [s]

3   Numerical Simulation

Parameters for the numerical simulation are considered based on the experiment measurements with some modifications. A sinusoidal cyclic movement is considered as follows for l_s in tension and x_e in bending, with a=1 [cm] for tension, 7.5 [mm] for bending without initial curvature, and 5 [mm] for the bending with initial curvature.

$$l_s \text{ or } x_e = a(\sin(2\pi t/t_f - \pi/2) + 1), \tag{19}$$

A simple spring-viscous damper model for the STIFF-FLOP module is considered as follows which should be added to τ_e in Eq. 18. The parameters are identified experimentally.

$$\tau_B = r k_B x_e + \sigma_B \dot{\theta}, \tag{20}$$

The simulation results for the fully jammed state show good agreement with the experimental results for the pure tension (Fig. 3), pure bending with initial curvature (Fig. 4) and bending with initial curvature (Fig.5). The simulation can predict the breaking point and hysteresis in the full cyclic actuation as well as the stiffness coefficient for the jammed and moving states with good accuracy (Fig. 3-5).

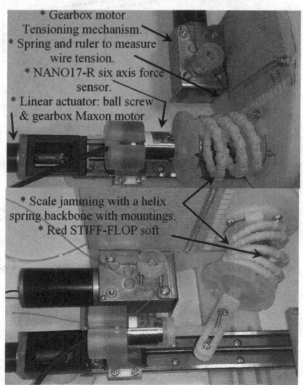

Fig. 3. Experiment setup, pure tension (up) and bending (down)

4   Experiments

A curved scale with circular cross sections is designed with a hole to pass the tension wire and 3D-printed. Both end surfaces jagged to increase the static friction while the coulomb friction is relatively low for a 3D printed part material. The experiment setup design is shown in Fig. 2 for the tension and bending experiments respectively. Fig. 3 and 4 show 40% increase in the blocking force in both tension and bending. There is a small change in stiffness constant for tension case but it is increased up to 40% in the bending test. Fig.5 shows the scales can fix the STIFF-FLOP module and block up to 0.6 [N] external force for r_0=30 [mm]. Fig. 6 shows up to 350% increase in stiffness and blocking force for antagonistic jammed state where the scales are jammed while the module chamber is pressurized which is significant. This is more than the linear summation of the system payload capacity without one of the jammed scales or internal pressure which emphasis on the antagonistic behaviour role. The results are filtered using a Moving Filter in Matlab software.

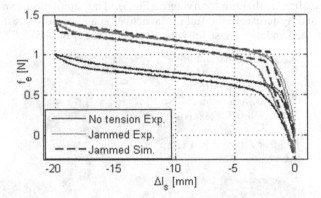

Fig. 4. Experimental and simulation results for pure tension with jammed (10N wire tension) and unjammed scales

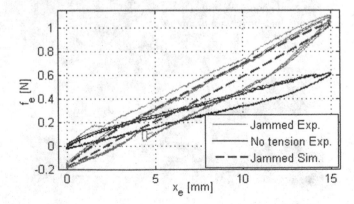

Fig. 5. Experimental and simulation results for pure bending of a straight module with jammed (1.5N wire tension) and unjammed scales

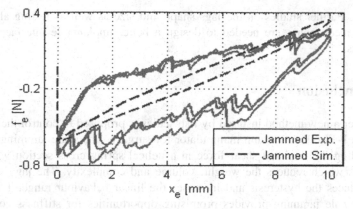

Fig. 6. Experimental and simulation results for pure bending of a bent module in jammed (1.5N wire tension) and unjammed scales

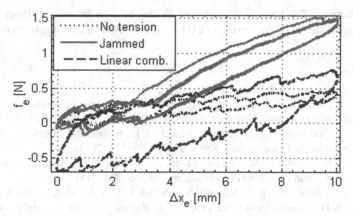

Fig. 7. Experiment results for antagonistic stiffening in pure bending for jammed (1.5N wire tension) and unjammed scales, and linear combination of scale jamming and pressurized modoule

5   Results and Discussion

The scale jamming in bending shows higher deformation rate ($\Delta x_e/l_{s0}$=0.428) than granular jamming by A. Jiang et. al. (=0.25) [5] and layer jamming by Kim et al. (=0.05) [9] and acts linearly in the whole deformation region while the deformation rate in the linear region for the later works are 0.18 and 0.01 respectively. The ratios of load capacity to initial length in the linear region are 28.6, 50 and 5 respectively which shows the scale jamming advantage over layer jamming. The granular jamming is better in terms of payload because of its bulky design. However, the scale jamming is better in terms of wearability, hysteresis and the ratio of achievable stiffness to device weight. The simulation results are in good agreement with the experiments; however, a more precise model considering the tensioning wire elongation and twist and the backlash between the jags is needed to achieve a model with a unified set of parameters. Further studies on the jags shape and size as well as testing alternative tensioning mechanisms are needed to design a better implantable interface for real applications.

6 Conclusion

In this paper a new method inspired by fish scales is proposed to control the stiffness of the STIFF-FLOP continuum manipulator for surgery. The scale jamming is based on controlling the torsional shear force in a helical spring cross section with small rigid scales which reduces the weight, volume and complexity. The jagged contact surface reduces the hysteresis and increases the linear behaviour range. The results show that scale jamming provides promising opportunities for stiffness control for soft robot manipulators for minimal invasive surgery.

Acknowledgement. This research is supported by the Seventh Framework Program of the European Commission in the framework of EU project STIFF-FLOP, grant agreement 287728.

References

1. Kuder, I.K., Arrieta, A.F., Raither, W.E., Ermanni, P.: Variable stiffness material and structural concepts for morphing applications. Prog. Aerosp. Sci. **63**, 33–55 (2013)
2. Laschi, C., Cianchetti, M., Mazzolai, B., Margheri, L., Follador, M., Dario, P.: Soft Robot Arm Inspired by the Octopus. Adv. Robot. **26**(7), 709–727 (2012)
3. Sornkarn, N., Howard, M., Nanayakkara, T.: Internal Impedance Control helps Information Gain in Embodied Perception, pp. 6685–6690 (2014)
4. Cheng, N.G., Lobovsky, M.B., Keating, S.J., Setapen, A.M., Gero, K.I., Hosoi, A.E., Iagnemma, K.D.: Design and Analysis of a Robust, Low-cost, Higly Articulated Manipulator Enabled by Jamming of Granular Media, pp. 4328–4333 (2012)
5. Jiang, A., Xynogalas, G., Dasgupta, P., Althoefer, K., Nanayakkara, T.: Design of a variable stiffness flexible manipulator with composite granular jamming and membrane coupling. In: 2012 IEEE/RSJ Int. Conf. Intell. Robot. Syst., pp. 2922–2927, October 2012
6. Jiang, A., Ataollahi, A., Althoefer, K., Dasgupta, P., Nanayakkara, T.: A Variable stiffness joint by granular jamming. In: 36th Mech. Robot. Conf. Parts A B, vol. 4, p. 267, August 2012
7. Santiago, J.L.C., Walker, I.D., Godage, I.S.: Continuum robots for space applications based on layer-jamming scales with stiffening capability. In: IEEE Aerospace Conference, pp. 1–13 (2015)
8. Ou, J., Yao, L., Tauber, D., Steimle, J., Niiyama, R., Ishii, H.: JamSheets: Thin Interfaces with Tunable Stiffness Enabled by Layer Jamming (2014)
9. Kim, Y.-J., Cheng, S., Kim, S., Iagnemma, K.: A Novel Layer Jamming Mechanism With Tunable Stiffness Capability for Minimally Invasive Surgery. IEEE Trans. Robot. **29**(4), 1031–1042 (2013)
10. Long, J.H., Hale, M.E., McHenry, M.J., Westneat, M.W.: Functions of fish skin: Flexural stiffness and steady swimming of longnose gar Lepisosteus osseus. J. Exp. Biol. **199**, 2139–2151 (1996)

11. Chang, J.H., Greenlee, A.S., Cheung, K.C., Slocum, A.H., Gupta, R.: Multi-turn, tension-stiffening catheter navigation system. In: 2010 IEEE Int. Conf. Robot. Autom., pp. 5570–5575, May 2010

12. Leech, A.R.: A study of the deformation of helical springs under eccentric loading. Naval Posigraduat1e School Monterey, California (1994)

13. Michalczyk, K.: Analysis of helical compression spring support influence on its deformation. Arch. Mech. Eng., vol. LVI (2009)

14. Cianchetti, M., Ranzani, T., Gerboni, G., De Falco, I., Laschi, C., Menciassi, A.: STIFF-FLOP surgical manipulator: Mechanical design and experimental characterization of the single module. In: IEEE Int. Conf. Intell. Robot. Syst., pp. 3576–3581 (2013)

15. Cianchetti, M., Ranzani, T., Gerboni, G., Nanayakkara, T., Althoefer, K., Dasgupta, P., Menciassi, A.: Soft Robotics Technologies to Address Shortcomings in Today's Minimally Invasive Surgery: The STIFF-FLOP Approach. Soft Robot. 1(2), 122–131 (2014)

16. Stilli, A., Wurdemann, H., Althoefer, K.: Shrinkable, stiffness-controllable soft manipulator based on a bio-inspired antagonistic actuation principle. In: IEEE/RJS International Conference on Intelligent Robots and Systems (2014)

17. Maghooa, F., Stilli, A., Althoefer, K., Wurdemann, H.: Tendon and pressure actuation for a bio-inspired manipulator based on an antagonistic principle. In: IEEE International Conference on Robotics and Automation (2015)

18. Song, X., Member, H.L., Member, K.A.: Efficient Break-Away Friction Ratio and Slip Prediction Based on Haptic Surface Exploration (2013)

Tuneable Stiffness Design of Soft Continuum Manipulator

Seri Mastura Mustaza[1](✉), Duale Mahdi[2], Chakravarthini Saaj[1],
Wissam A. Albukhanajer[1], Constantina Lekakou[2], Yahya Elsayed[2], and Jan Fras[3]

[1] Department of Electronic Engineering, Faculty of Engineering and Physical Sciences, University of Surrey, Guildford, Surrey GU2 7XH, UK
{s.mustaza,c.saaj,w.a.albukhanajer}@surrey.ac.uk
[2] Division of Mechanical, Medical and Aerospace Engineering, Faculty of Engineering and Physical Sciences, University of Surrey, Guildford, Surrey GU2 7XH, UK
{d.a.mahdi,c.lekakou,y.elsayed}@surrey.ac.uk
[3] Przemyslowy Instytut Automatyki i Pomiarow, Al. Jerozolimskie 202,
02-486 Warsaw, Poland
jfras@piap.pl

Abstract. Soft continuum robots are highly deformable and manoeuvrable manipulators, capable of navigating through confined space and interacting safely with their surrounding environment, making them ideal for minimally invasive surgical applications. A crucial requirement of a soft robot is to control its overall stiffness efficiently, in order to execute the necessary surgical task in an unstructured environment. This paper presents a comparative study detailing the stiffness characterization of two soft manipulator designs and the formulation of a dynamic stiffness matrix for the purpose of disturbance rejection and stiffness control for precise tip positioning. An empirical approach is used to accurately describe the stiffness characteristics along the length of the manipulator and the derived stiffness matrix is applied in real-time control to reject disturbances. Further, the capability of the two types of soft robots to reject disturbances using the dynamic control technique is tested and compared. The results presented in this paper provide new insights into controlling the stiffness of soft continuum robots for minimally invasive surgical applications.

Keywords: Stiffness control · Soft continuum robots · Robustness · Minimally invasive surgery

1 Introduction

There is an increasing interest to advance the state-of-the-art of soft robotics in medical applications. This is primarily driven by the need for a robot that can manoeuver in confined spaces and interact with organs and tissue safely, especially, in applications such as minimally invasive surgery (MIS). Numerous research studies have been conducted that demonstrate soft continuum robots for MIS applications [1, 2, 3, 4]. The highly articulated soft robot manipulators are advantageous compared

© Springer International Publishing Switzerland 2015
H. Liu et al. (Eds.): ICIRA 2015, Part III, LNAI 9246, pp. 152–163, 2015.
DOI: 10.1007/978-3-319-22873-0_14

with rigid manipulators because of their low modulus, making them compliant to compressive forces and hence, allowing to manoeuvre easily and safely through narrow orifices and tight passages between organs. However, flexibility of the manipulator also implies low payload capacity and poor distal stiffness. While it is desirable for the manipulator to be gentle when interacting and manoeuvring within its surroundings, significant stiffness is required once the robot is in place to perform useful surgical tasks.

For a fully operational soft robot, a key requirement is its ability to navigate in an unstructured environment in the presence of perturbation to avoid crude movements and instability of the distal tip. For these reasons, tuneable stiffness is required in order for the manipulator to reject perturbing forces anywhere in the workspace. This can be achieved by tuning the forces generated by the manipulator to compensate the effect of perturbations, thus ensuring safe manoeuvrability and manipulation of the soft robot.

The currently implemented tuneable stiffness methods for soft manipulators require a more complex mechanical design e.g. tendon driven mechanism [5, 6, 7] and granular jamming [7, 8, 9]. While prototypes based on these designs have been proven to provide the desired stiffness, they introduce additional problems. In the case of granular jamming, the granular matter used reduces steerability, while, tendon based mechanisms may introduce undesirable non-linear characteristic to the entire system as well as poor disturbance rejection [10]. The ideal solution is to design stiffness controller coupled with position estimation algorithm to produce a desired positional tip stiffness. However, assessing the stiffness of soft manipulator can be difficult and stiffness control for this type of manipulator is achieved indirectly from position control without sensing the forces or stiffness at the tip as proposed in [11]. Apart from the work in [11], stiffness control of soft manipulators has been poorly researched and developed.

The goal of this paper is to analyse and compare the stiffness characteristics of two different designs of soft continuum manipulators and use this information for real-time stiffness control by implementing a disturbance rejection algorithm. The empirical approach taken involves characterizing the stiffness of the manipulator through experiments and deriving a relationship between the force exerted by the manipulator and the joint displacement to generate a dynamic stiffness matrix. Disturbance rejection in this study is achieved by directly measuring the perturbing forces on the manipulator and compensating for any deflection using the dynamic stiffness matrix, in contrast to the work in [11] whereby no information on forces was utilized to execute the stiffness control.

This paper is organised into the following sections. Section 2 describes the prototypes of the soft manipulators. Section 3 details the stiffness analysis carried out. Section 4 discusses the relationship between the applied disturbance force at the tip and the change in chamber pressure to compensate for this tip displacement. Finally, Section 5 gives concluding remarks and suggests future work.

2   Soft Continuum Manipulator Prototype

This study investigates the stiffness characteristics of two soft continuum manipulator prototypes as shown in Fig. 1. Both of the prototypes were fabricated using Ecoflex TM 0050 silicone and are composed of three actuating pressure chambers displaced at 120o apart in symmetrical radial arrangement. Two different braiding structures are explored to maximize the longitudinal extension and bending of the manipulator in achieving the desired steerability. The structural differences between the two prototypes are explained below:

Prototype 1. The first prototype is composed of three equally spaced semi-cylindrical chambers in radial arrangement moulded directly into a single cylindrical unit as shown in Fig. 1(a) [12]. A crimped braided sheath is used around the module to constrain the inflation of the chamber and maximize bending under applied pressure. The diameter of each semi-cylindrical actuating chamber is 8mm and the diameter of the cylindrical unit is 30 mm.

Prototype 2. The second prototype is more compact in structure compared with Prototype 1. It comprises three full-cylindrical actuating pressure chambers without any braided sleeve around the module as illustrated in Fig. 1(b). The radial expansion of the pressure chambers are constrained by tightly wound nylon thread around each individual chamber to maximize the bending and longitudinal expansion. Similar to the first prototype, the actuating chamber is 8mm in diameter and with an overall smaller module diameter of 25mm [13].

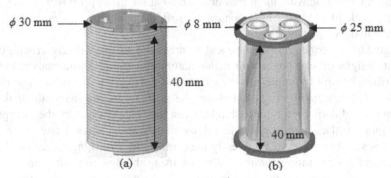

Fig. 1. (a) Prototype 1 (crimped braided sheath) and (b) prototype 2 (nylon threaded chambers)

Both braiding structures effectively minimize the radial expansion of the chambers under actuation, but their behaviours differ tremendously. Prototype 2 is less sensitive to changes in pressure compared to prototype 1. This is due to the internal braiding that restricts changes in chamber length and cross-sectional area. Furthermore, the prototype 2 eliminates cross-talk between chambers and friction between the outer surface of the silicone and the wall of the external braiding which prototype 1 suffers from. More details on the structural design of these prototypes can be found in [12, 13, 14].

3   Stiffness Analysis

This section investigates and compares the stiffness characteristics of the two proto-types presented in Section 2. The stiffness matrix for each module is derived and used to control the stiffness of the manipulator.

The stiffness matrix is defined as the relationship between the forces exerted to the displacements of all joints [15]:

$$F = Kd \tag{1}$$

where F is the force (n dimensional vector) at the tip in (N), d is the displacements of the joints (n dimensional vector) in (m) and K is stiffness matrix in (N/m). Nor-mally, this matrix is a constant throughout its operating workspace and configuration. However, due to the inherent non-linear behaviour of the manipulator, a constant ma-trix will not properly reflects the changes in the stiffness of the manipulator, hence, will not be effective in controlling the stiffness of the manipulator.

The forces exerted by each actuating chambers is symmetrical around 120 degree and the force sensor [16], [17] used has a symmetry of 90 degree on the x-y coordi-nate axes, therefore, the relationship between the displacement of each chamber and the forces acting on each chamber cannot be obtained directly. This relationship is obtained indirectly in two steps: first, the relationship between a change in force as a function of a change in pressure is obtained. The second step is to manipulate the relationship in the first step in terms of pressure and substitute this into the Euler-Bernoulli equation to obtain a relationship between displacement, stiffness and pres-sure.

In this study, the Euler-Bernoulli beam model equation [18] is used. This equation relates a change in length to a change in pressure:

$$\Delta dl_i = \frac{F_p}{EA} dl_i = \frac{dP_i A_{ch}}{EA} dl_i \tag{2}$$

where dl is the length of a small fragment of the chamber in (m), Δdl is the change in length of the chamber due to pressure in (m), $i \in \{1,2,3\}$ is an index used to indi-cate chamber number, dP is change in pressure in (N/m2), E is the material's Young's Modulus in (N/m2), and A is the cross sectional area of the module in (m2). The pressure in each chamber can be used to calculate the force applied on the mod-ule using (3):

$$F = PA_{ch} \tag{3}$$

where A_{ch} is the cross sectional area of the chamber. Table 1 shows the constants in (2) i.e. the Young's Modulus, the cross-sectional area of the module and the chamber cross-sectional area. Figure 2 shows a comparison between the beam model and the experimental data for the two prototypes investigated. The beam model and experi-mental data correlate for both prototypes and for this reason, the beam model is con-sidered a representative model.

Table 1. Beam model parameters

Parameter	Description	Prototype 1	Prototype 2
E	Young's modulus	40000 N/m^2	178000 N/m^2
A	Manipulator cross-sectional area	$4.22e\text{-}4 \text{ m}^2$	$4.52e\text{-}4 \text{ m}^2$
a	Chamber cross-sectional area	$2.21e\text{-}5 \text{ m}^2$	$4.42e\text{-}5 \text{ m}^2$

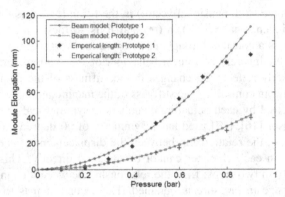

Fig. 2. Relationship between pressure and the change in length for one chamber

To obtain the relationship between change of force and change of pressure, the tip of the manipulator is constrained to emulate interactions with the environment. For a small change in pressure ΔP (see Fig. 3), the forces exerted at the tip of the module are measured using the force sensor developed by Noh et al [17]. This is done across the operational pressure range (up to 0.8 bar) and allows for the characterisation of the stiffness of the manipulator in all operational pressure range and configurations of interest. Thus, the behaviour and pattern between actuated pressure and force exertion for both prototypes is estimated. This analysis gives the relationship between force and pressure, for the three actuating chambers. For example, the change of force as a function of change of pressure relationship for chamber 1 of prototype 1 is governed by (4):

$$dF_x = -4.366dP_1 + 1.6368$$

$$dF_y = -1.7853dP_1 + 0.5973 \qquad (4)$$

$$dF_z = 1.6199dP_1 - 1.4472$$

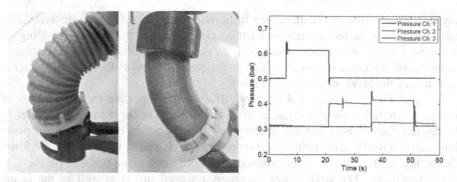

Fig. 3. Experimental setup to evaluate the relationship between pressure and force: (a) and (b) prototype 1 and 2 with constrained tip respectively, (c) input pressure supply to both prototypes

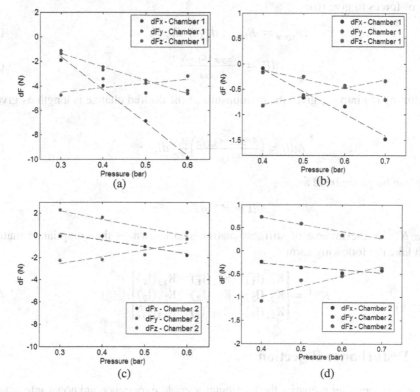

Fig. 4.   Linear fitting of the relationship between change in force and pressure: (a) and (c) for prototype 1, (b) and (d) for prototype 2

Figure 4 compares the relationships of the change in force as a function of change in pressure for the two prototype manipulators. The figure shows data for two chambers (out of three), i.e., chamber 1 and 2. As illustrated in these figures, both prototypes have a similar overall trend where the directions of the forces exerted are comparable. However, prototype 1 exerts a significantly higher force compared with prototype 2. This is due to the difference in braiding structure of the manipulator. For small pressure perturbations imposed within each chamber, the external braiding of prototype 1 allows it displace more (see Fig.2) compared with the second prototype. Furthermore, the cross-sectional area of prototype 1 changes as the chamber is actuated causing the resultant force to be higher.

The goal for stiffness control for surgical tasks is to maintain the tip position as forces are exerted at the distal end of the manipulator. The input to the controller is the desired displacement and this is compared with the actual displacement for position control through reducing the position error. The change in length(Δdl), the force (F_p) acting on the chamber is determined empirically at a given position (in the x, y and z directions). The straight line equations obtained in (4) as well as the beam model is utilized to generate the dynamic relationship between length and force as the chambers are pressurized at different levels. The linear equation in (5) is re-written in terms of forces to give (6):

$$dF_{x,y,z} = B_{i,x,y,z}\, dP_i + C_{i,x,y,z} \tag{5}$$

$$dP_i = \frac{dF_{x,y,z} - C_{i,xyz}}{B_{i,xyz}} \tag{6}$$

Substituting (6) into (2) gives the relationship of the desired change is length as given in (7):

$$\Delta dl_i = \left(\frac{dF_{x,y,z} - C_{i,xyz}}{B_{i,xyz}}\right)\frac{A_{ch}}{EA}\, dl_i \tag{7}$$

which can be generalized as:

$$\Delta dl = K^{-1} dF \tag{8}$$

where K^{-1} is the inverse of stiffness matrix also known as the compliance matrix which take the following form:

$$K^{-1} = \begin{bmatrix} K_{11}(l_1) & K_{12}(l_1) & K_{13}(l_1) \\ K_{21}(l_2) & K_{22}(l_2) & K_{23}(l_2) \\ K_{31}(l_3) & K_{32}(l_3) & K_{33}(l_3) \end{bmatrix}. \tag{9}$$

4   Disturbance Rejection

For a given operating scenario, the manipulator could experience unknown interaction forces at the tip which result in the deflection of the tip from its desired working point. This could have a detrimental effect in practice, possibly resulting in unintended interaction between the tip and its surroundings resulting in tissue damage. This deflection can be effectively compensated for by directly measuring the interaction forces in-situ and utilizing the compliance matrix derived in (9) to maintain the desired tip position.

The deflection and tip position of the manipulator (X_c, Y_c, Z_c) is measured using the NDI Aurora sensor [19]. To maintain the desired position in the presence of disturbance forces, the inverse kinematic model in [20] is utilized to calculate the desired joint position.

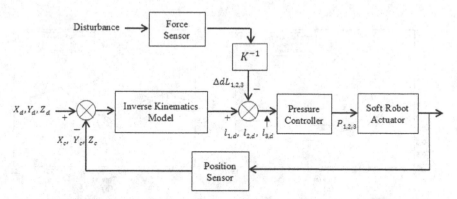

Fig. 5. Position control architecture through stiffness control and disturbance rejection

In the current test setup, disturbances are injected by applying arbitrary forces on the tip of the manipulator. Based on the measured forces, the system calculates the resulting $\Delta dl_{1,2,3}$ using (8) and re-calculates the desired chamber length $(l_{1,d}, l_{2,d}, l_{3,d})$ to overcome this disturbance. The controller then outputs the appropriate pressure, $P_{1,2,3}$, to achieve the desired length. The overall control architecture based on the inverse kinematic developed in [20] is as shown in Fig. 5.

Figure 6 illustrates the disturbance forces applied at the tip, the deflection of the tip due to this disturbance and the corresponding chamber pressure to counteract these disturbances for both prototypes. For prototype 1, disturbance forces in the range of 0.6 to 0.7 N are applied for a period of 7.5 seconds (Fig. 6(a)) which results in tip deflection in the range of 3 to 5 mm. For prototype 2, a steeper disturbance forces in the range of 0.8 to 1 N were applied for a period of 3.5 seconds (Fig. 6(b)) resulting in larger tip deflection between 2 to 5.5 mm. Larger disturbance forces were applied to prototype 2 because of its inherent structural stability resulting from the internal nylon threading. To counteract these disturbance forces, the controller utilizes the dynamic stiffness matrix to estimate the resultant change in length due to these disturbance forces and applies pressure in the relevant chambers to maintain the tip at the desired position. As can be seen in Fig. 6(e) and Fig. 6(f), the controller manages to keep the deflection of the tip to within 2 to 5.5mm and nullifies the displacement caused by the disturbances force applied at the tip of manipulator. Both prototypes successfully reject the disturbance but prototype 2 takes a longer time to achieve this. This was attributed to the difference in duration and magnitude of the disturbance force applied to the two prototypes. For both prototypes, corrective pressures in the range of 0.15 to 0.3 bar were applied. Figure 7 shows the results in the absence of the dynamic stiffness control without force feedback, whereby huge deflections in the range of 10 to 20 mm are observed. Comparing results in Fig. 6 and Fig. 7 confirms that a dynamic stiffness matrix is required in order to actively counteract disturbance forces for real-time applications. Further, the controller responds in real-time to maintain tip position at the desired position under perturbed conditions, thus proving the robustness of the stiffness controller.

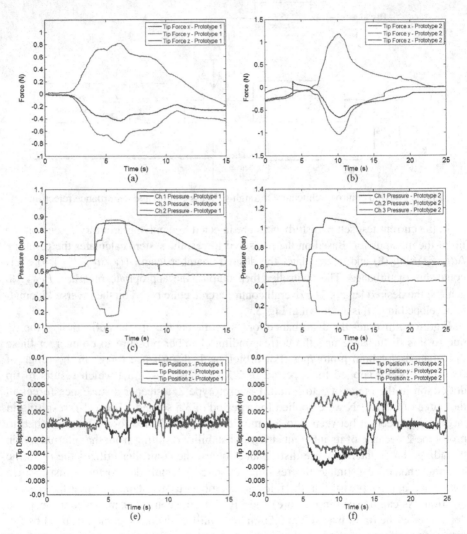

Fig. 6. Position control with force feedback for disturbance rejection and stiffness control (a) disturbance force at tip of manipulator, (c) pressure actuation to compensate the disturbance and (e) displacement of tip for prototype 1. Similarly (b, d and f) are the corresponding plots for prototype 2

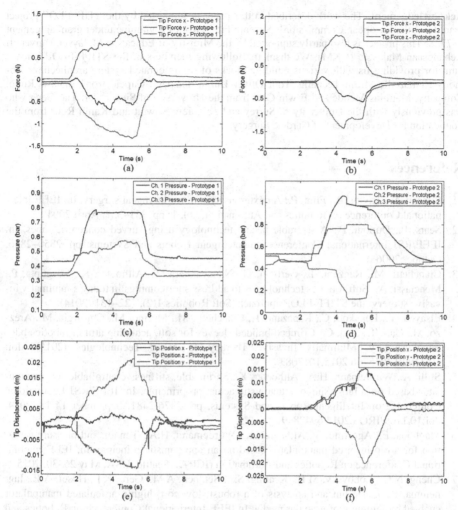

Fig. 7. Position control with no force-feedback: (a) disturbance force at tip of manipulator, (c) pressure actuation to compensate the disturbance and (e) displacement of tip for prototype 1. Similarly (b, d and f) are the corresponding plots for prototype 2

5   Conclusion and Future Work

The experimental results suggest that the two designs of the soft continuum robot modules have significantly different stiffness, re-enforcing the need for a separate dynamic stiffness matrix tailored for each system. The results covered in this paper show that the empirically derived stiffness matrix is suitable for disturbance rejection in real-time control. More investigation is required to develop a generalized stiffness matrix suitable for a class of soft continuum robots. Further research is under way to achieve a complete characterization that covers the entire workspace of interest for different designs of soft continuum robots.

Acknowledgement. The work described in this paper is supported by the STIFFFLOP project grant from the European Commission's Seventh Framework Programme under grant agreement 287728. This project is also partly supported by the Ministry of Education Malaysia, Universiti Kebangsaan Malaysia (UKM). We thank the following members of the STIFF-FLOP consortium for providing us with various building blocks of the integrated system architecture: Anthony Remazeilles from Tecnalia, Toni Oliver Duran and Ugo Cupcic from Shadow Robot Company, Matthias Mende and Erwin Gerz from the University of Siegen, Dr. Tao Geng who was previously with the University of Surrey and Zbigniew Nawrat and Kamil Rohr from the Foundation for Development of Cardiac Surgery.

References

1. Simaan, N., Taylor, R., Flint, P.: A dexterous system for laryngeal surgery. In: IEEE International Conference on Robotics and Automation, vol. 1, pp. 351–357, April 2004
2. Sears, P., Dupont, P.: A steerable needle technology using curved concentric tubes. In: IEEE/RSJ International Conference on Intelligent Robots and Systems, pp. 2850–2856, October 2006
3. Cianchetti, M., Ranzani, T., Gerboni, G., Nanayakkara, T., Althoefer, K., Dasgupta, P., Menciassi, A.: Soft robotics technologies to address shortcomings in today's minimally invasive surgery: the STIFF-FLOP approach. Soft Robotics **1**(2), 122–131 (2014)
4. Elsayed, Y., Lekakou, C., Ranazani, T., Cianchetti, M., Morino, M., Chirurgia, M., Arezzo, M., Gao, T., Saaj, C.: Crimped braided sleeves for soft, actuating arm in robotic abdominal surgery. Minimally Invasive Therapy & Allied Technologies (2015). doi: 10.3109/13645706.2015.1012083
5. Stilli, A., Wurdemann, H.A., Althoefer, K.: Shrinkable, stiffness-controllable soft manipulator based on a bio-inspired antagonistic actuation principle. In: IEEE/RSJ International Conference on Intelligent Robots and Systems, pp. 2476-2481, September 14-18, 2014. doi:10.1109/IROS.2014.6942899
6. Maghooa, F., Agostino, S., Althoefer, K., Wurdemann, H.A.: Tendon and pressure actuation for a bio-inspired manipulator based on an antagonistic principle. In: IEEE International Conference on Robotics and Automation (ICRA), Seattle, USA, May 26–30, 2015
7. Cheng, N.G., Lobovsky, M.B., Keating, S.J., Setapen, A.M., Gero, K.I., Hosoi, A.E., Iagnemma, K.D.: Design and analysis of a robust, low-cost, highly articulated manipulator enabled by jamming of granular media. In IEEE International Conference on Robotics and Automation, pp. 4328–4333, May 2012
8. Loeve, A.J., van de Ven, O.S., Vogel, J.G., Breedveld, P., Dankelman, J.: Vacuum packed particles as flexible endoscope guides with controllable rigidity. IGranular Matter **12**(6), 543–554 (2010)
9. Jiang, A., Xynogalas, G., Dasgupta, P., Althoefer, K., Nanayakkara, T.: Design of a variable stiffness flexible manipulator with composite granular jamming and membrane coupling. In: IEEE/RSJ International Conference on Intelligent Robots and Systems (IROS), pp. 2922–2927, October 7-12, 2012
10. Lee, Y.-T., Choi, H.-R., Chung, W.-K., Youm, Y.: Stiffness control of a coupled tendon-driven robot hand. IEEE Control Systems **14**(5), 10–19 (1994). doi:10.1109/37.320882
11. Mahvash, M., Dupont, P.: Stiffness control of surgical continuum manipulators. IEEE Transactions on Robotics **27**(2), 334–345 (2011)
12. Lekakou, C., Elsayed, Y., Geng, T., Saaj, C.M.: Skins and Sleeves for Soft Robotics: Inspiration from Nature and Architecture. Advanced Engineering Materials (2015). doi:10.1002/adem.201400406

13. Fraś, J., Czarnowski, J., Maciaś, M., Główka, J., Cianchetti, M., Menciassi, A.: New STIFF-FLOP module construction idea for improved actuation and sensing. In: IEEE International Conference on Robotics and Automation (ICRA), Seattle, USA, May 26-30, 2015
14. Cianchetti, M., Ranzani, T., Gerboni, G., De Falco, I., Laschi, C., Menciassi, A.: STIFF-FLOP surgical manipulator: mechanical design and experimental characterization of the single module. In: IEEE International Conference on Intelligent and Robotic Systems, pp. 3567-3581 (2014)
15. Carbone, G.: Stiffness analysis for grasping tasks: grasping in robotics. In: Carbone, G., (ed.) Mechanisms and Machine Science, vol. 10, pp. 17–55. Springer, London (2013)
16. Noh, Y., Secco, E.L., Sareh, S., Wurdemann, H., Faragasso, A., Back, J., Liu, H., Sklar, E., Althoefer, K.: A continuum body force sensor designed for flexible surgical robotic devices. In: 36th Annual International Conference of IEEE Engineering in Medicine and Biology Society (EMBC), pp. 3711–3714 (2014)
17. Noh, Y., Sareh, S., Back, J., Wurdemann, H.A., Ranzani, T., Secco, E.L., Faragasso, A., Liu, H., Althoefer, K.: A three-axial body force sensor for flexible manipulators. In: IEEE International Conference on Robotics and Automation (ICRA), pp. 6388–6393 (2014)
18. Fraś, J., Czarnowski, J., Maciaś, M., Główka, J.: Static Modeling of Multisection Soft Continuum Manipulator for Stiff-Flop project. Springer (2014)
19. NDI 3D measurement technology systems. http://www.ndigital.com/medical/products/aurora/
20. Calinon, S., Bruno, D., Malekzadeh, M.S., Nanayakkara, T., Caldwell, D.G.: Human-robot skills transfer interfaces for a flexible surgical robot. Computer Methods and Programs in Biomedicine 116(2), 81–96 (2014)

Lecture Notes in Computer Science:
An Antagonistic Actuation Technique
for Simultaneous Stiffness and Position Control

Helge A. Wurdemann[✉], Agostino Stilli, and Kaspar Althoefer

Centre for Robotics Research (CoRe), Department of Informatics,
King's College London, Strand, London, UK
{helge.wurdemann,agostino.stilli,kaspar.althoefer}@kcl.ac.uk
https://www.kcl.ac.uk, https://www.kings-core.com

Abstract. The application of soft robots can result in significant improvements within a number of areas where traditional robots are currently deployed. However, a challenging task when creating soft robots is to exert effective forces. This paper proposes to combine pneumatic and tendon-driven actuation mechanisms in an entirely soft outer sleeve realising a hybrid actuation principle, to realise a new type of robotic manipulator that can collapse entirely, extend along its main axis, bend along its main axis and vary its stiffness. The created robot arm is inherently flexible manufactured from sections that consist of an internal stretchable, air-tight balloon and an outer, non-stretchable sleeve preventing extension beyond a maximum volume. Tendons connected to the distal ends of the robot sections run along the outer sleeve allowing each section to bend in one direction when pulled. The results from our study show the capabilities of such a robot and the main advantages of the proposed technique compared to traditional, single-actuation type robot manipulators.

Keywords: Soft robot · Bio-inspiration · Hybrid actuation · Stiffness controllability

1   Introduction

Taking inspiration from nature, researchers have created new robotic systems to overcome limitations of traditional robots composed of rigid joints and links [4]. In particular, animals' appendages such as the elephant trunk or the octopus arm have become the focus of studies creating soft, hyper-redundant robots and aiming to achieve similar capabilities as their role models [7,9,17,19,27,28,30]. The application of these type of robots can result in significant improvements within a number of fields (such as navigation and manipulation in unstructured environments) where traditional robots are currently deployed [10,11,20]. However, a challenging task when creating soft robots is to exert effective forces [13].

© Springer International Publishing Switzerland 2015
H. Liu et al. (Eds.): ICIRA 2015, Part III, LNAI 9246, pp. 164–174, 2015.
DOI: 10.1007/978-3-319-22873-0_15

In recent years, researchers have investigated several solutions to the complex problem to change and control the stiffness of soft manipulators. A silicone-based, pneumatically actuated soft robot arm has been developed as part of the EU-funded project STIFF-FLOP. STIFF-FLOP focuses on exploring the mechanisms of the octopus and attempts to extract relevant biological features to develop medical robotic systems for Minimally Invasive Surgery (MIS) [14] with integrated sensors [22,23,25,31,32]. One segment of the current multi-segment manipulator prototype is equipped with three compressible chambers [8,24]. Stiffness variation is realised with an additional chamber within the silicone body filled with granular that can be jammed by applying a vacuum [13,14,18]. Hence, the control of the stiffness of the robot's body is achieved by extending the overall robot system through the introduction of an additional type of actuator. The concept of polymeric artificial muscles described in [5] to actuate a robot manipulator was furthered in [15] by integrating granule-filled chambers which when exposed to varying degrees of vacuum could actuate, soften and stiffen the manipulator's joints. A similar concept is proposed in [16]. A hollow snake-like manipulator consists of multiple overlapping layers of thin Mylar film. By applying vacuum pressure, the friction between the film layers increases which results in a tunable stiffness capability. In [6], the authors report on a thermally tunable composite for mechanical structures. This flexible open-cell foam coated in wax can change stiffness, strength, and volume. Altering between a stiff and soft state and vice versa introduces a time delay as the material does not instantly react to the heating-up or cooling-down process.

In this paper, we propose to combine pneumatic and tendon-driven actuation mechanisms in an entirely soft outer sleeve. The hybrid actuation mechanism and design of the manipulator result in a new type of robotic manipulator that can collapse entirely, extend and bend along its main axis, and vary its stiffness simultaneously. The robot arm is inherently flexible manufactured from sections that consist of an internal stretchable, air-tight balloon and an outer, non-stretchable sleeve preventing extension beyond a maximum volume. Tendons connected to the distal ends of the robot sections run along the outer sleeve allowing each section to bend in one direction when pulled. The results from our study show the capabilities of such a robot and the main advantages of the proposed technique compared to traditional, single-actuation type robot manipulators.

In Section 2, the antagonistic actuation principle is described and the scientific contributions of this paper are summarised. The mechanical design of the soft, stiffness-controllable robot arm is presented in Section 3 along with the overall control architecture. Sections 4 and 5 introduce the experimental setup to validate the tunable stiffness mechanism and present the achievements. The achievements are concluded in Section 6.

2 Contributions of the Antagonistic Actuation Principle

The role model for our work is the octopus [12,17]. Its soft arms virtually have an infinite number of Degrees of Freedom (DoFs). Studies by biologists show that the octopus arm has longitudinal and transversal muscles. Both sets of muscles can be activated in an antagonistic way so that it is possible for the octopus to control the stiffness of parts of its arm. This natural feature enables the octopus to catch fish, crawl across the seabed, or move obstacles.

We have here transferred the natural antagonistic principle of the longitudinal and transversal muscles into an antagonistic robotic actuation system. Our robot manipulator makes use of two fundamental actuation means - intrinsic, pneumatic actuation and extrinsic, tendon-based actuation - able to oppose each other. This cooperative hybrid system is capable of varying the arm's stiffness over a wide range combining the advantages of extrinsic and intrinsic actuation mechanisms at the same time [3]:

- Tendons are utilised to operate manipulators with a high payload due to high tensile strength [21].
- Due to the thin structure of tendons and the externally placed motors used to control the length of the tendons, tendon-driven manipulators can be easily miniaturised [1].
- Controlling the tip's position and orientation of tendon-actuated manipulators is fairly accurate [2].
- The payload of a tendon-driven robot depends on the tensile strength of the tendons and the maximum force that can be generated by the applied electrical drives.
- Pneumatically actuated manipulators are inherently compliant and suitable to share the working environment with humans [5,26].

Fusing these two sets of actuation mechanisms allows us creating a manipulation with enhanced capabilities above single-type actuation robots [29]. Hence, the significant contributions made by the creation of the actuation mechanism are as follows:

- The implementation of the hybrid actuation system together with the structure of the robot allows the manipulator to transform between being entirely shrunk and completely elongated. We are able to achieve an extension of factor 20 or more.
- The antagonistic actuation mechanism allows the manipulator to transform between a soft and stiff state: This can be achieved by inflating the robot and fastening the tendons at the same time. This collaborative principle results in a continuous stiffness controllability.
- Our developed robot has no backbone and a simple structure made of soft material such as fabric and latex. Hence, the manipulator can be miniaturised easily and squeezed through narrow openings being still fully functional.

3   Mechanical Structure and Design

3.1   Structure and Assembly of the Manipulator

A CAD drawing of the manipulator's structure is shown in Fig. 1. The structure of the prototype consists of three parts:

- The inner air-tight latex bladder is stretchable. The pneumatic actuation system is connected to the base of the bladder.
- The outer polyester fabric sleeve is non-stretchable but shrinkable. This sleeve restricts the expansion of the inner bladder when pressurised.
- Two sets of three nylon tendons are fixed at the tip and in the middle of the manipulator dividing the robot into two sections.

The cylindrical polyester sleeve is slipped over the latex bladder. The length of the inflated soft robot is 20 cm; the diameter is 23 mm. As mentioned earlier, the non-stretchable fabric material limits the expansion of the inner sleeve and prevents any ballooning in radial direction beyond the maximum diameter of 23 mm. Hence, the manipulator can only expand along its longitudinal axis (elongation) to a maximum length of 20 cm. In order to avoid the inner latex sleeve from being twisted whilst changing between an inflated and deflated state, the tip of the latex bladder and fabric sleeve are connected. Two sets of three tendons are inserted into guidance channels in the periphery of the manipulator along the outside of the polyester sleeve. These channels are spaced 120° apart. In our two-section prototype, three tendons are fixed to the tip of the manipulator and another set of three tendons are attached to the tip of the proximal section (i.e., the middle of the manipulator). Employing this approach, the robot's two sections can be independently controlled (see Figs. 1 and 2). Adjusting the length of the tendons appropriately also controls the stiffness of the arm - e.g., decreasing the tendon length at a given air pressure will reduce the manipulator's length and result in a stiffer state.

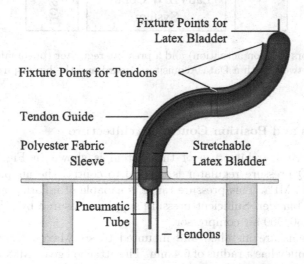

Fig. 1. The prototype of the manipulator consists of an inner air-tight, stretchable latex bladder, an outer, non-stretchable (but shrinkable) polyester fabric sleeve and three pairs of nylon tendons. The latex bladder is pneumatically inflated or deflated. The tendons are mounted on the outside at tip and in the middle of the manipulator allowing to control the robot's configuration [28].

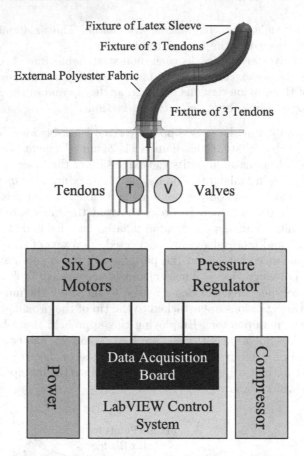

Tendons   Valves

Fig. 2. DC motors (tendon actuation) and a pressure regulator (pneumatic actuation) are interfaced between with a Data Acquisition Board and the bio-inspired manipulator [28].

3.2   Stiffness and Position Control Architecture

The overall control architecture of the system is shown in Fig. 2. A SMC ITV0010-3BS-Q pressure regulator is utilised to control the air pressure from 0.001 MPa to 0.1 MPa. This pressure range is capable of inflating and deflating the inner latex bladder. Sufficient pressure supply is ensured by a BAMBI MD Range Model 150/500 air compressor.

Pulley systems are installed and mounted to six Maxon RE-max 24 DC motors. Each pulley has a radius of 6.4 mm. The attached gear (Maxon Planetary Gearhead GP 22 C) provides a torque of up to 2 Nm resulting in a maximum force of 312.5 N considering the pulleys.

The DC motors and pressure regulators are interfaced via an NI USB-6211 DAQ card to LabVIEW software. The hybrid actuation system receives steering commands from a joystick (Logic 3 JS282 PC Joystick): The toggle controls the

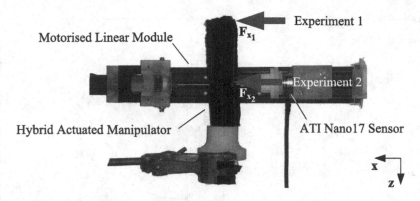

Fig. 3. Overview of the experimental setup to validate variable stiffness with a ATI Nano17 Force/Torque sensor and a motorised linear module [28].

angular position of the shaft drives which drive the tendon pulleys. The pulleys decrease or increase the length of the tendons resulting in bending of the manipulator. A button is integrated to regulate the air pressure inside the latex sleeve. Being able to steer the manipulator using the tendons and regulating the pressure at the same time allows keeping constant pressure when the deflatable arm shrinks, extends or bends on the one hand and to control the stiffness (increasing the pressure in the latex bladder) when pulling all tendons simultaneously on the other hand. In this way, a motion control architecture was implemented in order to conduct the experiments in Section 4.

4   Experimental Test Setup

In order to validate the manipulator's performance of changing stiffness, experiments have been conducted applying lateral forces F_x to the manipulator. Fig. 3 gives a top view of the overall experimental setup. The experiments consisted of loading and unloading the manipulator up to a 15 mm deflection. An ATI Nano17 Force/Torque sensor was mounted on a motorised linear module to monitor the stiffness variation. The motorised linear module was running at a speed of 0.25 $\frac{mm}{s}$. The unloading phases are identical to the loading phases, with the exception that the module is moving in the opposite direction to its initial position. Forces were applied at two different locations while the tendons were not actuated by the pulleys in order to evaluate the stiffness variation in a static condition.

Data from the force sensor and the linear module were recorded at 1 kHz using a DAQ card (NI USB-6211). Four trials were performed for each deflection location, and with the average and variability of each point plotted against the deflection distance. The results are reported in Section 5.1 and 5.2 respectively.

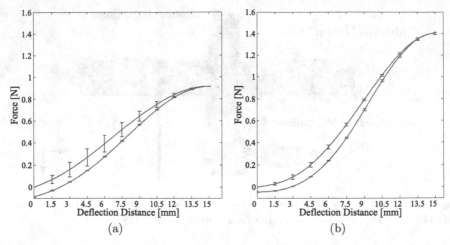

Fig. 4. Deflection versus force at the manipulator tip ($\boldsymbol{F}_{x_1}$) for pressures of (a) 0.0075 MPa and (b) 0.015 MPa [28].

5   Results

5.1   Lateral Forces (Experiment 1 and 2)

Measuring forces during the loading and unloading process of Experiments 1 and 2 are displayed in Figs. 4 and 5. The curves showing load forces (or forward motions) of the manipulator start in the origin of the coordinate frame $(0, 0)$ for all figures.

In Experiment 1, a displacement of 15 mm was applied to the manipulator's tip. The lateral forces $\boldsymbol{F}_{x_1}$ were recorded and can be seen in Fig. 4. In Fig. 4(a), the manipulator's internal bladder had a constant pressure of 0.0075 MPa; a pressure of 0.015 MPa was recorded in Fig. 4(b). From the graphs, the manipulator with lower pressure exhibited a fairly linear profile, reaching a peak of 0.9 N. When the pressure was increased to a higher value, the robot displayed a non-linear behaviour during the displacements test, reaching a maximum force $\boldsymbol{F}_{x_1}$ of 1.4 N. The hysteresis and average variability were significantly lower when applying a higher pressure.

Fig. 5 show the results of Experiment 2. Here, a lateral force $\boldsymbol{F}_{x_2}$ was recorded when deflecting the center of the manipulator. From Fig. 4(a), the maximum force achieved is 4.8 N for a pressure of 0.0075 MPa. When raising the value to 0.015 MPa, a force of 9.5 N was measured. Similar to the results of Experiment 2, the hysteresis is lower for a stiffer manipulator.

From Experiments 1 and 2, it can be noted that the amount of lateral force increases and the hysteresis decreases with higher pressures applied to the latex bladder during deflections.

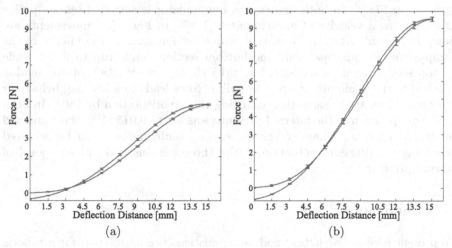

Fig. 5. Deflection versus force at the center of the manipulator (F_{x_2}) for pressures of (a) 0.0075 MPa and (b) 0.015 MPa [28].

Fig. 6. Single tendon actuation of the (a) bottom section and (b) top section [28].

5.2   Bending Behaviour

In order to show the working space and bending behaviour, experiments have been conducted actuating the tendons that are fixed at the tip and middle section of the manipulator. The results were captured for robot movements in one plane as shown in Fig. 6. In both figures, the manipulator is actuated by one pulley attached with a velocity of approximately $1 \frac{mm}{s}$. In Fig. 6(a), movements are shown that result from the actuation of one tendon fixed in the middle of the manipulator. In this case, only the bottom section bends (up to $90°$); while the top section remains straight. Fig. 6(b) shows the actuation of one tendon attached to the manipulator tip. The pulling force leads to a bending behaviour of both sections at the same time changing the tip orientation by $180°$. In both bending experiments, the internal pressure was set to 0.015 MPa and remained constant. Hence, an infinite range of bending configurations can be achieved combining simultaneous actuation of the three tendons and different level of internal pressure.

6   Conclusions

We present here a new hybrid and antagonistic actuation system for a robotic manipulator fusing pneumatic with tendon-driven actuation. Being inspired by the biological role model, the octopus, our antagonistic actuation system aims at modeling the octopus' way of using its longitudinal and transversal muscles in its arms: activating both types of muscles, the octopus can achieve a stiffening of its arms.

Our concept goes beyond state-of-the-art in the field of soft robotics: our robot is mainly made of thin sleeve-like components filled with air to achieve a fully-extended state, and thus can be shrunk to a considerably small size when entirely deflated. This capability to move between these two extreme states make the robot a particularly useful candidate for applications such as Minimally Invasive Surgery or search and rescue. Our experimental study shows that the our manipulators, indeed, is capable to bend, to morph from entirely inflated to completely shrunk as well as to squeeze through narrow openings.

Acknowledgments. The work described in this paper is partially funded by the Seventh Framework Programme of the European Commission under grant agreement 287728 in the framework of EU project STIFF-FLOP.

References

1. Ataollahi, A., Karim, R., Fallah, A.S., Rhode, K., Razavi, R., Seneviratne, L., Schaeffter, T., Althoefer, K.: 3-DOF MR-compatible multi-segment cardiac catheter steering mechanism. IEEE Transactions on Biomedical Engineering **99** (2013)
2. Bailly, Y., Amirat, Y.: Modeling and control of a hybrid continuum active catheter for aortic aneurysm treatment. In: IEEE International Conference on Robotics and Automation (2005)
3. Branicky, M., Borkar, V., Mitter, S.: A unified framework for hybrid control: Model and optimal control theory. IEEE Transactions on Automation Control **43**(1) (1998)

4. Buckingham, R.: Snake arm robots. Industrial Robot: An International Journal **29**(3), 242–245 (2002)
5. Caldwell, D., Medrano-Cerda, G., Goodwin, M.: Control of pneumatic muscle actuators. IEEE Control Systems **15** (1995)
6. Cheng, N., Gopinath, A., Wang, L., Iagnemma, K., Hosoi, A.: Thermally tunable, self-healing composites for soft robotic applications. Macromolecular Materials and Engineering **299**(11), 1279–1284 (2014)
7. Cianchetti, M., Arienti, A., Follador, M., Mazzolai, B., Dario, P., Laschi, C.: Design concept and validation of a robotic arm inspired by the octopus. Materials Science and Engineering C **31**, 1230–1239 (2011)
8. Cianchetti, M., Ranzani, T., Gerboni, G., de Falco, I., Laschi, C., Menciassi, A.: STIFF-FLOP surgical manipulator: Mechanical design and experimental charaterization of the single module. In: IEEE/RSJ International Conference on Intelligent Robots and Systems (2013)
9. Cieslak, R., Morecki, A.: Elephant trunk type elastic manipulator - a tool for bulk and liquid type materials transportation. Robotica **17**, 11–16 (1999)
10. Godage, I., Nanayakkara, T., Caldwell, D.: Locomotion with continuum limbs. In: IEEE/RSJ International Conference on Intelligent Robots and Systems (2012)
11. Gravagne, I., Rahn, C., Walker, I.D.: Large deflection dynamics and control for planar continuum robots. IEEE/ASME Transactions on Mechatronics **8**(2), 299–307 (2003)
12. Gutfreund, Y., Flash, T., Fiorito, G., Hochner, B.: Patterns of arm muscle activation involved in octopus reaching movements. The Journal of Neuroscience **18**(15), 5976–5987 (1998)
13. Jiang, A., Aste, T., Dasgupta, P., Althoefer, K., Nanayakkara, T.: Granular jamming with hydraulic control. In: ASME International Design Engineering Technical Conferences & Computers and Information in Engineering Conference (2013)
14. Jiang, A., Secco, E., Wurdemann, H., Nanayakkara, T., Dasgupta, P., Athoefer, K.: Stiffness-controllable octopus-like robot arm for minimally invasive surgery. In: Workshop on New Technologies for Computer/Robot Assisted Surgery (2013)
15. Jiang, A., Xynogalas, G., Dasgupta, P., Althoefer, K., Nanayakkara, T.: Design of a variable stiffness flexible manipulator with composite granular jamming and membrane coupling. In: IEEE/RSJ International Conference on Intelligent Robots and Systems (2012)
16. Kim, Y., Cheng, S., Kim, S., Iagnemma, K.: Design of a tubular snake-like manipulator with stiffening capability by layer jamming. In: IEEE/RSJ International Conference on Intelligent Robots and Systems (2012)
17. Laschi, C., Mazzolai, B., Mattoli, V., Cianchetti, M., Dario, P.: Design of a biomimetic robotic octopus arm. Bioinspiration and Biomimetics **4**, 1–8 (2009)
18. Li, M., Ranzani, T., Sareh, S., Seneviratne, L., Dasgupta, P., Wurdemann, H., Althoefer, K.: Multi-fingered haptic palpation utilising granular jamming stiffness feedback actuators. Smart Materials and Structures **23**(9), 095007 (2014)
19. Maghooa, F., Stilli, A., Althoefer, K., Wurdemann, H.: Tendon and pressure actuation for a bio-inspired manipulator based on an antagonistic principle. In: IEEE International Conference on Robotics and Automation (2015)
20. McMahan, W., Jones, B., Walker, I.D.: Design and implementation of a multisection continuum robot: Air-octor. In: IEEE/RJS International Conference on Intelligent Robots and Systems (2005)
21. Neppalli, S., Jones, B., McMahan, W., Chitrakaran, V., Walker, I., Pritts, M., Csencsits, M., Rahn, C., Grissom, M.: Octarm - a soft robotic manipulator. In: IROS (2007)

22. Noh, Y., Sareh, S., Back, J., Wurdemann, H., Ranzani, T., Secco, E., Faragasso, A., Liu, H., Althoefer, K.: A three-axial body force sensor for flexible manipulators. In: IEEE International Conference on Robotics and Automation (2014)
23. Noh, Y., Secco, E., Sareh, S., Wurdemann, H., Faragasso, A., Back, J., Liu, H., Sklar, E., Althoefer, K.: A continuum body force sensor designed for flexible surgical robotic devices. In: IEEE Engineering in Medicine and Biology Society (2014)
24. Ranzani, T., Gerboni, G., Cianchetti, M., Menciassi, A.: A bioinspired soft manipulator for minimally invasive surgery. Bioinspiration & Biomimetics 10(3), 035008 (2015)
25. Sareh, S., Jiang, A., Faragasso, A., Noh, Y., Nanayakkara, T., Dasgupta, P., Seneviratne, L., Wurdemann, H., Althoefer, K.: Bio-inspired tactile sensor sleeve for surgical soft manipulators. In: ICRA (2014)
26. Shin, D., Sardellitti, I., Khatib, O.: A hybrid actuation approach for human-friendly robot design. In: IEEE International Conference on Robotics and Automation, pp. 1747–1752 (2008)
27. Stilli, A., Maghooa, F., Wurdemann, H., Althoefer, K.: A new bio-inspired, antagonistically actuated and stiffness controllable manipulator. In: Workshop on Computer/Robot Assisted Surgery (2014)
28. Stilli, A., Wurdemann, H., Althoefer, K.: Shrinkable, stiffness-controllable soft manipulator based on a bio-inspired antagonistic actuation principle. In: IEEE/RJS International Conference on Intelligent Robots and Systems (2014)
29. Walker, I.D.: Robot strings: long, thin continuum robots. In: IEEE Aerospace Conference (2013)
30. Walker, I., Dawson, D., Flash, T., Grasso, F., Hanlon, R., Hochner, B., Kier, W., Pagano, C., Rahn, C., Zhang, Q.: Continuum robot arms inspired by cephalopods. In: UGVT VII (2005)
31. Wurdemann, H., Sareh, S., Shafti, A., Noh, Y., Faragasso, A., Liu, H., Hirai, S., Althoefer, K.: Embedded electro-conductive yarn for shape sensing of soft robotic manipulators. In: IEEE Engineering in Medicine and Biology Society (2015)
32. Xie, H., Jiang, A., Wurdemann, H., Liu, H., Seneviratne, L., Althoefer, K.: Magnetic resonance-compatible tactile force sensor using fibre optics and vision sensor. IEEE Sensors Journal 14(3), 829–838 (2014)

Surface Classification for Crawling Peristaltic Worm Robot

Thomas Manwell, Alexander Jupp, Kaspar Althoefer, and Hongbin Liu$^{(\boxtimes)}$

Centre for Robotic Reseach (CoRe), Department of Informatics,
King's College London, London, UK
{thomas.manwell,alexander.jupp,kaspar.althoefer,hongbin.liu}@kcl.ac.uk
http://nms.kcl.ac.uk/core/

Abstract. This paper represents a new application for existing classification techniques. A robotic worm device being developed for human endoscopy, fitted with a 3-axis accelerometer was driven over a variety of surfaces and the accelerometer data was used to identify, which surface the robot worm found itself. Within the Weka environment, three available classifiers, J48, LIBSVM and Perceptron were tested with both Fast Fourier Transform (FFT) and Mel-Frequency Cepstral Coefficients (MFCC) extraction techniques, frame sizes of 0.5 and 2 seconds. The highest testing accuracy demonstrated for this surface classification, was 83%. It is hoped that this machine learning will improve the operational use of the robot with the system identifying surface types and, later surface properties of hard to reach anatomical regions, both for locomotive efficiency and medical information.

Keywords: Classification · Accelerometer · Worm robot · Human endoscopy

1 Introduction

1.1 The Worm Robot

The KCL Robot Worm [9] is a small, minimally intrusive robot that has been inspired by multiple living organisms in nature, most notably earth worms. The potential future uses of this robot design are very broad, ranging from earth exploration to human medicine both endoscopy and enteroscopy.

To check for colorectal cancer, which is a commonly found cancer worldwide, endoscopy procedures are very necessary but have draw backs for both patient and doctor as a tube is forcefully moved into position. It requires a large amount of training, expertise and precision from the operator using the device. Enteroscopy procedures require a form of locomotion to reach deep into the small intestine. In search and rescue, rescuers have a very short time to maximise their

---

A. Jupp—Lead researcher for carrying out signal processing and classification processes.

© Springer International Publishing Switzerland 2015
H. Liu et al. (Eds.): ICIRA 2015, Part III, LNAI 9246, pp. 175–184, 2015.
DOI: 10.1007/978-3-319-22873-0_16

chances of finding survivors, and do not want to waste time in focusing on robotic parameters and often only narrow spaces are available.

The robotic worm is being developed to address these challenges; removing the pressures on operators, allowing them to focus on the task at hand.

In recent years, similar worm-inspired robots have been developed, [3],[10],[1], for a variety of applications. Most notably, [14], which used shape memory alloy, around a soft body, for actuation.

The KCL Robot Worm consists of segments, each containing a DC motor to actuate a simple tendon pulley system, which works antagonistically with the braided mesh structure by doubling up as a spring energy storage system to restore segments to resting position.

Several movement patterns are used to change the order in which segments elongate and shorten.

1.2   Surface Classification

It is hoped that surface classification will be used to distinguish the state of the biological tissue, most likely the gastrointestinal tract, for example how stiff and contracted the organ is, or how moist the surface is. If the robotic worm could analyse and diagnose its environment, it could make intelligent decisions regarding its power output, movement style and frequency etc. For example, in search and rescue, if the robot is traversing an area and can differentiate between a viscous surface, such as mud, and a flat surface, like plaster, then it could alter its output to the motors. This would save power to the worm, and allow experts to focus on what is necessary. It could also decide if the area is feasible for the robot to operate in, i.e. water is not a safe surface for the robot.

2   Methodology

2.1   The Accelerometer

The accelerometer used is a LilyPad Accelerometer ADXL335, a tri-axial accelerometer [11]. More importantly, it is approximately 2x2 centimetres, which

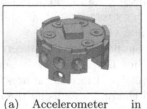

(a)   Accelerometer   in mounting

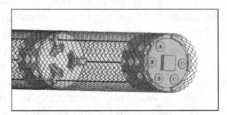

(b) Tip of Worm Robot with LilyPad Accelerometer

Fig. 1. LilyPad Accelerometer with the Worm Robot

fits inside the small body of the worm, and can operate at a very low voltage, which is ideal for the robotic worm.

A mounting for the accelerometer, Figure 1(a), was 3D printed to hold it securely in place at the tip of the robot worm, Figure 1(b). The fitting is pictured here, the accelerometer is coloured purple, and the fitting is coloured green. The accelerometer is fitted inside the fitting, with the wires soldered permanently into the accelerometer.

2.2 Testing on Multiple Surfaces

To obtain the data to test feature extraction techniques, and classifier; a number of different surfaces were put together in order to gather a wide range. Six surfaces were tested: wood, plastic, cardboard, silicon, sponge and polyester.

The cables for the worm robot were kept slack to avoid interference with the accelerometer readings. For the plastic sheet, this was suspended mid-air to give a different scenario to the wooden desktop. The silicon layer was approximately 5mm thick, with some variations.

2.3 Accelerometer Readings

The graphs in Figure 3 are samples from the data that was taken over multiple experiments on the surfaces.

3 Feature Extraction Techniques

Two feature extraction methods, Fast Fourier Transform (FFT) [6] and Mel-Frequency Cepstral Coefficients (MFCC) [8] were implemented to compare their effects.

3.1 Fast Fourier Transform (FFT)

The signal was broken into frames, with the FFT being performed on each individual frame, which is a standard method for this kind of signal analysis. The equation is:

$$Xk \equiv \sum_{n=0}^{N-1} x_n \cdot e^{\frac{-2\pi i k n}{N}} \tag{1}$$

The implementation of FFT can be seen in the example of the sponge surface in Figure 4(b). The extraction technique simplifies the sample, significantly reducing the number of features.

3.2 Mel-Frequency Cepstral Coefficients (MCFF)

The MCFF feature extraction technique [4] was inspired by voice recognition systems, which also process analogue signals [7],[2] and consists of the following steps:

1. Breakdown entire signal into frames and perform FFT on each frame using Equation 1.
2. Apply Mel-Filterbank to each individual frame, to create mel-frequencies. This was performed with triangular, overlapping frame, using Equation 2. The outcome is shown in Figure 4(c).

$$w(n) = 1 - \left| \frac{n - \frac{N-1}{2}}{\frac{L}{2}} \right| \tag{2}$$

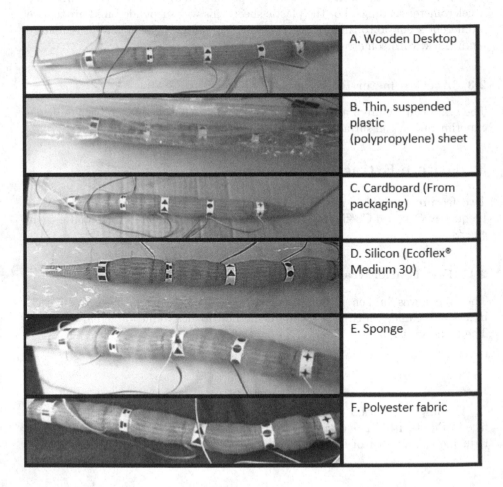

	A. Wooden Desktop
	B. Thin, suspended plastic (polypropylene) sheet
	C. Cardboard (From packaging)
	D. Silicon (Ecoflex® Medium 30)
	E. Sponge
	F. Polyester fabric

Fig. 2. The six surfaces used in the experiments

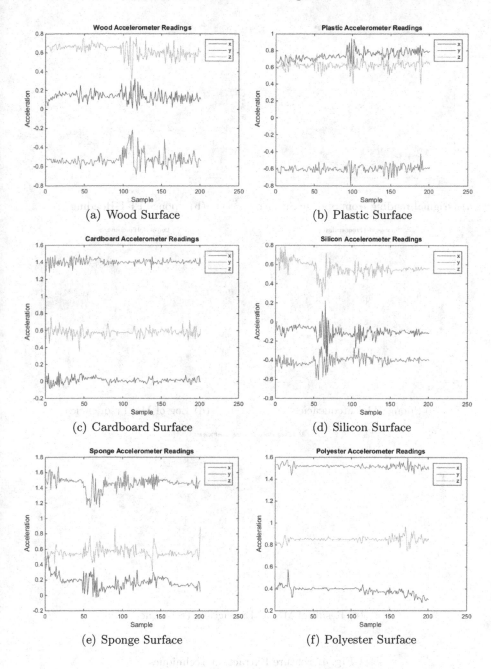

(a) Wood Surface

(b) Plastic Surface

(c) Cardboard Surface

(d) Silicon Surface

(e) Sponge Surface

(f) Polyester Surface

Fig. 3. Accelerometer Readings

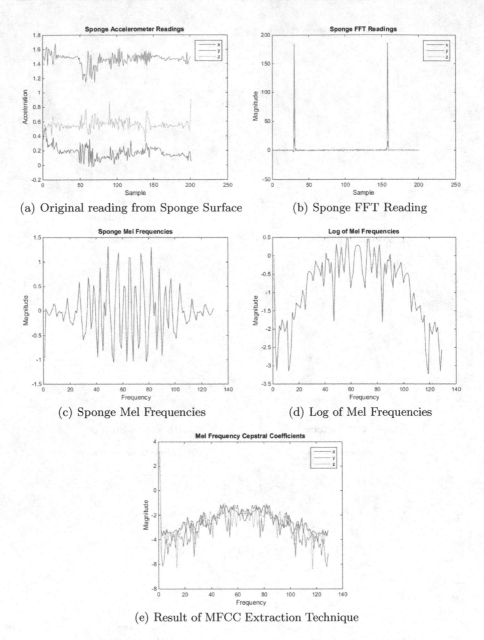

(a) Original reading from Sponge Surface

(b) Sponge FFT Reading

(c) Sponge Mel Frequencies

(d) Log of Mel Frequencies

(e) Result of MFCC Extraction Technique

Fig. 4. Feature Extraction Techniques

3. Take the log of these Mel frequencies. The outcome of which is shown in
   Figure 4(d).

4. Compute the Discrete Cosine Transformation (DCT) [5] of the mel-frequencies, using Equation 3. The outcome of which is shown in Figure 4(e).

$$Xk = \frac{1}{2}\left(x_0 + (-1)^k x_{N-1}\right) + \sum_{n=1}^{N-2} x_n \cos\left[\frac{\pi}{N-1}nk\right] \tag{3}$$

4   Classification

LIBSVM within MATLAB was used to classify between different surfaces and environments, with a 'binary decision tree' set-up. This structure was chosen due to similarity between the data and also to reduce computational time by reducing overall number of comparisons.

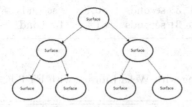

Fig. 5. Binary Decision Tree structure

To train and test models, the entire signal of 'x, y and z' values was split into frames. There was at least twenty-eight seconds of usable data for each surface, therefore twenty-two seconds was used to train up each model, and the remaining six available for testing.

To train and test other classifiers and look at the results, the popular library Weka was used. The classifiers that were tested, from within Weka, were J48, which is a java implementation of the C4.5 classification algorithm. LIBSVM, in Weka, to compare results and the Perceptron classifier, which is an artificial neural network.

4.1   Results

There were 26 seconds of good data gathered for each surface and movement pattern, 70% was used for training, and 30% for testing, i.e. 22 seconds for training and 4 seconds for testing. For the following results, the classifier was trained with two second chunks, and tested with two second chunks. In total, there were eleven two second frames for training, and two, two second frames for testing. To break up the signal into frames, the sampling frequency was 64, therefore there were 128 samples in a 2 second frame, and 11 frames in an individual pattern movement.

The default settings within LIBSVM in Weka, using a radial basis function kernel, were: gamma = 0, cost = 1.

Table 1. Classifying with LIMSVM Library

Classifier	Feature Extraction	Training Time	Training Accuracy	Testing Time	Testing Accuracy
Decision Tree: SVM	None	10 seconds	95%	1 second	66%
Decision Tree: SVM	FFT	130 seconds	70%	1 second	40%
Decision Tree: SVM	MFCC	120 seconds	66%	1 second	40%

Table 2. Classifying with Weka Library without Feature Extraction

Classifier	Training Time	Training Accuracy	Testing Time	Testing Accuracy
J48	2 seconds	88%	1 second	80%
LIBSVM	35 seconds	80%	2 seconds	83%
Perceptron	31 seconds	81%	1 second	83%

Table 3. Classifying with Weka Library with FTT Feature Extraction

Classifier	Training Time	Training Accuracy	Testing Time	Testing Accuracy
J48	5 seconds	67%	1 second	27%
LIBSVM	45 seconds	43%	2 seconds	40%
Perceptron	40 seconds	40%	1 second	39%

Table 4. Classifying with Weka Library with MFCC Feature Extraction

Classifier	Training Time	Training Accuracy	Testing Time	Testing Accuracy
J48	3 seconds	69%	2 second	33%
LIBSVM	120 seconds	46%	2 seconds	38%
Perceptron	30 seconds	43%	2 second	40%

Table 5. Classifying with Weka Library without Feature Extraction, (0.5 second frame size)

Classifier	Training Time	Training Accuracy	Testing Time	Testing Accuracy
J48	2 seconds	88%	1 second	80%
LIBSVM	40 seconds	82%	2 seconds	81%
Perceptron	30 seconds	79%	1 second	76%

Table 6. Classifying with Weka Library with FTT Feature Extraction, (0.5 second frame size)

Classifier	Training Time	Training Accuracy	Testing Time	Testing Accuracy
J48	8 seconds	65%	1 second	40%
LIBSVM	50 seconds	46%	2 seconds	41%
Perceptron	38 seconds	440%	1 second	37%

Table 7. Classifying with Weka Library with MFCC Feature Extraction, (0.5 second frame size)

Classifier	Training Time	Training Accuracy	Testing Time	Testing Accuracy
J48	3 seconds	72%	2 second	38%
LIBSVM	120 seconds	47%	2 seconds	41%
Perceptron	30 seconds	43%	2 second	39%

5   Conclusions

This system has been tested on 6 surfaces and was able to correctly identify the surface at up to 83% accuracy, using . The surfaces represent a number of scenarios where the robot might be used and the results indicate that they can be successfully differentiated. This also suggests that this methodology is a feasible way forward for inspecting the large intestine, as the surfaces give a wide range of friction, hardness and texture varieties.

We have also shown the results of 2 feature extraction techniques that transform the series into the frequency domain. Both of these actually yielded lower training and testing accuracies. This is likely due to the fact that many features that are found in the time domain are lost in the frequency domain. It might be that these features only appear after a certain portion of the locomotive cycle has passed.

We envisage this system being very useful during the implementation of the worm robot for human endoscopy for two reasons. The first is that this should help the operator with handling by making decisions for movement pattern or speed autonomously, and therefore benefit the efficiency and ease of use of the system. The second is to understand the state of the tissue in contact and provide that medical information to the Doctor. This can happen simultaneously as the doctor is using the device for optical inspection.

Currently the Robot Worm only has a vibration sensing capability. For medical applications ideally we need additional sensing modalities for deeper understanding of the interaction between the robot and human tissue. These include tactile sensing such as [12], shape sensing such as [15], curvature sensing such as [13] and body contact sensing such as [16].

References

1. Adachi, K., Yokojima, M., Hidaka, Y., Nakamura, T.: Development of multistage type endoscopic robot based on peristaltic crawling for inspecting the small intestine. In: 2011 IEEE/ASME International Conference on Advanced Intelligent Mechatronics (AIM), pp. 904–909, July 2011
2. Alcaraz Meseguer, N.: Speech analysis for automatic speech recognitione. Ph.D. thesis, Norwegian University of Science and Technology (2009)
3. Boxerbaum, A.S., Chiel, H.J., Quinn, R.D.: A new theory and methods for creating peristaltic motion in a robotic platform. In: 2010 IEEE International Conference on Robotics and Automation, pp. 1221–1227, May 2010
4. Weng, C.-W., Lin, C.-Y., Jang, J.S.R.: Music instrument identification using MFCC: Ehru as an example. Ph.D. thesis, TainanNationalCollege of the Arts, Taiwan (2004)
5. Govindaraju, N.K., Lloyd, B., Dotsenko, Y., Smith, B., Manferdelli, J.: High performance discrete fourier transforms on graphics processors. In: Proceedings of the 2008 ACM/IEEE Conference on Supercomputing, SC 2008, pp. 2:1–2:12. IEEE Press, Piscataway (2008)
6. Haque, A.F.: FFT and Wavelet-Based Feature Extraction for Acoustic Audio Classification. International Journal of Advance Innovations, Thoughts & Ideas 1(1) (2012)
7. Hsu, C.L., Jang, J.S.: On the improvement of singing voice separation for monaural recordings using the mir-1k dataset. IEEE Transactions on Audio, Speech, and Language Processing 18(2), 310–319 (2010)
8. Logan, B.: Mel frequency cepstral coefficients for music modeling. In: International Symposium on Music Information Retrieval (2000)
9. Manwell, T., Vitek, T., Ranzani, T., Menciassi, A., Althoefer, K., Liu, H.: Elastic mesh braided worm robot for locomotive endoscopy. In: 2014 36th Annual International Conference of the IEEE Engineering in Medicine and Biology Society (EMBC), pp. 848–851, August 2014
10. Menciassi, A., Gorini, S., Pernorio, G., Valvo, F., Dario, P.: Design, Fabrication and performances of a biomimetic robotic earthworm. In: 2004 IEEE International Conference on Robotics and Biomimetics, pp. 274–278. IEEE (2004)
11. Nickel, C., Busch, C.: Classifying accelerometer data via hidden markov models to authenticate people by the way they walk. In: 2011 IEEE International Carnahan Conference on Security Technology (ICCST), pp. 1–6, October 2011
12. Noh, Y., Sareh, S., Back, J., Wurdemann, H., Ranzani, T., Secco, E., Faragasso, A., Liu, H., Althoefer, K.: A three-axial body force sensor for flexible manipulators. In: 2014 IEEE International Conference on Robotics and Automation (ICRA), pp. 6388–6393, May 2014
13. Searle, T., Althoefer, K., Seneviratne, L., Liu, H.: An optical curvature sensor for flexible manipulators. In: 2013 IEEE International Conference on Robotics and Automation (ICRA), pp. 4415–4420, May 2013
14. Seok, S., Onal, C., Wood, R., Rus, D., Kim, S.: Peristaltic locomotion with antagonistic actuators in soft robotics. In: 2010 IEEE International Conference on Robotics and Automation (ICRA), pp. 1228–1233, May 2010
15. Wurdemann, H.A., Sareh, S., Shafti, A., Noh, Y., Faragasso, A., Chathuranga, D., Liu, H., Hirai, S., Althoefer, K.: Embedded electro-conductive yarn for shape sensing of soft robotic manipulators. In: IEEE Engineering and Biology Society (EMBC) (in press) (2015)
16. Xie, H., Liu, H., Noh, Y., Li, J., Wang, S., Althoefer, K.: A fiber-optics-based body contact sensor for a flexible manipulator. IEEE Sensors Journal 15(6), 3543–3550 (2015)

Learning the Stiffness of a Continuous Soft Manipulator from Multiple Demonstrations

Danilo Bruno[1](✉), Sylvain Calinon[1,2], Milad S. Malekzadeh[1],
and Darwin G. Caldwell[1]

[1] Department of Advanced Robotics, Istituto Italiano di Tecnologia (IIT),
Via Morego, 30, 16163 Genova, Italy
`danilo.bruno@iit.it`
[2] Idiap Research Institute, Rue Marconi 19, 1920 Martigny, Switzerland

Abstract. Continuous soft robots are becoming more and more widespread in applications, due to their increased safety and flexibility in critical applications. The possibility of having soft robots that are able to change their stiffness in selected parts can help in situations where higher forces need to be applied. This paper describes a theoretical framework for learning the desired stiffness characteristics of the robot from multiple demonstrations. The framework is based on a statistical mathematical model for encoding the motion of a continuous manipulator, coupled with an optimal control strategy for learning the best impedance parameters of the manipulator.

1 Introduction

The use of soft robots is becoming crucial to perform tasks where the contact with a delicate environment is needed. This is happening for surgical robots as well as manipulators that need to be put in contact with humans [1–4].

The possibility of changing the stiffness of soft robots can become a desired feature where the softness of the robot prevents it from performing some task, such as moving objects or carrying weights. In this case, the robot should be aware of the stiffness that is needed to perform the task and apply the correct values where needed.

Moreover, the choice of using soft materials usually comes together with a different kind of embodiment, that departs from rigid kinematic chains. Usually soft robots are described as continuous robots and their kinematic/dynamic description needs to take this characteristics into account.

In this paper we will present a mathematical framework that can be employed to learn the impedance characteristics of a soft continuous manipulator from demonstrations. The paper is based on the main idea that the robot needs to increase its stiffness whenever the task needs a higher precision. This is true, for instance, in transportation tasks, where the relative position between the carried

---

This work was partially supported by the STIFF-FLOP European project under contract FP7-ICT-287728.

H. Liu et al. (Eds.): ICIRA 2015, Part III, LNAI 9246, pp. 185–195, 2015.
DOI: 10.1007/978-3-319-22873-0_17

object and the end-effector of the robot is invariant during the crucial part of the task [5].

Within the learning from demonstrations scenario, the user providing demonstrations shows the task to perform several times: the consistency of the demonstrations is encoded into a statistical model and the information extracted back at reproduction time allows us to select the appropriate stiffness behaviour [6, 7].

In this paper, this feature is further enhanced by the use of a mathematical model handling the movement of a continuous robot. In this way, the model can be used to learn the correct stiffness behaviour of different parts along the continuous robot at different times.

The selection of the correct values of stiffness is performed by using optimal control techniques, that allow us to learn a positional controller along the manipulator that makes a trade off between minimizing the forces applied on the manipulator and tracking the demonstrations within the demonstrated variability [8].

The current paper is aimed at showing the general theoretical framework for learning the stiffness behaviour of a continuous robot and is organized as follows. In section 2 we present a generative model to learn motion skills from demonstrations. In particular, we show how a tracking controller can be learnt from demonstrations by using optimal control technique. Section 3 extends the model developed for Section 2 for encoding the motion of a continuum robot. Section 4 shows how those results can be effectively employed to learn the desired stiffness of the manipulator from demonstrations.

2   Learning Motion Skills from Demonstrations

2.1   Gaussian Mixture Models

The setup for learning motion skills from demonstrations presented in this section is based on the construction of a statistical model of observed movements (demonstrations) given by a user.

The observations $\{\boldsymbol{\xi}_n\}_{n=1}^{N}$ representing the points of the demonstrations are assumed to be independent realizations of a random vector and is assumed to be distributed as a linear combination of Normal distributions as:

$$P(\boldsymbol{\xi}_n) = \sum_{k=1}^{K} \pi_k \, \mathcal{N}(\boldsymbol{\xi}_n | \boldsymbol{\mu}_k, \boldsymbol{\Sigma}_k),$$

with

$$\mathcal{N}(\boldsymbol{\xi}_n | \boldsymbol{\mu}_k, \boldsymbol{\Sigma}_k) = \frac{1}{(2\pi)^{\frac{D}{2}} |\boldsymbol{\Sigma}_k|^{\frac{1}{2}}} \exp\left[-\frac{1}{2}(\boldsymbol{\xi}_n - \boldsymbol{\mu}_k)^{\top} \boldsymbol{\Sigma}_k^{-1}(\boldsymbol{\xi}_n - \boldsymbol{\mu}_k)\right].$$

The parameters of a *Gaussian mixture model* (GMM) with K components are thus defined by $\{\pi_k, \boldsymbol{\mu}_k, \boldsymbol{\Sigma}_k\}_{k=1}^{K}$, where π_k is the prior (mixing coefficient), $\boldsymbol{\mu}_k$ is the center, and $\boldsymbol{\Sigma}_k$ is the covariance matrix of the k-th mixture component.

The estimation of mixture parameters can be performed by maximizing the log-likelihood of the above distribution of the given dataset. This leads to an *expectation-maximization* (EM) process iteratively refining the model parameters to converge to a local optimum of the likelihood. These two steps are iteratively applied until a stopping criterion is satisfied. The two steps are described below.

E-step:

$$h_{n,i} = \frac{\pi_i \, \mathcal{N}(\xi_n | \mu_i, \Sigma_i)}{\sum_{k=1}^{K} \pi_k \, \mathcal{N}(\xi_n | \mu_k, \Sigma_k)}.$$

M-step:

$$\pi_i = \frac{\sum_{n=1}^{N} h_{n,i}}{N},$$

$$\mu_i = \frac{\sum_{n=1}^{N} h_{n,i} \xi_n}{\sum_{n=1}^{N} h_{n,i}},$$

$$\Sigma_i = \frac{\sum_{n=1}^{N} h_{n,i} (\xi_n - \mu_i)(\xi_n - \mu_i)^{\mathsf{T}}}{\sum_{n=1}^{N} h_{n,i}}.$$

The reproduction of an average movement or skill behavior can be formalized as a statistical regression problem. We demonstrated in previous work that Gaussian Mixture Models in combination with *Gaussian mixture regression* (GMR) offers a simple and elegant solution to handle encoding, recognition, prediction and reproduction in robot learning [5,9]. It provides a probabilistic representation of the movement, where the model can retrieve actions in real-time, within a computation time that is independent of the number of datapoints in the training set.

By defining which variables span for input and output parts (noted respectively by $\mathcal{I}$ and $\mathcal{O}$ superscripts), a block decomposition of the datapoint ξ_n, vectors μ_i and matrices Σ_i can be written as

$$\xi_n = \begin{bmatrix} \xi_n^I \\ \xi_n^O \end{bmatrix}, \quad \mu_i = \begin{bmatrix} \mu_i^I \\ \mu_i^O \end{bmatrix}, \quad \Sigma_i = \begin{bmatrix} \Sigma_i^I & \Sigma_i^{IO} \\ \Sigma_i^{OI} & \Sigma_i^O \end{bmatrix}.$$

The GMM thus encodes the joint distribution $\mathcal{P}(\xi^I, \xi^O) \sim \sum_{i=1}^{K} \pi_i \mathcal{N}(\mu_i, \Sigma_i)$ of the data ξ. At each reproduction step n, $\mathcal{P}(\xi_n^O | \xi_n^I)$ is computed as the conditional distribution

$$\mathcal{P}(\xi_n^O | \xi_n^I) \sim \sum_{i=1}^{K} h_i(\xi_n^I) \, \mathcal{N}\left(\hat{\mu}_i^O(\xi_n^I), \hat{\Sigma}_i^O\right), \tag{1}$$

$$\text{with} \quad \hat{\mu}_i^O(\xi_n^I) = \mu_i^O + \Sigma_i^{OI} \Sigma_i^{I^{-1}}(\xi_n^I - \mu_i^I), \tag{2}$$

$$\hat{\Sigma}_i^O = \Sigma_i^O - \Sigma_i^{OI} \Sigma_i^{I^{-1}} \Sigma_i^{IO},$$

$$\text{and} \quad h_i(\xi_n^I) = \frac{\pi_i \mathcal{N}(\xi_n^I | \mu_i^I, \Sigma_i^I)}{\sum_{k}^{K} \pi_k \mathcal{N}(\xi_n^I | \mu_k^I, \Sigma_k^I)}. \tag{3}$$

In the general case, eq. (1) represents a multimodal distribution. In problems where a single output is expected (single peaked distribution), eq. (1) can be approximated by a single normal distribution $\mathcal{N}(\hat{\boldsymbol{\mu}}_n^o, \hat{\boldsymbol{\Sigma}}_n^o)$ with parameters

$$\hat{\boldsymbol{\mu}}_n^o = \sum_i h_i(\boldsymbol{\xi}_n^I)\left[\boldsymbol{\mu}_i^o + \boldsymbol{\Sigma}_i^{OI}\boldsymbol{\Sigma}_i^{I^{-1}}(\boldsymbol{\xi}_n^I - \boldsymbol{\mu}_i^I)\right],$$

$$\hat{\boldsymbol{\Sigma}}_n^o = \sum_{i=1}^K h_i(\boldsymbol{\xi}_n^I)\boldsymbol{\Sigma}_i^o + \sum_{i=1}^K h_i(\boldsymbol{\xi}_n^I)\boldsymbol{\mu}_i^o(\boldsymbol{\mu}_i^o)^\top - \hat{\boldsymbol{\mu}}_n^o\hat{\boldsymbol{\mu}}_n^{o\top}. \tag{4}$$

Eq. (4) is computed in real-time from the model parameters. The retrieved signal encapsulates variation and correlation information in the form of a probabilistic flow tube, see e.g., [10].

2.2 LQR Representation for Motion Skills

One typical scenario consists in collecting time and position values of the demonstrations and then extract the resulting movement by choosing time as input. This allows us to reproduce the average behaviour extracted from the demonstrations. This approach implements an open-loop controller which is not robust to perturbations.

In order to cope with this drawback, the above approach can be complemented by coupling the reproduction step with a second order linear dynamical system, which improves robustness and smoothness of the reproductions [7]. Moreover the stiffness and damping parameters characterizing the second order system can also be estimated from the observation by following a reasoning similar to the development of linear quadratic regulators [11].

The approach exploits Gaussian conditioning to retrieve a reference trajectory in the form of a full distribution $\mathcal{N}(\hat{\boldsymbol{\xi}}_t^o, \hat{\boldsymbol{\Sigma}}_t^o)$ varying at each time step t, see [8] for an experiment with a 7 DOFs manipulator.

Similarly as the solution proposed by Medina et al. in the context of risk-sensitive control for haptic assistance [12], the predicted variability can be exploited to form a minimal intervention controller [13]. An acceleration command

$$\boldsymbol{u}_t = \hat{\boldsymbol{K}}_t^P(\hat{\boldsymbol{x}}_t - \boldsymbol{x}_t) - \hat{\boldsymbol{K}}_t^V\dot{\boldsymbol{x}}_t \tag{5}$$

is used to control the robot, with $\hat{\boldsymbol{x}}_t$ estimated by GMR. $\hat{\boldsymbol{K}}_t^P$ and $\hat{\boldsymbol{K}}_t^V$ are full stiffness and damping matrices estimated by a *linear quadratic regulator* (LQR) with time-varying weights. For a *finite horizon LQR*, this is achieved by minimizing the cost function

$$c^{(1)} = \sum_{t=1}^T(\hat{\boldsymbol{x}}_t - \boldsymbol{x}_t)^\top \boldsymbol{Q}_t(\hat{\boldsymbol{x}}_t - \boldsymbol{x}_t) + \boldsymbol{u}_t^\top \boldsymbol{R}_t\,\boldsymbol{u}_t, \tag{6}$$

subject to the constraints of a double integrator system. This cost function aims at finding an optimal feedback controller minimizing simultaneously the tracking

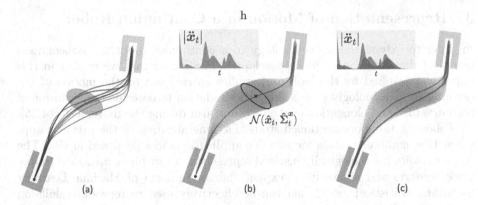

Fig. 1. Minimal intervention controller based on a linear quadratic regulator (LQR). *(a)* Multiple demonstrations. *(b)* Reproduction of the reference path defined as the centers of the Gaussians retrieved by GMR, with the corresponding acceleration profile. *(c)* Reproduction with a linear quadratic regulator, reducing the accelerations and jerks while keeping the movement within a boundary defined by the demonstrations.

errors and the control commands in proportion defined by weighting matrices Q_t and R_t.

The solution can be computed by backward integration of a *Riccati ordinary differential equation* with varying full weighting matrix $Q_t = (\hat{\Sigma}_t^x)^{-1}$ estimated by GMR and by setting R_t as a constant diagonal matrix. It provides a time-varying feedback control law in the form of Eq. (5) with full stiffness and damping matrices $\hat{K}_t^P$ and $\hat{K}_t^V$.

To solve the above minimization problem, a boundary condition needs to be set on the final feedback term, which can for example be set to zero. It is in this case assumed that the robot comes back to a compliant state after the task is fulfilled.

In some situations, it might be computationally expensive to recompute at each iteration t a prediction on the remaining movement. An approximation that we exploited in [8] can in this case be locally computed by considering an *infinite horizon LQR* formulation to estimate a feedback term at iteration t by considering only the current estimate $\hat{\Sigma}_t^x$. This corresponds to the estimation of a feedback controller that does not know in advance whether the precision at which it should track a target will vary. The corresponding cost function at iteration t corresponds to

$$c_t^{(2)} = \sum_{n=t}^{\infty} (\hat{x}_t - x_n)^{\top} Q_t (\hat{x}_t - x_n) + u_n^{\top} R_t u_n, \quad \forall t \in \{1, \ldots, T\} \qquad (7)$$

which can be solved iteratively through the *algebraic Riccati equation*, providing an optimal feedback controller in the form of Eq. (5) with full stiffness and damping matrices $\hat{K}_t^P$ and $\hat{K}_t^V$.

3   Representation of Motion in a Continuum Robot

In order to extend the above strategy to a continuous robot, a mathematical model of the latter needs to be developed. The model that we exploit in this paper was inspired by the biological studies carried out on the motion of the octopus. In fact, biologists use a similar formulation to describe the evolution of the curvature and elongation of the Octopus arm during the movement [14,15].

Following the aforementioned studies, a generalization of the previous approach that is more suitable for robotics applications was proposed in [16]. The setup exploits the biologically inspired representation within a statistical framework scenario and allows us to exploit the instruments of Machine Learning to build a statistical model that can be effectively used to reproduce skills on continuous robotics arms.

For this purpose, the continuum robotic arm is approximated as a robot with a high number of links: it can be described as a kinematic chain with a high number of revolute joints alternated with prismatic joints describing the local elongation. The joint space of the robot is described by a collection of 3 Euler angles and a scalar offset for each link, where the forward kinematics can be evaluated by using standard robotics techniques.

The motion of the system can be represented by assigning the set of 4 joint variables for each discrete position along the arm. This is performed by defining a continuous *arm-index* $s \in [0,1]$, representing the position of any point along an arm, with $s = 0$ and $s = 1$ representing respectively the base and the tip. We will rescale the duration of a movement so that it can be represented by a continuous *time index* $t \in [0,1]$. We thus have 3 Euler angles $\theta_x, \theta_y, \theta_z$ and an offset ΔL for each t and s, that can be represented as a set of surfaces as shown in Fig. 2.

The original raw data consist of noisy Cartesian positions of selected points along the arm. A preprocessing step is performed by resampling and smoothing through a two-dimensional polynomial fitting (surface fitting) with a 7-degree polynomial. The degree for the polynomial is set experimentally by testing different orders.

In order to represent the motion in the form of a statistical model, the preprocessed data are encoded into a Gaussian Mixture Model, representing the distribution of the joint variables for each value of s and t as shown in Fig.3.

Now, we can extend the dynamical system scenario to the continuous robot, by extending the concept of trajectory attractor to a surface attractor (generalization of the generic spring-damper system to a spatiotemporal dynamical system). We use both arm-index s and time t (rather than only time) as input variables, which enables the approach to encode the movement of the whole arm with a compact model (namely, by encoding the movement of all points along the arm).

A dynamical system can then be derived for the current situation in two different ways, either by integrating along the t index or along the s index. In the current scenario, the evolution is only semi bi-dimensional, in the sense that the surface is constructed by integrating either in the t direction from a given

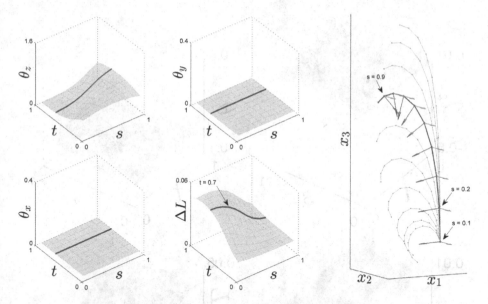

Fig. 2. Spatiotemporal representation. *Left:* The black lines represent slices in the Euler angles surfaces $\boldsymbol{\theta}$ and offset surface $\boldsymbol{\Delta L}$, corresponding to a static pose described by all the links $(0 \leq s \leq 1)$ at $t = 0.7$. *Right:* For the same time frame, some of the corresponding Frenet frames along the arm are depicted in a 3D Cartesian space.

curve at $t = 0$, representing the initial shape or, alternatively, by integrating in the s direction from a given curve at $s = 0$, representing the evolution of a single point (e.g. the base of the continuous robot).

The difference is illustrated in Fig.4. The surface attractor allows us to evaluate at each time step the desired pose of the continuum robot. The dynamical system in time describes the motion of each single link along with the impedance parameters of the positional controller. On the other hand, the dynamical system in s allows us to evaluate at each time step the static pose of the robot. The picture shows the learnt stiffness values with colors from white to black depicting increasing levels of stiffness

4   Learning Impedance Parameters

In the following section we will concentrate on the use of the above described scenario for learning impedance parameters from demonstrations. First of all a distinction should be made between the t-directed and the s-directed dynamical system. In the first case, the dynamical system implemented by eq.(5) represents the acceleration applied to the points of the continuous robot and can be interpreted as the impedance parameters of a positional controller implemented on the system.

On the other hand, the s-directed dynamical system has a different meaning. In this case, eq.(5) represents the second derivative with respect to a spatial

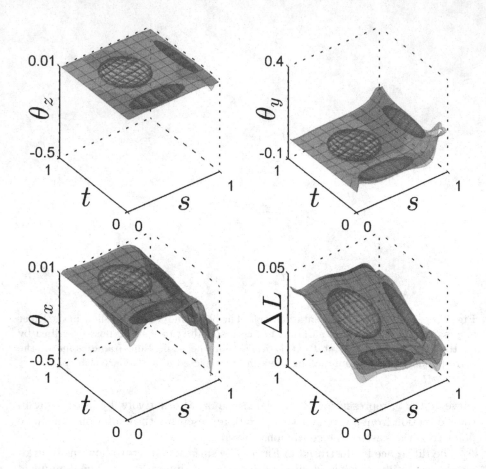

Fig. 3. The Gaussian Mixture Model encoding a movement for the joint variables representing the continuum robot. The number of components is chosen heuristically.

variable and can be related to the bending and elongation properties of the system. In particular the stiffness matrix represents the resistance to bending and elongation of the continuum robot. For this reason, the described scenario can be effectively employed to learn from demonstrations what is the optimal stiffness of the different parts of the manipulator along the task.

As an example of scenario, we show how this mechanism works for a simple motion. The demonstration phase consists of moving the end effector of the manipulator to reach a target in front of it. The end-effector of the manipulator is moved around in the first part of the movement with a high variability (exploration phase). In the second part of the task, the robot is instead moved directly towards the target. As a result, the robot learns a different controller for the different parts of the task, as shown in Fig.5.

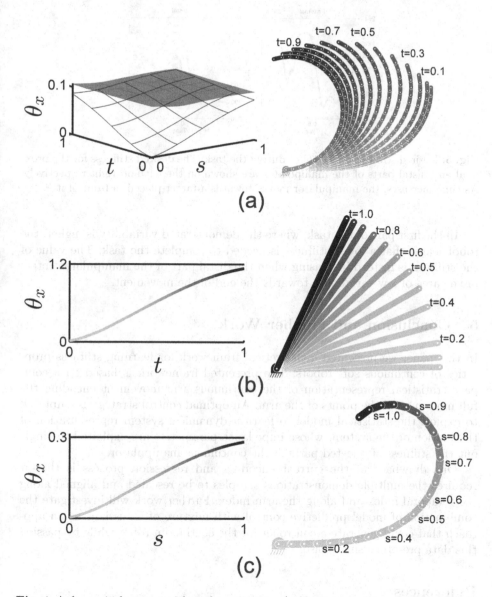

Fig. 4. A dynamical system with surface attractor (evolution over time and arm-index). *(a)* The grey surface on the left represents the attractor surface corresponding to the observed motion of the continuum arm, while the white surface is the reproduced motion of the arm. The right figure shows the arm configurations in 2D Cartesian space at different time steps. *(b)* The left figure shows the evolution in time of a link with a given arm-index s along the kinematic chain and the corresponding configurations (on the right). *(c)* The figures show the pose of the continuum arm for a given time step t. The evaluated stiffness values are represented by different colors (black means high stiffness).

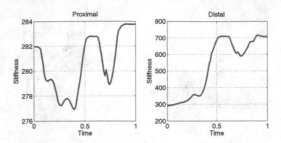

Fig. 5. Desired stiffness of the robot during the task. The desired stiffness for the proximal and distal parts of the manipulator are shown on the left and right respectively. As time increases, the manipulator moves towards a target placed in front of it.

In the first part of the task, where the demonstrated variability is higher, the robot learns that a lower stiffness is needed to complete the task. The value of the stiffness is instead increasing when the distal part of the manipulator enters in the area of low variability, towards the end of the movement.

5   Conclusion and Further Work

In this paper we presented a theoretical framework for learning stiffness properties of continuous soft robots. The presented framework is based on a compact statistical representation of the continuous arm movement encoding the full motion of all the points of the arm. An optimal control strategy is employed to exploit the statistical model to learn a dynamical system representation of the motion of the system, whose impedance parameters are exploited to single out the stiffness of selected parts of the continuous manipulator.

One drawback of the current encoding and regression process is that it requires the multiple demonstrations samples to be rescaled and aligned along the temporal index and along the arm index. Further work will investigate the combination of model predictive control with mixture of Gaussians as an approach that could generate a controller for the continuum robot while by-passing this data pre-processing phase.

References

1. Jiang, A., Xynogalas, G., Dasgupta, P., Althoefer, K., Nanayakkara, T.: Design of a variable stiffness flexible manipulator with composite granular jamming and membrane coupling. In: IEEE/RSJ International Conference on Intelligent Robots and Systems (IROS), pp. 2922–2927 (2012)
2. Jiang, A., Ataollahi, A., Althoefer, K., Dasgupta, P., Nanayakkara, T.: A variable stiffness joint by granular jamming. In: ASME Intl Design Engineering Technical Conf. & Computers and Information in Engineering Conf. (IDETC/CIE), pp. 267–275 (2012)

3. Cianchetti, M., Ranzani, T., Gerboni, G., De Falco, I., Laschi, C., Menciassi, A.: STIFF-FLOP surgical manipulator: mechanical design and experimental character-ization of the single module. In: IEEE/RSJ International Conference on Intelligent Robots and Systems (IROS), pp. 3567–3581 (2013)
4. Cianchetti, M., Ranzani, T., Gerboni, G., Nanayakkara, T., Althoefer, K., Dasgupta, P., Menciassi, A.: Soft robotics technologies to address shortcomings in today's minimally invasive surgery: the stiff-flop approach. Soft Robotics 1(2), 122–131 (2014)
5. Calinon, S., D'halluin, F., Sauser, E.L., Caldwell, D.G., Billard, A.G.: Learning and reproduction of gestures by imitation: An approach based on hidden Markov model and Gaussian mixture regression. IEEE Robotics and Automation Magazine 17, 44–54 (2010)
6. Ijspeert, A., Nakanishi, J., Pastor, P., Hoffmann, H., Schaal, S.: Dynamical move-ment primitives: Learning attractor models for motor behaviors. Neural Compu-tation 25(2), 328–373 (2013)
7. Calinon, S., Li, Z., Alizadeh, T., Tsagarakis, N.G., Caldwell, D.G.: Statistical dynamical systems for skills acquisition in humanoids. In: Proc. IEEE Intl Conf. on Humanoid Robots (Humanoids), Osaka, Japan, pp. 323–329 (2012)
8. Calinon, S., Bruno, D., Caldwell, D.G.: A task-parameterized probabilistic model with minimal intervention control. In: Proc. IEEE Intl Conf. on Robotics and Automation (ICRA), Hong Kong, China, pp. 3339–3344, May-June 2014
9. Calinon, S., Billard, A.G.: Recognition and reproduction of gestures usinga proba-bilistic framework combining PCA, ICA and HMM. In: Proc. Intl Conf. on Machine Learning (ICML), Bonn, Germany, pp. 105–112, August 2005
10. Lee, D., Ott, C.: Incremental kinesthetic teaching of motion primitives using the motion refinement tube. Autonomous Robots 31(2), 115–131 (2011)
11. Astrom, K.J., Murray, R.M.: Feedback Systems: An Introduction for Scientists and Engineers. Princeton University Press, Princeton (2008)
12. Medina, J.R., Lee, D., Hirche, S.: Risk-sensitive optimal feedback control for haptic assistance. In: IEEE Intl Conf. on Robotics and Automation (ICRA), pp. 1025–1031, May 2012
13. Todorov, E., Jordan, M.I.: Optimal feedback control as a theory of motor coordi-nation. Nature Neuroscience 5, 1226–1235 (2002)
14. Flash, T., Hochner, B.: Motor primitives in vertebrates and invertebrates. Current Opinion in Neurobiology 15(6), 660–666 (2005)
15. Zelman, I., Titon, M., Yekutieli, Y., Hanassy, S., Hochner, B., Flash, T.: Kinematic decomposition and classification of octopus arm movements. Frontiers in Compu-tational Neuroscience 7(60) (2013)
16. Malekzadeh, M.S., Calinon, S., Bruno, D., Caldwell, D.G.: Learning by imitation with the STIFF-FLOP surgical robot: A biomimetic approach inspired by octopus movements. Robotics and Biomimetics, Special Issue on Medical Robotics 1, 1–15 (2014)

Robot Mechanism and Design

A Method for Structure Coupling-reducing of Parallel Mechanisms

Huiping Shen[✉], Liangjie Yang, Xiaorong Zhu, Ju Li, and Hongbo Yin

School of Mechanical Engineering, Changzhou University, Changzhou 213164, China
shp65@126.com

Abstract. Structure coupling degree is used to describe the topological structure complexity of a mechanism. While structure coupling-reducing can reduces such complexity. But the topic is little being studied so far. This paper studies the principle, design method and application of the structure coupling-reducing of the mechanisms. Firstly, a novel concept is proposed which defines the structure coupling-reducing (SCR). The differences and relations of the structure coupling-reducing (SCR) and motion decoupling (MD) are also revealed. Then a SCR method is proposed, which requires to change the position and orientation characteristics chain into a driving chain and meanwhile remove one unconstrained driving chain. At last, two SCR examples are illustrated, which offers good application potential. The SCR principle and SCR method can be applied to all spatial mechanisms. They are of great significance in topological structure optimization, kinematics solution, real-time control.

Keywords: Parallel robot mechanisms · Parallel mechanisms · Topological structure optimization · Structure coupling-reducing · Motion decoupling

1 Introduction

As we all know, solution for forward position is one of the most basic important issues in parallel mechanism research, which ultimately attribute to a solving question of nonlinear position equations. Many scholars have contributed a lot to the issue using both algebra method and numerical method. When using the algebra method, by various elimination methods, Grobner bases method and other mathematical methods, the unknowns of position equations are eliminated and ultimately is expressed as solving of a quaternion algebra equation of higher degree [1-5]. When using numerical method, like chaotic iteration [6], successive approximation search method [7], continuous method [8], single open-chain method [9] or optimization method such as genetic algorithm [10], the position equations are transformed into solving of a optimization problem. The two methods all require for minimizing the dimensions of nonlinear position equations.

Yang proposed the concept of coupling degree of mechanism in 1983 [11]. The coupling degree is used to describe the dependence and coupling of position variables among the all independent loops and can represented by an index k ($k \geq 0$), which is an important index to describe the complexity of topological structure of a mechanism.

© Springer International Publishing Switzerland 2015
H. Liu et al. (Eds.): ICIRA 2015, Part III, LNAI 9246, pp. 199–210, 2015.
DOI: 10.1007/978-3-319-22873-0_18

It also reflects the complexity of solving mechanism kinematics and dynamics. k is constant once the topological structure of a mechanism is determined. It has been proved that the larger the value k is, the stronger the coupling of loop position variables of the mechanism and the higher of the complexity will be. Further, for a mechanism with $k=0$, it is a essentially statically determinated structure and its analytical forward position kinematics solution can be easily got. For a mechanism with $k>0$, its analytical forward position kinematics solution cannot be obtained directly, which needs be solved by simultaneous equations involved multiple loop position variables. The value k is just as the minimum dimension [12] of the simultaneous equations.

However, how to reduce the coupling degree k in order to reduce the dimension of nonlinear position equations of a mechanism? This problem is currently open and not yet studied.

The authors designed the ten weak coupling (i.e. with low coupling degree value) 6-*dof* parallel mechanisms [13,14], and eight weak coupling three-translational parallel mechanisms [15] in 2004 and 2005 by using designing of hybrid chains including one or two plane or space loops. In 2012, by using two kinds of hybrid branches such as double-sphere-pair hybrid branches, and triple-sphere-pair hybrid branches, six kinds of different structure 6-dof Steward-platform basic models (1-2-3, 3-1-1-1, 2-2-2, 2-2-1-1, 2-1-1-1-1, 1^6) are constructed. According to the calculated coupling degree k value, all 6-*dof* Steward parallel mechanisms can be divided into 4 categories, i.e., $k = 0, 1, 2, 3$. Further, one type with $k=0$, three types with $k=1$, one type with $k=2$ or 3 respectively. Then, the numerical forward position solutions [16-18] are obtained according to their k value. However, these examples mentioned above belong to the case study. Further, the general principle and practical design method of structure coupling-reducing are not yet presented.

This paper explores an important issue—how to reduce the coupling-degree of a parallel mechanism. We will propose the principle and one method for structure coupling-reducing, which can be applied to all spatial mechanisms.

The rest of this paper is organized as follows. In Section 2, on the basis of defining of a new concept of structure coupling-reducing of a parallel mechanism, the differences and relationship of the structure coupling-reducing and motion decoupling is revealed. In Section 3, a method for structure coupling-reducing is proposed. In Section 4, two novel mechanisms with the low coupling degree after being structure coupling-reduced are obtained by using the method for structure coupling-reducing. These novel mechanisms offer good application potential. In Section 5, we give the conclusions.

2   Structure Coupling-Reducing (SCR) and Motion Decoupling (MD)of Mechanisms

2.1   Structure Coupling-Reducing of a Mechanism and Its Significance

As mentioned above, the inherent dependence and correlation among the position variables within each basic loop, namely high coupling, is one of the most prominent features for multi-loop spatial parallel mechanisms. On the one hand, these characteristics make it have advantages such as strong bearing capacity, small accumulative error, large stiffness etc. On the other hand, configuration design, kinematics analysis and dynamics analysis, and the development of control system of a parallel mechanism have great difficulties, which make a great influence on its application ranges to some extent.

Keeping the DOF and POC (Position and Orientation Characteristics) remain unchanged, if the coupling degree can be reduced, not only the complexity of solving forward position solutions can be directly reduced, but also the calculation of workspace, error analysis and control algorithm etc will be easy. Further, that is preferable to make the coupling degree be $k=0$ because forward position solutions of a mechanism with coupling degree $k=0$ can be represented as an analytic expression. The value k is reduced from a larger value to a small value, or even zero, which means to reduce dimensions of the nonlinear position equations of the mechanism.

Therefore, in this paper it is called as "topological structure coupling-reducing (dimension-reducing)", For convenience, short for "structural coupling-reducing, written as SCR". By structure coupling-reducing, a mechanism could change from the strong structure coupling to the weak structure coupling while keeping the DOF and POC remain unchanged. The results bring to advantages to both forward position solution or even to obtaining analytical expressions and the real time control of kinematic parameters. Besides, it also makes the analysis process of kinematics and dynamics simple. This is the first important issue of topological structure optimization design for a mechanism.

2.2   Motion Decoupling of a Mechanism

Motion coupling of a mechanism refers to the dependence and correlation between the output parameters and the input parameters of the mechanism. Whether linear or nonlinear motions exists this relationship. Therefore, motion decoupling is the key issue of parametric design and motion control for a mechanism.

Many researchers have conducted in-depth research of motion decoupling design and analysis for parallel mechanisms. They put forward some parallel mechanisms with partial motion decoupling or full motion decoupling, like the mechanisms with two-rotation [19,20], one translation and two-rotation [21], three-translation[22], 2R-3R-4R spherical mechanism [23], 4-dof two-translation and two-rotation mechanisms [24], a decoupling spherical rotation mechanism [25]. 4-DOF 3T1R mechanism [26]. These works focused on the motion decoupling of a class of parallel mechanisms or special one.

In this paper, we call input-output motion completely decoupling or partial decoupling as motion decoupling, written as MD. Motion decoupling is advantageous to efficient control of position, velocity and acceleration for a mechanism. Therefore, it is the second important issue of topological structure optimization design for a mechanism.

2.3   Differences and Relations Between SCR and MD

As described above Section 2.1 and Section 2.2, structure coupling-reducing and motion decoupling of a mechanism are obviously two completely different concepts. The authors have a basic understanding on the differences and relations between SCR and MD as following.

(1) At present, "decoupling mechanism" that most of the literature referred to is actually MD mechanism. Regarding MD, there have being many case studies for both analysis and design at the parallel mechanism community. A general methodology for motion decoupling design needs to be put forward. On the contrary, there is few research on SCR. The two concepts are easy to confuse, which affects in-depth study on the performance and application of parallel mechanisms.

(2) MD of a mechanism reflects the dependence and the correlation between output parameters and input parameters, including the nonlinear position variables or linear velocity and acceleration variables. But coupling degree k of a mechanism reflects dependencies and correlation among nonlinear position variables or linear velocity and acceleration variables within each basic loops.

(3) MD of a mechanism is for the mechanism with more than two inputs [27, 28]. Generally speaking, the more the number of input joint is, the more likely complex the motion decoupling will be. But MD has nothing to do with the structure coupling degree k of the mechanism. For a mechanism with $k=0$, its input-output relationship can be the strong coupling, and even non MD. While strong structure coupling mechanism with $k=1, 2, 3$ may have MD, even completely decoupling or partial decoupling.

(4) Both strong structural coupling and strong motion coupling will affect real-time control of a mechanism. If structural coupling is strong it means the value k is larger. Then it is difficult to obtain real-time analytical forward position solutions, which will affect the position real-time control. If motion coupling is strong, it is difficult to design control algorithm of the position, velocity and acceleration, which will affect the actual application of parallel mechanisms.

This paper mainly focuses on the design method for structure coupling-reducing. Regarding a general methodology for motion decoupling will be discussed in another paper.

3   SCR Principle and Design Method of Parallel Mechanisms

3.1   The Coupling Degree and Its Analytical Expression

As what mentioned before, the coupling degree of mechanisms k refers to the complexity degree of position variables coupling among each basic loop in a Basic Kinematic Chain (in short, BKC) what mechanisms contain. This paper gives the definition of coupling degree according to *Ref.* [11,12] and presents a new analytical expression as follows.

1) The constraint degree of a single-open-chian [11,12,27,28]
    The constraint degree of the jth SOC_j in a kinematic chain is defined as

$$\Delta_j = \sum_{i=1}^{m_j} f_i - I_j - \xi_{L_j} = \begin{cases} \Delta_j^- = -5,-4,-3,-2,-1. \\ \Delta_j^0 = 0 \\ \Delta_j^+ = +1,+2,+3,\cdots \end{cases} \quad (1)$$

Here, m_j refers to the number of kinematic pair of the jth single-open-chian (for short, SOC_j). f_i refers to the degree of freedom of ith kinematic pair(excluding negative degree of freedom); I_j refers to the number of driving pair of the jth SOC_j; ξ_{L_j} refers to the number of independent displacement equations of the jth independent loop.

For a BKC, there must be:

$$\sum_{j=1}^{v} \Delta_j = 0$$

2) Coupling degree of a BKC [11,12,27,28]
    Coupling degree of a BKC is defined as

$$k = \frac{1}{2}\min\left\{\sum_{j=1}^{v} \Delta_j = 0\right\} \quad (2)$$

There may be several structure decomposition schemes when BKC is decomposed into v $SOC(\Delta_j)$, and we should choose the scheme what $\left(\sum |\Delta_j|\right)$ is the smallest.

Thus, number of BKC which mechanism contains and its coupling degree can be calculated. The coupling degree of the mechanism k is the max of the coupling degree of which contains BKC, which is

$$k = \max\{k_1, k_2, ..., k_i, ...\} \quad (3)$$

Here, v refers to the number of the basic loop, k_i refers to coupling degree of the ith BKC;

3) New analytical expression of a mechanism

This paper presents a new accurate expression based on the constitution of BKC of a mechanism, which is shown as follows:

$$PKM^K[\text{F,V}] = F - J_{in} + \sum_{j=1}^{v} p_k \cdot BKC^k(\Delta_1, \Delta_2, \cdots \Delta_j) \quad (4)$$

Here, $PKM^K[F,v]$—a parallel kinematic mechanism with degree of freedom F, baisc loop v and coupling degree k.

F-J_{in}— with F inputs

p_k — number of BKC whose coupling is k

$BKC^k(\Delta_1, \Delta_2, \cdots \Delta_j)$ —BKC with coupling degree k, constitute of SOC_1, $SOC_2, \cdots SOC_j$ whose constraint degree are $\Delta_1, \Delta_2, \cdots \Delta_j$ respectively.

Formula (4) reveals three information about the mechanism topological structure: ① coupling degree of the mechanism and the number of independent loops, ② the number of BKC_s and coupling degree k value of each BKC that the mechanism contains, ③ composition of each BKC, including the number of SOC and value of its constraint degree. So the meaning of the formula is more clear than that stated in the *Ref* [11,12,27,28] and suitable for any plan and spatial mechanisms.

3.2   SCR Principle and Method of Parallel Mechanisms

According to Section 2 stated above, SCR of a parallel mechanism refers to reducing the coupling degree k of the mechanism.

According to the formula (1) and (2), the value of coupling degree k of a BKC depends on the value of the constraint degree Δ of the first SOC. So we must reduce the constraint degree Δ of the first SOC in order to reduce the coupling degree k of the BKC. This is just SCR principle. According to this principle and the formula (1), we propose a SCR method based on changing POC chains into driving chains as follows.

Ref. [28] studied a class of 5-DOF parallel mechanisms with a linear platform and analyzed their topological characteristics, where a concept, i.e, the POC chain, is defined. That is, such a passive chain is called as the POC chain if the POC of the end link of a passive chain without driving joints is the same as that of the moving platform. Here, we use the symbol ☐ to express a POC chain. Obviously, the characteristics of the POC chain are as follows: ① It is a serial branched chain which has the least kinematic joints; ② the chain has no driving pair. The concept of the POC chain is more clear and concise than "passive chain", "redundant chain", "just constraint chain", "full constraints chain", "constraint chain" proposed in other literatures [19,29,30,31]. It reflects the output characteristics of the moving platform directly.

The coupling-reducing method we propose here based on changing POC chains into driving chains will contain two steps: ① change the POC chain into the driving chain; ② meanwhile, remove one unconstrained driving chain. The reason of two operations for reducing the coupling degree k can be described as follows.

For the parallel mechanisms which contain a POC chain, the POC chain has the smallest number of DOFs among all branched chains. We should choose the branched chains which include the POC chain as the first SOC, which will bring two benefits. On the one hand, the constraint degree Δ of the first SOC is 1 smaller than which before changing because the number of driving pair I increases by 1 according to formula (1). On the other hand, the possibility of coupling among basic loops will reduce because the number of basic loops of the mechanism decreases by 1. Therefore the method can reduce the coupling degree k effectively.

The mechanism like 3-SPS+$\boxed{\text{RPR}}$, 3-UPS+$\boxed{\text{UP}}$, 4-SPS+$\boxed{\text{SP}}$ and 5-SPS+$\boxed{\text{PRPU}}$ can be reduced the coupling degree k by this method. But we only illustrated the 3-SPS+$\boxed{\text{RPR}}$, 3-UPS+$\boxed{\text{UP}}$ mechanism for the limitation of the space of this paper.

4   Application of the SCR Method

4.1   Examples 1: *3-SPS+RPR Mechanism.*

① *The Original Mechanism.*

3-SPS+RPR mechanism is shown as Figure 1(a). Three branched chains of the mechanism are SPS unconstrained chains, and another branched chain $R_{41}P_{42}R_{43}$ is POC chain. Obviously the moving platform of the mechanism can realize the 2-translational and 1-rotational motion [31] which the POC chain can satisfy. DOF of the mechanism is 3. P_{42} is passive joint among them, and the coupling degree analysis is as follows.

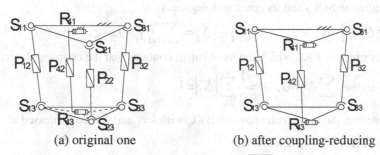

(a) original one   (b) after coupling-reducing

Fig. 1. Coupling-reducing design of 3-SPS+RPR mechanism

1) Determine SOC_1 (POC chain $R_{41}P_{42}R_{43}$ is included) and its constraint degree Δ_1

$$SOC_1\{-S_{11}-P_{12}-S_{13}-R_{43}-P_{42}-R_{41}-\}$$

$$\Delta_1 = \sum_{i=1}^{m_1} f_i - I_1 - \xi_1 = 9-1-6 = 2$$

2) Determine SOC_2 and its constraint degree Δ_2

$$SOC_2\{-S_{21}-P_{22}-S_{23}-\}$$

$$\Delta_2 = \sum_{i=1}^{m_2} f_i - I_2 - \xi_2 = 6-1-6 = -1$$

3) Determine SOC_3 and its constraint degree Δ_3

$$SOC_3\{-S_{31}-P_{32}-S_{33}-\} \quad \Delta_3 = \sum_{i=1}^{m_3} f_i - I_3 - \xi_3 = 6-1-6 = -1$$

4) Determine the BKC which the mechanism contains and the coupling degree κ

$$\sum_{j=1}^{3}\Delta_j=0, \quad \text{so} \quad \kappa=\frac{1}{2}\sum_{j=1}^{3}\left|\Delta_j\right|=2.$$

Therefore, the mechanism has one BKC with $k=2$ and can be expressed as
   $PKM^2=3\text{-}J_{in}+BKC^2(2,-1,-1)$

The coupling degree of the mechanism is $k=2$.

② *Coupling-Reducing Design*

We reduce its coupling degree in this way: (1) make P_{42} joint in the POC chain $\boxed{R_{41}P_{42}R_{43}}$ be driving joint, ② remove one SPS unconstrained driving chain, say $S_{21}\underline{P}_{22}S_{23}$. Thus we get the derived mechanism shown in Figure 1(b), and its coupling degree analysis is as follows:

1) Determine SOC_1 and its constraint degree Δ_1

$$SOC_1\{-S_{11}-P_{12}-S_{13}-R_{43}-P_{42}-R_{41}-\},$$

$$\Delta_1 = \sum_{i=1}^{m_1} f_i - I_1 - \xi_1 = 9\text{-}2\text{-}6 = 1$$

2) Determine SOC_2 and its constraint degree Δ_2

$$SOC_2\{-S_{31}-P_{32}-S_{33}-\}, \quad \Delta_2 = \sum_{i=1}^{m_2} f_i - I_2 - \xi_2 = 6\text{-}1\text{-}6 = -1$$

3) Determine the BKC which the mechanism contains and the coupling degree κ

Because $\sum_{j=1}^{2} \Delta_j = 0$, $\kappa = \dfrac{1}{2}\sum_{j=1}^{2} |\Delta_j| = 1$.

Therefore, the mechanism has one BKC with $k=1$ and can be expressed as

$$PKM^1 = 3\text{-}J_{in} + BKC^1(-1,1)$$

Then, the coupling degree of the mechanism is $k=1$. That is to say, the coupling degree has been reduced form 2 to 1. But DOF and POC of the mechanism stay the same.

In the same way, we make analysis of 4-SPS+$\boxed{SP}$, 3-UPS+$\boxed{UP}$ and 5-SPS+$\boxed{PRPU}$ mechanism [31]. To save space, we only given a brief results.

Examples 2: 3-UPS/UP Mechanism.

The famous Tricept 3-UPS+$\boxed{UP}$ mechanism is shown in Figure 2(a). It has three UPS unconstrained branched chains and one POC chain like $\boxed{UP}$. The DOF of the mechanism is 3. The moving platform of the mechanism can realize the 1-translational and 2-rotational motion which the POC chain $\boxed{U_{4142}P_{43}}$ (U_{4142} is expressed by R_{41} and R_{42}) can do. Coupling degree of the mechanism is 2 [32]. The mechanism has one BKC with $k=2$, composed of only three basic loops, and can be expressed as

$$PKM^2 = 3\text{-}J_{in} + BKC^2(2,-1,-1)$$

The coupling degree of the mechanism is $k=2$

The coupling-reducing procedure of the mechanism is as follows: ① make P_{43} joint in the POC chain $\boxed{U_{4142}P_{43}}$ into driving one; ② remove one UPS driving chain, say $SOC\{-\left(R_{11}\perp R_{12}\right)-P_{13}-S_{14}-\}$. Thus we get the derived TriVariant mechanism [35] shown in Figure 2(b). The mechanism has one BKC with $k=1$, composed of only two basic loops, and can be expressed as

$$PKM^1 = 3\text{-}J_{in} + BKC^1(1,-1)$$

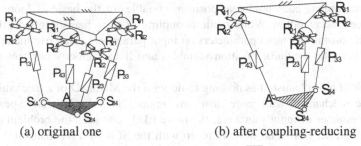

(a) original one   (b) after coupling-reducing

Fig. 2. Coupling-reducing design of 3-SPS+UP mechanism

Therefore the coupling degree of the mechanism after coupling-reducing is $k=1$. That is to say, the coupling degree has been reduced from 2 to 1while DOF is still 3 and POC stays the same as 1-translational and 2-rotational motion. But this moment the piston rod and oil cylinder, which constitute the driving joint, are subjected to a lateral force. That is the negative factor.

From the two examples above, DOF and POC of a mechanism stay unchanged but coupling degree of the mechanism decreases using coupling-reducing method based on changing POC chain into the driving chain. Coupling-reducing operation not only can make the forward position solutions more easy but also can simplify physical structure of the mechanism and be easily manufactured because the number of basic loops decreases. Therefore the method is an effective topological optimization method.

5   Conclusions

The paper proposes a method for coupling-reducing of the parallel mechanisms, which is a key issue of the topological structure optimization of parallel mechanisms.

Firstly, as a novel concept--SCR of a parallel mechanism is proposed and defined. The differences and relations between SCR and MD are also clarified.

Then, under the condition that the DOF and POC of the mechanism remain invariable, SCR method based on changing a POC chain into driving chain is proposed. Two SCR application examples are illustrated respectively. Two corresponding novel mechanisms with low coupling degrees are obtained, which offers potential applications of the optimizing configuration.

All conclusions obtained from this paper are as following:

1) Structure coupling means the correlation and dependence among the position variables of each loop in the parallel mechanism, which can be measured using an numerical index--structure coupling degree k. Reducing the structure coupling degree could reduce directly the complexity of solving forward position solutions for a mechanism.

2) Structure coupling-reducing (SCR) and motion decoupling (MD) of a mechanism are two important aspects in topological structure optimization. But they are two totally different concepts. The structure coupling degree reflects the correlation and

dependence among each nonlinear position variables of the basic   loops and their linear speed, acceleration. While motion coupling of a mechanism reflects the dependence relationship of output parameters on input parameters, each parameter of which includes both the nonlinear position variables and linear velocity or acceleration variables.

3) SCR of a mechanism has nothing to do with the MD. MD of a mechanism is only for the mechanism with more than two inputs [30, 31]. Generally speaking, the more the number of input joints are, the more likely complex the problem of motion coupling will be. But it has nothing to do with the SCR of a mechanism. Even for a mechanism with coupling degree $k=0$, its input-output relationship can be the strong coupling. And a mechanism with high coupling degree value, i.e., $k=1$, 2, 3, can have completely motion decoupling or partial motion decoupling.

4) Both strong structure coupling and strong motion coupling of a mechanism contribute effect on difficulty in real-time motion control. The former makes it difficult to obtain analytical forward position solutions, and real-time calculation is poor. The latter makes it difficult to design algorithm of the position, velocity and acceleration.

5) The SCR method can be applied to all spatial parallel mechanisms. They are of significance in topological structure optimization design, kinematics solution, real-time control and application.

Authors would like to thank Prof. Ting-li Yang for his fruitful discussion with us.

Acknowledgements. The financial support of National Natural Science Foundation of China (No.51375062, No.51405039, No.51475050) is gratefully acknowledged.

References

1. Innocenti, C.: Parenti-Castelli.V. Closed-form direct position analysis of a 5-5 Parallel Mechanism. J. Mech. Des., Trans. ASME **115**(3), 515–521 (1993)
2. Husty, M.L.: Algorithm for Solving the direct kinematic of Stewart-Gough-Type platforms. Mechanism and Machine Theory **31**(4), 365–380 (1996)
3. Dasgupta, B., Mruthyunjaya, T.S.: A constructive predictor-corrector algorithm for the direct position kinematics problem for a general 6-6 Stewart platform. Mechanism and Machine Theory **31**(6), 799–811 (1996)
4. Liang, C., Rong, H.: The forward displacement solution to a Stewart platform type manipulator. Chinese Journal of Mechanical Engineering **27**(2), 26–30 (1991)
5. Kong, X., Gosselin, C.M.: Generation and forward displacement analysis of RPR-PR-RPR analytic planer parallel manipulators. ASME Journal of Mechanical Design **124**(2), 294–300 (2002)
6. Luo, Y., Li, D.: Finding all solutions to forward displacement analysis problem of 6-SPS parallel robot mechanism with chaos-iteration method. Journal of Engineering Design **10**(2), 70–74 (2003)
7. Qü, Y., Huang, Z.: Three-dimensional search method for position analysis of space 6-dof parallel mechanism. Robot **3**(5), 25–29 (1989)
8. Liu, A., Yang, T.: Kinematic design of mechanical system. China Petrochemical Press, Beijing (1999)

9. Feng, Z., Zhang, C., Yang, T.: Direct displacement solution of 4-DOF spatial parallel mechanism based on ordered single-opened-chain. Chinese Journal of Mechanical Engineering **42**(7), 35–45 (2006)

10. Zheng, C., Jiao, L.: Forward kinematics of a general Stewart parallel manipulator using the genetic algorithm. Journal of Xidian University **30**(2), 165–173 (2003)

11. Yang, T.-L.: Structural analysis and number synthesis of spatial mechanisms. In: Proc. of the 6th World Cong., on the Theory of Machines and Mechanisms, New Delhi, vol. 1, pp. 280–283 (1983)

12. Yang, T.: Basic theory of mechanical system– mechanism, kinematics, dynamics. China Machine Press, Beijing (1996)

13. Shen, H.P., Yang, T.L., Ma, L.Z.: Methodology for synthesis of kinematic structure of 6-DOF weakly-coupled parallel mechanism. Chinese Journal of Mechanical Engineering **40**(7), 14–19 (2004)

14. Shen, H.P., Yang, T.L., Ma, L.Z.: Synthesis and Structure Analysis of Kinematic Structures of 6-dof Parallel Robotic Mechanisms. Mechanism and Machine Theory **40**(10), 1164–1180 (2005)

15. Shen, H.P., Ma, L.Z., Yang, T.L.: Kinematic structural synthesis of 3-translational weakly-coupled parallel mechanisms based on hybrid chains. Chinese Journal of Mechanical Engineering **41**(4), 22–27 (2005)

16. Yu, T., Shen, H., et al.: An easily manufactured structure and its analytic solutions for forward and inverse position of 1-2-3-SPS type 6-DOF basic parallel mechanism. In: Proceedings of 2012 IEEE International Conference on Robotics and Biomimetics, December, Guangzhou (2012)

17. Wang, Z., Shen, H., et al.: An easily manufactured 6-DOF 3-1-1-1 SPS type parallel mechanism and its forward kinematics. In: Proceedings of the 2nd IFToMM Symposium on Mechanism Design for Robotics, MEDER 2012 (2012)

18. Shen, H., Yin, H., Wang, Z., et al.: Research on forward position solution for 6-SPS parallel mechanisms based on topology structure analysis. Chinese Journal of Mechanical Engineering **49**(11), 14–20 (2013)

19. Zhen, H., et al.: On the degree of freedom of mechanisms. Science Press (2011). Hou, Y., Lu, W., Zeng, D., et al.: Motion decoupling of 2-DOF rotational parallel mechanism. Patent 201010617042 (2010)

20. Jin, Q., Yang, T.L.: Synthesis and Analysis of a Group of 3-Degree-of-Freedom Partially Decoupled Parallel Manipulators. ASME J. Mech. Des. **126**, 301–306 (2004)

21. Li, W., Gao, F., Zhang, J.: A three-DOF Translation Manipulator with Decoupled Geometry. Robtica **23**, 805–808 (2005)

22. Hang, L., Wang, Y., Wu, J., et al.: Research on decoupling conditions of spherical parallel mechanism based on the topological structure decoupling rule. Chinese Journal of Mechanical Engineering **41**(9), 28–32 (2005)

23. Zeng, Q., Fang, Y., Ehmann, K.F.: Design of a Novel 4-DOF Kinematotropic Hybrid Parallel Manipulator. J. Mechanical Desgin **133**(12) (2011)

24. Zhang, F., Zhang, D.: Structural synthesis of decoupled spherical parallel mechanism based on driven-chain principle. Transactions of the Chinese Society for Agricultural Machinery **42**(11), 195–199 (2011)

25. Richard, P.-L., Gosselin, C.M., Kong, X.: Kinematic analysis and prototyping of a partially decoupled 4-DOF 3T1R parallel manipulator. In: Proceeding of the 2006 ASME Design Engineering Conference, DETC2006-99570, USA (2006)

26. Yang, T.: Topology structure design of robot mechanisms. China Machine Press, Beijing (2004)

27. Yang, T., Liu, A., Luo, Y., et al.: Topological Structure Design for Robot mechanisms. Science Press, Beijing (2012)
28. Zhu, S.: Special 5-DOF Parallel Mechanisms and their Topological Analysis. Changzhou University (2014)
29. Tsai, L.W.: Multi-Degree-of-Freedom Mechanisms for Machine Tools and the Like. U.S. Patent No. 5656905 (1997)
30. Liu, L.: Research on synthesis and characterization study of 3-DOF parallel mechanism containing intermediate constraint branch. Yan Shan University, Qinhuangdao (2007)
31. Luo, J.: Research on little degrees of freedom parallel equipment configuration and constraint chains. Northeastern University, Shenyang (2006)
32. Huang, T., Liu, H., Li, M.: 5-DOF Robot. Patent. CN1709657 (2005)

Design of a Compact Finger-Integrated Mechanism with Robust Kinematics for Hand Exoskeleton

Baiyang Sun, Caihua Xiong(✉), Yaobin Huang, Wenbin Chen, and Wenrui Chen

State Key Lab of Digital Manufacturing Equipment and Technology,
Institute of Rehabilitation and Medical Robotics, Huazhong University of Science
and Technology, Wuhan 430074, China
chxiong@hust.edu.cn

Abstract. Compactness, accurate kinematic transmission, wearable convenience, safety and applicability are important factors for wearable hand exoskeletons. Currently, integrating all the factors into an exoskeleton is still difficult. Traditionally, matching center of rotations (CoR) between finger joints and exoskeletons is able to ensure accurate kinematic transmission, however lacks wearable easiness and applicability in some degree. Bypassing this difficulty, this paper introduces a novel compact wearable mechanism of a finger exoskeleton which does not need to match rotation centers to human joints, while possessing robust kinematic properties. The wearable structure presents compactness, wearable easiness and safety, while theoretical results prove movement and force transmission robustness.

Keywords: Hand exoskeleton · Human hand · Robotics · Mechanism design · Kinematic robustness

1 Introduction

As a wearable mechanism for an exoskeleton hand connects an emotionless device to an emotional person, it plays an important role in protecting people from hurting and transmitting accurate movement and force.

Currently, to assist a human hand performing a serious of desired motion and force, two kinds of mechanisms can be adopted: stand-alone mechanisms and integrated mechanisms[1]. The first ones are independent mechanical systems which can work autonomously without users, such as [2-5]. This kind of mechanisms is able to ensure accurate kinematics so that it can assistant a human hand to perform accurate motion as expected. But it needs to match the center of rotation or remote center of rotation to a human joint very carefully. Otherwise, the difference in the rotational axes may cause a collision between the user's hand and the device, resulting in damage to the user's hand [6]. The other ones, integrating finger phalanges and articulations into the system kinematic scheme, are considered to be able to bypass the problem of matching the center of rotation, such as [7, 8]. However, this method may cause issues in uncertain wear position so as to lead an inaccurate kinematic transmission. Though sensors could be used to measure human joint moments or angles,

© Springer International Publishing Switzerland 2015
H. Liu et al. (Eds.): ICIRA 2015, Part III, LNAI 9246, pp. 211–219, 2015.
DOI: 10.1007/978-3-319-22873-0_19

this paper aims to not use sensors in joint space to simply exoskeleton mechatronics design in future. From structural view, compactness, safety and applicability are important for finger exoskeleton considering its expansibility to a whole hand structure and suitable for different persons with various finger sizes. In a word, to design a wearable mechanism for hand exoskeleton, which possesses compactness, safety, wearable convenience and accurate kinematic transmission, still remains challenge.

In this paper, the latter approach is chosen considering its potential application in family or some other flexible vacations where easily wearing the exoskeletons is very important. Though uncertain wear position may lead to an inaccurate kinematic transmission, reasonable design method introduced in this paper could overcome the disadvantage. In the rest of this paper, we firstly introduce the analysis and selection of mechanism configuration considering compactness, wearable convenience and parameters variations robustness. And then develop a parameter optimization method for kinematic robustness. Finally we present the final compact wearable mechanism and theoretical results reflecting kinematic transmission robustness, force transmission robustness and force nonlinearity.

2   Mechanism Configuration

From the view of the wear mechanisms' placement, the mechanism placed externally to the hand backside is chosen in order to avoid interference with both the human counterparts and the grasped object [1] and make it easier to extend to a multi-finger structure. Among various integrated mechanisms placed above the finger, four-bar mechanism is considered as the simplest one. Two alternative configurations are listed in Fig. 1. The location of the black circle connecting phalanges expresses a standard position where the mechanism worn ideally. However, as there is not a clear reference for wearing the exoskeleton accurately, the human joint usually deviates from the ideal position. The purple circle indicates a possible arisen location when the mechanism is worn deviated from the ideal situation. The red frame shown in Fig. 1 (b) represents a boundary limiting all possible locations of the finger joint under various wear situations.

Actually, we can conclude that the larger the distance between lower pairs of mechanical parts and finger surface is, the more robust the parameter variations (length variation and angle deviation) of virtual links (the purple dotted lines shown in Fig. 1) will be. Therefore, mechanical joints must be far away from the surface of the finger to ensure the robust kinematics of the integrated system. Scheme of crank-rod mechanism which needs more space in the vertical direction may lead to burden in height. So we prefer the latter one. Although scheme of slider-crank mechanism has called for more space in the horizontal direction, reasonable design will avoid interference between adjacent mechanisms and save space in the vertical direction. Based on the analysis before, Slider-crank mechanism scheme which utilizes relative movement between slider and linkage driving the rotation of the human joint is finalized.

To form relative movement between slider and linkage, the mechanism consisting of pulley-tendon-leader shown clearly in Fig. 1 (c) is proposed. Rotating the pulley could realize desired relative linear movement through the tendon which is fixed at both ends of the leader and the pulley. The rotation of the pulley could be easily driven by actuators through the way of cable-driven.

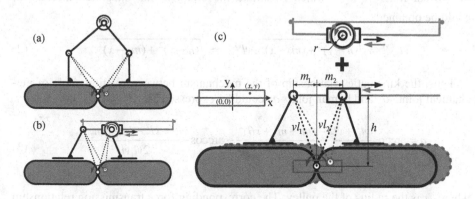

Fig. 1. Mechanism configurations

3   Parameters

3.1   Range of CoR

When the exoskeleton structure is inaccurately worn, finger joint position will deviate from its standard position. If the standard location of human joint is considered as origin of coordinates, the coordinates of any deflected position of rotation center which is caused by inaccurately wearing could be expressed as (x, y) (see Fig. 1). As the significant reference of mechanical joints when wearing, range of x is set as $x \in (-m_1, m_2)$ considering that the finger joint is always able to be ensured locating between the mechanical joints in the horizontal direction. In the vertical direction, range of y is set as $y \in (-y_m, y_m)$ considering motion errors produced either by possible extrusions between mechanical part and the skin of human finger or as the reason of a human finger phalange not rotating strictly around a fixed axis during motion. y_m is arbitrarily set as $2mm$ based on the analysis results from Zhang [9] that standard deviations (vertical distances from CoR to finger surface) of most finger joints' CoRs among different persons are less then $2mm$. Hence the following boundary condition is set:

$$x \in (-m_1, m_2), \ y \in (-2mm, 2mm) \tag{1}$$

3.2   Kinematic Relationship

Let h denote the vertical distance from the finger joint to the leader, m_1 the horizontal distance between the left mechanical joint and the finger joint, m_2 the horizontal distance between the right mechanical joint and the finger joint. Then the lengths of the virtual links vl_1, vl_2 which constitute the crank in the integrated mechanism could be obtained:

$$vl_1^{(x,y)} = \sqrt{(h-y)^2 + (m_1+x)^2}, \; vl_2^{(x,y)} = \sqrt{(h-y)^2 + (m_2-x)^2} \tag{2}$$

Thus, the kinematic relationship of the mechanism between rotation angle of mechanical joint θ and human joint q could be expressed as:

$$q = \arccos \frac{vl_1^2 + vl_2^2 - (m_1 + m_2 + r\theta)^2}{2vl_1vl_2} - \arccos \frac{vl_1^2 + vl_2^2 - (m_1 + m_2)^2}{2vl_1vl_2} \tag{3}$$

where r is the radius of the pulley. The corresponding force transmission relationship could be expressed as:

$$T = \frac{vl_1vl_2 \sin[\arccos(vl_1^2 + vl_2^2 - (m_1 + m_2 + r\theta)^2 / 2vl_1vl_2)]}{m_1 + m_2 + r\theta} T_0 \tag{4}$$

where T_0 is the input moment in mechanical joint and $T_{(x,y)}$ is the output moment in human joint.

There exist following function relationships which are easily derived from the relationships stated above:

$$q_{(x,y)} = f(x, y, q_{(0,0)}) \tag{5}$$

$$T_{(x,y)} = g(x, y, q_{(0,0)}) \tag{6}$$

3.3   Optimization Index

As the objective is to ensure robust kinematic under the premise of easily wearing the mechanism without matching its center of rotation to a human articulation, the following index could be considered as a means to evaluate robustness of motion transmission under a specific wear situation with the articulation position coordinate (x, y):

$$I_1^{(x,y)} = \frac{1}{q_{max}} \int_{q_{(0,0)}=0}^{q_{max}} (1 - \left| \frac{q_{(x,y)} - q_{(0,0)}}{q_{(0,0)}} \right|) \tag{7}$$

where q_{max} represents the limit rotation angle of human joint, set as 80 degrees. Considering all possible wear situations with the articulation position coordinate (x, y) locating in the range of $x \in (-m_1, m_2)$, $y \in (-y_m, y_m)$ based on the discussion before, the robustness of motion transmission under any possible wear situations could be expressed as:

$$I_1 = \frac{1}{2 y_m (m_1 + m_2)} \iint_{x,y} I_1^{(x,y)}$$ (8)

Similarly, the robustness index can be easily extended to force transmission. The criteria evaluating robustness of motion transmission under a specific wear situation with the articulation position coordinate (x, y) could be written as:

$$I_2^{(x,y)} = \frac{1}{\theta_{max}} \int_{q_{(0,0)}=0}^{q_{max}} (1 - \left| \frac{T_{(x,y)} - T_{(0,0)}}{T_{(0,0)}} \right|)$$ (9)

Hence:

$$I_2 = \frac{1}{2 y_m (m_1 + m_2)} \iint_{x,y} I_2^{(x,y)}$$ (10)

Another important kinematic characteristic of four-bar mechanism is the nonlinearity that influences the moment acting on the human joint isotropy. To reduce the nonlinearity to an extreme, an optimization index is considered to measure the nonlinear degree, expressed as follows:

$$I_3 = \frac{T_{min}}{T_{max}}$$ (11)

where T_{min} and T_{max} are respective the minimum and maximum moment in the whole rotation range of the human joint under constant force control mode. As $T_{(0,0)}$ is a continuous differentiable function of θ with only one extreme value existed, the index can be rewritten as:

$$I_3 = \frac{\min \left\{ T_{(0,0)} \big|_{\theta=0}, T_{(0,0)} \big|_{\partial T_{(x,y)}/\partial \theta=0}, T_{(0,0)} \big|_{\theta=\theta max} \right\}}{\max \left\{ T_{(0,0)} \big|_{\theta=0}, T_{(0,0)} \big|_{\partial T_{(x,y)}/\partial \theta=0}, T_{(0,0)} \big|_{\theta=\theta max} \right\}}$$ (12)

Taking above indices into account, a global index combining them is obtained:

$$I = I_1 I_2 I_3$$ (13)

3.4   Constraints

Kinematic Constraints

In order to avoid the dead-point problem that will lead to interference between human and mechanism, the following constraints must be satisfied in any possible wear situation:

$$0 < \frac{vl_1^2 + vl_2^2 - (m_1 + m_2)^2}{2vl_1vl_2} < 1 \tag{14}$$

$$\arccos \frac{vl_1^2 + vl_2^2 - (m_1 + m_2)^2}{2vl_1vl_2} + q_{max} < 180° \tag{15}$$

Size Constraints

In terms of finger sizes, the maximum vertical distance from CoR to finger or palm surface can reach about $16mm$ [9], and the length of a finger phalange is usually less than $40mm$. Considering these situations and manufacture factors, constraints of the structure parameters are set as follows:

$$\begin{aligned} 30mm \leq h \leq 50mm \\ 3mm \leq m_1 \leq 15mm \\ 3mm \leq m_2 \leq 15mm \\ l_{max} \leq 36mm \end{aligned} \tag{16}$$

where l_{max} represents the stoke of the relative displacement between slider and leader. It could be expressed as a function of h, m_1 and m_2.

3.5   Optimization Formulation

As the existence of absolute value in the integral term of indices I_1 and I_2, discrete method is used to overcome the difficulties. Step lengths of variables including decision parameters h, m_1, m_2 and process variables x, y are set as 1mm. Step length of process variable $q_{(0,0)}$ is set as 10 degrees. r is set as 7mm considering the tradeoff between structural limit and minimal input force in tendon around the pulley.

4   Results

The final mechanism parameters and movement ranges of key variables are obtained from above optimization process and shown in Table 1. The theoretical results are shown in Fig. 2. On the left of Fig. 2, the color reflects the robustness of movement transmission (upside) and force transmission (underside). The intensity of the color is decided by indices (7) and (9) which represent the robustness of movement and force

Table 1. Mechanism parameters and movement ranges of key variables

Parameters	Values (mm)	Variables	Ranges
h	30	q	$(0, 80°)$
m_1	6	θ	$(0, 295°)$
m_2	5	x	$(-6mm, 5mm)$
r	7	y	$(-2mm, 2mm)$

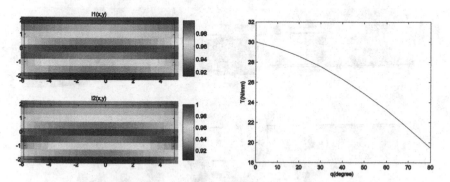

Fig. 2. Theoretical results with optimized parameters

transmission. The performances of robust indices for both movements and force are more than 90% in every possible wear situation. On the right, it represents the relationship between the moment and rotation angle of a human joint with the input tension being 1N with corresponding pulley radius equaling to 7mm (i.e., input torque equaling to 7 N*mm). Note that the range of CoR is determined considering different persons' data, of which the distance between the CoR to finger surface, in [9], hence, the robustness results are effective for different persons and prove its wide applicability.

5   Mechanism Structure

The final structure (for one finger) could be seen in Fig. 3. The whole size of the mechanism on each joint of the finger is 19mm(wideth of the mechanism) × 66mm(length of the whole leader) × 32mm(height from finger surface to the top of the mechanism). The structure is sufficiently small and compact for a wearable mechanism to fingers. And it allows to expanding to a whole hand exoskeleton structure because of its compact size. The finger exoskeleton can be divided into three parts: the driving part consisting of pulley and cable, the transmission part introduced in section 2 and the contacting part, i.e., soft nylon finger-cot. To avoid fingers moving over the natural range, a mechanical restriction on the leader is set.

In wearable process, the nylon finger-cot needs to satisfy the wear conditions that the finger articulation locates between the mechanical joints of the virtual links with the relative distance being 11mm ($(m_1 + m_2)$ obtained before). It is easy to be satisfied as the obvious reference of the mechanical joints and the larger distance. Also, it does not need to match CoR to the human joint. Hence, the wearable process is very easy and fast. To make it applicable for different persons with various finger sizes, a regulating mechanism is also set between two mechanical unites for PIP MCP joints of a finger.

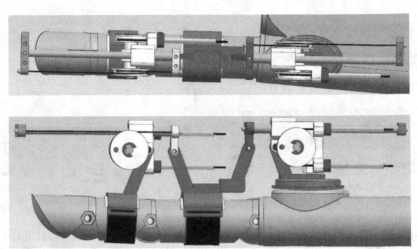

Fig. 3. Finger exoskeleton structure

6   Conclusion and Future Work

This paper presented a novel wearable mechanism design for hand exoskeleton. Compact structure ensures wearable safety, convenience and wide applicability, while mechanism parameters ensure robust kinematics with the theoretical results proving it. The advantages make it suitable for home rehabilitation or some other flexible vacations in future where wear convenience, structural simplicity and portability are required. The practical structure is under development and experimental tests will be taken in the future work. Meanwhile, we will expand the mechanism unit to a whole hand structure in the future.

7   Funding

This work was partly supported by the National Basic Research Program of China (973 Program, Grant No. 2011CB013301), National Natural Science Foundation of China (Grant No. 51335004, Grant No. 61301225, Grant No. 51305148).

References

1. Mozaffari Foumashi, M.: Synthesis of Hand Exoskeletons for the Rehabilitation of Post-Stroke Patients (2013)
2. Chiri, A., Vitiello, N., Giovacchini, F., Roccella, S., Vecchi, F., Carrozza, M.C.: Mechatronic design and characterization of the index finger module of a hand exoskeleton for post-stroke rehabilitation. IEEE/ASME Transactions on Mechatronics 17, 884–894 (2012)
3. Tong, K., Ho, S., Pang, P., Hu, X., Tam, W., Fung, K., et al.: An intention driven hand functions task training robotic system. In: 2010 Annual International Conference of the IEEE Engineering in Medicine and Biology Society (EMBC), pp. 3406–3409 (2010)
4. Li, J., Wang, S., Wang, J., Zheng, R., Zhang, Y., Chen, Z.: Development of a hand exoskeleton system for index finger rehabilitation. Chinese Journal of Mechanical Engineering 25, 223–233 (2012)
5. Arata, J., Ohmoto, K., Gassert, R., Lambercy, O., Fujimoto, H., Wada, I.: A new hand exoskeleton device for rehabilitation using a three-layered sliding spring mechanism. In: 2013 IEEE International Conference on Robotics and Automation (ICRA), pp. 3902–3907 (2013)
6. Heo, P., Gu, G.M., Lee, S.-J., Rhee, K., Kim, J.: Current hand exoskeleton technologies for rehabilitation and assistive engineering. International Journal of Precision Engineering and Manufacturing 13, 807–824 (2012)
7. Zhang, F., Hua, L., Fu, Y., Chen, H., Wang, S.: Design and development of a hand exoskeleton for rehabilitation of hand injuries. Mechanism and Machine Theory 73, 103–116 (2014)
8. Ueki, S., Kawasaki, H., Ito, S., Nishimoto, Y., Abe, M., Aoki, T., et al.: Development of a hand-assist robot with multi-degrees-of-freedom for rehabilitation therapy. IEEE/ASME Transactions on Mechatronics 17, 136–146 (2012)
9. Zhang, X., Lee, S.-W., Braido, P.: Determining finger segmental centers of rotation in flexion–extension based on surface marker measurement. Journal of Biomechanics 36, 1097–1102 (2003)

Structure Optimization and Implementation of a Lightweight Sandwiched Quadcopter

Qiaoyu Zhang[1], Jie Chen[1], Luo Yang[2], Wei Dong[1], Xinjun Sheng[1(✉)], and Xiangyang Zhu[1]

[1] Institute of Robotics, School of Mechanical Engineering, Shanghai Jiao Tong University, 800 Dongchuan Road, Shanghai 200240, People's Republic of China
{owenqyzhang,chenjiesjtu14,dr.dongwei}@gmail.com,
{xjsheng,mexyzhu}@sjtu.edu.cn
[2] Sinsun Co., Ltd, 351 Jinzang Rd., Shanghai 201206, People's Republic of China
yangluo@sinsun.com

Abstract. A three-layered sandwiched structure of quadcopter was proposed to lower the weight and rotary inertia, resulting in an increase in endurance time and payload in this present work. The framework was optimized with two carbon fiber layers on the surface and balsa in the middle. The weight was reduced to 148 g *via* the options of aluminum alloy, balsa and carbon fiber reinforced polymer (CFRP). Stress analysis shows that the stress and strain of this structure were within the safety range even when all four rotors are at maximum thrust with maximum payload which guarantees enough stiffness of the structure. A prototype controlled by an open source controller was used to run the tests. The flight tests indicated that endurance time was 29 min and the payload was 700 g, respectively.

Keywords: Quadcopter · Structure optimization · Sandwiched structure

1 Introduction

Research and development of unmanned aerial vehicle (UAV) and micro aerial vehicle (MAV) are getting high encouragement nowadays due to its broad potential applications in civilian and military areas [1]. Civilian applications of UAVs were initially considered for D3 (dirty, dull and dangerous) operations. For instance, the use of UAVs in radioactive contamination was documented after the Fukushima reactor damage [2, 3]. The use of UAVs for dull operations includes their use in frontier surveillance [4] and digital elevation model (DEM) creation [5]. UAV-based applications in dangerous situations include monitoring hurricanes and wildfire situations [6, 7]. Apart from the applications mentioned above, a number of remote sensing operations have tested the use of UAVs in the monitoring of wildlife, ice cover, weather phenomena, climate change, etc. [8]. The role of UAVs in daily life is predicted to grow in the next decade [9], which suggests the increased demand for the specialists in the fields of miniature UAV design and implementation. Scientific studies have been mainly concerned with the precursors of remote sensing flights using UAVs. They have shown the feasibility of UAVs and the advantages of using such platforms.

© Springer International Publishing Switzerland 2015
H. Liu et al. (Eds.): ICIRA 2015, Part III, LNAI 9246, pp. 220–229, 2015.
DOI: 10.1007/978-3-319-22873-0_20

Quadcopter (often also referred to as quadrotor) aircraft is one of the UAV that are major focuses of active researches in recent years. Design and implementation of a typical quadrotor system imposes some challenges, among of which are limited payload and flight time, fast and unstable dynamics, and tight integration of control electronics and sensors. However, the drift of inertial sensors leads to errors during time-discrete integration, making a steadily accurate estimation of the absolute pose nearly impossible. Recent development in quadcopters has made some achievements in control algorithm. In our previous work, control strategy based on the optimal control and subspace stabilization approach [10] and a flight controller with disturbance observer (DOB) [11] were developed to solve the two-point boundary value problem of a highly under-actuated quadrotor.

Although the control algorithms and sensors on quadcopters have made great progress, researches on structural design are rarely found in the existing literature. Structure of the quadcopter can be an essential factor in control model. Numerous aspects concerning the performance of the machine including endurance, maximum speed and acceleration are determined by the structural parameters of the frame.

In fact, the primary consideration is weight and rotary inertia. The quadcopter designs currently available in the market are almost identical. Most of them are made of plastic, aluminum alloy, and sometimes carbon fiber to lower the weight. But almost none of them managed to strike a balance between weight and strength.

The structure layout of a quadcopter is optimized with three layers based on the kinematics and dynamic models in this present work. To lower the weight and the rotary inertia of the framework, different materials are chosen, including carbon fiber reinforce polymer (CFRP) and aluminum alloy and a novel sandwiched structured material. Later the stress and strain of the optimized structure are analyzed, respectively. A prototype is built based on our design, which is controlled with a stable open source control board. The real time tests illustrate the flight time and the payload of the optimized structure.

2   Dynamic Model of the Quadcopters

Kinematics Model

The quadcopter structure is presented in Fig.1 including the corresponding angular velocities, torques and forces created by the four rotors.

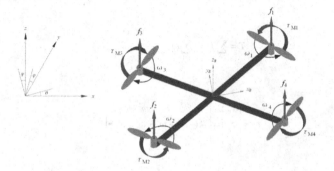

Fig. 1. The inertial and body frames of a quadcopter

The linear position of the quadcopter is defined in the inertial frame x, y, z-axis with a vector ξ consisting three linear coordinates. The attitude (angular position) is defined in the inertial frame with a vector η consisting three angles. Angle θ, φ, ψ determines the rotation of the quad around the x, y, z-axis respectively. Vector p contains linear position vector ξ and angular position vector η.

$$\xi = \begin{pmatrix} x \\ y \\ z \end{pmatrix}, \quad \eta = \begin{pmatrix} \theta \\ \varphi \\ \psi \end{pmatrix}, \quad p = \begin{pmatrix} \xi \\ \eta \end{pmatrix} \tag{1}$$

The origin of the quadcopter is in the mass center of the frame. The linear velocity and the angular velocity are defined by vector V and ω respectively.

$$V = \begin{pmatrix} v_x \\ v_y \\ v_z \end{pmatrix} = \begin{pmatrix} \dot{x} \\ \dot{y} \\ \dot{z} \end{pmatrix}, \quad \omega = \begin{pmatrix} p \\ q \\ r \end{pmatrix} \tag{2}$$

The rotation matrix between the body frame and the inertial frame is

$$R = \begin{pmatrix} \cos\psi\cos\varphi & \cos\psi\sin\varphi\sin\theta - \sin\psi\cos\theta & \cos\psi\sin\varphi\cos\theta + \sin\psi\sin\theta \\ \sin\psi\sin\varphi & \sin\psi\sin\varphi\sin\theta + \cos\psi\cos\theta & \sin\psi\sin\varphi\cos\theta - \cos\psi\sin\theta \\ -\sin\theta & \cos\varphi\sin\theta & \cos\varphi\cos\theta \end{pmatrix} \tag{3}$$

The angular velocity of rotor i creates force f_i in the direction of the rotor axis. The angular velocity and acceleration of the rotor also create torque τ_M around the rotor axis

$$f_i = k\omega_i^2, \quad \tau_{M_i} = b\omega_i^2 + I_M\dot{\omega}_i \tag{4}$$

In which k is the lift constant, b is the drag constant, and I_M is the inertia moment of the rotor. The effect of $\dot{\omega}_i$ is usually small and can be neglected.

The combined force of the four rotors create thrust T in the direction of the body z-axis. Torque τ_B consists of torque τ_θ, τ_φ and τ_ψ in the corresponding directions in the body frame.

$$T = \sum_{i=1}^{4} f_i = k\sum_{i=1}^{4} \omega_i^2, \quad T = \begin{pmatrix} 0 \\ 0 \\ T \end{pmatrix} \tag{5}$$

$$\tau_B = \begin{pmatrix} \tau_\theta \\ \tau_\varphi \\ \tau_\psi \end{pmatrix} = \begin{pmatrix} lk(\omega_1^2 - \omega_2^2 + \omega_3^2 - \omega_4^2) \\ lk(-\omega_1^2 + \omega_2^2 + \omega_3^2 - \omega_4^2) \\ \sum_{i=1}^{4} \tau_{M_i} \end{pmatrix} \tag{6}$$

In the equations above, l is the distance between the rotor and the center of the mass. Thus the attitude of the quadcopter can be controlled by controlling the angular velocity of the four rotors respectively.

Control Model

Controlling the quadcopter based on the physical model discussed above is one of the most important jobs in designing the system. The design of the control model must consider its responding time, reusability and interoperability.

In this present work, the quad rotor employs a Proportional-Integral-Derivative control system. The PID controller is a closed-loop feedback system which sends a control signal and receives a feedback from the inertial sensors. The controller then calculates the difference between the set position and attitude and adjusts the output accordingly.

To test the performance of our designed framework, we choose Pixhawk, an open source controller. The control model of Pixhawk is shown in Fig.2.

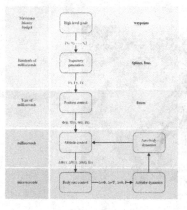

Fig. 2. Control architecture of Pixhawk. Different levels have different update and latency requirements [16].

3   Proposed Optimized Framework

The thrust T and the torque τ_B can be controlled by rotors. The above-mentioned model showed that the acceleration matrix $\ddot{\xi}$ is determined by the mass of the frame m, and the angular acceleration $\ddot{\eta}$ is determined by the rotary inertia I. To acquire greater agility, the mass and the rotary inertia of the frame have to be lowered.

A critical problem of quadcopter is its endurance. Usually, a quadcopter can fly about 20 min, and significantly less if they are carrying high payload. Thus, the lighter weight is, the longer battery life and the more payloads.

A typical quadrotor utilizes a four-spar method, with each spar anchored to the central hub. It consists of four rotors and electric speed controllers (ESCs), a control board, a battery, a radio transmitter and numerous sensors including GPS, compass, gyroscope, accelerometer, etc. The frame has to provide enough room for the equipment. The weight of the equipment attached cannot be lightened, so the only way to lower the rotary inertia is to move them as close to the center of mass as possible.

At present, mainstream quadrotors available in the market, such as DJI F450 and Hummingbird, have a single-layered popular frame. In the frame of DJI F450, all the equipment is jammed into a limited space. The battery is placed beneath the frame, which causes the frame not to stand on the ground without a high landing gear, resulting in the increase of the weight and rotary inertia. An overview of the DJI F450 is shown in Fig.3 (a).

Hummingbird from Ascending Technology, which is widely employed type for indoor flights of quadcopter, is made of carbon fiber and magnesium alloy. Although Hummingbird weighs only 500g, its maximum payload is only 200g, which means it can carry nothing other than the battery. In addition, it does not have a GPS module. That is to say, it is not able to locate itself outdoors. These limitations greatly confine its application. The overview of Hummingbird is shown in Fig.3 (b).

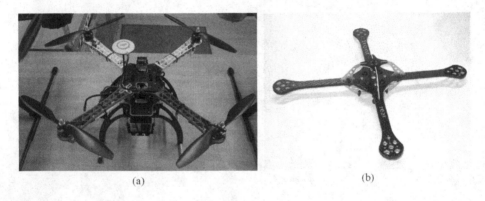

(a)   (b)

Fig. 3. Overview of DJI F450 and Hummingbird. (a) Actual physical picture of DJI F450 quadrotor, and (b) Actual physical picture of Hummingbird framework

Unlike them, our optimized design has three-layered sandwiched structure. In this sandwiched structure, the control board and most of the sensors can be placed on the middle layer. The top layer is designed for GPS, and the bottom layer is for battery. The battery can be locked with two carbon fiber panels to reduce oscillation. Compared to other conventional single-layered frames, the three-layered sandwiched structure makes it possible to spread the sensors to avoid electromagnetic interruption. The optimized design is also flattened so that it can access some narrower places. The overview of the optimized frame is shown in Fig. 4.

When designing an autonomous quad rotor, material options must be considered according to the durability, machinability, and price. And in this case, weight is the ultimate concern. The materials in consideration for our design include aluminum, plastic, and carbon fiber. The thrust of the rotors and the vibration requires high strength. To make sure the control model keeps working, the deformation has to be small enough to be neglected. Based on the two factors, carbon fiber and aluminum alloy are chosen for their high strength, low density and acceptable price.

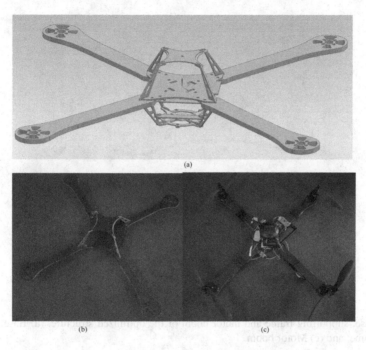

Fig. 4. Overview of our three-layered sandwiched structure. (a) The design model of framework; (b) Actual picture of our optimized framework; and (c) Actual picture of our quadrotor

The base is the center of the frame, which is made of carbon fiber reinforced polymer (CFRP). The control board is placed in the center of the base, with all the sensors around it and the GPS on the top. The base is large enough to avoid electromagnetic interference among the chips and onboard devices. This is essential because electric current can affect the accuracy of the sensors. To make the installation of the frame easier, several holes and slots are made on the base to allow wires pass through. Base of optimized design of the quadcopter is shown in Fig. 5 (a).

The supporting frame supports the panels on the top and bottom. To increase the strength of the frame with a relatively lower weight, the supporting frame is made of aluminum alloy. Several holes are designed to make the frame lighter. Supporting frame of optimized design of the quadcopter is illustrated in Fig. 5 (b).

The motor boom of the quadcopter is one of the most important parts of the frame. It has to stand great moment. Most of the quadcopters are using engineering plastics or carbon fiber to build the wings of the frame. The problem is that they are either too heavy or too crispy. In our design, the motor boom is made of composite sandwiched panel. The panel consists of two layers of carbon fiber on the surface and a wood layer in the center to increase the stiffness. The motor boom design is shown in Fig. 5 (c).

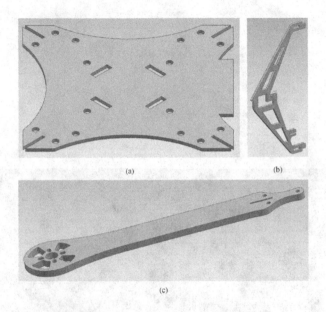

Fig. 5. Base, supporting frame and motor boom of the optimized structure. (a) Base; (b) Supporting frame; and (c) Motor boom

4   Analysis and Test

The mass of the frame weighs 148g. The comparison between our design and some mainstream frame designs is shown in Table 1. The weight of our frame was reduced by 45% to 51%, compared with the weight of some mainstream frame designs (with same wheelbase).

Table 1. Comparisons between our design and some mainstream frame designs (with same wheelbase)

Model	DJI F450	Q450 V3 Glass Fiber	Hobbyking SK450	Our Design
Weight (g)	282	270	300	148

Stress Analysis

Stress analysis of the optimized design using ANSYS is shown in Fig.6.

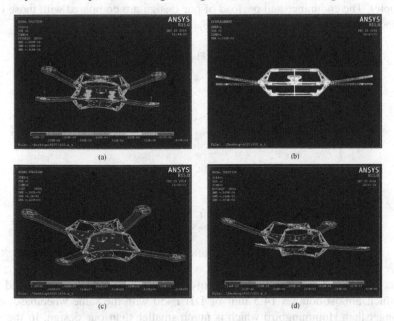

Fig. 6. Stress and strain analysis. (a) Abridged view of the simulation; (b) Deformation distribution; (c) Stress Distribution; and (d) Strain Distribution

With four rotors all at maximum thrust, the maximum deformation is small enough to be neglected. The maximum stress and strain are also within safe range.

The stiffness of the optimized framework is illustrated in Fig.7. The stiffness is approximately 74222N/m under the force of one rotor, and the stiffness is approximately 145608N/m under the forces of four rotors. When controlling a quadcopter, the forces are relatively small, and the stiffness is big enough to ensure the safety of the vehicle.

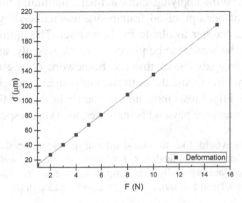

Fig. 7. The stiffness of the optimized framework

Endurance and Payload Test

The main goal of our optimization is to increase the endurance and payload of the quadcopter. The endurance and payload of our design are compared with those of DJI F450 and Hummingbird shown as in Fig. 8.

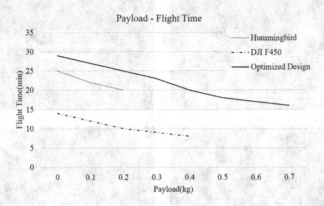

Fig. 8. Payload vs. flight time for different designs. (With same battery)

As is shown in Fig.8, the flight time of our optimized model is prolonged to 29 min, which almost doubles 14.5 min of DJI F450 with the same wheelbase. It also flies longer than Hummingbird which is much smaller than our design. In the viewpoint of payload, our optimized design can carry as much as 700 g, while the maximum payload of Hummingbird is only 200 g, and that of DJI F450 is not more than 400 g. At maximum payload, the optimized design can still fly longer than DJI F450 without any payload.

5   Conclusion

In summary, the design of a lightweight quadcopter is optimized by employing a sandwiched structure. With applying carbon fiber, aluminum alloy, and composite materials, the weight of our optimized framework was reduced by 45% to 51% compared with typical quadcopter available in the market. The optimized design allows the control board and the sensors to be placed more dispersedly, and the frame can be flattened. The stress analysis shows that the framework is safe even when all four rotors are at maximum thrust. And the deformation is small enough so that the control model can be utilized. Flight tests indicate that the flight time of the optimized model is prolonged to 29 min and the payload is increased to 700 g, respectively.

Acknowledgments. We would like to thank all our partners for their great cooperation concerning the vehicle design, the feedback for improvements as well as the experiments performed together. These are especially Wu Tong, Fang Jiahao, Ye Xin and Wu Junjie from Shanghai Jiaotong University and all others who have participated in the experiment.

References

1. Williamson, W.R., Abdel-Hafez, M.F., Rhee, I., Song, E.J., Dolfe, W.J., Chichka, D.F., Speyer, J.L.: An instrumentation system applied to formation flight. IEEE Trans. Control Syst. Technol. **15**(1), 75–85 (2007)
2. Ackerman, E.: Japan earthquake: Global Hawk UAV may be Able to Peek inside Damaged Reactors. http://spectrum.ieee.org/automaton/robotics/military-robots/global-hawk-uav-may-be-able-to-peek-inside-damaged-reactors (accessed on February 13, 2015)
3. Reavis, B., Hem, B.: Honeywell T-Hawk Aids Fukushima Daiichi Disaster Recovery: Unmanned Micro Air Vehicle Provides Video Feed to Remote Monitors. http://honeywell.com/News/Pages/Honeywell-T-Hawk-Aids-Fukushima-Daiichi-Disaster-Recovery.aspx (accessed on February 13, 2015)
4. Baker, R.E.: Combining micro technologies and unmanned systems to support public safety and homeland security. J. Civ. Eng. Archit. **6**, 1399–1404 (2012)
5. Turner, D., Lucieer, A., Watson, C.: An automated technique for generating georectified mosaics from ultra-high resolution unmanned aerial vehicle (UAV) imagery, based on structure from motion (SfM) point clouds. Remote Sens. **4**, 1392–1410 (2012)
6. Ambrosia, V., Buechel, S., Wegener, D., Sullivan, F., Enomoto, E., Zajkowski, T.: Unmanned airborne systems supporting disaster observations: Near-Real-Time data needs. Int. Soc. Photogramm. Remote Sens. **144**, 1–4 (2011)
7. Salamí, E., Pedre, S., Borensztejn, P., Barrado, C., Stoliar, A., Pastor, E.: Decision support system for hot spot detection. Intell. Environ. **2**, 277–284 (2009)
8. Watts, A.C., Ambrosia, V.G., Hinkley, E.A.: Unmanned aircraft systems in remote sensing and scientific research: Classification and considerations of use. Remot. Sens. **4**, 1671–1692 (2012)
9. Devalla, V., Prakash, O.: Developments in unmanned powered parachute aerial vehicle: A review. IEEE Aerospace and Electronic Systems Magazine **29**(11), 6–20 (2014)
10. Achtelik, M., Zhang, T., Kuhnlenz, K., Bus, M.: Visual tracking and control of a quadcopter using a stereo camera system and inertial sensors. In: International Conference on Mechatronics and Automation, ICMA 2009, pp. 2863–2869. IEEE (2009)
11. Dong, W., Gu, G.Y., Zhu, X.Y., Ding, H.: Solving the Boundary Value Problem of an Under-Actuated Quadrotor with Subspace Stabilization Approach. Journal of Intelligent & Robotic Systems (2014). doi:10.1007/s10846-014-0161-3
12. Dong, W., Gu, G.Y., Zhu, X.Y., Ding, H.: High-performance trajectory tracking control of a quadrotor with disturbance observer. Sensors and Actuators A: Physical **211**, 67–77 (2014)
13. Meier, L., Honegger, D., Pollefeys, M.: PX4: A Node-Based Multithreaded Open Source Robotics Framework for Deeply Embedded Platforms. http://people.inf.ethz.ch/lomeier/publications/px4_autopilot_icra2015.pdf (accessed on April 13, 2015)

Design of a Passive Gait-Based Lower-Extremity-Exoskeleton for Supporting Bodyweight

Kok-Meng Lee[1,2(✉)], Donghai Wang[2,3], and Jingjing Ji[1]

[1] State Key Laboratory of Digital Manufacturing Equipment and Technology,
Huazhong University of Science and Technology, Wuhan 430074, Hubei, China
[2] Woodruff School of Mechanical Engineering, Georgia Institute of Technology,
Atlanta, GA 30332, USA
kokmeng.lee@me.gatech.edu
[3] State Key Laboratory of Fluid Power Transmission and Control, Zhejiang University,
Hangzhou 310027, Zhejiang, China

Abstract. This paper presents the design of a bodyweight-supporting lower-extremity-exoskeleton (LEE) with compliant joints to relieve compressive load in human knees during walking. Based on experimental measurements that relate plantar forces with gait phase, the design of a gait-based LEE is divided into BW-supporting and free-swinging and realized by means of built-in compliant mechanisms in its exoskeleton-knees. Design considerations to accommodate human knee geometry and adapt walking gaits are highlighted. The snap-fit mechanisms for human gait-based operations are illustrated and analyzed numerically. The effects of several different exoskeleton-knee designs on reducing plantar force are experimentally compared validating the effectiveness and light-weight advantages of LEE in reducing plantar force in walking.

Keywords: Lower-Extremity-Exoskeleton · Body-support · Mechanical knee · Compliant mechanism

1 Introduction

People with knee osteoarthritis (OA) suffer severe knee pain [1]. The mechanism between OA and mechanical or psychosocial factors has been an important research topic; and knee load has been well recognized as a significant factor. Body mass index has a large effect on knee pain intensity for people with knee OA [2]. There is an obvious direct association between body mass and knee load [3]. From an in-vivo study [4], the rate of OA progression is associated with increased load during ambulation. However, knee joints are highly loaded during walking, stair descending, stair ascending, standing up, etc. [5] in daily life.

To help early prevention from OA and release pain which prohibits human walking, a possible solution is to use a well-designed lower-extremity-exoskeleton to support human bodyweight (BW) and accommodate any kinematic differences between the exoskeleton joint and human knee in walking. A good example is Honda BW support system [6] which uses two actuators to reduce ground reaction force.

© Springer International Publishing Switzerland 2015
H. Liu et al. (Eds.): ICIRA 2015, Part III, LNAI 9246, pp. 230–242, 2015.
DOI: 10.1007/978-3-319-22873-0_21

Exoskeletons are widely studied in mechatronics in recent years, and potentially have a broad spectrum of rehabilitation and human-motion assistance. Some examples include an adaptive wearable ankle robot for physical rehabilitation [7], and an active leg-exoskeleton for walking [8]. Although an exoskeleton can improve human athletic ability, there are possibilities for discomfort and injury risk if it does not match well with the wearer. A human knee has complicated kinematics with femur rolling and sliding on tibia. However, traditional rigid knee exoskeleton joints not adapting to this complicated character would tend to create residual forces if the design does not have sufficient degrees of freedom (DOFs) [9]. Thus, the coupling between engineering exoskeleton with natural knee must be compliant.

The ergonomic design for a lower-extremity-exoskeleton is important for human comfort. To connect an exoskeleton with soft human body, compliance must be incorporated in its joint design [10]. Compliant mechanism has been widely applied in engineering design such as snap-fits [11], flexure hinges [12] and compliant joints [13] [14]. It offers some advantages in interacting highly deformable biological body over rigid mechanisms in terms of simple structure, low cost and high performance.

This paper presents a BW-supporting lower-extremity-exoskeleton (LEE) to relieve knee compressive load. The remainder of this paper offers the following:

a) As a basis for developing the design concept, the plantar forces with gait phase are measured with an instrumented shoe. The results verify the needs to support human BW using LEE in some gait phases.
b) The principle and theoretical basis for designing a LEE to support human bodyweight during walking are discussed. Design considerations to adapt human knee geometry and walking gaits are highlighted. The methods to design built-in snap-fit mechanisms to adapt human walking gaits are detailed. The compliant mechanisms for the BW-supporting LEE are numerically analyzed.
c) The effects of several different exoskeleton-knee-joints on reducing plantar force during walking are experimentally compared.

2   Gait-Based Body-Support Exoskeleton

A lower-extremity-exoskeleton (LEE) based on human gait phases is designed to support bodyweight (BW) during walking. Unlike an engineering pin-joint that has a center-of-rotation, human joints not only rotate but also roll, slide and contain clearances between moving components. The design requirements include

1) The LEE have all six DOFs of the natural legs, and its operations are fail-safe.
2) The joints of the LEE must be compliant to accommodate bio-joint variations in the closed-chain mechanism formed by the human-leg and exoskeleton.
3) For safety and ease-of-operation, the LEE is operated without fastening to the user, capable of self-standing and requires no external actuation.
4) The LEE conserves energy created in some gait phases to compensate for the additional weight introduced by exoskeleton.

2.1   Experimental Gait Analysis

Normal human gaits generally follow a universal law [15] with plantar-force distribution depending on feet-ground contact condition when walking.   Fig. 1(a) illustrates a typical five sequential phases of a gait cycle:

IC (initial contact) occurs from swing to stance of the lower extremity. It begins when the heel just touches the ground.

MS (mid stance) is a transition moving the BW from the heel to forefoot.

During **TS** (terminal stance), the heel rises and hallux touches the ground as the body moves ahead of the forefoot.

PS (pre-swing) starts with initial contact of the opposite extremity, and ends with ipsilateral toe off.

During **SP** (swing phase), the foot lifts from the ground and the dorsiflexion angle θ_a increases preparing for the heel strike.

In the following discussion, the left leg will be used for illustration.

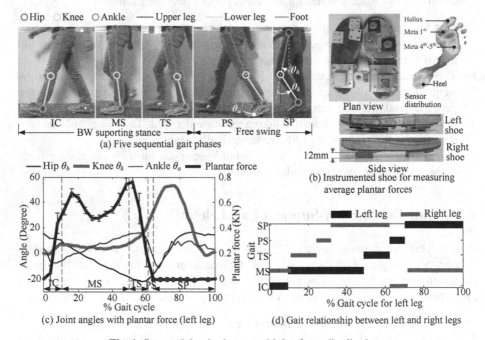

Fig. 1. Sequential gait phases and joint-force distribution

To design a LEE that adapts human gaits, an experiment was performed to measure average plantar forces using an instrumented shoe consisting of four force sensors on the shoe-sole as shown in Fig. 1(b). Four same-size succedaneums were similarly placed on the other shoe. For relating the plantar force to the joints (hip, knee and ankle), markers were placed on the subject (28 year-old male; 63.5kg; 1.69m) to locate the joint positions by a digital camera. For the following analysis, the joint angles in sagittal plane are defined in Fig. 1(a): The hip angle θ_h is measured from a vertical line; the knee angle θ_k of $0°$ corresponds to a fully extended knee; and the ankle θ_A is measured from the foot position at which the foot is $90°$ from the lower leg. Fig. 1(c) relates measured plantar forces (averages and standard deviations of 5 trials) to gait phases showing that human knees bear a large load pressure as the left and right knees alternatively transfer the human BW to the ground (Fig. 1d). Some physically-intuitive insights for design optimization of a LEE can be revealed from Fig. 1:

a) The knee is mostly in $\theta_k \approx 0°$ in the stance phases while the LEE supports the BW, and flexes in the swing phases. Mathematically, the BW (with the supporting knee fully extended) is rotated about the heel during stances which make up more than half of the gait cycle. Thus, the mechanical knee is designed such that it fully extends at the IC phase, maintains at $\theta_k \approx 0°$.
b) The LEE with its knee supports the BW throughout the MS phase, and is smoothly unlocked during TS to prepare for the swing phases.
c) At the end of stance, the body moves forward relative to the left leg. The hip angle θ_h becomes negative and continues to increase more negatively in PS (while the BW is supported by the other leg). As in Fig. 1(c), the measured plantar force is zero in SP during which the hip angle θ_h flexes from $-20°$ to $+20°$. Thus, the hip-joint is designed to conserve energy created by the forward-motion of the body during TS and convert it to useful work done for flexing the hip joint in SP.

Design and Operation of the BW-Supporting LEE
The joints for the six-DOFs of a typical human lower-extremity (anatomical figure from ZygoteBody, adult male) are illustrated in Fig. 2(a). The LEE designed to support human BW during walking is shown in Fig. 2(b). The knee and ankle joints of the LEE are designed to rest at $-3°$ and $5°$ respectively so that LEE leans forward by $8°$ taking advantages of the gravity for self-standing as shown in Fig. 2(c). When not in motion, the static LEE behaves like a "chair" facilitating the user to "ride" on it. Because LEE does not require external actuators and batteries, it has an obvious weight advantage (in the order of 1kg).

DOF Distribution of the Compliant Joints
The six DOFs of the leg are distributed among the three mechanical (hip, knee and ankle) joints as illustrated in Fig. 2(d~f). Unlike a 3DOF ball-joint-like human hip joint, the mechanical hip-joint has two consecutive pin-joints adapting the flexion/extension and abduction/adduction rotations. To prevent mechanical hip-joint to rotate the knee, the circumduction (secondary in human hip) is transferred to the mechanical ankle (where the lower link is rotatable about its axis). The torsion-spring in Fig. 2(d) harvests hip energy (negative φ_h) during TS/PS phases, and releases the stored energy (positive φ_h) lifting the leg in SP to prepare for IC of the next gait cycle. In IC and MS, the torsion spring is free as illustrated in Fig. 2(d). The harvesting torque τ_h is designed based on

$$\tau_h = \begin{cases} 0 & \text{if } \varphi_h \geq 0 \\ k_h \varphi_h & \text{if } \varphi_h < 0 \end{cases} \tag{1}$$

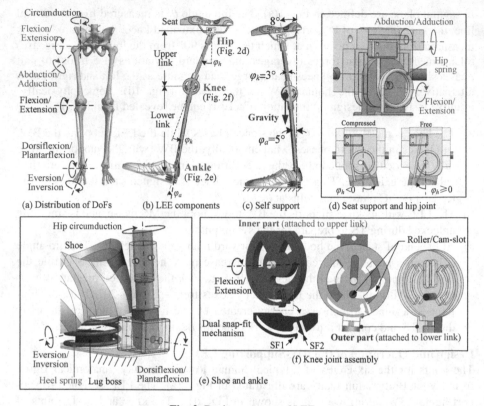

(a) Distribution of DoFs   (b) LEE components   (c) Self support   (d) Seat support and hip joint

(e) Shoe and ankle   (f) Knee joint assembly

Fig. 2. Design concept of LEE

Example: For harvesting 1 Joule of energy in TS/PS ($\varphi_k=-20°$) and release it in SP in each gait cycle, the hip-spring is designed to have stiffness $k_h = 2 / \varphi_k^2 = 0.286$ Nm/°.

As shown in Fig. 2(e), the mechanical ankle consists of a pin-joint to allow for dorsiflexion/plantarflexion, and a heel-spring to accommodate eversion/inversion (secondary in human ankle). The heel-spring increases the shoe heel-height in free-state and is compressed by the BW in stance phases. The heel-spring compression lowers the human body relative to the seat which transfers the BW to the exoskeleton.

A human knee has complicated kinematics, which not only flexes but extends with femur and also rolls/slides on tibia [9]. As shown in Fig. 2(f), the compliant knee consists of an inner part and outer part attached to the upper link and lower link respectively, where the scallop holes in the two parts are materials removed for lightweight design. The knee flexion/extension is designed to roll in the cam-slot to reduce negative effects of closed-chain knee-exoskeleton interaction on the human knee [9]. The compliant knee is mechanically operated in the range $\varphi_k \geq -3°$ by a dual snap-fit mechanism as shown in Fig. 2(f). The 1st snap-fit (SF1) maintains the knee in full extension ($-3° \leq \varphi_k < 0°$) in IC, and is unlocked by a trigger-spring in MS while the BW is supported and rotated forward about the supporting heel. The small negative φ range avoids singularity at 0° and maintains full knee extension in MS. The 2nd snap-fit (SF2) holds SF1 allowing the leg to flex in TS, PS and SP.

2.2 Bodyweight (BW) Supporting Mechanism

The BW is supported by a BW-support spring between the two links of the lower leg mechanism. As shown in Fig. 3(a), the BW-support spring is attached to an expansion screw for calibrating the length of the link to that of the human leg. Similar expansion screw is used to calibrate the upper link. During stance, the BW compresses the spring transmitting the force from the outer link to the inner link. The depression exposes the trigger spring to release the lower link from SF1 as shown Fig. 3(b).

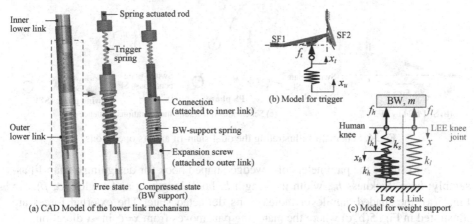

Fig. 3. Schematics illustrating the BW-supporting mechanism

The relationship between the human knee and mechanical knee is best illustrated by Fig. 3(c), where the human body and the upper links of LEE above the knees is modeled as mass m. As seen in Fig. 2(e), the human heel is supported by the heel-spring (soft stiffness k_s) and lug-boss (hard stiffness k_h) in parallel but the two springs are separated by an initial distance l_h in free-state. When the heel-spring and/or lug-boss are compressed, the BW is shared between human knee and the BW-support spring force f_l (Fig. 3c):

$$mg = f_h + f_l \text{ where } f_l = k_l x \text{ and } f_h = \begin{cases} k_s x & \text{if } x < l_h \\ k_s x + k_h(x - l_h) & \text{if } x \geq l_h \end{cases} \quad \text{(2a, b, c)}$$

For a specified percentage BW (W_s) to be supported by the LEE, the spring constant of the BW-support spring can be determined from (3):

$$k_l = W_s(mg)/l_h \quad (3)$$

Example: For LEE to support 20% of m=63.5kg; $k_l = 6.223$ N/mm for $l_h = 20$mm.

Design of Compliant Knee for BW-Supporting

The operational principle of the dual snap-fit compliant mechanism is illustrated in Fig. 4 where the switch-plate is attached to the lower link through the outer part of the mechanical knee. As shown in Fig. 4, the SF1 hook prevents the lower link from rotating anticlockwise in IC and is pushed into SF2 by the trigger spring in MS until the switch-plate releasing it in PS. At the end of SP, the switch-plate bends SF1 to return the knee in full extension to prepare for the next IC.

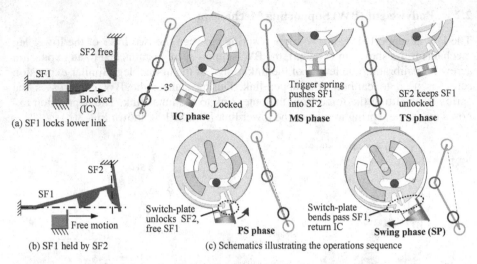

Fig. 4. Schematics illustrating the dual snap-fit and its operations

Fig. 5(a) shows the parameters of a wedge-shaped hook for designing a snap-fit assembly; base thickness h_o, width w_b, length l_t, hook length l_m and angles (α, β). In Fig. 5(a), the shaded cantilever indicates its deflected state. The external forces are illustrated in Fig. 5(b, c) where the matching part moves from $x=\ell_t$ in $-x$ direction.

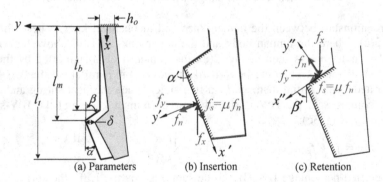

Fig. 5. Schematics illustrating the parameters and variables of a snap-fit

The general design procedure and governing equations for design a wedged hook for snap-fit assembly and disassembly can be found in [11]. The three snap-fit processes are insertion, dwelling and retention:

– During insertion (Fig. 5b), the matching part that advances to snap moves up along the front surface of the inclined wedge (against downward friction). The cantilever is deflected by the matching part with maximum deflection δ at $x=\ell_m$.
– The matching part slides on the tip while maintaining its maximum deflection during <u>dwelling</u>.

- The elastic deformation gradually returns to zero as the matching part moves down the inclined surface completing the <u>retention</u> process (Fig. 5c). Unlike insertion, the frictional force is upward during retention.

The forces can be calculated from Eq. (4) where μ is the friction coefficient between the two sliding surfaces:

$$f_y = \frac{3w_b h_o^3 \delta E}{4\ell_b^3 Q} \text{ and } \frac{f_x}{f_y} = \begin{cases} \dfrac{\mu + \tan\alpha'}{1 - \mu\tan\alpha'} & \text{for insertion} \\[2mm] \mu & \text{for dewelling in MS} \\[2mm] \dfrac{-\mu + \tan\beta'}{1 + \mu\tan\beta'} & \text{for retention in TS} \end{cases} \tag{4a,b}$$

where $\alpha' = \alpha + \tan^{-1}(\delta/\ell_m)$ for insertion; and $\beta' = \beta - \tan^{-1}(\delta/\ell_m)$ for retention. In (4a), E is the Young modulus of the material; and Q is an empirical coefficient for a specific snap-fit type (l_b/h_o ratio) [11].

3   Results and Discussion

Two sets of results are presented to illustrate the key components of the LEE: The *first* is a numerical analysis on the built-in dual-snap-fit mechanism for the gait-based exoskeleton-knee. The *second* is an experimental study comparing the effects of different knee designs on reducing plantar force.

3.1   Numerical Analysis of Built-in Snap-Fits for Gait-Based LEE

The design and operation of the built-in dual-snap-fit mechanism for the gait-based exoskeleton-knee are best illustrated numerically. Equations (4a, b) are utilized to calculate the forces required to insert or retract the matching part. The detailed deformed shapes and stresses induced in the materials of the dual snap-fit mechanism are analyzed by COMSOL (Commercial software) which also provides a basis for comparison. The parametric values used in this analysis are listed in Table 1, where l_b, l_m and l_t are the effective moment-arm lengths.

Table 1. Specification of the dual snap-fit mechanism

Snap fit	l_b (mm)	l_m (mm)	l_t (mm)	h_b (mm)	h (mm)	α (°)	β (°)	w_b (mm)	Q
SF1	36.8	36.8	39.8	2.5	2.5	44	95	7.62	2.10
SF2	8.82	10.0	14.6	2	2.5	30	70	7.62	2.65

Material (Delrin): Density $\rho=1.39\times10^3$ Kg/m^3; Young Modulus $E=2.62$ GPa; Poisson ratio $\upsilon=0.39$; Permissible stress $\sigma = 65.5$ MPa.

The results are summarized in Table 2 and Fig. 6. The 2nd subscript i (=1, 2) in *face$_{ri}$*, *face$_{di}$* and *face$_{ii}$* indicates that the surfaces for the retention, dwelling and insertion correspond to SF1 and SF2 respectively. Both SF1 and SF2 are free in IC and most of the initial SP. Simulation results from COMSOL are discussed as follows:

- In MS, maximum deflections occur when the trigger-spring pushes SF1 hook into SF2. For the designed deflections of 4.0mm and 0.45mm (perpendicular to $face_{d1}$ and $face_{d2}$), the corresponding root stresses are 18.2MPa and 29.8MPa respectively, which are well below the permissible stress of 65.5MPa. The force f (applied vertically at x_1=29.22mm) to snap SF1 into SF2 can be calculated from the moment equation $f_{y1} = -\left(f_{y2}x_2 + f_{x2}y_2 \right) / x_1$ where the force $\mathbf{f_2}$ (f_{x2}, f_{y2}) acting by SF1 on SF2 at the contact $\mathbf{x_2}$ between two hooks were determined numerically from the COMSOL; $\mathbf{f_2}$ $(-6, -5)$ N at $\mathbf{x_2}$ $(41.63, 16.34)$ mm. The actuating force from the trigger-spring is approximately 14.3N; using a safety factor of 2, k_t=1.91N/mm.
- In TS, SF1 is designed to store low strain energy so that it can be positively held by SF2 to prevent accidentally locking the lower leg for safe operation. In other words, SF1 is deflected with a low force f_d whereas SF2 must be released by a high force f_r. From COMSOL, f_d=2.7N is approximately an order lower than f_r=26N.
- In PS, SF1 is unlocked. This requires the switch-plate to bend SF2 anticlockwise with a force (f_a=10.6N at $face_{i2}$) to release SF1 from SF2.
- When transiting from SP to the next IC, the switch-plate bends SF1 clockwise with an "insertion" force (f_c=5.5N perpendicular to $face_{i1}$) to fully extend the knee.

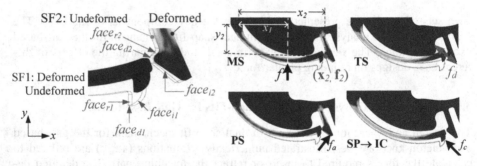

Fig. 6. Numerical results simulated by COMSOL

Table 2. Summary for the simulated results

Gait Phase	Design requirements	Results
MS (SF1, SF2)	Specified deflection δ (4.0, 0.45)mm:	Maximum stress σ_{max} (18.2, 29.8) MPa
	Required force f = 14.3N:	Trigger spring stiffness k_t : 1.91 N/mm
TS (SF1, SF2)	$f_d \ll f_r$, SF1 is positively held by SF2:	(f_d=2.7N, f_r = 26N)
PS	Release SF1 from SF2:	f_a =10.6N at $face_{i2}$
SP→IC	Fully extend the knee in transition:	f_c=5.5N perpendicular to $face_{i1}$

3.2   Experimental Study of Mechanical Knee Design

Fig. 7 shows three knee designs (adaptive, multi-centric hinge and compliant) for analyzing the kinematic effects on reducing the plantar force:

- The adaptive knee [10] (Fig. 7a) uses a roller/cam-slot mechanism to approximate human knee motion [9]. The cam profile that relates the incremental distance to the flexion angle φ is given by

$$\Delta C_a = 0.648\varphi^4 - 6.708\varphi^3 + 15.924\varphi^2 - 0.495\varphi \tag{5}$$

- The multi-centric hinge (Fig. 7b) is a commercial knee brace (BLEDSOE BRACE SYSTEMS) using a dual pin/cam-slot mechanism to accommodate human knee motion. The effective cam profile for the multi-centric hinge is given by

$$\Delta C_m = 0.029\varphi^4 - 5.860\varphi^3 + 17.355\varphi^2 - 1.392\varphi \tag{6}$$

- The compliant knee [14] (Fig. 7c) that uses a pair of deformable rings to adapt for the human knee kinematics was designed to support human BW during MS and TS. Unlike an adaptive knee or a multi-centric knee where an effective rotation center follows a specific cam profile, the effective rotation-center of a complaint knee moves over a region. The two bounds ($\pm f_{\mathrm{LEE}}$) were determined experimentally:

$$\Delta C_c = \pm\left(1.947\varphi^4 - 4.041\varphi^3 + 10.638\varphi^2 - 3.857\varphi\right) \tag{7}$$

The cam profiles of adaptive joint and multi-centric hinge, which are nearly identical, are compared against that of the compliant knee in Fig. 8(a). The compliant knee kinematics is similar to the cam-based designs in low φ ($<10°$) and high φ ($>85°$). Some differences occur in flexion ($10°<\varphi<85°$) due to conservative stiffness design for supporting BW and to avoid compromising the rigidity in stance phases.

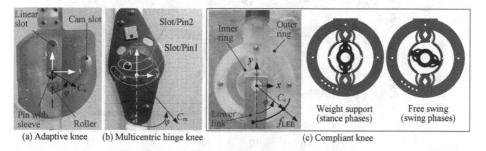

(a) Adaptive knee   (b) Multicentric hinge knee   (c) Compliant knee

Fig. 7. Different knee-joint exoskeleton designs

Experiments were performed on the same subject as for Fig. 1. The mean and standard deviation values (over five trials) of the measured plantar forces for the LEE with multi-centric hinge knee) and compliant knee are plotted in Fig. 8(b), where results are compared against those without LEE. The comparisons confirm that the LEEs with multi-centric hinge and compliant knees reduce mean plantar forces in MS by (21.9%, 20.4%) respectively, validating their effectiveness in supporting the human BW. When using a multi-centric hinge knee, human percentile gait does not change significantly between with and without the LEE. However, delays in the two peaks can be seen in the plantar-force plot; from (16%, 52%) without LEE to (12%, 46%) for the design where compliant knees are used in LEE.

Table 3 compares the LEE (adaptive knee) design with three published BW-supporting exoskeletons; Honda BW-supporting system, Moon-walker, LEE (complaint knee). While all four designs effectively reduce plantar force, the passive LEE offers several advantages (light-weight and better mobility) because no external actuator and battery are needed. With built-in compliance, the ergonomic comfort is improved for the safety of human body. Unlike active exoskeletons, the reduced percentage depends largely on the spring stiffness specification for the passive LEE.

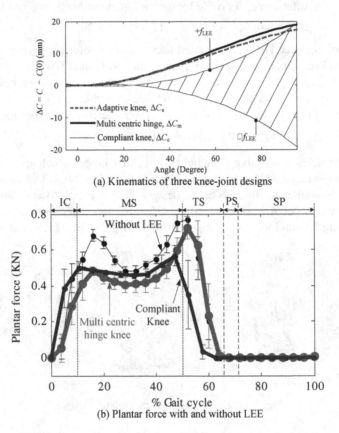

(a) Kinematics of three knee-joint designs

(b) Plantar force with and without LEE

Fig. 8. Comparison among three different knee-joint designs

Table 3. Performance index for four LEEs

Parameters	Honda [6]	Moon Walker [16]	LEE gait-based knee	
			Compliant [14]	Adaptive [10]
Actuator	Active	Quasi-active	Passive	
Power	Battery	External power	no external power is needed	
Mass (Kg)	6.5	Heavy (NR)	2.357	1.14
Mechanism	Rigid	Compliant	Compliant	Compliant
Mean reduced plantar force (%)	≈ 20	33~50	20.4	20

4   Conclusions

Based on an experimental study on measured plantar force, a gait-based lower-extremity-exoskeleton (LEE) to accommodate biological joint geometry/motion and support human bodyweight during walking has been designed and analyzed. Using a dual snap-fit compliant mechanism built-in the exoskeleton-knees, the gait-based LEE effectively decouples BW-supporting and free-swinging phases. Results are presented to illustrate the key components of the LEE. Experimental results compare three published designs validating the effectiveness of LEE in reducing plantar force.

Acknowledgements. This work was supported in part by the National Basic Research Program of China (973 Program, Grant No. 2013CB035803), China Postdoctoral Science Foundation (Grant No. 2014M562012), Guangdong Innovative Research Team Program (No. 2011G006), and China Scholarship Council.

Abbreviation

LEE	lower-extremity-exoskeleton	IC	initial contact	SP	swing phase
OA	osteoarthritis	MS	mid stance	SF	snap-fit
BW	bodyweight	TS	terminal stance		
DOF	degree of freedom	PS	pre-swing		

References

1. Maly, M.R., Costigan, P.A., Olney, S.J.: Determinants of self-report mobility outcome measures in people with knee osteoarthritis. Arch. Phys. Med. Rehabil. **87**(1), 96–104 (2006)
2. Maly, M.R., Costigan, P.A., Olney, S.J.: Mechanical factors relate to pain in knee osteoarthritis. Clin. Biomech. **23**(6), 796–805 (2008)
3. Messier, S.P., Gutekunst, D.J., Davis, C., DeVita, P.: Weight loss reduces knee-joint loads in overweight and obese older adults with knee osteoarthritis. Arthritis & Rheumatism **52**(7), 2026–2032 (2005)
4. Andriacchi, T.P., Mündermann, A.: The role of ambulatory mechanics in the initiation and progression of knee osteoarthritis. Current Opinion in Rheumatology **18**(5), 514–518 (2006)
5. Kutzner, I., Heinlein, B., Graichen, F., Bender, A., Rohlmann, A., Halder, A., Beier, A., Bergmann, G.: Loading of the knee joint during activities of daily living measured in vivo in five subjects. J. Biomechanics **43**(11), 2164–2173 (2010)
6. Ikeuchi, Y., Ashihara, J., Hiki, Y., Kudoh, H., Noda, T.: Walking assist device with body-weight support system. In: Proc. IEEE/RSJ Int. Conf. Intell. Robots Syst., Louis, USA, pp. 4073–4079 (2009)
7. Jamwal, P.K., Xie, S.Q., Hussain, S., Parsons, J.G.: An adaptive wearable parallel robot for the treatment of ankle injuries. IEEE/ASME Trans. Mechatronics **19**(1), 64–75 (2014)
8. Banala, S.K., Agrawal, S.K., Kim, S.H., Scholz, J.P.: Novel gait adaptation and neuromotor training results using an active leg exoskeleton. IEEE/ASME Trans. Mechatronics **15**(2), 216–225 (2010)

9. Lee, K.-M., Guo, J.: Kinematic and dynamic analysis of an anatomically based knee joint. J. Biomech. **43**(7), 1231–1236 (2010)
10. Wang, D., Lee, K.-M., Guo, J., Yang, C.-J.: Adaptive knee joint exoskeleton based on biological geometries. IEEE/ASME Trans. Mechatronics **19**(4), 1268–1278 (2014)
11. Ji, J., Lee, K.-M., Zhang, S.: Cantilever snap-fit performance analysis for haptic evaluation. ASME J. Mech. Design **133**(12), 121004-1–121004-8 (2011)
12. Yi, B.-J., Chung, G.B., Na, H.Y., Kim, W.K., Suh, I.H.: Design and experiment of a 3-dof parallel micromechanism utilizing flexure hinges. IEEE Trans. Robot. and Autom. **19**(4), 604–612 (2003)
13. Ji, J., Lee, K.-M., Guo, J., Zhang, S.: Discrete deformation models for real-time computation of compliant mechanisms in two and three dimensional space. IEEE/ASME Trans. Mechatronics **19**(5), 1268–1278 (2014)
14. Lee, K.-M., Wang, D.: Design analysis of a passive weight-support lower-extremity-exoskeleton with compliant knee-joint. In: Proc. IEEE Int. Conf. Robot. Autom., Settle, USA (2015) (accepted)
15. Perry, J.: Gait analysis: normal and pathological function. Slack Incorporated, Thorofare (1992)
16. Krut, S., Benoit, M., Dombre, E., Pierrot, F.: Moonwalker, a lower limb exoskeleton able to sustain bodyweight using a passive force balancer. In: Proc. IEEE Int. Conf. Robot. Autom., Alaska, USA, pp. 2215–2220 (2010)

The Effect of Asymmetrical Body-Mass Distribution on the Stability and Dynamics of Quadruped Bounding with Articulated Spine

Hua Nie, Ronglei Sun[(⊠)], and Caihua Xiong

State Key Laboratory of Digital Manufacturing Equipment and Technology,
Huazhong University of Science and Technology, Wuhan 430074, China
D201177095@hust.edu.cn

Abstract. This paper examines the dynamics and stability of quadruped bounding with asymmetrical mass distribution between front and rear body. A sagittal-plane model with an asymmetrical articulated torso and two compliant legs is introduced to capture the dynamics of robot bounding. Numerical return map studies of the system in dimensionless setting reveal that the speed of this model increases when the ratio of front body-mass to rear body-mass decreases. Next, the stability properties of this model are investigated. The results indicate that the stability of robot doesn't vary with the decreased ratio, but it will decrease with the increased ratio. In conclusion, we conclude that appropriate decreasing the ratio of front body-mass to rear body-mass can enhance the locomotion performance of quadruped bounding with articulated spine without scarfing robot's stability.

Keywords: Asymmetry · Bounding · Dynamics · Quadruped · Stability

1 Introduction

Bionic legged robots have been rapidly developed for decades because they can adapt to much rougher land than other mobile robots, for example the grassland, the snowfield, the muddy road, etc. As we all know, legged robot is a multi-joints series-parallel institution, which is very complicated to build its complete dynamic model and precise control system. The well-known Spring Loaded Inverted Pendulum (SLIP) model for legged locomotion is proposed from biomechanical, as it's very effective in the description of dynamic behaviors of animals of many kinds. Raibert [1] proposed a control method of dynamical legged locomotion based on this model which is considered to be the most effective and the simplest. His machines including one-leg, two-leg and four-leg designed by his team could be seen as great extension of this idea in robotics. The success of the SLIP model in simplifying structure of the robot inspired many researchers' interest. In order to accurately predict and control the SLIP model, a series of approximate solution were proposed to analyze the existence of self-stable running orbits of the SLIP model by many researchers (such as Koditschek [2], Schwind [3], Geyer [4] and Arslan [5]). Although the SLIP model

© Springer International Publishing Switzerland 2015
H. Liu et al. (Eds.): ICIRA 2015, Part III, LNAI 9246, pp. 243–253, 2015.
DOI: 10.1007/978-3-319-22873-0_22

can capture the basic properties of the animal, it ignores the pitching oscillation of the torso which is one of the dominant modes of animal.

As was mentioned above, in order to overcome the deficiency of the SLIP model, the planar quadruped model with rigid torso and two massless legs were proposed for analyzing the stability of the system where torso pitching was the main factor to influence locomotion performance of the robot. Mur [6] phy studied this model and found that the system could run passively without any control when a dimensionless moment of inertia of the robot is less than unity. Li [7] and Berkemeier [8] utilized mathematic analysis to prove and enrich Murphy's results. Furthermore, Poulakakis [9] found that the two variations of the bounding gait which had been experimentally observed on Scout II, could be passively generated with appropriate initial conditions based on the analysis of numerically derived return maps. Studies above are based on the torso whose mass is symmetric, but animal body is not uniformly distributed. Zou [10] found that asymmetrical body-mass distribution had a great influence on the stability and dynamics of quadruped bounding. He concluded that the dimensionless moment of inertia of the robot was less than $1 - \beta^2$, where β was a dimensionless measure of the symmetry. However, the torso of animal is not rigid, which becomes more flexible when speed increases.

Most recently, more attention were drawn to Cheetah robot, such as Boston Dynamics Cheetah robot, MIT Cheetah robot [11], ALSP Cheetah robot [12], because of its ability to achieve high-speed running. The primary difference between the Cheetah robot model and previous quadruped robots is the flexible torso. Biology studies found that stride frequency was one way to increase the robot's speed, but it worked mainly at a low-speed. At high-speed, the robot system relied on the extension the limbs and flexible torso. Khoramshahi [13] found that active spine supported actuation led to faster locomotion, with less foot sliding on the ground, and a higher stability to go straight forward through large number of experiments with a small compliant quadruped robot. To investigate the role of flexible torso during the robot's flexion and extension, many researchers regard it as a sagittal-plane model with articulated spine which composed of front body, rear body and elastic articulated spine. Through analyzing this model's passive dynamics, Qi [14] presented the motion of spine in passive quadrupedal bounding and its effect on leg behavior. Cao [15] further analyzed the inherent stability of this model and concluded that the combination of the stiffness properties of the legs and the torso was the dominant factor to influence the self-stability of the robot. However, animal structure parameters between front body and rear body are not uniform judging from animal skeleton. To the best of the authors' knowledge, the mass of front body and rear body were all equal in previous research. So far, the model of asymmetrical front and rear body is still lack of research.

In this study, we will investigate the effect of asymmetrical front and rear body-mass distribution on the stability and dynamics of quadruped bounding with articulated spine and propose optimal design method of the corresponding model. This paper is structured as follows. Section 2 presents quadruped bounding model with articulated spine and asymmetrical front body and rear body. In section 3, we will research the bounding model's locomotion performance and inherent stability when asymmetrical front body and rear body's mass varies, and put forward the body's optimal design method. Discussions and Results are summarized in section 4 and 5.

2   Quadruped Bounding Model with Articulated Spine

2.1   Model Description

In this paper, a planer model (Fig 1) is built based on the model of Poulakakis et al., where the mass of the front body and rear body is not equal. It's well known that legged robots belong in the category of hybrid systems. A full stride of the dynamic model, shown in Figure 2, can be divided in four or five different phases in a complete bounding cycle. Depending on the state of a leg which is stance or flight, the four phases are: the back leg stance, double leg stance, front leg stance, double leg flight. Based on the robot's running speed, there are two different bounding cycle. At low speed, the robot state transfers from front leg stance to double leg stance and then to back leg stance (the dotted line in Figure 2). When the speed increased, double leg stance phase will gradually disappear. The bounding cycle of the robot will transfer from front leg stance to double leg flight and then to back leg stance (the solid line in Figure 2).

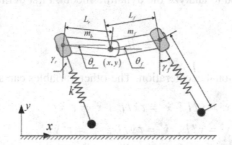

Fig. 1. Sagittal quadruped model with articulated spine

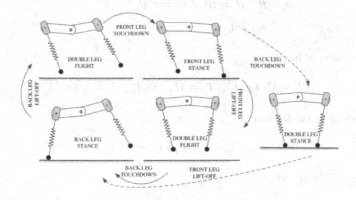

Fig. 2. Bounding phases and events

2.2   Dynamics in Non-dimensional Form

To derive a theoretical mathematical model, the following two simplifying hypotheses were taken into consideration: first, the mass of the legs are negligible with respect to the entire mass of the body. Second is that both ends of the leg, one in contact with the ground, the other attached to the body, are treated as a frictionless pin joint. Based on the above simplification, the equations of motion are obtained using the Lagrangian approach and could be stated in the unified form.

$$\dot{x} = \frac{d}{dt}[_q^q] = [_{-M(q)(F(q)+G(q))}^{\dot{q}}] = f(x) \tag{1}$$

where $q = [x, y, \theta]^T$ is Cartesian coordinates shown in Figure 4, x is state-space $[q, \dot{q}]^T$, M represents the mass matrix, and F and G stands for the vectors of the elastic and the gravitational forces, respectively. It is important to note that the equations of motion depend on the following nine physical parameters $L_r, L_f, m_r, m_f, I_r, I_f, l_0, k_{troso}, k_{leg}$. To simplify the physical parameters needed to be analyzed, and make subsequent analysis become more general, dimensional analysis will be applied to analyze the dynamic model. First define characteristic time scale τ as

$$\tau = \sqrt{l_0 / g} \tag{2}$$

where g is the gravitational acceleration. The other variables can show as

$$x^* = x/l_0, \ \dot{x}^* = \tau\dot{x}/l_0, \ \ddot{x}^* = \tau^2\ddot{x}/l_0$$
$$y^* = y/l_0, \ \dot{y}^* = \tau\dot{y}/l_0, \ \ddot{y}^* = \tau^2\ddot{y}/l_0$$
$$\theta_r^* = \theta_r, \ \dot{\theta}_r^* = \tau\dot{\theta}_r, \ \ddot{\theta}_r^* = \tau^2\ddot{\theta}_r \tag{3}$$
$$\theta_f^* = \theta_f, \ \dot{\theta}_f^* = \tau\dot{\theta}_f, \ \ddot{\theta}_f^* = \tau^2\ddot{\theta}_f$$
$$q_m = m_f / m_r,$$

Relative moment of inertia

$$I_r^* = 4I_r / m_r L_r^2, \ I_f^* = 4I_f / m_f L_f^2, \ I_r = I_f \tag{4}$$

Relative hip-to-COM distance:

$$L_r^* = L_r / l_0, \ L_f^* = L_f / l_0, \ L_r = L_f \tag{5}$$

Relative leg stiffness and torso stiffness

$$k_r^* = k_r l_0 / m_r g, \ k_f^* = k_f l_0 / m_b g, \ k_{torso}^* = k_{torso} / m_b g l_0 \tag{6}$$

where the superscript '*' denotes a dimensionless quantity, the subscript n indicates the stride number, td denotes touchdown, f and r denotes front and rear respectively. Additionally, Froude number is a dimensionless number, which enables size independent comparison of animals and robots in terms of speed and is used to characterize the forward speed.

$$Fr = v / \sqrt{gl} \tag{7}$$

where v is forward velocity, g is acceleration of gravity and l is the hip joint height of the standing robot. By substituting (3)-(6) into (1), the continuous-time dynamics of the system become

$$\dot{x}^* = f^*(x^*) \tag{8}$$

where x^* are the dimensionless counterparts of the states. After dimensionless analysis, the system reduces nine parameters to five dimensionless parameters $q_m, L_f^*, I_f^*, k_{troso}^*, k_{leg}^*$. As previously analyzed, the combination of the stiffness properties of the legs and the torso were the dominant factor to decide self-stability of the robot in the flexible-torso case. Here, our model's basic parameters are chosen from the study of Cao. Tables 1 show the variables and parameters of the model. So far q_m has not yet been determined. This paper is aim to optimize q_m through dynamics and stability of the robot.

Table 1. Non-dimensional physical parameters

Parameter	Value
Relative moment of inertia (I_f^*)	2.0121
Relative hip-to-COM distance(L_f^*)	0.7667
Relative leg spring stiffness(k_{leg}^*)	23.6
Relative torso spring stiffness(k_{troso}^*)	4

2.3   Poincaré Bound Gait and Fixed Points

In a periodic bound gait shown in Figure 2, the dynamics of the bounding gait can be built by concatenating the continuous-time phases. Referring to Altendorfer et al and Poulakakis's research, the method of Poincaré map could be used to study the existence and stability properties of periodic motions. We define Poincaré map P as mapping x_n^* to x_{n+1}^*, which is given by

$$x_{n+1}^* = P(x_n^*, u_n) \tag{9}$$

where x_n^* is the apex height states of the model, $u_n = [\lambda_f \ \lambda_r]$. If a point x comes rear to the same point after a full stride cycle, this point is called fixed point. The existence of this important fixed point enforced the passive motion in cycles so as to reach the most energy efficient status. The goal of this section is to search fixed points corresponding to the steady state cyclic bounding motions of the dynamic model. This is equivalent to solving the equation as follow

$$x_n^* - P(x_n^*, u_n) = 0 \qquad (10)$$

To simplify the calculations, the apex $[y, \theta_r, \theta_f, \dot{x}, \dot{\theta}_r, \dot{\theta}_f]^T$ in the double leg flight phase is considered as initial point in a full stride cycle because of reducing one-order dimension. At the same time, fixed points solved in this paper are at the same energy state. Initial state variables change to $[y, E, \theta_f, \dot{x}, \dot{\theta}_r, \dot{\theta}_f]^T$. Since the energy coordinate E can be mapped to itself, the form of fix point reduce order to $[y, \theta_f, \dot{x}, \dot{\theta}_r, \dot{\theta}_f]^T$. The search for fixed points is conducted numerically using Newton-Raphson method. Differential equations of motion are solved by using MATLAB's ode45 function with 1e-6 and 1e-7relative and absolute tolerances respectively. All fixed points for E=7.2 are presented in Figure 3(1) in which horizontal axis denotes touchdown angle, vertical axis denotes torso oscillation $\theta_f - \theta_r$ and color bar denotes velocity. The results indicate that increased speed results in an increase on touchdown angle and torso oscillation. Figure 3(2) shows energy distribution of each fix point at E=7.2. For all fixed points at one energy state, we can conclude that rotational kinetic and vertical kinetic are negligible compared to other energy, gravitational potential remain basically unchanged, forward kinetic and torso spring potential transform each other when touchdown angle vary from the figure 3.

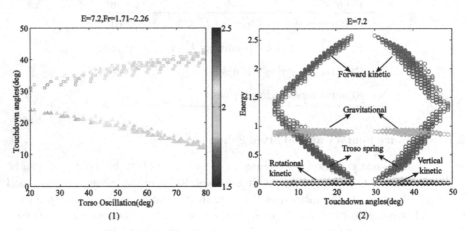

Fig. 3. Bounding fixed points and energy distribution

3   Body Parameters Optimization Analysis

As we all known, asymmetrical body-mass distribution had a great influence on the stability and dynamics of quadruped bounding with rigid spine. Moreover, q_m needs to be determined with the dimensionless analysis. Motivated by these two points, the effect of asymmetrical body-mass distribution (q_m) on locomotion performance of quadruped bounding with articulated spine will be researched in this section.

3.1   Locomotion Performance Optimization

Faster speeds and higher altitudes indicate better locomotion performance. Our aim is to improve the velocity and height of all fixed points through optimizing q_m. Without loss of generality, different energy states are chosen for calculation. For every energy state, range of speed and height of all fixed points are recorded. Here, any two energy states with E=7.9 and E=4.5 are used for analyzing. Bounding fixed points for q_m varying from 0.6 to 1.1 at E=7.9 and E=4.5 are presented in Fig. 4.

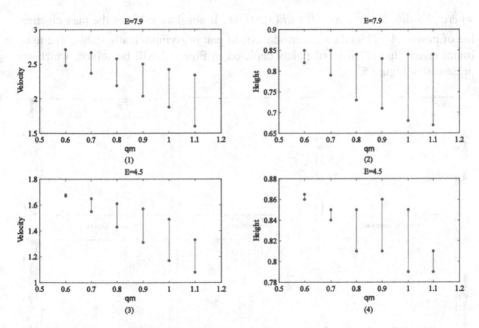

Fig. 4. Range of speed and height are obtained under q_m varying from 0.6 to 1.1

It can be seen from Figure 4(1,3) that the velocity change exists certain rules along with the change of q_m. When q_m becomes larger, the speed of the robot decreases. In contrast, when q_m becomes smaller, the speed of the robot increases. The velocity of fixed points increase 20% at q_m = 0.7 or 0.8, compared with q_m = 1. If q is further reduced, velocity of the robot grows slowly, and the number of fixed points sharp decline. Therefore, possible decreasing q_m resulted in an increase in robot's speed. Then, Figure 4(2,4) show that range of height present small fluctuations. The simulation results are well consistent with the energy distribution analysis in section 2.3 that gravitational potential remain basically unchanged under any cases. As height of the fixed points is essentially the same, range of height will be not considered later.

3.2   Stability Analysis

Previous section tell us that decreasing q_m is beneficial to enhance the locomotion performance of the robot. However, it's not advisable if this design sacrifices the stability of the robot. Stability will be researched with parameter optimization. The existence of fixed points shows that the robot can run passively without any control. However, in real situations robot is continuously disturbed, if a fixed point is unstable, the robot can't maintain its periodic motion without any control. The maximum eigenvalue of Jacobian matrix of Poincaré map is used to estimate the stability of a periodic bound gait. Δx denotes perturbation from x, as follow.

$$\Delta x_{n+1} = A\Delta x_n + B\Delta u_n \qquad (11)$$

where $A = \partial P(x,u)/\partial x$ and $B = \partial P(x,u)/\partial u$. If absolute value of the max eigenvalue of matrix A is less than 1, periodic bound gait is asymptotically stable. The maximum eigenvalue of all fixed points depicted in Figure 3 will be solved, which are presented in Figure 5.

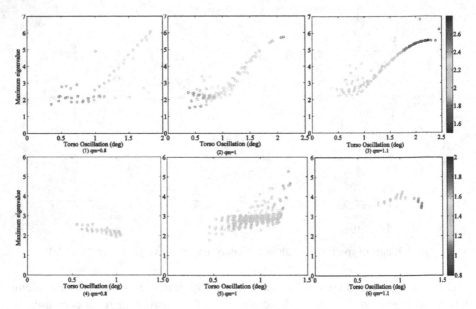

Fig. 5. The maximum eigenvalue of all fixed points with varying q_m and E=7.9 and 4.5

It can be seen from Figure 5 that the maximum eigenvalue of fixed point decreases with decreasing torso oscillation when torso oscillation larger than 1, and the maximum eigenvalue of fixed point remains unchanged when torso oscillation less than 1. Meanwhile the red fixed points become more with decreasing q_m. This is to say that higher speeds of the fixed points which own to decreased torso oscillation and q_m correspond to better inherent stability. The inherent stability of fixed point at the highest speed doesn't become better when q_m decreases, as the minimum eigenvalue in Figure 5(1) and (2) are basically the same. But the inherent stability of fixed point at the highest speed reduces when q_m increases, since the minimum eigenvalue with $q_m =1.1$ is bigger than the minimum eigenvalue with $q_m =1$ or 0.8.

Taken together, the stability of the robot doesn't vary with decreased q_m, but it will reduce with increased q_m. In consequence, reasonable decreasing q_m is a feasible design method.

4   Discussion

The main work of our study is to explore the effect of asymmetrical front and rear body-mass distribution on the stability and dynamics of quadruped bounding with articulated spine. In the bounding model with rigid torso, many researchers developed an analytical stability criterion through accurate mathematical derivation under the assumption of small pitch angle. However, it's very difficult to use analytical method to study the dynamics and stability of the bounding model with articulated spine, as model became further complicated and pitching angles of front and rear body couldn't be ignored compared with rigid torso model. Therefore, q_m described in this paper was optimized through solving range of velocity and height of the fixed points.

In order to make the conclusions of this article more general, dimensional analysis was adapted to build dynamic model of the robot. Through massive computations of the system's Poincaré map, a number of fixed points with varying q_m were found at different energy states. The simulation results showed that reasonable decreasing q_m could enhance locomotion performance of the robot. On the contrary the robot's speed decreased with increased q_m. We also concluded that robot's forward kinetic and torso spring potential transform each other with varying touchdown angle at the same energy state. Hence, torso oscillation of the model reduces with decreased q_m, resulting an increase in the model's speed.

As one of the most important properties, self-stability is entitled to be the prominent criterion for the design of a cheetah robot which enables its smooth motion in variously complex surface. Cao, who was the first to analyze the stability of bounding model with articulated spine, concluded that the combination of the stiffness properties of the legs and the torso was the dominant factor to influence the self-stability of the robot. However, the front and rear body parameters of the bounding model adopt by Cao were the same. This inspired us to study stability of the bounding model with asymmetrical front and rear body. We concluded that0020the robot's stability doesn't vary with decreased q_m, but it would reduce with increased q_m.

5   Conclusions

In conclusion, reasonable decreasing the ratio of front body-mass to rear body-mass of quadruped bounding model with articulated spine can increase robot's speed with without sacrificing stability. In the near future, we will develop a cheetah robot using the optimization method proposed in this article, and study the control method of the cheetah robot with asymmetrical torso structure.

6   Funding

This work is supported by the National Basic Research Program of China (grant 2013CB035805) and the Graduates' Innovation Fund of Huazhong University of Science & Technology (grant 01-09-070092).

References

1. Raibert, M.H.: Trotting, pacing and bounding by a quadruped robot. Journal of Biomechanics **23**, 79–98 (1990)
2. Full, R.J., Koditschek, D.E.: Templates and anchors: neuromechanical hypotheses of legged locomotion on land. Journal of Experimental Biology **202**, 3325–3332 (1999)
3. Schwind, W.J., Koditschek, D.E.: Approximating the stance map of a 2-DOF monoped runner. Journal of Nonlinear Science **10**, 533–568 (2000)
4. Geyer, H., Seyfarth, A., Blickhan, R.: Spring-mass running: simple approximate solution and application to gait stability. Journal of theoretical biology **232**, 315–328 (2005)
5. Saranlı, U., Arslan, Ö., Ankaralı, M.M., Morgül, Ö.: Approximate analytic solutions to non-symmetric stance trajectories of the passive spring-loaded inverted pendulum with damping. Nonlinear Dynamics **62**, 729–742 (2010)
6. Raibert, M.H., Brown Jr., H.B., Chepponis, M., Koechling, J., Hodgins, J.K., Dustman, D., et al.: Dynamically Stable Legged Locomotion (September 1985-Septembers1989) (1989)
7. Neishtadt, A., Li, Z.: Stability proof of Raibert's four-legged hopper in bounding gait. New York State University, New York (1991)
8. Berkemeier, M.D.: Modeling the dynamics of quadrupedal running. The International Journal of Robotics Research **17**, 971–985 (1998)
9. Poulakakis, I., Papadopoulos, E., Buehler, M.: On the stability of the passive dynamics of quadrupedal running with a bounding gait. The International Journal of Robotics Research **25**, 669–687 (2006)
10. Zou, H., Schmiedeler, J.P.: The effect of asymmetrical body-mass distribution on the stability and dynamics of quadruped bounding. IEEE Transactions on Robotics **22**, 711–723 (2006)
11. Hyun, D.J., Seok, S., Lee, J., Kim, S.: High speed trot-running: Implementation of a hierarchical controller using proprioceptive impedance control on the MIT Cheetah. The International Journal of Robotics Research **33**, 1417–1445 (2014)

12. Spröwitz, A., Tuleu, A., Vespignani, M., Ajallooeian, M., Badri, E., Ijspeert, A.J.: Towards dynamic trot gait locomotion: Design, control, and experiments with Cheetah-cub, a compliant quadruped robot. The International Journal of Robotics Research **32**, 932–950 (2013)
13. Khoramshahi, M., Sprowitz, A., Tuleu, A., Ahmadabadi, M.N. Ijspeert, A.: Benefits of an active spine supported bounding locomotion with a small compliant quadruped robot. In: Proceedings of 2013 IEEE International Conference on Robotics and Automation (2013)
14. Deng, Q., Wang, S., Xu, W., Mo, J., Liang, Q.: Quasi passive bounding of a quadruped model with articulated spine. Mechanism and Machine Theory **52**, 232–242 (2012)
15. Cao, Q., Poulakakis, I.: Quadrupedal bounding with a segmented flexible torso: passive stability and feedback control. Bioinspiration & Biomimetics **8**, 046007 (2013)

Modeling and Control of a Wheelchair Considering Center of Mass Lateral Displacements

Víctor H. Andaluz[1,2(✉)], Paúl Canseco[2], José Varela[2], Jessica S. Ortiz[3],
María G. Pérez[4], Vicente Morales[2], Flavio Roberti[5], and Ricardo Carelli[5]

[1] Universida de Las Fuerzas Armadas ESPE, Sangolquí, Ecuador
vhandaluz1@espe.edu.ec
[2] Universida Técnica de Ambato, Ambato, Ecuador
jvmorales99@uta.edu.ec
[3] Universida Tecnológica Indoamérica, Ambato, Ecuador
jsortiz@uti.edu.ec
[4] Escuela Politécnica Nacional, Quito, Ecuador
maria.perez@epn.edu.ec
[5] Universida Nacional de San Juan, San Juan, Argentina
{rcarelli,froberti}@inaut.unsj.edu.ar

Abstract. This work presents the kinematic and dynamic modeling of a human-wheelchair system where it is considered that its mass center is not located at the wheels' axis center of the wheelchair. Furthermore, it is presents a new motion controller for human-wheelchair system that is capable of performing positioning and path-following tasks. The proposed controller has the advantage of simultaneously performing the approximation of the robot to the proposed path by the shortest route and limiting its velocity. This controller design is based on two cascaded subsystems: a kinematic controller with command saturation, and a dynamic controller that compensates the dynamics of the robot.

Keywords: Wheelchair · Dynamic modeling · Cascade control · Lyapunov's method

1 Introduction

In recent years, robotics research has experienced a significant change. The research interests are moving from the development of robots for structured industrial environments to the development of autonomous mobile robots operating in unstructured and natural environments [1-3,7,8,9]. The integration of robotic issues into the medical field has become of great interest in recent years. Service, assistance, rehabilitation and surgery are the more benefited human health-care areas by the recent advances in robotics. Specifically, autonomous and safe navigation of wheelchairs inside known and unknown environments is one of the important goals in assistance robotics [1-7,16].

A robotic wheelchair can be used to allow people with both lower and upper extremity impairments or severe motor dysfunctions overcome the difficulties in driving a wheelchair. The robotic wheelchair system integrates a sensory subsystem, a navigation and control module and a user-machine interface to guide the wheelchair in

© Springer International Publishing Switzerland 2015
H. Liu et al. (Eds.): ICIRA 2015, Part III, LNAI 9246, pp. 254–270, 2015.
DOI: 10.1007/978-3-319-22873-0_23

autonomous or semi-autonomous mode [4-6]. In autonomous mode, the robotic wheelchair goes to the chosen destination without any participation of the user in the control. This mode is intended for people who have great difficulties to guide the wheelchair. In the semi-autonomous mode the user shares the control with the robotic wheelchair. In this case only some motor skills are needed from the user.

Hence, a trajectory will be automatically generated and a trajectory tracking control will guide the wheelchair to the desired target. As indicated, the fundamental problems of motion control of wheelchair robots can be roughly classified in three groups [10-11]: 1) *point stabilization*: the goal is to stabilize the wheelchair at a given target point, with a desired orientation; 2) *trajectory tracking*: the wheelchair is required to track a time parameterized reference; and 3) *path following*: the wheelchair is required to converge to a path and follow it, without any time specifications; this work is focused to resolve the path following problem.

The path following problem has been well studied and many solutions have been proposed and applied in a wide range of applications. Let $\mathcal{P}_d(s) \in \Re^2$ be a desired geometric path parameterized by the curvilinear abscissa $s \in \Re$. In the literature is common to find different control algorithms for path following where is conseder $s(t)$ as an additional control input [12-15]. Furthermore, it is important to consider the wheelchair's dynamics in addition to its kinematics because wheelchairs carry relatively heavy loads. As an example, the trajectory tracking task can be severely affected by the change imposed to the wheelchair dynamics when it is carrying a person, as shown in [16]. Hence, some path following control architectures already proposed in the literature have considered the dynamics of the wheelchair robots [17,18].

In such context, is important to indicate that the wheelchair's center of gravity changes due to postural issues, limb amputations, or obesity [19]. Therefore, in the present work a dynamic model of the human-wheelchair system is developed considering lateral deviations of the center of mass originated in user's movement, limb amputations, or obesity. Useful properties of the model are proved in the paper. The obtained dynamic model has an adequate structure and properties for designing a control law and the control input can be given in terms of linear and angular reference velocities, as usually found in commercial mobile robots. This latter characteristic is an advantage when evaluating the control experimentally [17]; in the literature is common to find that they use wheel torques as control signals, which cannot be directly manipulated in most of the real vehicles. Furthermore, in this work it is proposed a new method to solve the path following problem and positioning for a wheelchair robot to assist persons with severe motor diseases. The proposed control scheme is divided into two subsystems, each one being a controller itself: i) the first on is a kinematic controller with saturation of velocity commands, which is based on the wheelchair robot's kinematic. The path following problem is addressed in this subsystem. It is worth noting that the proposed controller does not consider $s(t)$ as an additional control input as it is frequent in literature; and ii) an dynamic compensation controller that considered the human-wheelchair system dynamic model, which are directly related to physical parameters of the system. In addition, both stability and robustness properties to parametric uncertainties in the dynamic model are proven through Lyapunov's method. To validate the proposed control algorithm, experimental results are included and discussed.

The paper is organized as follows: Section II shown the build of the wheelchair and kinematic and dynamic modeling of the human-wheelchair system, while Section 3 describes the path following's formulation problem and it also presents the controllers design. Furthermore, the analysis of the system's stability is developed. Next, experimental results are presented and discussed in Section 4, and finally the conclusions are given in Section 5.

2   Human – Wheelchair System

In this work presents a wheelchair robot which was developed at the Technical University Ambato (see Fig. 1). The wheelchair has two independently driven wheels by two direct current motors (in the center part), and four caster wheel around the central axis conferring greater stability to the human-wheelchair system (two in the rear part and two in the front part). Encoders installed on each one of the motor shafts allow knowing the relative position and orientation of the wheelchair.

Fig. 1. Autonomous wheelchair robotic

Information provided by the encoders is used by the PID controllers responsible for getting an independent velocity control of the left and the right wheel.

The hardware architecture of the robotic wheelchair consists of: i) a commercial powered wheelchair from which only the mechanical structure and motors are used; the power card and joystick were discarded; ii) two encoders directly connected to the motors; iii) a microcontroller where the low-level velocity controller is implemented; iv) a power card that amplifies PWM signals obtained from the microcontroller and sends them to the motors; v) a computer where thehigh-level control algorithms and signal processing of the human–machine interface are implemented. The high-level controller is implemented under the Windows operating system using Microsoft Visual C++. The block diagram of the low-level velocity controller is shown in Fig. 2.

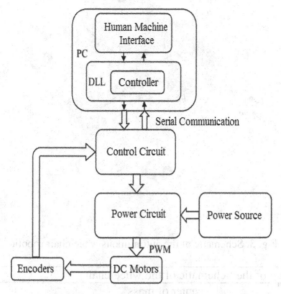

Fig. 2. Mechatronics architecture of the wheelchair robot

The wheelchair robot used in this work presents similar characteristics to that of a unicycle-like mobile robot, because it has two driven wheels which are controlled independently by two direct current motors and four caster wheel to maintain balance, while the unicycle-type mobile robots have a caster wheel to maintain stability. The kinematic and dynamic modeling of the human-wheelchair system is developed in the next subsection, considering a horizontal work plane where the wheelchair moves. The wheelchair type unicycle-like mobile robot presents the advantages of high mobility, high traction with pneumatic tires, and a simple wheel configuration.

2.1 Kinematic and Dynamic Modeling

Based on what was written in previous paragraphs, this work is based on unicycle-like wheelchair. A unicycle wheelchair is a driving robot that can rotate freely around its axis. The term unicycle is often used in robotics to mean a generalized cart or car moving in a two-dimensional world; these are also often called unicycle-like or unicycle-type vehicles.

It is assumed that the human-wheelchair system moves on a planar horizontal surface. Let $\mathcal{R}(X, Y, Z)$ be any fixed frame with Z vertical. Traditionally, in the motion control of wheelchair robots, the wheelchair is considered as a point located at the middle of the virtual axle. However, in this work, the point that should follow a predetermined trajectory is located in front of the virtual axle (point $h(x, y)$ of Fig. 3). Such point is herein after named as the point of interest. Fig.3 illustrates the wheelchair considering in this work.

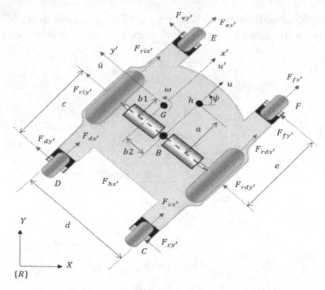

Fig. 3. Schematic of the autonomous wheelchair robotic

Nomenclature of the Schematic of the wheelchair	
G	center of mass
B	the center axis of the two wheels
$h(x, y)$	point that is required to track a path in $\mathcal{R}$
x, y	position in the axes of the wheelchair
ψ	orientation of the wheelchair
ω	angular velocity
$u', \bar{u}$	longitudinal and lateral velocities of the center of mass
$F_{rrx'}, F_{rry'}$	longitudinal and lateral tire forces of the right wheel
$F_{rlx'}, F_{rly'}$	longitudinal and lateral tire forces of the left wheel
$F_{hx'}, F_{hy'}$	longitudinal and lateral force exerted on H by the human
τ_h	moment exerted by the human
$F_{cx'}, F_{cy'}, F_{dx'}, F_{dy'}$	longitudinal and lateral tire forces exerted on C and D by the rear castor wheels
$F_{ex'}, F_{ey'}, F_{fx'}, F_{fy'}$	longitudinal and lateral tire forces exerted on E and F by the front castor wheels
d, b_1, b_2, a, c	Distances

The force and moment equations for the mobile robot are:

$$\sum F_{x'} = m\left(\dot{u}' - \bar{u}\omega\right) = F_{rlx'} + F_{rrx'} + F_{cx'} + F_{dx'} + F_{ex'} + F_{fx'} \tag{1}$$

$$\sum F_{y'} = m\left(\dot{\bar{u}} - u'\omega\right) = F_{rly'} + F_{rry'} + F_{cy'} + F_{dy'} + F_{ey'} + F_{fy'} \tag{2}$$

$$\sum M_{z} = I_{z}\dot{\omega} = \tfrac{d}{2}\left(F_{rrx'} - F_{rlx'}\right) + b_{2}\left(F_{rrx'} + F_{rlx'}\right) - b_{1}\left(F_{rly'} + F_{rry'}\right) -$$
$$- \left(c - b_{1}\right)\left(F_{dy'} + F_{cy'}\right) + \left(e - b_{1}\right)\left(F_{fy'} + F_{ey'}\right) + \left(\tfrac{d}{2} + b_{2}\right)\left(F_{cx'} + \tag{3}$$
$$+ F_{fx'}\right) - \left(\tfrac{d}{2} - b_{2}\right)\left(F_{dx'} + F_{ex'}\right)$$

where $m = m_{h} + m_{w}$ is the human-wheelchair system mass in which m_{h} is the human mass and m_{w} is the wheelchair mass; and I_{z} is the human-wheelchair system moment of inertia about the vertical axis located in G.

According to [19], velocities u, ω and $\bar{u}$, including the slip speeds, are given by:

$$u = \tfrac{r}{2}\left(\omega_{r} + \omega_{l}\right) \; ; \; \omega = \tfrac{r}{d}\left(\omega_{r} - \omega_{l}\right) \tag{4}$$

$$\bar{u} = b_{1}\omega \tag{5}$$

where, r is the right and left wheel radius; d is the distance between wheels; ω_{r} and ω_{l} are the angular velocities of the right and left wheels, respectively.

It is considered that motors of the wheelchair are identical, thus, the motor models attained by neglecting the voltage on the inductances are:

$$\tau_{r} = \frac{k_{a}\left(v_{r} - k_{b}\omega_{r}\right)}{R_{a}} \; ; \; \tau_{l} = \frac{k_{a}\left(v_{l} - k_{b}\omega_{l}\right)}{R_{a}} \tag{6}$$

where v_{r} and v_{l} are the input voltages applied to the right and left motors; k_{b} is equal to the voltage constant multiplied by the gear ratio; R_{a} is the electric resistance constant; τ_{r} and τ_{l} are the right and left motor torques multiplied by the gear ratio; and k_{a} is the torque constant multiplied by the gear ratio. The dynamic equations of the motor-wheels are:

$$I_{e}\dot{\omega}_{r} + B_{e}\omega_{r} = \tau_{r} - F_{rrx'}r \tag{7}$$

$$I_{e}\dot{\omega}_{l} + B_{e}\omega_{l} = \tau_{l} - F_{rlx'}r \tag{8}$$

where, I_{e} and B_{e} are the moment of inertia and the viscous friction coefficient of the combined motor rotor, gearbox, and wheel.

In general, most market-available robots have low level PID velocity controllers to track input reference velocities and do not allow the motor voltage to be driven directly. Therefore, it is useful to express the mobile robot model in a suitable way by considering rotational and translational reference velocities as input signals. For this purpose, the velocity controllers are included into the model. To simplify the model, a

PD velocity controller has been considered which is described by the following equations:

$$v_u = k_{PT}\left(u_{ref} - u_{me}\right) + k_{DT}\left(\dot{u}_{ref} - \dot{u}_{me}\right) \tag{9}$$

$$v_\omega = k_{PR}\left(\omega_{ref} - \omega_{me}\right) + k_{DR}\left(\dot{\omega}_{ref} - \dot{\omega}_{me}\right) \tag{10}$$

where, k_{PT}, k_{DT}, k_{PR} and k_{DR} are gain positive constants of the PD controllers.

From (1 – 10) the following dynamic model of the mobile robot is obtained:

2.1.1   Kinematic Model

The configuration instantaneous kinematic model of the holonomic wheelchairis is defined as,

$$\begin{cases} \dot{x} = u\cos\psi - a\omega\sin\psi \\ \dot{y} = u\sin\psi + a\omega\cos\psi \\ \dot{\psi} = \omega \end{cases} \tag{11}$$

also the equation system (11) can be written in compact form as

$$\dot{\mathbf{h}} = \mathbf{J}(\psi)\mathbf{v}$$
$$\dot{\psi} = \omega \tag{12}$$

where $\dot{\mathbf{h}} = [\dot{x} \quad \dot{y}]^T \in \mathfrak{R}^2$ represents the vector of axis velocity of the $\mathcal{R}(X,Y,Z)$ system; $\mathbf{J}(\psi) = \begin{bmatrix} \cos\psi & a\sin\psi \\ \sin\psi & a\cos\psi \end{bmatrix} \in \mathfrak{R}^{2x2}$ is a singular matrix; and the control of maneuverability of the wheelchair is defined $\mathbf{v} \in \mathfrak{R}^n$ and $\mathbf{v} = [u \quad \omega]^T \in \mathfrak{R}^2$ in which u and ω represent the linear and angular velocities of the wheelchair, respectively.

On the other side, of (11) is determined the non-holonomic velocity constraint of the wheelchair robotic which determines that it can only move perpendicular to the wheels axis,

$$\dot{x}\sin\psi - \dot{y}\cos\psi + a\omega = 0 \tag{13}$$

2.1.2   Dynamic Model

Without including disturbances can by defined as,

$$u_{ref} = \varsigma_1\dot{u} - \varsigma_2\dot{\omega} + \varsigma_3 u - \varsigma_4\omega^2 \tag{14}$$

$$\omega_{ref} = -\varsigma_5\dot{u} + \varsigma_6\dot{\omega} + \varsigma_7\omega + \varsigma_8 u\omega \tag{15}$$

where the dynamic parameters of the model of the human-wheelchair system are defined as,

$$\varsigma_1 = \frac{\frac{2l_e}{r} + \frac{2k_a}{R_a}k_{DT} + rm}{2\frac{k_a}{R_a}k_{PT}} ; \qquad \varsigma_2 = \frac{rmb_2}{2\frac{k_a}{R_a}k_{PT}} ;$$

$$\varsigma_3 = \frac{\frac{2B_e}{r} + \frac{k_a}{rR_a}k_b + 2\frac{k_a}{R_a}k_{PT}}{2\frac{k_a}{R_a}k_{PT}} ; \qquad \varsigma_4 = \frac{rmb_1}{2\frac{k_a}{R_a}k_{PT}} ;$$

$$\varsigma_5 = \frac{\frac{rmb_2}{d}}{\frac{k_a k_{PR}}{R_a}} ; \quad \varsigma_6 = \frac{\frac{dl_e}{r} + 2\frac{k_a k_{DR}}{R_a} + 2r\frac{m(b_2^2+b_1^2)+I_z}{d}}{2\frac{k_a k_{PR}}{R_a}} ; \varsigma_7 = \frac{\frac{dB_e}{r} + \frac{dk_a k_b}{2rR_a} + \frac{2k_a k_{PR}}{R_a}}{2\frac{k_a k_{PR}}{R_a}} ; \quad \text{and} \quad \varsigma_8 = \frac{\frac{rmb_1}{d}}{\frac{k_a k_{PR}}{R_a}} .$$

Now, the dynamic equations (14) and (15) can be represented as follows,

$$\mathbf{M}(\varsigma)\dot{\mathbf{v}} + \mathbf{C}(\varsigma,\mathbf{v})\mathbf{v} = \mathbf{v}_{\text{ref}} \tag{16}$$

where, $\mathbf{M}(\varsigma) \in \mathfrak{R}^{n \times n}$ with $n = 2$ and $\mathbf{M}(\varsigma) = \begin{bmatrix} \varsigma_1 & -\varsigma_7 \\ -\varsigma_8 & \varsigma_2 \end{bmatrix}$ represents the human-wheelchair system's inertia; $\mathbf{C}(\varsigma,\mathbf{v}) \in \mathfrak{R}^{n \times n}$ and $\mathbf{C}(\varsigma,\mathbf{v}) = \begin{bmatrix} \varsigma_4 & -\varsigma_3 \omega \\ \varsigma_5 \omega & \varsigma_6 \end{bmatrix}$ represents the components of the centripetal forces; $\mathbf{v} \in \mathfrak{R}^n$ and $\mathbf{v} = \begin{bmatrix} u & \omega \end{bmatrix}^T$ is the vector of system's velocity; $\mathbf{v}_{\text{ref}} \in \mathfrak{R}^n$ and $\mathbf{v}_{\text{ref}} = \begin{bmatrix} u_{\text{ref}} & \omega_{\text{ref}} \end{bmatrix}^T$ is the vector of velocity control signals for the wheelchair; and $\varsigma \in \mathfrak{R}^l$ with $l = 8$ and $\varsigma = \begin{bmatrix} \varsigma_1 & \varsigma_2 & \cdots & \varsigma_l \end{bmatrix}^T$ is the vector of sistem's dynamic parameters. Hence the properties for the dynamic model with reference velocities as control signals as:

Property 1. Matrix $\mathbf{M}(\varsigma)$ is positive definite, additionally it is known that $\|\mathbf{M}(\varsigma)\| < k_M$, where, k_M is a positive constant;

Property 2. Furthermore, the following inequalities are also satisfied $\|\mathbf{C}(\varsigma,\mathbf{v})\mathbf{v}\| < k_C \|\mathbf{v}\|$, where, k_C denote a positive constants.

Property 3. The dynamic model of the mobile manipulator can be represented by $\mathbf{M}(\varsigma)\dot{\mathbf{v}} + \mathbf{C}(\varsigma,\mathbf{v})\mathbf{v} = \Omega(\mathbf{v})\varsigma$, where, $\Omega(\mathbf{v}) \in \mathfrak{R}^{n \times l}$ and $\varsigma = \begin{bmatrix} \varsigma_1 & \varsigma_2 & \cdots & \varsigma_l \end{bmatrix}^T \in \mathfrak{R}^l$ is the vector of l unknown parameters of the human-wheelchair system, *i.e.*, mass of the human, mass of the wheelchair, physical parameters of the wheelchair, motors, velocity, and others.

Hence, the full mathematical model of the human-wheelchair system is represented by: (12) the kinematic model and (16) the dynamic model, taking the reference velocities of the robot as input signals.

3   Path Following and Positioning Controller

As represented in Fig. 4, the path to be followed is denoted as $\mathcal{P}(s)$, where $\mathcal{P}(s) = \left(x_{\mathcal{P}}(s), y_{\mathcal{P}}(s) \right)$; the actual desired location $P_d = \left(x_{\mathcal{P}}(s_D), y_{\mathcal{P}}(s_D) \right)$ is defined as the closest point on $\mathcal{P}(s)$ to the human-wheelchair system, with s_D being the curvilinear abscissa defining the point P_d; the unit vector tangent to the path in the point P_d is denoted by $\mathbf{T}$; θ_T is the orientation of $\mathbf{T}$ with respect to the inertial frame $\mathcal{R}(X, Y, Z)$; $\tilde{x} = x_{\mathcal{P}}(s_D) - x$ is the position error in the X direction; $\tilde{y} = y_{\mathcal{P}}(s_D) - y$ is the position error in the Y direction; ρ represents the distance between the wheelchair position $h(x, y)$ and the desired point P_d, where the position error in the ρ direction is $\tilde{\rho} = 0 - \rho = -\rho$, i.e., the desired distance between the wheelchair position $h(x, y)$ and the desired point P_d must be zero; and θ_ρ is the orientation of the error $\tilde{\rho}$ with respect to the inertial frame $\mathcal{R}(X, Y, Z)$.

According to Fig. 2, the path-following problem is solved by a control law capable of making the point of interest to assume a desired velocity equal to

$$V = \boldsymbol{v}_P(s_D, h) = |\boldsymbol{v}_P(s_D, h)| \angle \theta_T \tag{17}$$

besides making the robot to stay on the path, that is, $\tilde{x} = 0$ and $\tilde{y} = 0$. Therefore, if $\lim_{t \to \infty} \tilde{x}(t) = 0$ and $\lim_{t \to \infty} \tilde{y}(t) = 0$ then $\lim_{t \to \infty} \rho(t) = 0$ and $\lim_{t \to \infty} \tilde{\psi}(t) = 0$, being $\tilde{\psi}$ the orientation error of the wheelchair, defined as $\tilde{\psi} = \theta_T - \psi$.

Worth noting that the reference desired velocity $\boldsymbol{v}_P(s_D, h)$ of the wheelchair during the tracking path need not be constant, with is common in the literature [1,3,13-16],

$$\boldsymbol{v}_P(s_D, h) = f\left(k, s_D, \rho(t), \omega(t), \dots \right) \tag{18}$$

the wheelchair's desired velocity can be expressed as: constant function, curvilinear abscissa function of the path, position error function, angular velocities function of the wheelchair; and the others consideration.

Remark 1. Notice that the positioning problem is a particular case of path-following, for which $\boldsymbol{v}_P(s_D, h) = 0$.

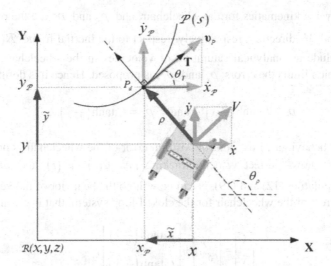

Fig. 4. The orthogonal projection of the point of interest over the path

3.1 Structure of the Controller

The structure of the controller here proposed to the path following and positioning problems, shown in Fig. 5, the design of the controller is based mainly on two cascaded subsystems: 1) *Kinematic controller* where the control errors $\rho(t)$ and $\tilde{\psi}(t)$ may be calculated at every measurement time and used to drive the mobile robot in a direction which decreases the errors; and 2) *Dynamic compensation controller*, which main objective is to compensate the dynamics of the human-wheelchair system, thus reducing the velocity tracking error.

3.1.1 Kinematic Controller

The proposed kinematic controller is based on the kinematic model of the wheelchair (12), i.e., $\dot{\mathbf{h}} = f(\psi)\mathbf{v}$. Hence following control law is proposed,

$$\begin{bmatrix} u_c \\ \omega_c \end{bmatrix} = \mathbf{J}^{-1}\left(\begin{bmatrix} \dot{x}_p \\ \dot{y}_p \end{bmatrix} + \begin{bmatrix} \rho_x \\ \rho_y \end{bmatrix} \right) \tag{19}$$

with

$$\dot{x}_p = |v_P|\cos(\theta_T) \quad \text{and} \quad \dot{y}_p = |v_P|\sin(\theta_T) \tag{20}$$

where u_c and ω_c are the velocities outputs of the kinematic controller, v_P is the reference velocity input of the wheelchair for the controller, $\dot{x}_R$ is the projection of v_P in the X direction, $\dot{y}_P$ is the projection of v_P in the Y direction, $\mathbf{J}^{-1}$ is the matrix of inverse kinematics for the wheelchair, and ρ_x and ρ_y are the position error in the X and Y direction, respectively, respect to the inertial frame $\mathcal{R}(X,Y,Z)$, In order to include an analytical saturation of velocities in the wheelchair, the **tanh(.)** function, which limits the errors ρ_x and ρ_y, is proposed. Hence it is defined as,

$$\rho_x = l_x \tanh\left(\tfrac{k_x}{l_x}\tilde{x}\right) \quad \text{and} \quad \rho_y = l_y \tanh\left(\tfrac{k_y}{l_y}\tilde{y}\right). \tag{21}$$

Now, the behaviour of the control position error of the wheelchair is now analysed assuming -by now- perfect velocity tracking *i.e.,* $u(t) \equiv u_c(t)$ and $\omega(t) \equiv \omega_c(t)$. Hence manipulating (12) and (19), is can be written the behavior of the velocity of the point of interest of the wheelchair for the closed-loop system, that is given by

$$\begin{bmatrix} \dot{x} \\ \dot{y} \end{bmatrix} = \begin{bmatrix} \dot{x}_P \\ \dot{y}_P \end{bmatrix} + \begin{bmatrix} l_x \tanh\left(\tfrac{k_x}{l_x}\tilde{x}\right) \\ l_y \tanh\left(\tfrac{k_y}{l_y}\tilde{y}\right) \end{bmatrix}. \tag{22}$$

The analysis of the stability of the closed-loop system is started getting the following relations from Fig 4,

$$\tilde{x} = -\tilde{\rho}\sin(\theta_T) \quad \text{and} \quad \tilde{y} = \tilde{\rho}\cos(\theta_T), \tag{23}$$

furthermore that

$$\dot{\rho} = -\dot{x}\sin(\theta_T) + \dot{y}\cos(\theta_T). \tag{24}$$

Now considering $\tilde{\rho} = -\rho$, its time derivative is,

$$\dot{\tilde{\rho}} = -\dot{\rho} \tag{25}$$

and substituting (24) in (25), it is obtained,

$$\dot{\tilde{\rho}} = \dot{x}\sin(\theta_T) - \dot{y}\cos(\theta_T). \tag{26}$$

On the other hand, introducing (20) and (22) in (26), one gets

$$\dot{\tilde{\rho}} = l_x \tanh\left(\tfrac{k_x}{l_x}\tilde{x}\right)\sin(\theta_T) - l_y \tanh\left(\tfrac{k_y}{l_y}\tilde{y}\right)\cos(\theta_T). \tag{27}$$

Finally, the behavior of $\tilde{\rho}$ for the closed-loop system it is obtained substituting (23) into (27) and it's can be written as,

$$\begin{aligned} \dot{\tilde{\rho}} = &\ l_x \tanh\left(-\tilde{\rho}\tfrac{k_x}{l_x}\sin(\theta_T)\right)\sin(\theta_T) \\ &- l_y \tanh\left(\tilde{\rho}\tfrac{k_y}{l_y}\cos(\theta_T)\right)\cos(\theta_T) \end{aligned} \tag{28}$$

Remark 2. From (28) it's can be conclude that such a system has only one equilibrium point, which is $\tilde{\rho} = 0$.

For the stability analysis the following Lyapunov candidate function is considered $V(\tilde{\rho}) = \frac{1}{2}\tilde{\rho}^2 > 0$. Its time derivative on the trajectories of the system is $\dot{V}(\tilde{\rho}) = \tilde{\rho}\dot{\tilde{\rho}}$, a sufficient condition for the stability of the equilibrium of the close-loop system is that $\dot{V}(\tilde{\rho})$ to be negative definite. Then introducing the closed-loop system (28) in $\dot{V}(\tilde{\rho})$, one gets

$$\dot{V}(\tilde{\rho}) = \tilde{\rho}l_x \tanh\left(-\tilde{\rho}\tfrac{k_x}{T_x}\sin\left(\theta_T\right)\right)\sin\left(\theta_T\right)$$
$$- \tilde{\rho}l_y \tanh\left(\tilde{\rho}\tfrac{k_y}{T_y}\cos\left(\theta_T\right)\right)\cos\left(\theta_T\right) \tag{29}$$

that is $\dot{V}(\tilde{\rho}) < 0$, guaranteeing the stability of the closed-loop system if the gain constants of the controller that weigh the control error are: $l_x > 0$, $k_x > 0$, $l_y > 0$ and $k_y > 0$.

Hence, from (29) it can now be concluded that $\lim_{t \to \infty} \tilde{\rho}(t) \to 0$, *i.e.*, $\tilde{x}(t) \to 0$ and $\tilde{y}(t) \to 0$ with $t \to \infty$ asymptotically. Therefore, from (22) it can be concluded that the final velocity of the point of interest will be $V = |v_P(s_D, h)| \angle \theta_T$ hence $\tilde{\psi}(t) \to 0$ for $t \to \infty$ asymptotically.

Remark 3. Notice that for positioning tasks, *i.e.*, $v_P(s_D, h) = 0$, when the robot reaches the target point it stays in that position (but, without the control of its final orientation).

3.1.2 Dynamic Compensation Controller

If not considered the perfect velocity tracking in kinematic controller design, i.e., $u(t) \neq u_c(t)$ and $\omega(t) \neq \omega_c(t)$. This velocity error motivates to design of an dynamic compensation controller; the objective of this controller is to compensate the dynamic of the human and of the wheelchair, thus reducing the velocity tracking error, hence the following control law dynamic model based (16) is proposed,

$$\begin{bmatrix} u_{ref} \\ \omega_{ref} \end{bmatrix} = \mathbf{M}\left(\begin{bmatrix} \dot{u}_c \\ \dot{\omega}_c \end{bmatrix} + \begin{bmatrix} \sigma_u \\ \sigma_\omega \end{bmatrix}\right) + \mathbf{C}\begin{bmatrix} u \\ \omega \end{bmatrix} \tag{30}$$

with

$$\sigma_u = l_u \tanh\left(\tfrac{k_u}{l_u}\tilde{u}\right) \quad \text{and} \quad \sigma_\omega = l_\omega \tanh\left(\tfrac{k_\omega}{l_\omega}\tilde{\omega}\right)$$

where $\tilde{u}(t) = u_c(t) - u(t)$ *and* $\tilde{\omega}(t) = \omega_c(t) - \omega(t)$ are the linear and angular velocity errors, respectively; $l_u > 0$, $k_u > 0$, $l_\omega > 0$ and $k_\omega > 0$ are positive gain constants that weigh the control error.

Now manipulating (16) and (30), have the behavior of the velocity errors of the human-wheelchair for the closed-loop system,

$$
\begin{bmatrix} \dot{\tilde{u}} \\ \dot{\tilde{\omega}} \end{bmatrix} = \begin{bmatrix} \dot{\tilde{u}}_c \\ \dot{\tilde{\omega}}_c \end{bmatrix} + \begin{bmatrix} l_u \tanh\left(\frac{k_u}{l_u}\tilde{u}\right) \\ l_\omega \tanh\left(\frac{k_\omega}{l_\omega}\tilde{\omega}\right) \end{bmatrix}.
\tag{31}
$$

Next, a Lyapunov candidate function and its time derivative on the system trajectories are introduced in order to consider the corresponding stability analysis $V(\tilde{u},\tilde{\omega}) = \frac{1}{2}(\tilde{u}^2 + \tilde{\omega}^2) > 0$; the time derivative of the Lyapunov candidate function is,

$$
\dot{V}(\tilde{u},\tilde{\omega}) = \tilde{u}\dot{\tilde{u}} + \tilde{\omega}\dot{\tilde{\omega}}
\tag{32}
$$

After introducing the derivate of (31) in (32), the time derivative $\dot{V}(\tilde{u},\tilde{\omega})$ is now

$$
\dot{V}(\tilde{u},\tilde{\omega}) = -\tilde{u}l_u \tanh\left(\frac{k_u}{l_u}\tilde{u}\right) - \tilde{\omega}l_\omega \tanh\left(\frac{k_\omega}{l_\omega}\tilde{\omega}\right) < 0
\tag{33}
$$

Hence, from (33) it can now be concluded that $\tilde{u}(t) \to 0$ and $\tilde{\omega}(t) \to 0$ with $t \to \infty$ asymptotically.

4   Experimental Results

In this section several experiments were executed for show the performance of the proposed controller and dynamic modelling of the human-machine system. Some of the results experiments using the wheelchair presented on Fig. 1 are reported in this section. In order to assess and discuss the performance of the proposed modelling and controller it is developed an analysis and simulation platform wheelchair with LabView interface.

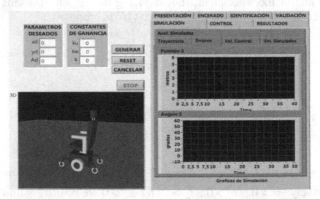

Fig. 5. Human Machine Interface developed for the analysis of the model and the performance of the controller proposed

First, the identification and validation of the proposed dynamic model is shown. The identification of the human-wheelchair system was performed by using least squares estimation [21] applied to a filtered regression model [22]. The identified of dynamic parameters of the human-wheelchair system with a human of 73kg are: $\chi_1 = 0.4903$, $\chi_2 = 0.2112$, $\chi_3 = -0.0011$, $\chi_4 = 1.0203$, $\chi_5 = 0.0346$, $\chi_6 = 0.9766$, $\chi_7 = -0.0016$ and $\chi_8 = -0.1001$. Fig. 6, shown the validation the proposed dynamic model, where it can be seen the good performance of the obtained dynamic model.

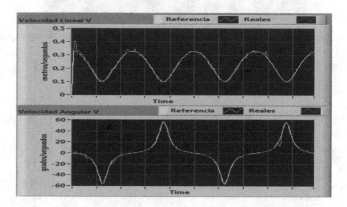

Fig. 6. Validation of the proposed dynamic model. The white line represents the proposed dynamic model signals and the red lines are the velocities experimental signals of human-wheelchair system.

Next, the experiment corresponds to the performance of the proposed controller (control laws (19) and (30)) for the positioning and path following problems. Note that for the positioning problem the desired velocity is zero, while for the path following problem the desired velocity of the wheelchair will depend on the task, the control error, the angular velocity, etc. For this case, it is consider that the reference velocity depends on the control errors and the angular velocity. It is defined as, $\left| v_P(s_D,h) \right| = v_P / \left(1 + k_\rho \rho + k_\omega \omega \right)$ where, k_ρ and k_ω are positive constants that weigh the control error and the angular velocity, respectability. Figures 7-8 show the results of the experiment. Fig. 7 shows the movement of the wheelchair on the X-Y space of the point stabilization problem experiment, and finally the Fig. 9 present the control errors of the path following problem. It can be seen that the proposed controller works correctly.

Fig. 7. Stroboscopic movement of the wheelchair in the point stabilization experiment.

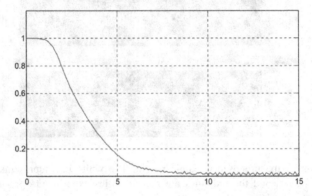

Fig. 8. Distance between the wheelchair and the closest point on the path

5   Using the Template

In this paper the modeling and control of the human-wheelchair system considering lateral deviations of the center of mass was proposed, which has reference velocities as control signals to the system. The controller proposed resolved the path following problem for wheelchair robot, which is also capable of positioning the robot. The design of controller was based on two cascaded subsystems: a kinematic controller which complies with the task objective (path following and positioning), and a dynamic controller that compensates the dynamics of the human-wheelchair system. Finally, the stability and robustness are proved by considering the Lyapunov's method, and the performance of the proposed controller is shown through real experiments.

Acknowledgment. The authors would like to thanks to Universidad Técnica de Ambato for financing the project "Robotic Assistance for Persons with Disabilities" (Resolution 1151-CU-P-2012). Also to the Universidad de las Fuerzas Armadas ESPE, Universidad Tecnológica Indoamérica, Escuela Politécnica Nacional and of the Instituto de Automática, Universidad Nacional de San Juan - Argentina, for the support to develop this work.

References

1. Bastos-Filho, T., et al.: Towards a New Modality-Independent Interface for a Robotic Wheelchair. IEEE Transactions on Neural Systems and Rehabilitation Engineering **22**(3) (2014)
2. Wang, Y., Chen, W.:Hybrid map-based navigation for intelligent wheelchair. In: IEEE International Conference on Robotics and Automation, China, pp. 637–642 (2011)
3. Cheng, W.-C., Chiang, C.-C.:The development of the automatic lane following navigation system for the intelligent robotic wheelchair. In: IEEE International Conference on Fuzzy Systems, Taiwan, pp. 1946–1952 (2011)
4. Mazo, M.: An integral system for assisted mobility. IEEE Robotics and Automation Magazine **8**(1), 46–56 (2001)
5. Zeng, Q., Teo, Ch.L., Rebsamen, B., Burdet, E.: A collaborative wheelchair system. IEEE Transactions on Neural Systems and Rehabilitation Engineering **16**(2), 161–170 (2008)
6. Parikh, S.P., Grassi, V., Kumar, V., Okamoto, J.: Integrating human inputs with autonomous behaviors on an intelligent wheelchair platform. IEEE Intelligent Systems **22**(2), 33–41 (2007)
7. Biswas, K., Mazumder, O., Kundu, A.S.: Multichannel fused EMG based biofeedback system with virtual reality for gait rehabilitation. In: IEEE Proceedings in International Conference on Intelligent Human Computer Interaction, India (2012)
8. Munakata, Y., Tanaka, A., Wada, M.: A five-wheel wheelchair with an active-caster drive system. In: IEEE International Conference on Rehabilitation Robotics, USA, pp. 1–6 (2013)
9. Yuan, J.: Stability analyses of wheelchair robot based on human-in-the-loop control theory. In: IEEE International Conference on Robotics and Biomimetics, Thailand, pp. 419–424 (2009)
10. Soeanto, D., Lapierre, L., Pascoal., A.: Adaptive non-singular path-following, control of dynamic wheeled robots. In: Proceedings of 42nd IEEE Conference on Decision and Control, Hawaii, USA, 9-12 December, pp. 1765–1770 (2003)
11. Andaluz, V., Roberti, F., Toibero, J.M., Carelli, R.: Adaptive unified motion control of mobile manipulators. Control Engineering Practice **20**, 1337–1352 (2012)
12. Yuhua, X., Chongwei, Z., Wei, B., Lin, T.: Dynamic sliding mode controller based on particle swarm optimization for mobile robot's path following. In: International Forum on Information Technology and Applications, pp. 257–260 (2009)
13. Wangmanaopituk, S., Voos, H., Kongprawechnon, W.: Collaborative nonlinear model-predictive collision avoidance and path following of mobile robots. In: ICROS-SICE International Joint Conference 2009, Japan, pp.3205–3210(2009)
14. Tanimoto, Y., Yamamoto, H., Namba, K., Tokuhiro, A., Furusawa, K., Ukida, H.: Imaging of the turn space and path of movement of a wheelchair for remodeling houses of individuals with SCI. In: IEEE International Conference on Imaging Systems and Techniques (2012)
15. Chen, S.-H., Chou, J.-J.:Motion control of the electric wheelchair powered by rim motors based on event-based cross-coupling control strategy. In: IEEE/SICE International Symposium on System Integration (2011)
16. Martins, F.N., Celeste, W., Carelli, R., Sarcinelli-Filho, M., Bastos-Filho, T.: An Adaptive Dynamic Controller for Autonomous Mobile Robot Trajectory Tracking. Control Engineering Practice **16**, 1354–1363 (2008)
17. De La Cruz, C., Bastos, T.F., Carelli, R.: Adaptive motion control law of a robotic wheelchair. Control Engineering Practice, 113–125 (2011)

18. Yahaya, S.Z., Boudville, R., Taib, M.N., Hussain, Z.: Dynamic modeling and control of wheel-chaired elliptical stepping exercise. In: IEEE International Conference on Control System, Computing and Engineering, pp. 204–209 (2012)
19. Sapey, B., Stewart, J., Donaldson, G.: Increases in wheelchair use and perceptions of disablement. Disability & Society **20**(5), 489–505 (2005)
20. Zhang, Y., Hong, D., Chung, J.H., Velinsky, S.A.: Dynamic model based robust tracking control of a differentially steered wheeled mobile robot. In: Proceedings of the American Control Conference, Philadelphia, Pennsylvania, pp. 850–855 (1998)
21. Aström, K.J., Wittenmark, B.: Adaptive Control. Addison-Wesley (1995)
22. Reyes, F., Kelly, R.: On parameter identification of robot manipulator. In: IEEE International Conference on Robotics and Automation, pp. 1910–1915 (1997)

Robotic Vision, Recognition and Reconstruction

Vision-Based Automatic Hair Follicular Unit Separation

Bohan Yang[1], Hesheng Wang[1,2(✉)], Weidong Chen[1,2(✉)], and Yusen Liang[1]

[1] Department of Automation, Shanghai Jiao Tong University, Shanghai 200240, China
[2] Key Laboratory of System Control and Information Processing,
Ministry of Education of China, Shanghai 200240, China
{wanghesheng,wdchen}@sjtu.edu.cn

Abstract. In this paper, a vision-based method is proposed to automatically recognize the hair follicles and plan the cutting path to separate them into units. By using color information and machine learning, hair area in the image can be recognized. And the interferences such as adipose shadows and scalpel parts will be eliminated by texture and area information. In order to recognize single piece of hair, a curve detection method is proposed which combine the linear Hough transform and the quadratic curve fitting method to detect hair pieces with follicles on them. After determining the location and distribution of hair follicles, based on the hair growth direction and the minimum external rectangle of hair area, cutting path will be planned to separate each follicular unit. Compared with the traditional artificial hair follicular unit separation, this method not only ensures the fitting accuracy, but also speeds up the processing speed.

Keywords: Follicular unit transplant · Curve detection · Machine learning · Robot

1   Introduction

Alopecia refers to the phenomenon of hair loss which can be divided into normal hair loss and pathological hair loss [1][2]. There has been so many people across the world suffering from the alopecia for a long time.

In order to effectively combat this harried alopecia, hair transplant technology has been widely developed. The commonly used hair transplant methods are FUT (Follicle Unit Transplant) [3] and FUE (Follicle Unit Extraction) [4]. The extraction and separation of hair follicle tissue in FUT can be separated at the same time. The FUE technology won't leave obvious scar after surgery basically, and the recovery cycle is short [5].

Overall, whether FUT or FUE, for patients, since the anesthesia is local, they need to lie on the bed consciously and wait for the hair follicles separation process to be complete for hours. For the doctor, the time-consuming process of hair follicle separation under a magnifying glass or microscope is very tired. So the artificial hair follicles separation is tough for both patients and doctors. If automatic mechanical arm can achieve rapid separation of hair follicles, operation time will be greatly shortened. Not only can relieve patients' pain, and reduce doctors' burden, but also avoid the

© Springer International Publishing Switzerland 2015
H. Liu et al. (Eds.): ICIRA 2015, Part III, LNAI 9246, pp. 273–284, 2015.
DOI: 10.1007/978-3-319-22873-0_24

damage of hair follicles and the risk of infection caused by the long time exposure of the tissue, and thus improve the effect of treatment a lot.

Recently, the rise of technology on the basis of minimally invasive surgery robot improves the accuracy and feasibility of surgery operation to a new level. A new era of surgery operated by advanced robot and fast signal processing technology is coming. Surgery robot such as the da Vinci [6] has been worked well in cardiothoracic surgery, urinary surgery, gynecology and abdominal surgery and so on [7][8][9][10].

As for hair transplant surgery, as early as in 2002, Houston neurosurgeon Philip Gildenberg came up with the idea of using robot instead of a doctor to transplant hair on the scalp more quickly and accurately. The technology mainly depends on the medical 3d image simulation technology. According to the 3D image of the patient's head, automatic hair transplant program is designed by the doctor and the robot can operate hair transplant surgery afterward [11].

At present, the FUE based automatic hair transplant surgery has had some mature application, of which the most remarkable one is ARTAS system. The system uses computer as an aid of the extraction of hair follicles, which includes the manipulator and the FUE based image processing techniques [12]. In specific applications, the doctor paints the scope of the hair follicle tissue on the patient's bald scalp and the robot takes pictures from different views which will be sent to the computer afterward to reconstruct the 3D image of the patient's head. This 3D image can be used to compute the angel, direction and depth information which will indicates the robot to extract hair follicle tissue. A study shows that the success rate of ARTAS system to extract hair follicle tissue is about 95%, and for the successfully extracted hair follicle tissue, the hair cutting off probability is 4.9%, and the system can extract around 600 pieces of hair within an hour [13].

The paper proposes a method to help improve the efficiency and accuracy of the FUT surgery, which visually recognize hair pieces and plan the cutting path to separate them into follicular units. To distinguish hair area from its surroundings in the image, for every pixel, its color information is used to train the SVM. Interferences need to be eliminated to get the hair binary image. Texture analysis is used to eliminate the adipose shadow interferences and convex hull detection is used to exclude the scalpel area. Based on hair growth direction estimated by Tamura texture feature and the minimum external rectangle of hair area, cutting path is planned to separate parts of scalp into hair clusters. Aiming at recognize single hair piece, we combine the linear Hough transform and the quadratic curve fitting method as our curve detection method. After getting the curve of single hair piece, we compute the cutting point and cutting direction referring to the point connecting adipose tissue and its tangent direction. This planning algorithm can separate hair clusters into follicular units without cutting hair pieces off.

The paper is organized as follows: Section 2 will introduce the process and the main idea of the hair follicular units recognizing and cutting algorithm; Section 3 will discuss experimental results; and Section 4 will present the conclusion.

2   Recognition and Cutting Path Planning of Scalps

The main process of the FUT method is to cut off the scalp tissue taken from the normal scalp skin, from big parts to small ones, and finally separate the hair into pieces which will be cultured afterward. The paper proposes a method of vision-based process of follicular unit recognition and cutting path planning, which can improve the efficiency and accuracy of the FUT surgery.

The algorithm proposed here can be divided into three main parts. Firstly, large scalps need to be recognized and a cutting path should be planned to separate large scalps into smaller parts; Secondly, detect the small scalps and plan to cut them into clusters of hair; Finally, use curve detection method to recognize each single piece of hair and analysis its direction, cutting along which and then separated hair follicular units can be obtained. This process is shown in Fig. 1.

| Large Scalps | Small Scalps | Hair Clusters | Follicular Units |

Fig. 1. General algorithm process

2.1   Recognition and Cutting Path Planning of Large Scalps

This part aims to recognize large parts of the scalp, and meanwhile to figure out the cutting position based on specific area and orientation.

The algorithm process is as follows.

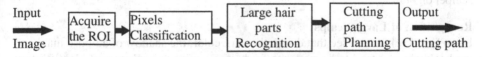

Input → Image | Acquire the ROI | Pixels Classification | Large hair parts Recognition | Cutting path Planning | Output → Cutting path

Fig. 2. Process of large scalps separation

Fig. 3. Large scalps separation

ROI Area Acquisition

The ROI area is extracted using template based matching method. Preset template image of key region in the separating algorithm, after which calculating the similarity between the template image and different regions of the current image based on color information to extract the ROI area for subsequent image processing.

The template image is acquired by set four endpoints coordinates of image ROI area of one frame based on empirical knowledge.

Pixels Classification

According to Fig. 3, color is the main difference between the hair and other parts. In order to recognize the large scalps, here, this paper uses SVM to define a classification standard according to color information, to distinguish the hair from other parts of the image preliminarily.

Image pixels can be divided into four categories, namely, blocks, adipose tissue, scalpel and hair. We collect sample points of the four categories, extracting gray scale channel, three channels of RGB space and three channels of HSV space, a total of seven features, of each pixel. After marked on label, these samples are input to the training support vector machine, SVM. The classification standard will be acquired after training.

Fig. 4. (a)The 18th frame (b)ROI area (c)SVM classification result

As shown in Fig. 4, the hair is marked on red, and adipose tissue on green, and scalpel on blue.

Recognition of Large Scalps

From Fig. 4 (c) we can see that due to the dark color, the shadow of adipose mass and the backlight part tend to be mistaken for hair, affecting the subsequent identification of hair. This paper proposes a method to eliminate this interference by using texture information.

Since the hair area is coarser, the difference of grey value between the near point is large than the shadow area. So we analyze the gray scale variance of the 8-neighbor of pixels, if it is greater than the threshold value, the pixel is identified as hair.

In order to analyze the identified discrete hair pixels as a whole, we operate connected components analysis on the binary image. If the connected component's area is greater than the threshold, we define it to be the large scalp. After this process, mistaken parts brought by the adipose tissue have turned into small bits and pieces. Then we operate connected components analysis on the image, and save the connected components whose area is greater than the threshold. The process is shown in Fig. 5.

Fig. 5. (a)Excluding the shadow interferences ; (b)Recognition result

Cutting Path Planning

The recognized scalps need to be cut into strips of uniform size while preventing each piece of hair to be cut off.

In order to avoid cutting off the hair, the direction of the cutting path should be along the direction of hair growth .We use Tamura texture feature [15] to obtain the orientation. For each hair pixel, we compute the difference of its 8-neighbor gray level in both horizontal Δ_V and vertical direction Δ_H.

$$\Delta_V = \begin{bmatrix} i-1,j-1 & i,j-1 & i+1,j-1 \\ i-1,j & i,j & i+1,j \\ i-1,j+1 & i,j+1 & i+1,j+1 \end{bmatrix} * \begin{bmatrix} -1 & 0 & 1 \\ -1 & 0 & 1 \\ -1 & 0 & 1 \end{bmatrix} \tag{1}$$

$$\Delta_H = \begin{bmatrix} i-1,j-1 & i,j-1 & i+1,j-1 \\ i-1,j & i,j & i+1,j \\ i-1,j+1 & i,j+1 & i+1,j+1 \end{bmatrix} * \begin{bmatrix} 1 & 1 & 1 \\ 0 & 0 & 0 \\ -1 & -1 & -1 \end{bmatrix} \tag{2}$$

$$\theta = tan^{-1}(\Delta_V / \Delta_H) + \frac{\pi}{2} \tag{3}$$

Using the value θ of every hair pixels, we can statistic the pixel direction histogram. The maximum direction of the histogram is defined as the direction of hair growth.

Then compute the minimum circumscribed rectangle of the hair pixels, and choose a place which is at fixed distance up from the bottom of the rectangle, at where start cutting along the hair growth direction, as shown in Fig. 6.

Fig. 6. The final cutting path

2.2   Recognition and Cutting Path Planning of Small Scalps

This part aims to separate the small strips of scalp, after completing large scalp separation, into small clusters of hair, as shown in Fig. 8.

Overall, this process is similar to the process of large scalps. The difference lies on the extra parts of identifying the scalpel and eliminating the scalpel interferences.

Since the area of the scalpel is similar to the area of hair, it may produce interference to the recognition of hair. So the scalpel part in the image should be excluded. In order to exclude the scalpel area out of the ROI area, we remove the convex hull of the largest connected component of scalpel pixels in the SVM classification results out of the ROI area. The result is shown in Fig. 7.

Fig. 7. (a)Scalpel and its convex hull (b) The exclusion result

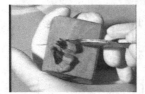

Fig. 8. Small scalp   **Fig. 9.** Cutting path

After that, operate Tamura texture analysis on the area of hair to be separated, and get the hair growth direction. Compute its minimum circumscribed rectangle, and cut along the growth direction of hair, as shown in Fig. 9.

2.3   Recognition and Cutting Path Planning of Hair Follicular Units

This part aims to subdivide the small clusters of hair separated from the small scalp, as shown in Fig. 10, by cutting off the adipose tissue around the hair roots. After that, separate them into pieces of single hair follicles for subsequent culturing. The process is shown in Fig. 12.

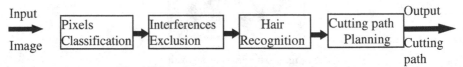

Fig. 10. Process of hair follicular unit separation

The main difficulty is the recognition of hair. Since hair piece is thin, the above mentioned method cannot be used. Moreover, the single piece of hair is so slender that the effect of the wrong recognition will be more obvious. So more accurate algorithm need to be designed to identify the hair in the image.

Fig. 11. Hair follicular unit separation   **Fig. 12.** SVM classification result

Pixels Classification

In this part the requirement of the accuracy of the classification is higher than the former two parts, so we resample 653 background pixels and 531 hair pixels, 664 adipose pixels, 450 scalpel pixels as four kinds of sample set to train the SVM to get new classification criteria. The accuracy of training is 92.2%.

Then recognize the pixels in the ROI area using the new classification standard. As shown in Fig. 11, the hair is marked on red, and adipose tissue on green, and scalpel on blue.

Interferences Exclusion

There are shadows around the adipose tissue and the scalpel, as mentioned above, and also unsegregated scalp parts, which all, will affect the subsequent identification of hair. For scalpel interferences and unsegregated scalp interferences, we exclude the largest connected component out of the image, as shown in Fig. 13(a).

Fig. 13. (a)Scalpel (b)Adipose (c)Hair without interferences

Fig. 14. (a)SVM result of classification (b)Hair without large part of interferences

For the separated adipose parts, on the contrary, we exclude the connected components whose area is less than the threshold. Then operate the dilation on the recognized scalpel and adipose area to eliminate the shadow. The resulting hair binary image is shown in Fig. 13.

Large pieces of hair which hasn't been cut will seriously affect the subsequent Hough transform. So we need to exclude the connected components whose area is greater than the threshold and the slender length is less than the threshold. The result is shown in Fig. 14.

Fig. 15. (a) Hair binary image (b)Hair recognition result 1 (c) Hair recognition result 2

Hair Recognition

After getting the hair binary image, operate curve detection on the image to recognize each single hair. There are a variety of different curve detection methods, such as quadratic [16] and random Hough transforms [17], and methods that track the contour and then compound image edges to get the curve information [18].

Aiming at hair recognition, we propose a method to detect the curve. We map every pixel in the hair binary image to the Hough space, and classify them into different groups corresponding to specific curves.

Yet there may be a curve recognized as two because of the nature of linear Hough transformation. We need to use edge connection to combine two straight lines with the head-tail connection into one. So operate 8-connected components analysis on the results to classify the points in the area. And apply quadratic curve fitting to each class of points [19], so the functional expression of each piece of hair can be computed. The process is shown in Fig. 15.

Cutting Path Planning

For every pieces of the hair recognized in the above part, we start from the right endpoint probing along the tangent direction within the length of six pixels. If adipose tissue is detected, we assume that the point is the starting point of cutting path and the curve is the hair need to be cut. The cutting paths are straight lines rayed above and below from the starting point whose lengths are 10. And we cut the hair along tangent at the starting point. The path is shown in Fig. 16.

Fig. 16. Cutting path planning

3   Experiment and Analysis

We experiment our separating method on the video taken from Shanghai fifth people's hospital of a hair follicle tissue transplant operation. The video consists of three main sections, which are respectively large parts of hair follicle separation, small parts of hair follicle and hair follicular unit separation.

Analyze and process the image in the video by frames, and compare the results obtained from the separation algorithm and the artificial results, we evaluate the feasibility of the algorithm by the accuracy of recognition.

We can see from Fig. 17, Fig. 19, Fig. 21 that our recognized the object and excluded most of the interferences. And the cutting path planned according to the hair growth direction computed by texture feature is basically in accordance with the cutting path of the surgeon. The recognition accuracy is 95.5%, 90%, 94.08% respectively.

3.1   Analysis of Large Scalp Experiment

The cutting path of large scalps is shown in Fig. 17.

Fig. 17. Cutting path of large scalps

Analyzing the recognition error, we find out that the main cause is the interferences of shadow, as shown in Fig. 18. Because of their similar color, it is hard to distinguish these two parts. And this problem also causes error in computing the minimum external rectangle and hair growth direction. In practice, shadow-less lamp can be used to avoid the interferences of shadow and to improve the accuracy.

Fig. 18. (a) SVM classification result (b) Hair recognition result (c) Cutting path planning

3.2   Analysis of Small Scalp Experiment

The cutting path of small scalps is shown in Fig. 19.

Fig. 19. Cutting path of small scalps

There are 14 frames of inaccurate cutting paths caused by the misjudgment of the hair growth direction. This is due to two clusters of hair which haven't been separated in time as shown in Fig. 20. In practice, we can preset the program separate clusters for some distance to avoid such errors.

Fig. 20. (a)The original frame (b)Hair recognition result (c)Cutting path planning

3.3   Analysis of Hair Follicular Unit Experiment

The cutting path of hair follicular units is shown in Fig. 21.

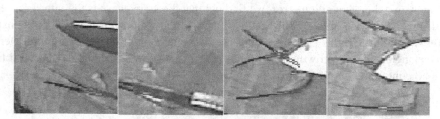

Fig. 21. Cutting path of follicular units

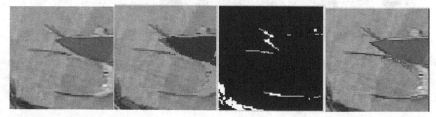

Fig. 22. (a)The original frame (b)SVM classification result (c)Hair recognition result (d)Cutting path planning

Recognition error mainly caused by the interferences of the scalpel and background wood. As shown in Fig. 22, part of the scalpel tip is mistakenly identified as hair, and the subsequent dilation failed to completely remove this interference, resulting in the inaccurate cutting path. In practice, we can select scalpel and wooden plate in different color with hair to improve the accuracy of the SVM classification.

4   Conclusion

The paper proposes a method to visually recognize the follicular unit and plan the cutting path, which can be used to help improve the efficiency and accuracy of the FUT surgery. We use the 7 channels value of color information to distinguish the hair and the adipose tissue and the surroundings, and use Tamura texture feature to estimate the cutting slope, and use Hough transform, edge connections, and curve fitting method to extract every pieces of hair and plan the cutting path. And the algorithm is experimented on the video taken from Shanghai fifth people's hospital of a hair follicle tissue transplant surgery. After comparing the results obtained from the separation algorithm and the artificial results compared with the results of the artificial recognition, we conclude that the accuracy of the separating method is more than 90%.

In the future, we will work on improving the accuracy and robustness of the recognition algorithm. Moreover, we consider designing a robot-assistant system to realize the automatic separation of hair follicular units.

Acknowledgement. This work was supported in part by Shanghai Rising-Star Program under Grant 14QA1402500, in part by the Natural Science Foundation of China under Grant 61105095 and 61473191.

References

1. Dou, G.: Reasons of hair loss, and methods of hair protection (in Chinese). J. Family Medicine **3**, 17 (2014)
2. Zhao, D.: Reasons and solutions of hair loss (in Chinese). J. Health Expo. **12**, 46–47 (2014)
3. Bernstein, R.M., Rassman, W.R.: The aesthetics of follicular transplantation. J. Dermatol. Surg. **23**, 785–799 (1997)
4. Rassman, W.R., Jones, R., Worton, E., Uyttendaele, H.: Follicular unit extraction: minimally invasive surgery for hair transplantation. J. Dermatologic Surgery **28**(8), 720–728 (2002)
5. Comparing, F.U.T.: Pre-making recipient sites to increase graft survival in manual and robotic FUE procedures. J. Hair Transplant Forum Int. **4**, 128–131 (2012)
6. Ji, W., Li, N., Li, J.: Application progress of Da Vinci surgical robot (in Chinese). J. Southeast of the National Defense Medicine **12**, 427–430 (2010)
7. Paul, M., Rodriguez, E., Chitwood, W.R.: Robotics in cardiac surgery. J. Interactive cardiovascular and thoracic Surgery **9**(3), 500–505 (2009)
8. Lee, E.K., Baack, J., Duchene, D.A.: Survey of practicing urologists: robotic versus open radical prostatectomy. J. The Canadian Journal of Urology **17**(2), 5094–5098 (2010)

9. Cho, J.E., Nezhat, F.R.: Robotics and gynecologic oncology: review of the literature. J. Journal of Minimally Invasive Gynecology **16**(6), 669–681 (2009)
10. Roukos, D.H.: The era of robotic surgery for colorectal cancer. J. Annals of Surgical Oncology **17**(1), 338–347 (2010)
11. Li, J.: Robots plant hair as fast as plant seeds. J. China's High-Tech Enterprises **1**, 72–73 (2004)
12. Robert, M., Bernstein, M.D., Zingaretti, G.: Robotic Recipient Site Creation in Hair Transplantation. J. Hair Transplant Forum Intl. **24**(3), 95–97 (2014)
13. Shin, J.W., Kwon, S.H., Kim, S.A., Kim, J.Y., Im, N.J., Park, K.C., Huh, C.H.: Characteristics of robotically harvested hair follicles in Koreans. J. Journal of the American Academy of Dermatology **72**(1), 146–150 (2015)
14. Cortes, C., Vapnik, V.: Support-vector network. J. Machine learning **20**(3), 273–297 (1995)
15. Tamura, H., Mori, S., Yamawaki, T.: Textural features corresponding to visual perception. J. IEEE Transactions on Systems, Man and Cybernetics **8**(6), 460–473 (1978)
16. Huang, Z., Guo, S., Liao, J.: Ellipse Hough transform based detection of red cells in the urinary sediments (in Chinese). J. The Computer System Application **9**, 74–77 (2009)
17. Xu, L., Oja, E., Kultanen, P.: A new curve detection method: randomized Hough transform (RHT). J. Pattern Recognition Letters **11**(5), 331–338 (1990)
18. Parent, P., Zucker, S.W.: Trace inference, curvature consistency, and curve detection. J. IEEE Transactions on Pattern Analysis and Machine Intelligence **11**(8), 823–839 (1989)
19. Akima, H.: A new method of interpolation and smooth curve fitting based on local procedures. J. Journal of the ACM **17**(4), 589–602 (1970)

The Decreasing of 3D Position Errors in the System of Two Cameras by Means of the Interpolation Method

Tadeusz Szkodny[✉] and Adrian Łęgowski

Institute of Automatic Control, Silesian University of Technology,
Akademicka 16 St., 44-100 Gliwice, Poland
Tadeusz.Szkodny@polsl.pl

Abstract. In this paper the original method of 3D position errors decreasing in the system of two cameras is proposed. The analysis of accuracy of determining the 3D coordinates of points on the plane template in the shape of rectangle is presented. The points lie at the corners of squares with side of 22 *mm*. Images of these points were obtained using Edimax IC-7100P cameras from two different points of view and analyzed. Position and orientation coordinates of cameras relatively to the reference system were calculated. Coordinates of points on the ideal image (without optical distortions) were determined. After reading from the image real coordinates, optical distortion model coefficients of the camera were calculated. After that, errors caused by optical distortion were determined. Coordinates read from the image were corrected and coordinates of observed points in the reference system were calculated. Next to decreasing computed 3D position errors the interpolation method was proposed. In this method the interpolation of the real coordinates of image points was used. The calculated coordinates were compared to them real values and them maximal differences were determined. Finally, the maximal position errors caused by finite dimensions of pixels were computed.

Keywords: Computer vision · Soft computing · Robot intelligence

1   Introduction

One of the basic component of computer intelligence of robots is software, that calculates coordinates of position and orientation of manipulated objects, seen by the cameras. Designing of such software must take into account the errors of coordinates read from the camera matrix. These errors cause inaccuracies of calculations of points coordinates in reference system, associated with technical station.

In order to determine accuracy of calculated coordinates of observed points, analysis of errors is needed. These errors are caused by: reading errors of coordinates from the matrix of the camera, optical distortions of the camera, errors of parameters that describes optical system of the cameras, and errors of calculations. During designing of the vision system, minimization of mentioned errors is needed.

To Edimax IC-7100P camera study, template with points surrounded by circles (Fig.1) is used. These points lie in the corners of squares with side of 22 *mm*. In the figure, x and y axis of reference system are marked in the Fig. 1.

© Springer International Publishing Switzerland 2015
H. Liu et al. (Eds.): ICIRA 2015, Part III, LNAI 9246, pp. 285–298, 2015.
DOI: 10.1007/978-3-319-22873-0_25

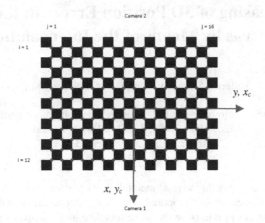

Fig. 1. The template used to camera Edimax IC-7100P study.

User can read points coordinates directly from picture in pixels using Microsoft Paint program, but this way is connected with possibility of making mistakes. To decrease risk of making such mistakes, reading of these coordinates in this work has been made automatically using image processing algorithm [1] implemented in C# language, in Microsoft Visual C# 2010 Express environment.

To compensate optical distortion errors, correction of the read coordinates has been used. Mathematical model of optical distortions used here is presented in works [2,3].

Parameters which describe the coordinate system of camera $x_c y_c z_c$ in reference system xyz are coordinates of position and orientation, focal length and size of the pixel. Calculation of every parameter can be made using iteration methods, which minimize square form of errors [4-5]. Errors of each mentioned parameters occurs in this form. However fundamental disadvantage of these methods is large number of calculations which cause great numerical errors and long time of calculations. In this work, coordinates of position and orientation was calculated using fast and accurate *Camera* algorithm [6]. Precision of calculations of this algorithm amounted to 10^{-6} *mm*. Focal length and size of the pixels were taken from camera datasheet.

In this work, precision of calculation of 192 points from Fig. 1 was analyzed in two different settings of cameras above this template. In second chapter, results of calculations of cameras position and orientation coordinates are presented. Third chapter contains calculations of mathematical model coefficients of optical distortions. Fourth chapter is about analysis of 3D coordinates calculations errors in reference system, using of the mathematical model of optical distortion. In the fifth chapter the coordinate interpolation method is described. The position errors caused by finite dimensions of pixels presents the sixth chapter. Seventh chapter summarizes all of the studies.

2   Position and Orientation of Camera

Here, calculation of position and orientation coordinates of cameras in two different angles with respect to plane of template from Fig.1 is performed. Beginning point O_c of camera coordinates system $x_c y_c z_c$ has been associated with the center of camera matrix. Cameras settings are presented in Fig. 2.

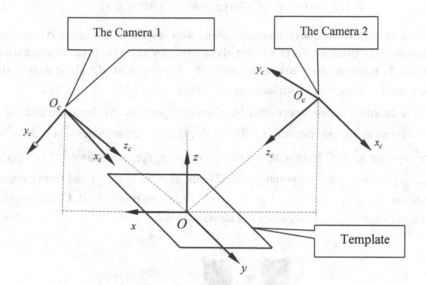

Fig. 2. Cameras settings.

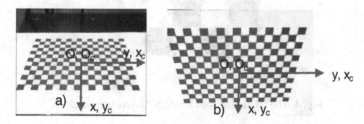

Fig. 3. The template image of: a) Camera1, b) Camera 2.

Images of template from the *Camera* 1 and the *Camera* 2 are presented in the Figures 3a-b.

Coordinates of all 192 points from the template are read in pixels with usage of algorithm of image processing [1] implemented in C# language in Microsoft Visual C# 2010 Express environment. From Edimax IC-7100P camera data sheet, can be read, that size of the pixel is equal to $2.8 \cdot 10^{-3} mm$ x $2.8 \cdot 10^{-3} mm$ and focal length $f_c = 5.01$ mm. After multiplication of coordinates in pixels by the pixel size, we obtain coordinates of points read from the image in *mm*, in $x_c y_c$- system. Coordinates of points in reference system *xy* are easy to determine. These points are placed in the corners of 22*mm* by 22*mm* square.

Position and orientation of camera system $x_c y_c z_c$ with respect to reference system *xyz* are described by the homogenous matrix $\mathbf{T}_c$ of transformation, like in the equation (1).

$$\mathbf{T}_c = Trans(d_x, d_y, d_z)Rot(z, \gamma)Rot(y, \beta)Rot(x, \alpha) . \qquad (1)$$

It is notation of successive transformations with respect to reference system *xyz*. Mentioned transformations are: rotation about axis *x* by angle α, rotation about axis *y* by angle β, rotation about axis *z* by angle γ, displacement d_z along axis *z*, displacement d_y along axis *y*, displacement d_x along axis *x* [7,8].

These coordinates are determined by *Camera* algorithm [6]. Input parameters of these algorithms are coordinates α, β, γ, d_x, d_y, d_z; coordinates $^c z_A$, $^c z_B$, $^c z_C$ of three points A, B, C from template in the system $x_c y_c z_c$; coordinates $^c x_{Ac}$, $^c y_{Ac}$, $^c x_{Bc}$, $^c y_{Bc}$, $^c x_{Cc}$, $^c y_{Cc}$ of points A, B, C in the system $x_c y_c z_c$ (read from camera); coordinates x_A, y_A, z_A, x_B, y_B, z_B, x_C, y_C, z_C of points A, B, C in the system *xyz*; focal length f_c; and accuracy of calculations *delta*.

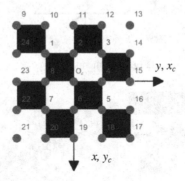

Fig. 4. The 24 points from which 5056 sets were created.

Calculations of coordinates of cameras from Fig.2 was made by means of program *Oject_Coordinates* written in C# language, in Microsoft Visual C# 2010 Express environment. The program created 5056 sets of points A, B, C, next calculated the camera coordinates for each of these sets, and finally averaged these coordinates. These sets were created from the 24 points lying beyond the beginning of coordinate system O_c, but closest to this beginning were chosen. These points are shown in Fig. 4. Read coordinates of these points have small errors caused by optical distortions. These points are located on the squares with side length of 44 *mm* and 22 *mm*, which centre approximately coincide with point O_c. Coordinates $z_A = z_B = z_C = 0$ *mm* .

For initial values of input parameters α, β, γ, d_x, d_y, d_z, $^c z_A$, $^c z_B$, $^c z_C$, x_A, y_A, z_A, x_B, y_B, z_B, x_C, y_C, z_C calculated roughly using geometrical dependencies and $^c x_{Ac}$, $^c y_{Ac}$, $^c x_{Bc}$, $^c y_{Bc}$, $^c x_{Cc}$, $^c y_{Cc}$ read from camera, camera coordinates α, β, γ, d_x, d_y, d_z were calculated with accuracy $delta=10^{-6}$ mm [6].

Screen in Fig.5 illustrates matrices $\mathbf{T}_{c1}$ and $\mathbf{T}_{c2}$ of the *Camera* 1 and *Camera* 2.

Fig. 5. The screen of program Camera_Coordinates.

3 The Optical Distortions Errors

For simplification, rows and columns are introduced. Rows are consist of points, lying on lines, which are parallel to axis y on Fig.1. Each row consist of 16 points. Number of rows is equal to 12, according to Fig.1. Columns are consist of points, lying on lines, which are parallel to axis x on Fig.1. Number of columns is equal to 16. P_{ij} is the point of i-th row and j-th column. Coordinates $^c x_c(i, j)$ and $^c y_c(i, j)$ of image of points P_{ij} of template, read from the camera system $x_c y_c$, have errors $\Delta_* {}^c x_c(i, j)$ and $\Delta_* {}^c y_c(i, j)$, caused by optical distortions. These errors can be calculated from mathematical description of distortions by means of coefficients k_1, k_2, k_3, p_1 and p_2 [2,3]. Equations (2a) and (2b) describes these errors. Errors $\Delta^c x_c(i, j)$ and $\Delta^c y_c(i, j)$ can be determine from coordinates read from camera. These errors are described by equation (2c).

$$\Delta_* {}^c x_c(i,j) = {}^c x_{ci}(i,j)[k_1 {}^c r_{ci}(i,j)^2 + k_2 {}^c r_{ci}(i,j)^4 + k_3 {}^c r_{ci}(i,j)^6] + 2p_1 \cdot {}^c x_{ci}(i,j) \cdot {}^c y_{ci}(i,j)$$
$$+ p_2 [{}^c r_{ci}(i,j)^2 + 2 \cdot {}^c x_{ci}(i,j)^2], \tag{2a}$$

$$\Delta_* {}^c y_c(i,j) = {}^c y_{ci}(i,j)[k_1 {}^c r_{ci}(i,j)^2 + k_2 {}^c r_{ci}(i,j)^4 + k_3 {}^c r_{ci}(i,j)^6] + 2p_2 \cdot {}^c x_{ci}(i,j) \cdot {}^c y_{ci}(i,j)$$
$$+ p_1 [{}^c r_{ci}(i,j)^2 + 2 \cdot {}^c y_{ci}(i,j)^2], \tag{2b}$$

$$\Delta^c x_c(i,j) = {}^c x_c(i,j) - {}^c x_{ci}(i,j), \quad \Delta^c y_c(i,j) = {}^c y_c(i,j) - {}^c y_{ci}(i,j), \tag{2c}$$

$${}^c r_{ci}(i,j)^2 = {}^c x_{ci}(i,j)^2 + {}^c y_{ci}(i,j)^2.$$

In equations (2a-c) occurs ideal coordinates of points ${}^c x_{ci}(i,j)$ and ${}^c y_{ci}(i,j)$, with no optical distortions. These coefficients can be calculated from homogenous form $\mathbf{r}(i,j)$ of vector that describes point P_{ij} in reference system. Coordinates of point P_{ij} in reference system can be note using indexes as follows: $x(i,j) = (i-7) \cdot 22\,mm$, $y(i,j) = (j-9) \cdot 22\,mm$. Point P_{79} is the origin O of the reference system. All points are lying on the plane xy, so $z(i,j) = 0$. The method of the calculations the ideal coordinates ${}^c x_{ci}(i,j)$ and ${}^c y_{ci}(i,j)$ is presented in the work [1].

Using ideal coordinates we can calculate the errors $\Delta^c x_c(i,j)$, $\Delta^c y_c(i,j)$ and ${}^c r_{ci}(i,j)^2$ occurring in equations (2a-c). Since these values are known, it allows to apply equations (2a) and (2b) for calculation of coefficients k_1, k_2, k_3, p_1 and p_2. If following sum is created

$$S = \sum_{i=1}^{12} \sum_{j=1}^{16} \{ [\Delta^c x_c(i,j) - \Delta_* {}^c x_c(i,j)]^2 + [\Delta^c y_c(i,j) - \Delta_* {}^c y_c(i,j)]^2 \},$$

unknown coefficients can be calculated by using minimally square method. Results of these calculations for two cameras from the Fig.2 are shown in screen from the Fig.5. On this screen n is the number of the coefficients occurring in equations (2a,b). For $n=1$, $k_1 \neq 0$, and the remaining coefficients k_2, k_3, p_1 and p_1 are equal to zero. For $n=2$ $k_1, k_2 \neq 0$, for $n=3$ $k_1, k_2, k_3 \neq 0$, for $n=5$ all coefficients different from zero.

4   The Calculation Coordinates Errors

Coefficients $k_1 \div k_3$, p_1 and p_2 shown in the screen from Fig.5 can be applied to calculate errors $\Delta_* {}^c x_c(i,j)$ and $\Delta_* {}^c y_c(i,j)$, described by equations (2a,b). After calculating these errors correction of coordinates of points ${}^c x_c(i,j)$ and ${}^c y_c(i,j)$ (read from camera matrix in the coordinate system $x_c y_c$) can be made. By correction it means subtraction of errors $\Delta_*{}^c x_c(i, j)$ and $\Delta_*{}^c y_c(i, j)$ from the coordinates $^c x_c(i, j)$ and $^c y_c(i, j)$. Coordinates after this correction are note by $^c x_{ccor}(i, j)$ and $^c y_{ccor}(i, j)$. Equation (3) describes corrected coordinates.

$$^c x_{ccor}(i, j) = {}^c x_c(i, j) - \Delta_*{}^c x_c(i, j), \quad {}^c y_{ccor}(i, j) = {}^c y_c(i, j) - \Delta_*{}^c y_c(i, j). \tag{3}$$

From corrected coordinates $^c x_{ccor}(i, j)$ and $^c y_{ccor}(i, j)$ of two cameras, coordinates $x_{cor}(i, j)$, $y_{cor}(i, j)$ and $z_{cor}(i, j)$ of points P_{ijcor} in the reference system xyz can be calculated. These coordinates are shown in the Fig.6.

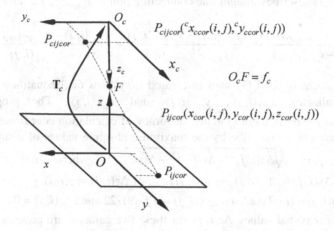

Fig. 6. The coordinates x of points P_{ijcor} and P_{cijcor}.

In order to calculate coordinates $x_{cor}(i, j)$, $y_{cor}(i, j)$ and $z_{cor}(i, j)$, coordinates $x_{ccor}(i, j)$, $y_{ccor}(i, j)$, $z_{ccor}(i, j)$ of point P_{cijcor} and x_F, y_F, z_F of focal F in reference system are necessary. Equations (4) and (5) describes that coordinates by means of the coordinates $^c x_{ccor1}(i, j)$, $^c y_{ccor1}(i, j)$, $^c x_{ccor2}(i, j)$ and $^c y_{ccor2}(i, j)$. The $^c x_{ccor1}(i, j)$, $^c y_{ccor1}(i, j)$ are coordinates of *Camera* 1, and $^c x_{ccor2}(i, j)$, $^c y_{ccor2}(i, j)$ - coordinates of *Camera* 2 from Fig. 2. The matrices $\mathbf{T}_c$, others coordinates, and lengths f of the two cameras were marked similarly.

$$
\begin{bmatrix} x_{ccor1}(i, j) \\ y_{ccor1}(i, j) \\ z_{ccor1}(i, j) \\ 1 \end{bmatrix} = \mathbf{T}_{c1} \begin{bmatrix} {}^c x_{ccor1}(i, j) \\ {}^c y_{ccor1}(i, j) \\ 0 \\ 1 \end{bmatrix}, \quad
\begin{bmatrix} x_{ccor2}(i, j) \\ y_{ccor2}(i, j) \\ z_{ccor2}(i, j) \\ 1 \end{bmatrix} = \mathbf{T}_{c2} \begin{bmatrix} {}^c x_{ccor2}(i, j) \\ {}^c y_{ccor2}(i, j) \\ 0 \\ 1 \end{bmatrix}, \tag{4}
$$

$$\begin{bmatrix} x_{F1} \\ y_{F1} \\ z_{F1} \\ 1 \end{bmatrix} = \mathbf{T}_{c1} \begin{bmatrix} 0 \\ 0 \\ f_{c1} \\ 1 \end{bmatrix}, \quad \begin{bmatrix} x_{F2} \\ y_{F2} \\ z_{F2} \\ 1 \end{bmatrix} = \mathbf{T}_{c2} \begin{bmatrix} 0 \\ 0 \\ f_{c2} \\ 1 \end{bmatrix}. \tag{5}$$

Equations (6) describes straight line connecting points $P_{cijcor1}$, $F1$ and P_{ijcor}.

$$\frac{x_{cor}(i,j) - x_{ccor1}(i,j)}{x_{F1} - x_{ccor1}(i,j)} = \frac{y_{cor}(i,j) - y_{ccor1}(i,j)}{y_{F1} - y_{ccor1}(i,j)} = \frac{z_{cor}(i,j) - z_{ccor1}(i,j)}{z_{F1} - z_{ccor1}(i,j)}. \tag{6}$$

Equations (7) describes straight line connecting points $P_{cijcor2}$, $F2$ and P_{ijcor}.

$$\frac{x_{cor}(i,j) - x_{ccor2}(i,j)}{x_{F2} - x_{ccor2}(i,j)} = \frac{y_{cor}(i,j) - y_{ccor2}(i,j)}{y_{F2} - y_{ccor2}(i,j)} = \frac{z_{cor}(i,j) - z_{ccor2}(i,j)}{z_{F2} - z_{ccor2}(i,j)}. \tag{7}$$

From equations (6) and (7) we can created 6 systems of 3 equations with 3 unknown coordinates $x_{cor}(i,j)$, $y_{cor}(i,j)$ and $z_{cor}(i,j)$. The program *Object_Coordinates* solves these systems. Accuracy of calculation coordinates of points in the template can be describe by the maximum absolute values of distance differences $\Delta r(i,j) = \sqrt{\Delta x(i,j)^2 + \Delta y(i,j)^2 + \Delta z(i,j)^2}$, where $\Delta x(i,j) = | x(i,j) - x_{cor}(i,j)|$, $\Delta y(i,j) = | y(i,j) - y_{cor}(i,j)|$, and $\Delta z(i,j) = | z(i,j) - z_{cor}(i,j)|$. As a reminder: $x(i,j) = (i-7) \cdot 22 \, mm$, $y(i,j) = (j-9) \cdot 22 \, mm$, $z(i,j) = 0$. Results of calculation of maximal values $\Delta r(i,j)$ for these two cameras are presented below in Table1. In the table n is number of coefficients distortions shown in screen from Fig.5.

Table 1. Maximum values of $\Delta r(i,j)$ result from the distortion model.

n	0	1	2	3	5
i	10	1	1	1	11
j	3	1	1	1	6
max $\Delta r(i,j)$	1.8590 mm	3.2259 mm	5.2110 mm	4.3349 mm	1.5773 mm

For $n=0$ calculations were done without corrections, i.e. for coordinates $^{c}x_{ccor}(i,j)$ and $^{c}y_{ccor}(i,j)$ respectively equal to $^{c}x_{c}(i,j)$ and $^{c}y_{c}(i,j)$. The greatest error appears for $n=2$. The smallest error is obtained when all five coefficients k_1, k_2, k_3, p_1 and p_2 are applied.

From calculations for coordinates $^{c}x_{ccor}(i,j)$ and $^{c}y_{ccor}(i,j)$ respectively equal to ideal coordinates $^{c}x_{ci}(i,j)$ and $^{c}y_{ci}(i,j)$ results max $\Delta r(i,j) = $ max $\Delta r(1,1) = 1.3782 \cdot 10^{-11} \, mm$! So small error indicates that the mathematical model of the optical distortion (1a, b) is poorly correct for local nature of these distortions. This indicates the possibility of a greater reduction of the calculation error by taking into account their local character.

In the next section 5 a method for the better reducing errors, based on their local character, is presented.

5   The Interpolation Method of the Coordinates Calculations Error Decreasing

From the fourth point results a very small calculation errors of the coordinates $x_{cor}(i, j)$, $y_{cor}(i, j)$ and $z_{cor}(i, j)$ for the ideal coordinates ${}^c x_{ci}(i, j)$ and ${}^c y_{ci}(i, j)$. We can calculate the ideal coordinates for each point of the template. Therefore, to calculate the position of each point on the template is very simple, because for each of these points can be calculate the ideal coordinates [1].

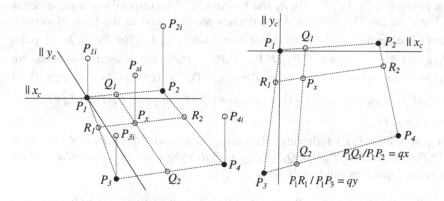

Fig. 7. Illustration of the point P_x and P_{xi}.

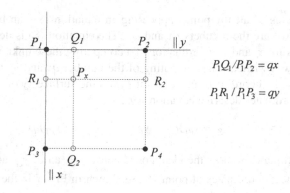

Fig. 8. The points P_x interpolated from the $x_c y_c$ surface on template surface xy.

So far, we considered only the coordinates $^c x_c(i, j)$ and $^c y_c(i, j)$ of image points located in the corners of the tetragon $P_1P_2P_3P_4$ (see Fig. 7), and squares $P_1P_2P_3P_4$ of the template (see Fig. 8). In general, the points P_x can lie outside of these corners, e.g. within these tetragons. Then we can read from the camera only $^c x_c$ and $^c y_c$ coordinates of these points. Coordinates of these points in the reference system we do not know, therefore the calculating of the corresponding ideal coordinates is impossible.

The local nature of the cameras optical error for points P_x we consider using corrected coordinates of these points on the $x_c y_c$ surface. The corrected coordinates are ideal coordinates $^c x_{ci}$ and $^c y_{ci}$. To calculation of these ideal coordinates we are using the coordinates x and y of the points P_x, interpolated from the surface $x_c y_c$ to the template surface xy. Figure 8 shows the interpolation. Are computed the ideal coordinates because they give very little calculation errors of coordinates in the reference system xyz.

The points P_x, $P_1 \div P_4$, Q_1, Q_2, R_1 and R_2 from Fig.7 correspond to the same points from Fig.8. In the Fig.7 the ideal coordinates are indicated in the form of vertical sections. The section P_1P_{1i} is the ideal coordinate $^c x_{ci}(i, j)$ or $^c y_{ci}(i, j)$ of point $P_1 = P_{ij}(i,j)$. Likely sections P_2P_{2i}, P_3P_{3i}, P_4P_{4i}, P_xP_{xi} are respectively equal to: $^c x_{ci}(i, j+1)$ or $^c y_{ci}(i, j+1)$ of point $P_2 = P_{ij}(i,j+1)$, $^c x_{ci}(i+1, j)$ or $^c y_{ci}(i+1, j)$ of point $P_3 = P_{ij}(i+1,j)$, $^c x_{ci}(i+1, j+1)$ or $^c y_{ci}(i+1, j+1)$ of point $P_4 = P_{ij}(i+1,j+1)$, $^c x_{ci}$ or $^c y_{ci}$ of point P_x.

For points from Fig.7 following relationships are valid $P_1Q_1/P_1P_2 = R_1P_x/R_1R_2 = P_3Q_2/P_3P_4 = qx$ and $P_1R_1/P_1P_3 = Q_1P_x/Q_1Q_2 = P_2R_2/P_2P_4 = qy$. From the relationships result vector equations valid on the surface $x_c y_c$.

$$\vec{P}_x = \vec{P}_1 + (\vec{P}_3 - \vec{P}_1)qy + (\vec{R}_2 - \vec{R}_1)qx, \quad \vec{R}_2 - \vec{R}_1 = \vec{P}_2 + (\vec{P}_4 - \vec{P}_2)qy - (\vec{P}_1 + (\vec{P}_3 - \vec{P}_1)qy).$$

$$\rightarrow \vec{P}_x = \vec{P}_1 + (\vec{P}_3 - \vec{P}_1)qx + (\vec{P}_2 + (\vec{P}_4 - \vec{P}_2)qy - (\vec{P}_1 + (\vec{P}_3 - \vec{P}_1)qy)qx. \tag{8}$$

The coordinates of all the points appearing in equation (8) can be read from the camera. Unknown are the numbers qx and qy. The equation (8) is description of two scalar equations for x_c and y_c coordination. To computing these unknown numbers the program $odwz$ was written. The computing of the number qx and qy allows the calculation of the x and y coordinates of the point P_x, on the surface xy of the template (see Fig. 8). The coordinates describe equations (9).

$$x = x_{P1} + (x_{P3} - x_{P1})qy, \quad y = y_{P1} + (y_{P2} - y_{P1})qx. \tag{9}$$

For this coordinates x and y the ideal coordinates $^c x_{ci}$ and $^c y_{ci}$ are computed [1]. So computed ideal coordinates of point P_x are shown in Fig. 7 in the form of vertical section P_xP_{xi}.

For the analysis of the accuracy of the interpolation method calculations we will create a set of points P_x within the area template image. An image area template S_1 of *Camera1* is described by extreme coordinates of the points template $x_{1\min} = \min({}^c x_c(i,j))$, $x_{1\max} = \max({}^c x_c(i,j))$, $y_{1\min} = \min({}^c y_c(i,j))$, $y_{1\max} = \max$ $({}^c y_c(i,j))$. An image area template S_2 of *Camera2* is described similarly, i.e. $x_{2\min} = \min({}^c x_c(i,j))$, $x_{2\max} = \max({}^c x_c(i,j))$, $y_{2\min} = \min({}^c y_c(i,j))$, $y_{2\max} = \max$ $({}^c y_c(i,j))$. The area $S_2 = (x_{2\max} - x_{2\min})\cdot(y_{2\max} - y_{2\min}) = 2.0541\,mm^2$, $S_1 = (x_{1\max}$ $-x_{1\min})\cdot(y_{1\max} - y_{1\min}) = 2.4340\,mm^2$, so we will create a set of points P_x covering a smaller area, i.e. S_2. The numbers M and N of dividing the relatively $dx_2 = x_{2\max} - x_{2\min}$ and $dy_2 = y_{2\max} - y_{2\min}$ sides must be less than the limits M_l and N_l, resulting from the size of a pixel. The size of the pixel is equal to $2.8\cdot10^{-3}mm$ x $2.8\cdot10^{-3}mm$, so $M_l=dx_2/2.8\cdot10^{-3}mm=677$ and $N_l=dy_2/2.8\cdot10^{-3}mm=387$.

We assumed $M = 339$ and $N = 194$ at which the distance between neighboring points of the set is approximately equal to the length of two sides of the pixel. The program *doklB1B2odwz* was calculated of maximal errors $\Delta r(i,j)$ for these dividing number M and N. Now the indexes $1 \le i \le M$ and $1 \le j \le N$. The errors $\Delta r(i,j) =$

$\sqrt{\Delta x(i,j)^2 + \Delta y(i,j)^2 + \Delta z(i,j)^2}$, where $\Delta x(i,j) = |\, x_{Px}(i,j) - x_{Px\,cor}(i,j)\,|$, $\Delta y(i,j) =$ $|\, y_{Px}(i,j) - y_{Px\,cor}(i,j)\,|$, and $\Delta z(i,j) = |\, z_{Px}(i,j) - z_{Px\,cor}(i,j)\,|$. $x_{Px}(i,j) \div z_{Px}(i,j)$ are coordinates of points P_x illustrated in Fig.8. $x_{Px\,cor}(i,j) \div z_{Px\,cor}(i,j)$ are coordinates of points $P_{x\,cor}$ computed from Eq. (6,7).

From these calculations results $\max \Delta r(i,j) = 3.6824\cdot10^{-12}\,mm$! So small position error indicates a high efficiency interpolation method proposed in this work. This method perfectly takes into account the local character errors. However, in this method, we used linear interpolation, which is equivalent to the assumption of a linear distribution of optical errors along the coordinate x_c and y_c camera. In fact, the error distribution is not linear. The closer the images template points are, the more the distribution of optical errors is closer to linear. In our case, this distance is not greater than the $0.2\,mm$. Therefore, our assumption is justified and real calculation errors of this method can be considered small and negligible in comparison with the image quantization errors, resulting from the finite size of pixels. The calculation method of the image quantization errors is presented in next chapter.

6   The Calculation of Image Quantization Errors

Position of the image points on the camera matrix can be read with an accuracy of one pixel, which has finite dimensions. Thus, the coordinates of the position are quantized respective lengths of the sides of the pixel. With the physics laws results that this quantization errors cause the camera resolution errors. The pixels are squares with side $px = 2.8\cdot10^{-3}mm$. In Fig.9 and 10 are illustrated the error vectors $\vec{r}_c$ and $\overrightarrow{PR}$.

The vector $\vec{r}_c$ results from maximal image quantization error in point P_c on the camera matrix. The vector $\overrightarrow{PR}$ resulting from vector $\vec{r}_c$ describes the camera resolution error in point P on the template.

The vector $\vec{r}_c$ should have a length of the diagonal of a square pixel. Let's assume that the greater among segments PR and PQ is a measure of the camera resolution error. From Fig.9 and 10 results the equation (10)

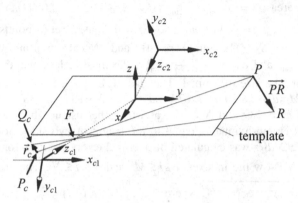

Fig. 9. The quantization error vector $\vec{r}_c$ and camera resolution error vector $\overrightarrow{PR}$.

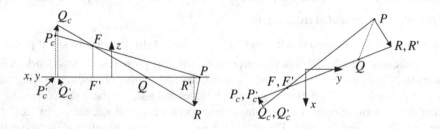

Fig. 10. The horizontal and vertical view of the vectors $\vec{r}_c$ and $\overrightarrow{PR}$.

$$\overrightarrow{PR} = -\vec{r}_c \left(PF / P_c F \right). \tag{10}$$

This equation shows that the largest section PR will be for the template point P farthest from the camera focal F.

Thanks T_{c1} array shown in Figure 5, we can calculate from equations (4,5) the coordinates of all the vectors and the points in Fig. 9 and 10, in the reference system xyz. The focal of the *Camera 1* has following coordinates: F(507.5924 *mm*, 8.0218 *mm*, 412.2506 *mm*). For this focal coordinates the template point P_{ij} (1,1) is farthest from the camera focal F. To determine the direction of the vector $\vec{r}_c$ we will use image coordinates of the two points of the template. There are points $P_{ij}(1,1)$ and $P_{ij}(2,2)$. The point $P_{ij}(2,2)$ is the closest camera focal F among the points $P_{ij}(1,1)$, $P_{ij}(1,2)$, $P_{ij}(2,2)$ and $P_{ij}(2,1)$ lying in the corners of a template square. Differences image coordinates of these points in the camera system are as follows: $\Delta^c x_c = {}^c x_c(2,2) - {}^c x_c(1,1) = 1.0217mm - 1.1421mm < 0$, $\Delta^c y_c = {}^c y_c(2,2) - {}^c y_c(1,1) = 0.4679mm - 0.5499mm < 0$. Direction of the vector $\vec{r}_c$ describes the signs of these differences, as follows: $\vec{r}_c = sign(\Delta^c x_c) \cdot px \cdot \vec{a} + sign(\Delta^c y_c) \cdot px \cdot \vec{b} = -px \cdot \vec{a} - px \cdot \vec{b}$, where $\vec{a}$ and $\vec{b}$ are unit vectors respectively of axis x_c and y_c of the camera system $x_c y_c z_c$. For such a vector $\vec{r}_c$ and focal coordinates of the *Camera 1* we obtain $PQ = 0.9671$ *mm* and $PR = 0.5999$ *mm*. Thus, the measure of the resolution errors of the *Camera 1* will be length of a larger sections, that is $PQ = 0.9671$ *mm*.

After a similar calculation for *Camera 2*, using the array $\mathbf{T}_{c2}$ shown in Figure 5, we get following coordinates the camera focal: $F(-449.9927$ *mm*, -17.8738 *mm*, 461.3084 *mm*). For this focal coordinates the template point $P_{ij}(12,16)$ is farthest from the camera focal F. The point $P_{ij}(11,15)$ is the closest camera focal F among the points $P_{ij}(12,15)$, $P_{ij}(12,16)$, $P_{ij}(11,16)$ and $P_{ij}(11,15)$ lying in the corners of a template square. Differences image coordinates of these points in the camera system are as follows: $\Delta^c x_c = {}^c x_c(11,15) - {}^c x_c(12,16) = -0.9385mm - (-1.10703)mm > 0$, $\Delta^c y_c = {}^c y_c(11,15) - {}^c y_c(12,16) = -0.4307mm - (-0.5291)mm > 0$. Direction of the vector $\vec{r}_c$ describes the signs of these differences, as follows: $\vec{r}_c = sign(\Delta^c x_c) \cdot px \cdot \vec{a} + sign(\Delta^c y_c) \cdot px \cdot \vec{b} = px \cdot \vec{a} + px \cdot \vec{b}$. For these calculations we obtain the following lengths $PQ = 0.8212$ *mm* and $PR = 0.5733$ *mm*. The measure of the resolution error of the *Camera 2* will be length of a larger sections, that is $PQ = 0.8212$ *mm*.

The longer section PQ of *Camera 1* and *Camera 2* determines the resolution error of the position measurement by means of these cameras. For cameras from Fig.2 the resolution error is section length of the *Camera 1* i.e. $PQ = 0.9671$ *mm*.

7   Summary

This work presents a method for calculating the position coordinates of points in the reference system, based on the coordinates of the images of these points, obtained from two cameras. Also shown the analysis of the effectiveness of the decreasing of the calculation position error of the template points.

Calculation errors resulting from optical distortion depends on the number n of coefficients k_1, k_2, k_3, p_1, p_2, taken into account in the equations (2a-2b). Table 1 shows that for $n = 1,2,3$ coefficients calculation errors are larger than omitting the optical distortion (for $n = 0$). Taking into account all the factors ($n = 5$) causes a small decrease calculation errors, from $1.8590\,mm$ to $1.5773\,mm$. Can be concluded that equation (2a-2b) weakly take into account the local nature of the optical errors.

To achieve greater accuracy of the calculations it can be used the original interpolation method of the coordinates proposed in this work. This method allows to achieve increasing the accuracy of calculations. For this method the calculations error is very small, order 10^{-12} mm ! So small position error indicates a high efficiency interpolation method proposed in this work. This method perfectly takes into account the local character errors. The application of the interpolation method of ideal coordinates proposed here is more effective than mathematical model of optical errors cameras.

The estimation of the position errors calculation should consider the quantization error of the image of two cameras. The resolution error of two cameras presented here is equal to 0.9671 mm.

Studies presented here should be treated as preliminary step of designing vision system with two cameras. In this system, the interpolation method proposed here should be used instead of the mathematical model of optical distortion. Then, the maximum the position error calculations of the system 2 cameras proposed here is equal to the resolution error of these cameras i.e. 0.9671 mm.

From equation (10), and Fig.10 results that the cameras resolution error can be reduced by: a) reduction of the vector $\vec{r}_c$, that is using cameras with smaller pixel dimensions, b) reduction of FP, that is close-template cameras c) reducing the ratio of PF/P_CF e.g. by increasing the focal length, d) increase the angle of FPF' e.g. by reducing the distance FP for a constant height of the camera above the template.

References

1. Szkodny, T., Meller, A., Palka, K.: Accuracy of determining the coordinates of points observed by camera. In: Zhang, X., Liu, H., Chen, Z., Wang, N. (eds.) ICIRA 2014, Part II. LNCS, vol. 8918, pp. 273–284. Springer, Heidelberg (2014)
2. Beyer, H.A.: Geometric and Radiometric Analysis of CCD-Camera Based Photogrammetric Closed-Range System. Dissertation, ETH, Zurich, No. 9701 (1992)
3. Google: The Calibraction of Camera. etacar.put.poznan.pl/marcin Kiełczewski (2013) (in Polish)
4. Chesi, G., Garulli, A., Vicino, A., Cipolla, R.: On the estimation of the fundamental matrix: a convex approach to constrained least-squares. In: Vernon, D. (ed.) ECCV 2000. LNCS, vol. 1842, pp. 236–250. Springer, Heidelberg (2000)
5. Golub, G.H., Van Loan, C.F.: Matrix Computations, 3-th edn. The Johns Hopkins University Press (1996)
6. Szkodny, T.: Calculation of the location coordinates of an object observed by a camera. In: Gruca, A., Czachórski, T., Kozielski, S. (eds.) Man-Machine Interactions 3. AISC, vol. 242, pp. 143–156. Springer, Heidelberg (2014)
7. Szkodny, T.: Foundation of Robotics. Silesian University of Technology Publ. Company, Gliwice (2012)
8. Craig, J.J.: Introduction to Robotics. 2-nd edn. Addison Wesley Publ. Comp. (1986)

A Quaternion-Vector Edge Detecting Approach of Height-Measurement in Complex Environment

Dajiang Hong[1,2], Juan Du[1(✉)], and Yueming Hu[1]

[1] School of Automation Science and Engineering, South China University of Technology,
Guangzhou 510640, Guangdong, People's Republic of China
djh_hong@163.com, {Dujuan,auymhu}@scut.edu.cn
[2] Department of Computer Science and Engineering, Hanshan Normal University,
Chaozhou 521041, Guangdong, People's Republic of China

Abstract. In the image processing of 3D precision measurement field, the traditional edge detection algorithms cannot find all the relevant edges which hide the height information. A quaternion-vector edge detecting approach of color image is presented to realize the micro-height-measurement by using the color fringes encoding structure light. The proposed edge detection procedure consists of two steps: firstly, the color similarity function is designed based on the quaternion-vector theory; secondly, the horizontal and vertical edges are distinguished based on the function presented in the first step. The characteristic points, as well as their height fringes, are then determined, and the height values are thus obtained. Finally, experiments are performed to show the effectiveness of the proposed approach.

Keywords: 3D height measurement · Edge detection · Encoding structure light · Quaternion-vector

1 Introduction

According to the complex physical environment constraint such as the precision advanced IC packaging production, the appearance detection can only be done by machine vision based image processing technique. For most mainstream IC chips, such as PQFP or QFP, the horizontal detection precision is getting higher and higher, at the same time, the speed of its appearance inspection is getting faster and faster, too. Especially, the detection precision and speed of the z-axis on the premise of maintaining the horizontal detection quality are limited by a number of factors, such as the image resolution between horizontal and vertical directions, the detecting precision of z-axis and its detecting speed. Most measuring systems can only inspect distances with high precision of one direction, may be horizontal or vertical. It takes much more measuring time, such as the method of laser scanning, the phase profilometry

This work was partially supported by the National 863 Program of China under Grant 2012AA041312; Project on the Integration of Industry, Education and Research of Guangdong Province (2012B091100039); National Science and Technology Major Project (2014ZX02503).

H. Liu et al. (Eds.): ICIRA 2015, Part III, LNAI 9246, pp. 299–309, 2015.
DOI: 10.1007/978-3-319-22873-0_26

approaches and the other structure light measurement systems. The method of laser scanning has high measuring precision, but takes much more processing time.

Recently, many researchers have paid more and more attention to the application of quaternion theory[1-4] in image processing, such as color image edge detector [5-9],color texture segmentation[10],color image registration[11],motion estimation[12-14] and so on. Some invariant principles based on quaternion theory are also introduced, e.g. quaternion Fourier-Mellin moments[15] and quaternion Zernike moments[16]. However, there is very rare successful application in micro-distance detecting system based on quaternion theory.

As to the color image processing, most traditional methods are only based on gray or vector. These two processing methods cannot uncover the spectrum relations among the R,G,B channels, so that it will produce bad effects on the detection speed and reliability. The color image processing approach based on quaternion handles R,G,B channels as a whole, therefore won't lose any details during processing and lead to high quality data support for the continued processing steps.

Image edge detection is often applied to object identification and segmentation in image processing. In this paper, according to 3D physical characteristics of QFP chips, a kind of detection system based on color fringes encoding structure light is used to conduct precise defect detection of 3D micro-height on-line. By using effective region segmentation to get closer to the characteristic points rapidly, the quaternion-vector edge detection algorithm is then applied to determine the feature point's position precisely and figure out the feature fringe boundary with the hidden height information. This method can improve the accuracy of the measurement system without increasing its processing time. Firstly, with the quaternion-vector approach, the relevant edges can be distinguished from the information hidden in the color pixels. Secondly, this approach only processes the points of ROI (region of interest) with image segmentation. The processing time will be shortened. Furthermore, the image segmentation is conducted based on the erosion and dilation methods of morphology in Reference [17-19] to get the rough coordinates of corners for image segmentation.

In order to inspect distances with high precision of both horizontal and vertical directions at the same time, and shorten the computation time to meet the real-time requirements, a new measurement system is developed by using the quaternion-vector (QV) theory in this paper. The remained parts of this paper are organized as follows. A brief introduction of quaternion is given in section 2. The quaternion-vector color similarity function is then provided in section 3. The quaternion-vector edge detection algorithm is also presented in section 4. The simulations and results analysis are finally presented in section 5.

2   Preliminaries

The notation of quaternion q is shown as below:

$$\begin{cases} q = q_r + q_i i + q_j j + q_k k \\ i^2 = j^2 = k^2 = ijk = -1 \\ ij = k, jk = i, ki = j \\ ji = -k, kj = -i, ik = -j \end{cases} \tag{1}$$

where q_r, q_i, q_j, q_k are real numbers and i, j, k are orthogonal imaginary units, and the quaternion representation $f_q(x,y)$ of color pixel is given as follows:

$$f_q(x, y) = 0 + q_R i + q_G j + q_B k = [0 \quad q_R \quad q_G \quad q_B] \tag{2}$$

where q_R, q_G, q_B are the red, green and blue components of the color pixel.

3   Quaternion-Vector Hybrid Color Similarity Function

To determine the edges between each two color fringes around the feature point, the quaternion-vector hybrid color similarity function [20,21] is proposed. According to the characteristic of the color image, each two color pixels is compared by the following formula.

$$f_{similarity}(x_1, y_1, x_2, y_2)$$
$$= \begin{cases} f_Q(x_1, y_1, x_2, y_2), & f_Q > T \\ f_V(x_1, y_1, x_2, y_2), & f_Q < T \end{cases} \tag{3}$$

where $f_{similarity}(x_1, y_1, x_2, y_2)$ is the similarity function between pixel (x_1, y_1) and pixel (x_2, y_2), $f_Q(x_1, y_1, x_2, y_2)$ is the quaternion color similarity function, its result will be used as the similarity degree if its value is bigger than the given threshold T, and the vector color similarity function $f_V(x_1, y_1, x_2, y_2)$ will be used if the quaternion color similarity degree value is smaller than T. The definition of $f_Q(x_1, y_1, x_2, y_2)$ is described as follows:

$$\begin{cases} f_Q(x_1, y_1, x_2, y_2) = \left\| f_q(x_{12}, y_{12}) - f_{q, gray}(x_{12}, y_{12}) \right\|_2 \\ f_q(x_{12}, y_{12}) = Rf_q(x_1, y_1)\overline{R} + f_q(x_2, y_2) \\ f_{q, gray}(x_{12}, y_{12}, i) = \frac{1}{\sqrt{3}} \sqrt{\sum_{k=2,3,4} f_q^2(x_{12}, y_{12}, k)}, i = 2,3,4 \\ R = e_n \\ \overline{R} \text{ is the conjugate quaternion of } R. \\ e_n = (i + j + k)/\sqrt{3} \end{cases} \tag{4}$$

function $f_V(x_1, y_1, x_2, y_2)$ is defined as below:

$$f_V(x_1, y_1, x_2, y_2) = \left\| f(x_1, y_1) - f(x_2, y_2) \right\|_2 \tag{5}$$

where $f_q(x_1, y_1)$ with the subscript "q" is the quaternion representation of color pixel (x_1, y_1), and $f(x_1, y_1)$ is its vector notation, and same as referring to other color pixels, $f_{q, gray}(x_{12}, y_{12})$ is the nearest approximate "gray" line of $f_q(x_{12}, y_{12})$, $f_q(x_{12}, y_{12}, i)$ is the number i element of quaternion $f_q(x_{12}, y_{12})$, $\|\|_2$ is the Euclidean distance.

As it is compared with the cosine similarity function, the quaternion-vector one can distinguish different colors better. For example, as to color pixel-pairs {[0,255,255],

[0,50,50]}, or {[0,255,255], [0,0,0]}, the results of the cosine method all get 1. It shows that all these colors are similar to each other, but it is a mistake obviously. As to quaternion-vector similarity function, the results are 83.6909 and 104.1033, and all the colors can be distinguished easily.

Different color pixel can be distinguished by quaternion similarity function except the neighboring pixels are almost gray, so that the vector similarity function is used to handle this particular case. For example, as to color pixel-pairs {[180,100,80], [100,100,100]} and {[255,255,255], [100,100,100]}, the results of quaternion method are 37.4166 and 0. It is obvious that the latter pair cannot be distinguished from each other by this way; therefore the vector one is used to calculate and get the answer of 268.4679.Using these two algorithms, all the color pixels can be distinguished by the similarity degree compared to the reasonable threshold value.

4   Quaternion-Vector Edge Detection

The color fringes' direction of the encoding structure light in the developed system is horizontal. The pixels' values of these horizontal direction fringes are thus almost the same within the stripe, and only change suddenly on the boundary between two different color stripes in ideal condition. In height measurement of IC chips, the sudden change of value will be happened with the change of height and then maintain the new value of the new height. According to the features of the IC subjects, the edge detection will be required to conduct from both horizontal and vertical directions.

4.1   Get the Horizontal Edges

There is a white line between each two color fringes normally in the encoding structure light system in use in this paper. If the pixel's color is similar to white or gray, it shows that this pixel is located in the boundary.

$$f_Q(x_1, y_1, x_g, y_g) = \left\| f_q(x_{1g}, y_{1g}) - f_{q, gray}(x_{1g}, y_{1g}) \right\|_2 \tag{6}$$

where $f_Q(x_1, y_1, x_g, y_g)$ is the similarity function between pixel (x_1, y_1) and its nearest approximate "*gray*" line (x_g, y_g). If the pixel in comparison is not similar to white or gray, compare the pixel with the former and the latter pixel in vertical direction using the similarity function introduced in section 3. If the two answers of the similarity function are all bigger than the given threshold $T_{horizontal}$, it shows that the pixel in comparison is in the horizontal boundary, or else it does not. Record all the horizontal edge points in the *H-array*.

$$\begin{cases} f_{similarity}(x, y, x-1, y) > T_{horizontal} \\ \quad and \ f_{similarity}(x, y, x+1, y) > T_{horizontal} \ , yes \\ f_{similarity}(x, y, x-1, y) < T_{horizontal} \\ \quad or \ f_{similarity}(x, y, x+1, y) < T_{horizontal} \ , no \end{cases} \tag{7}$$

4.2   Determine the Vertical Edges

Firstly, compare the pixel with white or gray as introduced in section 4.1 Then, if the pixel in comparison is not similar to white or gray, compare the pixel with the latter pixel in horizontal direction by using the similarity function introduced in section 3. If the answer of the similarity function is bigger than the given threshold $T_{vertical}$, it shows that the pixel in comparison is in the vertical boundary, or else it doesn't. Usually, the value of $T_{vertical}$ is bigger than $T_{horizontal}$. Record all the vertical edge points in the V-array.

$$\begin{cases} f_{similarity}(x, y, x, y+1) > T_{vertical}, yes \\ f_{similarity}(x, y, x, y+1) < T_{vertical}, no \end{cases} \tag{8}$$

If the value of vertical and horizontal thresholds get closer to zero, the more accurate results will be gotten.

4.3   Identify the Characteristic Points

Compared with the H-array and V-array, the cross-points are the coordinates of the characteristic points in need, and the adjoining horizontal boundaries are the characteristic fringes in which the height information of the feature points is hidden. With the characteristic points and their feature fringes, the height can be worked out and the defect type can also be identified.

$$\begin{cases} h = \dfrac{L}{d} \Delta row \cdot uPixHeight \\ \Delta row = row_{center, left} - row_{center, right} \end{cases} \tag{9}$$

where L is the vertical distance between reference plane and the plane of the camera, d is the distance between the center of camera lens and the center of the encoding structure light, Δrow is the pixel-shifting of the height fringe, and uPixHeight is the actual height of 1 pixel shifting. And $row_{center, left}$ and $row_{center, left}$ are the values of the left and right center of the height fringes.

5   Simulation and Analysis

By using the quaternion-vector edge detection algorithms mentioned in section 4, the images of the conditions with no rotating, left rotating and right rotating are processed in this section. The minimum resolution of the designed system is 0.01 mm, and the relevant experiments will be performed below with the reference height of 0.1 mm.

Fig. 1 shows the images with no rotating, comparing the quaternion-vector(QV) method with the canny and sobel ones. With the characteristic points and its relevant horizontal fringes, all the heights of the points have been calculated and partly showed in table 1. Data in table 1 show that the max absolute error of the QV method is less than 0.01 mm (0.005 21 mm), and far less than the one of the canny method(0.045 28 mm) and the one of the sobel method (0.101 85 mm).

Table 1. The results of Height with no rotating

Edge	Row No.	Height	Abs. Err	Max Abs. Err
QV	10	0.10077	0.00077	0.00521
	16	0.10031	0.00031	
	38	0.09815	0.00185	
	43	0.10463	0.00463	
	65	0.10521	0.00521	
	71	0.09925	0.00075	
Canny	9	0.10542	0.00542	0.04528
	16	0.10031	0.00031	
	36	0.07816	0.02184	
	39	0.10916	0.00916	
	39	0.13502	0.03502	
	44	0.09306	0.00694	
	64	0.14528	0.04528	
	71	0.09777	0.00223	
Sobel	10	0.10077	0.00077	0.10185
	17	-0.00185	0.10185	
	37	0.09818	0.00182	
	39	0.10916	0.00916	
	42	0.08492	0.01508	
	44	0.10571	0.00571	
	64	0.09028	0.00972	
	71	0.11469	0.01469	

As to the phase profilometry methods, e.g. the four-step phase shifting method, at least eight pictures should be taken, and after phase extraction, the difficulty of phase unwrapping will increase the occurrence probability of unknown errors, yield great measurement errors, and even result in measurement mistakes. The profilometry cannot be used in the on-line detection of precision electronic packing field due to its bad stability of detection accuracy and non-real-time processing feature.

Fig. 2 shows images with left rotating, and table 2 shows partial results of the heights. Data in table 2 show that the accuracy of the QV method does not getting worse with left rotations and the errors of canny and sobel are still much bigger than the QV ones.

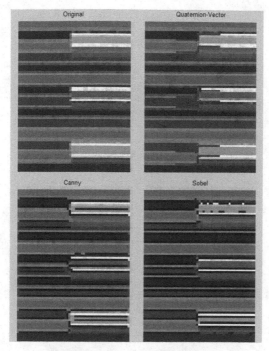

Fig. 1. Examples of no rotating

Table 2. The results of Height with left rotating (mm)

Edge	Row No.	Height	Abs. Err	Max Abs. Err
QV	16	0.10038	0.00038	0.00056
	21	0.09944	0.00056	
	43	0.10042	0.00042	
	71	0.10043	0.00043	
Canny	16	0.10028	0.00028	0.14358
	20	0.08394	0.01606	
	42	0.12400	0.02400	
	47	0.24358	0.14358	
	69	0.09565	0.00435	
Sobel	15	0.08390	0.01610	0.01610
	17	0.09054	0.00946	
	21	0.10592	0.00592	
	42	0.09448	0.00552	
	66	0.11401	0.01401	
	70	0.11633	0.01633	

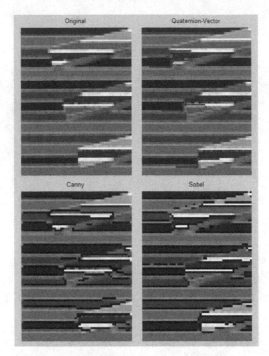

Fig. 2. Examples of left rotating

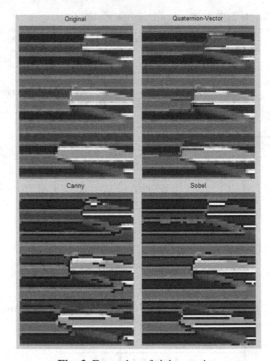

Fig. 3. Examples of right rotating

Fig. 3 is images with right rotating, and table 3 shows partial results of the heights. Data in table 3 show that same results of comparison as table 2.

Table 3. The results of Height with right rotating (mm)

Edge	Row No.	Height	Abs. Err	Max Abs. Err
QV	9	0.10522	0.00522	
	15	0.09920	0.00080	
	37	0.09866	0.00134	0.00522
	42	0.10501	0.00501	
	64	0.09960	0.00040	
Canny	8	0.10040	0.00040	
	15	0.09873	0.00127	
	35	0.09905	0.00095	0.01902
	42	0.10458	0.00458	
	63	0.11902	0.01902	
Sobel	8	0.09300	0.00700	
	13	0.08062	0.01939	
	15	0.09920	0.00080	
	36	0.12088	0.02088	0.02088
	43	0.10767	0.00767	
	63	0.08494	0.01506	

6   Conclusions

The experimental results show that the quaternion-vector edge detection algorithm which has been proposed in this paper can find all the relevant fringe edges of the characteristic points, and accurately locate the coordinates of the feature points. Compared with the other edge detecting approach, the QV method has higher accuracy. Compared with the laser scanning methods and phase profilometry approaches, the quaternion-vector edge detection algorithm can get the high detection precision as laser scanning methods and much shorter processing time. The future work will be focused on the improvement of the precision and processing speed of the algorithms.

References

1. Leiba, R.: Stability of Invariant Subspaces of Quaternion Matrices. Complex Analysis and Operator Theory **6**, 1069–1119 (2012)
2. He, Z.-H., Wang, Q.-W.: A real quaternion matrix equation with applications. Linear & Multilinear Algebra **61**, 725–740 (2013)
3. Haiyan, Z., Wenzhu, X.: Tame Kernels of Quaternion Number Fields. Communications in Algebra **42**, 2496–2501 (2014)
4. Benjamin, L.: Selectivity in quaternion algebras. Journal of Number Theory **132**, 1425–1437 (2012)
5. Florian, S., Andreas, K., Dietmar, H.: FPGA-Accelerated color edge detection using a geometric-algebra-to-verilog compiler. In: 15th Annual International Symposium on System-on-Chip (SoC). IEEE Press, Tampere (2013)
6. Xu, J., Ye, L., Luo, W.: Color Edge Detection Uing Multiscale Quaternion Convolution. International Journal of Imaging Systems and Technology **20**, 354–358 (2010)
7. Xin, G., Ke, C., Xiaoguang, H.: An improved Canny edge detection algorithm for color image. In: 10th IEEE International Conference on Industrial Informatics (INDIN), pp. 113-117. IEEE Press, Beijing (2012)
8. Gao, C.B., Zhou, J.L., Hu, J.R., et al.: Edge detection of colour image based on quaternion fractional differential. IET Image Processing **5**, 261–272 (2011)
9. Giuseppe, P., Nicolai, P.: Edge and line oriented contour detection: State of the art. Image and Vision Computing **29**, 79–103 (2011)
10. Li, X., Jin, L., Liu, H., et al.: A quaternion-based spectral clustering method for color image segmentation. In: 7th Symposium on Multispectral Image Processing and Pattern Recognition (MIPPR) - Automatic Target Recognition and Image Analysis, p. 800303-(1-8). SPIE Press, Guilin (2011)
11. Wang, Q., Wang, Z.: Color image registration based on quaternion Fourier transformation. Optical Engineering **51**, 057002-10 (2012)
12. Parker, J.G., Mair, B.A., Gilland, D.R.: Respiratory motion correction in gated cardiac SPECT using quaternion-based rigid-body registration. Medical Physics **36**, 4742–4754 (2009)
13. To, G., Mahfouz, M.R.: Quaternionic Attitude Estimation for Robotic and Human Motion Tracking Using Sequential Monte Carlo Methods With von Mises-Fisher and Nonuniform Densities Simulations. IEEE Transactions on Biomedical Engineering **60**, 3046–3059 (2013)
14. Alexiadis, D.S., Sergiadis, G.D.: Estimation of Motions in Color Image Sequences Using Hypercomplex Fourier Transforms. IEEE Transactions on Image Processing **18**, 168–187 (2009)
15. Guo, L.-Q., Zhu, M.: Quaternion Fourier-Mellin moments for color images. Pattern Recognition **44**, 187–195 (2011)
16. Chen, B.J., Shu, H.Z., Zhang, H., et al.: Quaternion Zernike moments and their invariants for color image analysis and object recognition. Signal Processing **92**, 308–318 (2012)
17. Jun, C., Anzheng, G.L., Lu, W.: Chessboard corner detection under image physical coordinate. Optics and Laser Technology **48**, 599–605 (2013)
18. Zhao, Q.-J., Zhao, D.-B.: New Algorithm for Corner Detection with Regulated Morphology. Journal of University of Electronic Science and Technology of China **39**, 886–890 (2010)

19. Senel, H.G., Cihan, I.K.: Topology based corner detection. In: 15th IEEE Signal Processing and Communications Applications, pp. 798-801. IEEE Press, Eskisehir (2007)
20. Amir, K., Orly, Y.-P.: Quaternion Structural Similarity: A New Quality Index for Color Images. IEEE Transactions on Image Processing 21, 1526–1536 (2012)
21. Alvarado-Cervantes, R., Felipe-Riveron Edgardo, M., Sanchez- Fernandez Luis, P.: Color Image segmentation by means of a similarity function. In: 15th Iberoamerican Congress on Pattern Recognition, pp. 319-328. SPRINGER-VERLAG, Sao Paulo (2010)

Dajiang Hong was born in the county of Chaozhou in Guangdong province, P.R.China. Hong received his M.S. degree in control theory and control engeering from South China University of China, Guangzhou, Guangdong, China, in 2003. He is currently studying for the Enn.D. degree in College of Automation Science and Engineering of SCUT. His research is about analysis and synthesis of control system.

Juan Du (Corresponding Author) was born in the county of Kaifeng in Henan province, P.R.China. Du got her Doctor's degree in 2010. Since 2003, She has been working in College of Automation Science and Engineering of SCUT. She was appointed as vice professor in 2012. Her research is about image recognition in electronic manufacturing Equipment.

Yueming Hu was born in the county of Jixi in Anhui province, P.R.China, professor Hu got his Bachelor's degree in 1982, Master's degree in 1985 and doctor's degree in June 1991 respectively. Since 1991, he has been working in College of Automation Science and Engineering of SCUT. Now he is the dean of Engineering Research Center of Precision Electronic Manufacturing Equipments, Ministry of Education, P.R.China.

Professor Hu has taken charge of and accomplished over 30 projects supported by National 863 Program, National Natural Science Foundation, etc. He has published over 150 journal papers such as Automatica, IEEE Trans. on AC, Acta Automatica Sinica. Now Professor Hu is devoting the basic research and development of the automatic electronic manufacturing equipments based on computer vision and intelligent motion control.

Parallel Stereo Matching Based on Edge-Aware Filter

Fan Bu[✉] and Chunxiao Fan

Beijing Key Laboratory of Work Safety Intelligent Monitoring,
School of Electronic Engineering, Beijing University of Posts and Telecommunications,
Beijing 100876, People's Republic of China
stbbff99@gmail.com, fcxg100@163.com

Abstract. This paper presents a novel parallel stereo matching algorithm based on edge-aware filter with good performance in accuracy and speed. The initial matching cost is built with census transform and sobel operator. Then the aggregated cost is computed by rolling guidance filter and guided filter. The final disparity is computed by rolling guidance filter and weighted median filter. The key idea is to eliminate the influence of small scale structures when computing weights in aggregation step and post-processing step. The proposed method ranks 17th on Middlebury benchmark and the results cost 52.5ms on one GPU and 33.8ms on two GPUs.

Keywords: Stereo matching · Edge-aware filter · Cost aggregation · Disparity

1 Introduction

Stereo matching seeks the correspondence in stereo images. The method can be classified into two categories: local methods and global methods [1]. Global methods regard the problem as energy function and minimize the function with belief propagation, graph cuts and so on. They can produce accurate results but they consume too much time. Local methods compute disparity within a support region. The traditional representative local methods are AdaptWeight [2, 3], GeoSup [4] and so on.

Traditional local methods are not well paralleled. So they can't achieve real-time performance on GPU. Nowadays, with the development of edge-aware filter, some local methods utilize edge-aware filter to do cost aggregation step. In article [5, 8-10], joint bilateral filter, guided filter and domain-transform filter are used to compute the weights in cost aggregation step. These methods achieve good performance in both accuracy and speed.

But, there exists a serious issue in the above methods. When computing weights in the aggregation step, some small scale structures will influence the weights' distribution. Zhang proposed rolling guidance filter [11] in 2014. This filter has an edge-aware property while removing small scale structures. The filter is simple in implementation and consumes little time. It gives us some clues to solve the issue.

In this paper, we propose a novel stereo matching algorithm based on edge-aware filter. We adopt a Census-Sobel measure to compute initial matching cost. Then we use rolling guidance filter and guided filter to do cost aggregation. Finally we utilize

© Springer International Publishing Switzerland 2015
H. Liu et al. (Eds.): ICIRA 2015, Part III, LNAI 9246, pp. 310–321, 2015.
DOI: 10.1007/978-3-319-22873-0_27

rolling guidance filter and weighted median filter to get the disparity. Our method is well-paralleled and we give an efficient implementation on CUDA. The results on Middlebury benchmark demonstrate the great performance in accuracy and speed.

The rest of the paper is organized as follows: Section 2 provides a summary of rolling guidance filter. Section 3 gives a full description of our algorithm. Section 4 gives the key of the implementation on CUDA. Section 5 shows the experimental results. Section 6 gives the conclusion.

2   Rolling Guidance Filter

Fig. 1 illustrates the work flow of rolling guidance filter [11]. This filter has two steps. Step 1 uses Gaussian filter to remove small scale structures. Step 2 uses edge-aware filter to do edge recovery, such as joint bilateral filter, guided filter and so on. Step 2 is an iterative step. The final J^n is the output image. Here we take guided filter for example. The equations are listed in Eq. (1-4). Step 1 uses Eq. (1-2) and step 2 uses Eq. (3-4). Guided filter takes one-channel image as input. But in Fig. 1, the input image I can be a 3-channel image (RGB). The solution is to handle each channel separately. Then we get three 1-channel images: J_R^n, J_G^n and J_B^n. Finally we combine the three 1-channel images to a 3-channel image. The detailed information about rolling guidance filter and guided filter can be found in [6, 7, 11].

$$J^1(p) = \frac{1}{K_p} \sum_{q \in N(p)} \exp\left(-\frac{\|p-q\|^2}{2\sigma_s^2}\right) I(q) \tag{1}$$

$$K_p = \sum_{q \in N(p)} \exp\left(-\frac{\|p-q\|^2}{2\sigma_s^2}\right) \tag{2}$$

$$J^{n+1}(p) = \sum_q W_{pq}(J^n) I(q) \tag{3}$$

$$W_{pq}(J^n) = \frac{1}{|w|^2} \sum_{k:(p,q) \in w_k} \left(1 + \frac{(J_p^n - \mu_k)(J_q^n - \mu_k)}{\sigma_k^2 + \varepsilon}\right) \tag{4}$$

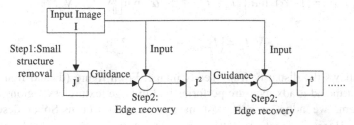

Fig. 1. Work flow of rolling guidance filter

Fig. 2 gives an example of rolling guidance filter. Figure (a) shows the original image while figure (b) shows the result image after rolling guidance filter. The small scale structures like small dots are well removed. And the edges of large scale structures like number 8 and number 2 are well preserved.

(a)   (b)

Fig. 2. (a) Input image. (b) Output image after rolling guidance filter.

3   Stereo Matching

Fig. 3 illustrates the work flow of our method. Like most local methods, our method is organized by four parts: matching cost computation, cost aggregation, disparity selection and post-processing [1]. Here we do rolling guidance filter on original stereo images and utilize the filtered images to do cost aggregation and post-processing.

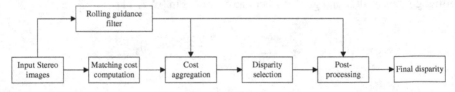

Fig. 3. Work flow of stereo matching

3.1   Matching Cost Computation

Rhemann et al. introduced a simple but effective measure in [8]. The cost is combined of differences of color and gradient, and we name it ad-gradx. It is formally expressed in Eq. (5). $C(p,d)$ is the matching cost. $\nabla_x(I_p)$ in Eq. (5) represents the x-direction gradient of pixel p in image I and the gradient is computed as Eq. (6).

$$C(p,d) = (1-\alpha) \cdot \min\left[\left\|I_p^L - I_q^R\right\|, \tau_1\right] + \alpha \cdot \min\left[\left\|\nabla_x I_p^L - \nabla_x I_q^R\right\|, \tau_2\right] \tag{5}$$

$$\nabla_x I_p = \left(I_{p+(1,0)} - I_{p-(1,0)}\right)/2 \tag{6}$$

Fig. 4 shows the disparity from Teddy image. From figure (d) and figure (e), we can see that the ad-gradx measure performs bad in large textureless regions. To solve this problem, we adopt a new cost measure named Census-Sobel described in Eq. (7-10). Here, $census(p)$ is the census transform of pixel p. $sobelx(p)$ is the x-direction gradient of pixel p computed by sobel operator and $sobely(p)$ is the y-direction gradient. Pixel q is the corresponding pixel in the right image when the disparity of pixel p is d. *Ham* means hamming distance. γ_{census}, γ_{sobelx} and γ_{sobely} are parameters. Figure (c) proves the effectiveness of our Census-Sobel measure.

$$C(p,d) = C_{census}(p,d) + C_{sobelx}(p,d) + C_{sobely}(p,d) \tag{7}$$

$$C_{census}(p,d) = 1.0 - \exp\left(-\frac{Ham[census(p), census(q)]}{\gamma_{census}}\right) \tag{8}$$

$$C_{sobelx}(p,d) = 1.0 - \exp\left(-\frac{\|sobelx(p) - sobelx(q)\|}{\gamma_{sobelx}}\right) \tag{9}$$

$$C_{sobely}(p,d) = 1.0 - \exp\left(-\frac{\|sobely(p) - sobely(q)\|}{\gamma_{sobely}}\right) \tag{10}$$

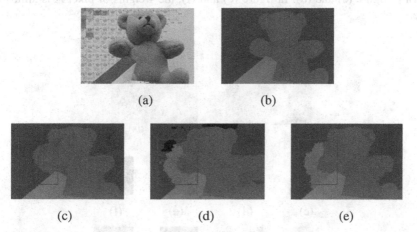

(a)   (b)

(c)   (d)   (e)

Fig. 4. (a) A part of Teddy image. (b) Groundtruth. (c) Our disparity with census-sobel cost. (d) Our disparity with ad-gradx cost. (e) Disparity in article [8], using ad-gradx cost.

3.2   Cost Aggregation

Rhemann et al. utilized guided filter to do cost aggregation [8] in Eq. (11-12). Here $C_2(p,d)$ is the aggregated cost of pixel p at disparity d. $W_{pq}(I)$ is the weight of pixel p and q in image I. I is the left image in the stereo images and $C(p,d)$ is the initial matching cost in Eq. (7). Details of Eq. (12) are included in article [6-9].

$$C_2(p,d) = \sum_q W_{pq}(I) C(p,d) \tag{11}$$

$$W_{pq}(I) = \frac{1}{|w|^2} \sum_{k:(p,q)\in w_k} \left(1 + \left(I_p - \mu_k\right)^T \left(\Sigma_k + \varepsilon U\right)^{-1} \left(I_q - \mu_k\right)\right) \tag{12}$$

When pixel p and q locate on the same side of the edge, $W_{pq}(I)$ computed by Eq. (12) is larger than the case locating on the opposite side of the edge. Locating on the same side means that pixel p and q have more opportunity to locate on the same disparity plane while locating on the opposited side of the edge means that pixel p and q have less opportunity. So the aggregation way in Eq. (11-12) is effective.

However, small scale structures may influence the weights' distribution. In Fig. 5-(a), all the pixels in the red square centered at pixel B locate on the same disparity plane so the weights in the red square should be almost the same. But in fact, pixel B is surrounded by some different color pixels and these pixels are small scale structures in figure (g). Here small scale structures mean they only occupy few pixels. Figure (i) shows the effects of these structures, the weights' distribution is disordered. Figure (h) shows the zoom of pixel B after rolling guidance filter. The small scale structures are removed while the weights are computed correctly in figure (j). Meanwhile, rolling guidance filter doesn't destroy the edges of large scale structures, as shown in figure (c) and (d). In figure (e) and (f), the weights of pixel A is almost the same.

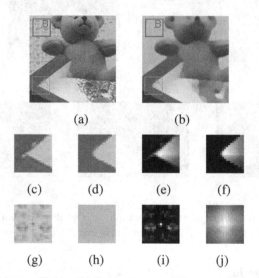

(a)   (b)

(c)   (d)   (e)   (f)

(g)   (h)   (i)   (j)

Fig. 5. (a) A part of Image Teddy. (b) Output image of figure a after rolling guidance filter. (c) Zoom of pixel A in figure a. (d) Zoom of pixel A in figure b. (e) Weights of pixel A in figure c. (f) Weights of pixel A in figure d. (g) Zoom of pixel B in figure a. (h) Zoom of pixel B in figure b. (i) Weights of pixel B in figure g. (j) Weights of pixel B in figure h.

In conclusion, we take Eq. (12-13) for cost aggregation step. Image J in Eq. (13) is the output image of image I after rolling guidance filter and the weight $W_{pq}(J)$ is calculated by Eq. (12). Namely, we use image J instead of image I in Eq. (11).

$$C_2(p,d) = \sum_q W_{pq}(J)C(p,d) \tag{13}$$

3.3   Disparity Selection and Post-Processing

We apply the Winner-Takes-all method (WTA) to choose the disparity for pixel p [1]. In Eq. (14), d_p is the disparity for pixel p, S is the set of the possible disparities. $C_2(p,d)$ is the aggregated cost.

$$d_p = \arg\min_{d \in S} C_2(p,d) \tag{14}$$

Once we get the initial disparity, we apply two strategies to do post-processing: left-right check and weighted median filter.

Left-Right Check

We compute two disparity images for this step. Disparity image 0 chooses left image as the reference view and disparity image 1 chooses right image as the reference view. Then we test the pixel's disparity by Eq. (15). Pixels which don't fulfill Eq. (15) will be marked as error pixels, the others will be marked as correct pixels. Here $d_L(x,y)$ represents the disparity for the pixel which coordinate is (x,y) in the left image.

The error pixels' disparities in disparity image 0 will be corrected by the disparity of the closest correct pixel [9] while the correct pixels' disparities keep unchanged. Detailed information is stated in article [8-9].

$$d_L(x,y) = d_R(x - d_L(x,y), y) \tag{15}$$

Weighted Median Filter

The disparity after left-right check will have some streak-like artifacts. Ma et al. proposed a simple but effective solution for the above problem in article [12]. They use weighted median filter to do post-processing in Eq. (16). Here I is the left image. The weight $W_{xy}(I)$ is computed by Eq. (12). Details about Eq. (16) refer to article [12].

$$h(x,i) = \sum_{y \in N(x)} w_{xy}(I)\delta(V(y) - i) \tag{16}$$

But there exists the same problem described in section 3.2. Small scale structures will influence the weights' values. Here we use Eq. (17) instead of Eq. (16). In Eq. (17), we use image J instead of image I. Image J is the output image of I after rolling guidance filter. Here image J is the same as the image J in section 3.2.

$$h(x,i) = \sum_{y \in N(x)} w_{xy}(J)\delta(V(y) - i) \tag{17}$$

4   Implementation on CUDA

Guided filter can be transformed into several box filters and box filter can be computed by integral image [14]. Thus, the efficient implementation of integral image is the key to reduce running time. Here, we take NVIDIA Tesla K20c GPU for example and state the implementation of integral image on CUDA.

4.1   Prefix Sum(Scan)

The definition of integral image is listed in Eq. (18). Here $I_s(x,y)$ is the integral image of image I. Eq. (19-20) are another definitions of integral image. So we can implement integral image in two steps. The first step computes the prefix sum in row-major order. The second step computes the prefix sum in column-major order.

$$I_s(x,y) = \sum_{i=0}^{x}\sum_{j=0}^{y} I(i,j) \tag{18}$$

$$I_X(x,j) = \sum_{i=0}^{x} I(i,j) \tag{19}$$

$$I_s(x,y) = \sum_{j=0}^{y} I_X(x,j) \tag{20}$$

Traditional prefix sum methods utilized shared memory to do threads' communication [15]. NVIDIA Kepler architecture GPUs give us a more efficient communication way using warp shuffle functions like __shfl(), __shfl_up() and so on. These functions can exchange a variable between threads within a warp without using shared memory. The example of 32 elements' scan is listed below [19].

```
__global__ void scan(void) {
  int laneId = threadIdx.x & 0x1f;
  int value = 31 - laneId;
  for (int i=1; i<32; i*=2) {
    int n = __shfl_up(value, i, 32);
    if (laneId >= i) value += n;
  }
}
```

Eq. (19) computes each row's scan in the image. We can use one warp to do one row's scan [16] as Fig. 6 shows. When warp 0 finishes the computation of the first 32 pixels' scan (Pixel 0-31), we use a temporary value to store the 31th pixel's prefix sum. When warp 0 finishes the second 32 pixels' scan (Pixel 32-63), we add the temporary value to the second 32 pixels' scan. Then we update the temporary value with the 63th pixel's prefix sum. Repeat the above procedures until all the elements' prefix sum is finished. Eq. (20) is implemented by the same way.

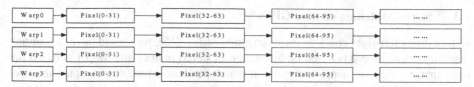

Fig. 6. Scan in row-major order

4.2   Novel Implementation of Integral Image

The two-dimension array is stored by row-major order in CUDA. So when computing scan in column-major order, the accesses of memory can't be coalesced. Thus much time is wasted in the accesses of memory. Bilgic et al. utilized matrix transpose to handle this problem in [17]. The method of integral image in [17] is listed below:

1. Compute the row-major scan
2. Matrix transpose
3. Compute the row-major scan
4. Matrix transpose

However, the matrix transpose procedure is not necessary. We propose a novel method for integral image. In Fig. 7, we will compute A's integral image. A is a 2-dimension array. The height of A is H and the width of A is W. B is the transpose of A. First we combine A's memory and B's memory with CUDA's texture memory. Texture memory has cache optimized for 2D spatial locality, so threads of the same warp that read texture addresses that are close together will achieve the best performance [19]. Firstly, we compute the column-major scan of A as section 4.1 shows but we write the results into B. Thus, we use one warp to compute A's one column's prefix sum and write results into one row of B. The row's number in B and the column's number in A should be the same. Secondly, we do the same procedure to array B. we use one warp to compute B's one column's prefix sum and write the result into one row of array C. Here C is another array which has the same size of A. Finally, C is the integral image of A. Our method only takes two column prefix sum, avoiding matrix transpose steps. So our method has better performance than the method in [17].

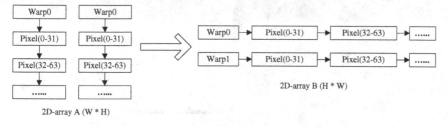

Fig. 7. Implementation of integral image

4.3   Stereo Matching on Two GPUs

Before doing left-right check as the description of Eq. (15), we should have two initial disparity maps. Map 0 is based on left image in stereo images and Map 1 is based on right image. The two maps' computation is independent, so we can compute the two maps on two GPUs to save about 1/2 time. Thus, we compute map 0 on device 0 and compute map 1 on device 1. The key of the implementation is minimizing memory's copy time between two GPUs. When the program is run as a 64-bit process, devices of Tesla GPU can use peer-to-peer access feature to accelerate the copy process [19]:

1. Call function cudaDeviceEnablePeerAccess() to enable peer-to-peer access.
2. Call function cudaMemcpyPeer() and so on to do the memory copy.

5   Experimental Results

We test our method with the Middlebury benchmark [18]. We use a server with Intel E5-2620-v2 2.1GHz CPU and NVIDIA Tesla K20c graphics card. The parameters are listed in Table 1. r_1 is the radius of Gaussian filter and r_2 is the radius of guided image filter. *Iters* are the times of iteration in rolling guidance filter.

Table 1. The parameters in our method.

σ_s	r_1	ε	r_2	*Iters*
1.06	1	0.000215	7	4
γ_{census}	γ_{sobelx}	γ_{sobely}	r_{census}	
7.5	2.5	10.3	1	

Fig. 8 shows the results of our method. Figure (a) shows the final disparity maps and figure (b) shows the error maps under pixel error threshold 1.

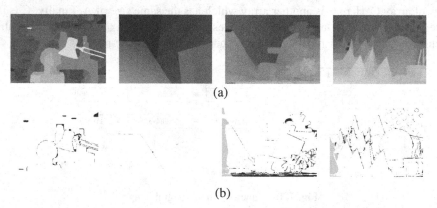

(a)

(b)

Fig. 8. (a) Final disparity maps. (b) Error maps under pixel error threshold 1.

Table 2 lists the average error ratio of several stereo matching methods on the Middlebury datasets. Our method has better performance over the other methods in the table. Up to May 18, 2015, our method ranks 17th of all the 162 methods. The key of our method is eliminating small structures' effect when computing weights.

Table 2. Evaluation for several local methods on Middlebury stereo images.

Algorithms	Avg. Error (%)	Rank
Ours	4.80	17
Guided agg+WM[12]	5.50	n/a
CostFilter[8]	5.55	50
GeoSup[4]	5.80	60
AdaptWeight[2]	6.67	98

For further proving our method's innovation, Table 3 shows contrast between our method and three other experiments. Experiment 1 only takes ad-gradx cost in Eq. (5) instead of Census-Sobel cost in Eq. (7) while other steps keep the same. Experiment 2 only takes Eq. (11-12) for cost aggregation instead of Eq. (12-13) and experiment 3 only takes Eq. (16) for weighted median filter instead of Eq. (17). From table 2, our method has better performance over the other three experiments.

Table 3. The contrast between our method and three experiments.

Algorithm	Avg. Error (%)
Ours	4.80
Ex. 1	5.77
Ex. 2	5.00
Ex. 3	4.84

Our method is easily implemented on CUDA. Table 4 gives the running time of our method on one GPU and on two GPUs. The running time includes all the steps in our method. The average time is 52.5ms on one GPU and 33.8ms on two GPUs.

Table 4. Running time of our method.

Image	Max Dis	Resolution	Time on 1 GPU (ms)	Time on 2 GPUs (ms)
Tsukuba	15	384*288	18.75	11.63
Venus	19	434*383	30.80	19.43
Teddy	59	450*375	80.24	52
Cones	59	450*375	80.24	52

Table 5 lists several other methods' average running time on Middlebury image sets. The data is selected from article [8, 12]. Our method is a little faster than the others on 1 GPU. The main reason is our fast implementation of integral image on GPU. The methods which rank higher than our method usually utilize Graph cuts, belief propagation and so on to get more accurate results, which will cost much time. So balancing the disparity's accuracy and running time, our method is a great choice.

Table 5. Evaluation of running time of several methods.

Algorithm	Avg. Time on GPU (ms)	Avg. Error (%)
Ours(1 GPU)	52.5	4.80
Ours(2 GPUs)	33.8	4.80
CostFilter(1 GPU)	65	5.55
Guided agg+WM(1 GPU)	54	5.50

Table 6 gives the running time of the integral image. Here the testing image's resolution is 1M (1024 * 1024) and the type of each pixel is float4. We can see our implementation described in section 4 has the best performance.

Table 6. Running time of integral image

Algorithm	Avg. Time (ms)
Ours	0.72
Our prefix scan + matrix transpose	2.1
Method in [17]	4.0

Fig. 9 gives the photos captured by Logitech C270 consumer cameras in real scenes. We can notice that our approach can produce accurate disparity image.

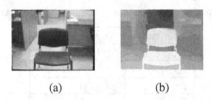

(a)   (b)

Fig. 9. (a) Left input image. (b) Disparity image.

6   Conclusion

This paper has presented a parallel stereo matching algorithm based on edge-aware filter and fast implementation on CUDA. Our method is based on three key points: using Census-Sobel cost measure, utilizing rolling guidance filter and guided filter to do cost aggregation and utilizing rolling guidance filter and weighted median filter to do post-processing. The effect of small scale structures is eliminated when computing weights, thus the disparity quality is improved. The results on Middlebury stereo images show the great performance in accuracy and speed. As heterogeneous computing technology develops, we will explore more efficient implementation on GPU.

Acknowledgement. The work was supported by the National Natural Science Foundation of China (Grants No. NSFC-61402046, NSFC-61170176), Fund for the Doctoral Program of Higher Education of China (Grants No.20120005110002), Fund for Beijing University of Posts and Telecommunications (No.2013XZ10, 2013XD-04).

References

1. Scharstein, D., Szeliski, R.: A taxonomy and evaluation of dense two-frame stereo correspondence algorithms. International Journal of Computer Vision **47**(1–3), 7–42 (2002)
2. Yoon, K.J., Kweon, I.S.: Locally adaptive support-weight approach for visual correspondence search. In: IEEE Computer Society Conference on Computer Vision and Pattern Recognition, CVPR 2005, vol. 2, pp. 924–931. IEEE (2005)
3. Yoon, K.J., Kweon, I.S.: Adaptive support-weight approach for correspondence search. IEEE Transactions on Pattern Analysis and Machine Intelligence **28**(4), 650–656 (2006)
4. Hosni, A., Bleyer, M., Gelautz, M., et al.: Local stereo matching using geodesic support weights. In: 16th IEEE International Conference on Image Processing (ICIP), pp. 2093–2096. IEEE (2009)
5. Mattoccia, S., Giardino, S., Gambini, A.: Accurate and efficient cost aggregation strategy for stereo correspondence based on approximated joint bilateral filtering. In: Zha, H., Taniguchi, R.-i., Maybank, S. (eds.) ACCV 2009, Part II. LNCS, vol. 5995, pp. 371–380. Springer, Heidelberg (2010)
6. He, K., Sun, J., Tang, X.: Guided image filtering. In: Daniilidis, K., Maragos, P., Paragios, N. (eds.) ECCV 2010, Part I. LNCS, vol. 6311, pp. 1–14. Springer, Heidelberg (2010)
7. He, K., Sun, J., Tang, X.: Guided image filtering. IEEE Transactions on Pattern Analysis and Machine Intelligence **35**(6), 1397–1409 (2013)
8. Rhemann, C., Hosni, A., Bleyer, M., et al.: Fast cost-volume filtering for visual correspondence and beyond. In: 2011 IEEE Conference on Computer Vision and Pattern Recognition (CVPR), pp. 3017–3024. IEEE (2011)
9. Hosni, A., Rhemann, C., Bleyer, M., et al.: Fast cost-volume filtering for visual correspondence and beyond. IEEE Transactions on Pattern Analysis and Machine Intelligence **35**(2), 504–511 (2013)
10. Pham, C.C., Jeon, J.W.: Domain transformation-based efficient cost aggregation for local stereo matching. IEEE Transactions on Circuits and Systems for Video Technology **23**(7), 1119–1130 (2013)
11. Zhang, Q., Shen, X., Xu, L., Jia, J.: Rolling guidance filter. In: Fleet, D., Pajdla, T., Schiele, B., Tuytelaars, T. (eds.) ECCV 2014, Part III. LNCS, vol. 8691, pp. 815–830. Springer, Heidelberg (2014)
12. Ma, Z., He, K., Wei, Y., et al.: Constant time weighted median filtering for stereo matching and beyond. In: 2013 IEEE International Conference on Computer Vision (ICCV), pp. 49–56. IEEE (2013)
13. Zabih, R., Woodfill, J.: Non-parametric local transforms for computing visual correspondence. In: Eklundh, J.-O. (ed.) ECCV 1994. LNCS, vol. 801, pp. 151–158. Springer, Heidelberg (1994)
14. Crow, F.C.: Summed-area tables for texture mapping. ACM SIGGRAPH Computer Graphics **18**(3), 207–212 (1984)
15. Harris, M., Sengupta, S., Owens, J.D.: Parallel prefix sum (scan) with CUDA. GPU Gems **3**(39), 851–876 (2007)
16. Li, J.: High performance edge-preserving filter on GPU. NVIDIA GTC (2015)
17. Bilgic, B., Horn, B.K.P., Masaki, I.: Efficient integral image computation on the GPU. In: 2010 IEEE Intelligent Vehicles Symposium (IV), pp. 528–533. IEEE (2010)
18. http://vision.middlebury.edu/stereo
19. NVIDIA C Programming Guide Version 7.0

Object Tracking of the Mobile Robot
Using the Stereo Camera

Hyun-Uk Ha[1(✉)], Ha-Neul Yoon[1], Yun-Ki Kim[1], Dong-Hyuk Lee[1],
and Jang-Myung Lee[2]

[1] Department of Electrical and Computer Engineering,
Pusan National University, Busan, South Korea
hyunuk.ha@gmail.com, {haneul1696,mecha8404,ldh0917}@pusan.ac.kr
[2] Department of Electronic Engineering, Pusan National University, Busan, South Korea
jmlee@pusan.ac.kr

Abstract. This paper proposes a visual servoing algorism for the object tracking by a mobile robot with the stereo camera. The mobile robot performs an object recognition and object tracking using the SIFT and CAMSHIFT algorism
for the visual servoing. The CAMSHIFT algorism has been used to obtain the
three-dimensional position and orientation of the mobile robot. With the visual
servoing, a stable balance control has been realized by a control system which
calculates a desired angle of the center of gravity whose location depends on
variations of link rotation angles of the manipulator. To demonstrate the control
performance of the visual servoing, real experiments ware performed using the
mobile manipulator system developed for this research.

Keywords: Object tracking · Mobile robot · Stereo camera · CAMSHIFT ·
SIFT · Visual servoing

1 Introduction

The robot is widely used in the industrial sites as well as daily life because the technic
and performance have been developed constantly in the robot industry[1].
The performance of robot has been improving with the development of robotics technology like as Computer technology, Sensor, Actuator[2]. Recently, intelligent robot
is equipped the camera and acquire the vision information. And then the acquired
vision information has been used a system like as the object recognition, real-time
monitoring.

The robot platform equipped the vision is generally fixed model and moving model. The fixed model is designed to be operated in the industry cite for faulty inspection. So, It is easy to equip the vision system to the fixed robot. Otherwise, the
moving robot is changed frequently the environment because the robot platform is
moving. So, It is difficulty to equip the vision system to the moving robot. According
to the application field, a variety of vision technique is needed. But moving robot
platform is needed suitable vision technique because of the special conditions[3].

© Springer International Publishing Switzerland 2015
H. Liu et al. (Eds.): ICIRA 2015, Part III, LNAI 9246, pp. 322–330, 2015.
DOI: 10.1007/978-3-319-22873-0_28

Fig. 1. Visual Servoing for Mobile Manipulator

The object estimation technique using the vision system is divided into a Position Based Visual Servoing(PBVS) and Image Based Visual Servoing(IBVS)[4-5]. Recently, the research has been studied fusing two technique. Two visual control methods are switched based on a certain boundary. Also, Camera viewing angle is adjusted to use the IBVS to compensate disadvantage of PBVS through the panning and tilting[6]. But, the additional device must be equipped for the panning and tilting. And more complex algorism must be designed because the equipped robot model has a decisive effect on the mobile robot control.

In this paper, hybrid visual servoing control technique is suggested for object estimation of mobile manipulator. Overall system is composed of SIFT and CAMSHIFT. SHIF algorism is used for object estimation and CAMSHIFT algorism is used for object tracking[7-8]. First, The center point of object is acquired to apply CAMSHIFT to the cognized object using SIFT. And then left and right image are applied the same algorism using the acquired center point. Finally, 3-dimension position information can be acquired through calculating a Disparity. Also, the rotation information of object can be acquired using a SIFT and CAMSHIFT information. 3-dimension position information can be used on visual servoing based on location of mobile manipulator[9]. Visual servoing based on visual information conjugates an error on the image. So, it can reduce control error and can operate successive moving despite of parameter error[10].

First, the object recognition using SIFT and CAMSHIFT, object tracking are explained in chapter 2. Second, the problem of PBVS, IBVS are explained in chapter3. And then the experiment result applying the suggested algorism is explained in chapter 4. Finally, the paper is concluded in chapter 5.

2   Vision System

The object detecting needs 3-dimension stereo image-related data for following and carrying out smoothly on a robot. For visual servoing, object detecting has to be ahead. And it will use target's special point.

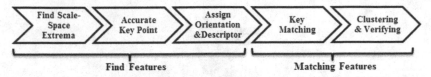

Fig. 2. SIFT algorism flow chart

2.1   SIFT(Scale Invariant Feature Transform)

SIFT algorism has two steps. First step is finding Key point and attaching descriptor. Second step is Matching step that compare natural DB image and target image using descriptor. In other words, in first step, after finding image Features(at this time we decide Extrema to Feature) on various scale we determine direction and amplitude data that can be solved at each point. Second step is comparing feature calculated in previous step's DB and distance of Feature, matching and sorting inlier and outlier for exact matching. Finally, we demonstrate matching that using Least Mean Square[7].

In the study we use SIFT algorism basically for object detecting. However, target's 3-dimension position data that using SIFT algorism is unsuitable on visual servoing, because it is not correct and has a lot of noise. So, improvingly we use CAMSHIFT(Continuously Adaptive Meanshift) algorism on object detecting.

2.2   CAMSHIFT(Continuously Adaptive Mean Shift)

CAMSHIFT algorism is tracing algorism using object's color data. It traces similar color part continuously base on object's color-histogram. At first selecting object, and after converting subject's RGB data to fig1's HSV color model, RGB data will be Hue data on histogram that represent color. It will be used at tracing model and be selection reason for object's boundary. Detecting window be set by interesting area selection will trace object until converged at this boundary by position and amplitude[8]. By applying CAMSHIFT algorism to object that recognized SIFT, we can find central point and decide central point to main special point, apply same algorism to right & left image base on main special point, and get Disparity between special points. So, we can get very precious and powerful 3-dimension data. CAMSHIFT has feature getting central point by using object's blob extent, so it is strong to light and noise. By the way, object's rotation information uses mutually affine parameter and CAMSHIFT's direction vector. In conclusion we made mixed vision system by using two algorisms.

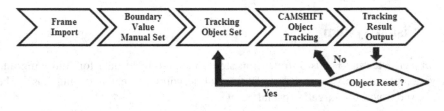

Fig. 3. CAMSHIFT algorism flow chart

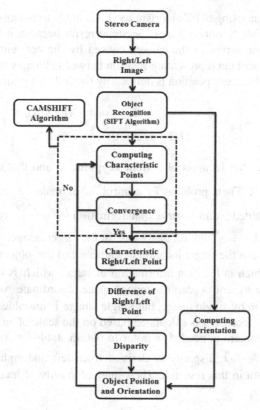

Fig. 4. Block diagram of the vision system

3   HVS(Hybrid Visual Servoing)

The existing visual servoing methods each have advantages and disadvantages. In this paper, in order to take the best advantage of the existing visual servoing method, combination of IBVS and PBVS is utilized to organize HVS system. By default, the HVS system is configured by cooperatively implementing CAMSHIFT algorism for object tracking and SIFT algorism for object recognition, separately.

Table 1. Kinds and characteristics of stereo vision

Stereo Vision with Position Based Visual Servoing (PBVS)	Stereo Vision with Image Based Visual Servoing (IBVS)
Optical parallel	Optical cross
Coordinate measuring by fixel	Coordinate measuring by rotation angle
Calibration is necessary	Calibration is unnecessary
Existing always Calibration error	Needing Angle measurement motor

Object Recognition using SIFT algorism has been made to position of camera more precisely. It is possible to obtain a more accurate result because it behaves robust to parameters calibration errors of the camera caused by the conventional IBVS techniques. The image Jacobian representing relation between changes the image's feature points and change of camera position is defined by the following formula (1).

$$L(s,Z) = \begin{bmatrix} -1/Z & 0 & x/Z & xu & -(1+x^2) & u \\ 0 & -1/Z & u/Z & 1+u^2 & -xu & -x \end{bmatrix} \tag{1}$$

The error between the location of current keypoint s and that of target keypoint s^* is $e = C(s - s^*)$. Then problem of control is to make e zero. Matrix C of weights can be regarded as an inverse transformation $L(s^*, Z^*)^+$ of the image Jacobian that $s = s^*$ and $z = z^*$ are set as a value in the target image. z^* is the approximate distance between the target location of camera and the object. The space coordinate (X, Y, Z) which is location information of target which is needed for PBVS. Z value which represent a depth among space coordinate was derived from CAMSHIFT algorism by single image. but single image is unstable and insensitive to movement of object because it is calculated based on the scale of an object. So disparity value of stereo camera is needed in order to obtain stable Z value. The point of center of gravity (x_c, y_c) disparity is derived from left and right image by using CAMSHIFT algorism in this research. The center of gravity of image can be derived as follows.

$$M_{00} = \sum_x \sum_y I(x, y) \tag{2}$$

$$M_{10} = \sum_x \sum_y x I(x, y) \tag{3}$$

$$M_{01} = \sum_x \sum_y y I(x, y) \tag{4}$$

$$x_c = \frac{M_{10}}{M_{00}}, \quad y_c = \frac{M_{01}}{M_{00}} \tag{5}$$

M_{00} in Eq.2 is the value of 0-th moment, Eq.3 and 4 are the value of first moment. x_c, y_c represent a point of center of mass. If these points are converged within some range, x_0, y_0 will be point of center of mass. This point less affected for noise and illumination changes, because it has averaged information of the target object. And, it's possible to simply determine the distance of mass by obtain Disparity of a point. Eq. 6 shows z which is established distance using Disparity.

$$z = \frac{fb}{(x_l - x_r)} = \frac{fb}{d} \tag{6}$$

Each value of f, b represents focal distance and distance value of left and right lens in stereo camera. z is the distance to the object from stereo camera. d indicates x-coordinate difference matching point of between the left and right image, namely Disparity. Because Orientation is known independently by the CAMSHIFT and SIFT algorism, it used to compare the value. Essentially, calculated each frame for 2-dimension Orientation that is generated by CAMSHIFT algorism is used to Visual Servoing. Orientation of SIFT algorism is not used to a real-time Visual Servoing because it is calculated by object recognition. If the error between Orientation of SIFT algorism and CAMSHIFT algorism is exceed a certain value, it recognize the Orientation value for stable and accuracy of Orientation value of CAMSHIFT algorism[11].

Proposed algorism in this paper is similar to existing PBVS except that 3D information of target object isn't used. Target location of camera which is used for PBVS was predicted ahead of time by the earlier proposed method for estimating a location. So, the camera can reach target location by selecting the component of velocity derived from equation (7) by using difference between current location and the location of target object.

$$w = \lambda_r r, \quad v = \lambda_t t \tag{7}$$

λ_r and $\lambda_r \in R$ are positive value of control gain. In this paper, control vector of existing PBVS wasn't used as it is and we modified the renewal of component of angular velocity in order to solve above-mentioned problem about the existing method for visual servoing by using equation (8).

$$w = \lambda_r r' = \lambda_r (\alpha r + (1 - \alpha) r^{object}) \tag{8}$$

Here, r' was derived as a weighted sum of r solved by $r^{current} - r^{t\,\arg et}$ and r^{object} which shows an orientation of an object. The renewal of component of location and velocity was determined by error t of location component is consist of $r^{current} - r^{t\,\arg et}$ which is difference between current location and target object.

4   Experiment Result

The experiment result is acquired the center point of object is acquired to apply CAMSHIFT to the cognized object using SIF. And then apply to the mobile manipulator.

The experiment was performed in the following procedure. First, two objects(=golf ball) are located from camera about 0.98[m], 2.01[m]. And then distance error is calculated about 10[sec] using HVS and PBVS, IBVS methods. Finally, three result using HVS, PBVS, IBVS method is compared for accuracy of three methods. The FRAME speed is 125[FPS]. It operates continually stereo calibration.

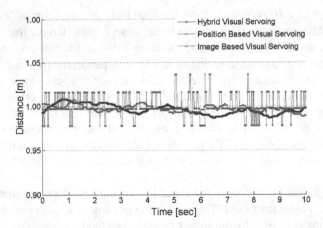

Fig. 5. Comparison of the distance error measuring 1.0m

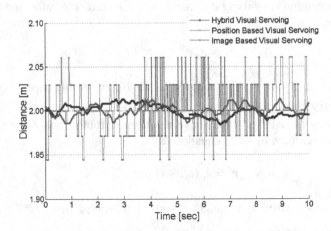

Fig. 6. Comparison of the distance error measuring 2.0m

Figures 5-6 are the result of the measurement three methods(HVS and PBVS, IBVS). In this figure, PBVS method has the largest error. And, HVS method is more stable than IBVS.

Error of three methods is increasing in case distance grows from object. But, HVS and IBVS have more improved result than PBVS about 0.02 ~ 0.06[m]. Also, PBVS and IBVS have a limitation to limited viewing angle and Calibration error. So, It is difficult to apply visual servoing separately. In order to solve stated shortcomings, this paper proposes HVS, which compensate limited FOV and singularity problem by PBVS and IBVS, respectively, and realize reliable visual servoing algorism for mobile manipulator based on stereo camera.

5   Conclusion

In this paper, HVS(Hybrid Visual Servoing) for object estimation is proposed using the Stereo Camera. For HVS, SIFT and CAMSHIFT algorism are used. It can estimate correct object recognition. CAMSHIFT is used for acquiring 3-dimension location and orientation using Stereo camera. HVS is applied to Mobile Manipulator using acquired image data. The experiment in this paper is that real-time image and mobile Manipulator is used. And research grapping the object is doing now.

Acknowledgments. This work was supported by the National Research Foundation of Korea(NRF) Grant funded by the Korean Government(MSIP) (NRF-2013R1A1A2021174).

This research was supported by the MOTIE (Ministry of Trade, Industry & Energy), Korea, under the Industry Convergence Liaison Robotics Creative Graduates Education Program supervised by the KIAT (N0001126).

References

1. Kim, D.Y., Park, J.W., Lee, C.W.: Object-Tracking System Using Combination of CAMshift and Kalman filter Algorithm. Journal of Korea Multimedia Society **16**(5), 619–628 (2013)
2. Tak, M.H., Cho, Y.J., Joo, Y.H.: Behavior Control for Cooperative Exploration for Swarm Robots. In: Proceedings of KIIS Fall Conference, Korea, vol. 21(2), pp.169–170, December 2011
3. Kim, S.H.: The robot vision technology trend for the intelligent mobile robot. Journal of Korea Robotics Society **9**(1), 26–35 (2012). (in Korean)
4. Wang, X., Li, X.: Study of movingtarget tracking based on kalman - CAMshift in the video. In: 2010 2nd International Conference Information Science and Engineering (ICISE), pp. 1–4 (2010)
5. Wang, Y., Lang, H., de Silva, C.W.: A Hybrid Visual Servo Controller for Robust Grasping by Wheeled Mobile Robots. IEEE/ASME Trans. of Mechatronics **15**(5), 757–769 (2010)
6. Low, E.M.P., Manchester, I.R., Savkin, A.V.: A biologically inspired method for vision-based docking of wheeled mobile robots. International Journal of Robotics and Autonomous systems **55**(10), 769–784 (2007)
7. Lowe, D.G.: Distinctive Image Features from Scale-Invariant Keypoints. International Journal of Computer Vision **60**(2), 91–100 (2004)
8. Lee, D.Y., Seo, B.C., Kim, S.S., Park, S.H.: Verification of Camera-Image-Based Target-Tracking Algorithm for Mobile Surveillance Robot Using Virtual Simulation. The 10th International Trans. Korea. Soc. Mech. Eng. A **36**(11), 1463–1471 (2012)
9. Lippiello, V., Siciliano, B., Vilani, L.: Position Based Visual Servoing in Industrial Multirobot Cells Using a Hybrid Camera Configuration. IEEE Trans. of Robotics **23**(1), 73–86 (2007)

10. Horaud, R., Dornaika, F., Espiau, B.: Visually Guided Object Grasping. IEEE Trans. of Robotics and Automation 14(4), August 1998

11. Kim, C.O., Choi, S., Cheong, J.N., Yang, G.W., Kim, H.S.: Robust Position Tracking for Position-Based Visual Servoing and Its Application to Dual-Arm Task. Journal of Korea Robotics Society 2(2), 129–136 (2007). (in Korean)

12. Kim, D.H., Chung, M.J.: A Switched Visual Servoing Technique Robust to Camera Calibration Errors for Reaching the Desired Location Following a Straight Line in 3-D Space. Journal of Korea Robotics Society 1(2), 125–134 (2006). (in Korean)

Robust Visual Tracking Based on Improved Perceptual Hashing for Robot Vision

Mengjuan Fei[1], Jing Li[2(✉)], Ling Shao[3], Zhaojie Ju[4], and Gaoxiang Ouyang[5]

[1] Institute of Cyber-Systems and Control, Zhejiang University, HangZhou 310027, China
feimegnjuan@zju.edu.cn
[2] School of Information Engineering, Nanchang University, Nanchang 330031, China
jingli@ncu.edu.cn
[3] Department of Computer Science and Digital Technologies, Northumbria University,
Ellison Place 2, Newcastle-upon-Tyne NE1 8ST, UK
ling.shao@ieee.org
[4] Intelligent Systems and Biomedical Robotics Group, School of Computing,
University of Portsmouth, Portsmouth PO1 3HE, UK
zhaojie.ju@port.ac.uk
[5] National Key Laboratory of Cognitive Neuroscience and Learning,
Beijing Normal University, Beijing, China
ouyang@bnu.edu.cn

Abstract. In this paper, perceptual hash codes are adopted as appearance models of objects for visual tracking. Based on three existing basic perceptual hashing techniques, we propose Laplace-based hash (LHash) and Laplace-based difference hash (LDHash) to efficiently and robustly track objects in challenging video sequences. By qualitative and quantitative comparison with previous representative tracking methods such as mean-shift and compressive tracking, experimental results show perceptual hashing-based tracking outperforms and the newly proposed two algorithms perform the best under various challenging environments in terms of efficiency, accuracy and robustness. Especially, they can overcome severe challenges such as illumination changes, motion blur and pose variation.

Keywords: Visual tracking · Perceptual hashing · AHash · PHash · DHash

1 Introduction

In robotics, robots mimic human behaviors through environmental information collected from different kinds of sensors for various applications. Visual tracking aims at generating the movement trajectories of a robot or an object by calculating its state information such as position, speed and the relationships between frames. As an important topic in robot vision, not only does visual tracking provide reliable data for motion analysis and scene understanding, but also underpins a wide range of applications, including military navigation guidance [1], autonomous driving [2], activity analysis [3]. Recently, researchers have proposed numerous algorithms to this

© Springer International Publishing Switzerland 2015
H. Liu et al. (Eds.): ICIRA 2015, Part III, LNAI 9246, pp. 331–340, 2015.
DOI: 10.1007/978-3-319-22873-0_29

fundamental subject, most of which fall into three main categories, i.e., kernel tracking, point tracking and silhouette tracking.

Recent tracking algorithms proposed in the literature perform well in terms of accuracy, such as compressive sensing for tracking [4], sparse representation [5] and semi-supervised learning [6]. However, the above-mentioned algorithms require high-computational costs owing to the complexity of appearance models. To this end, we apply perceptual hashing to visual tracking and propose two perceptual hashing algorithms which build a fast and robust image representation for object tracking. Firstly, three appearance models are built for three basic perceptual hashing methods, namely average hash (aHash), perceptive hash (pHash), and difference hash (dHash), respectively. To deal with challenging video sequences with more complex scenarios, we propose two improved perceptual hashing methods for object tracking, namely Laplace-based hash (LHash) and Laplace-based difference hash (LDHash). Consequently, visual tracking based on perceptual hashing can run in real time and outperform some state-of-the-art trackers, while the two newly proposed perceptual hashing algorithms perform better than the basic perceptual hashing methods on most video sequences in terms of accuracy and robustness.

The remainder of this paper is organized as follows. Section 2 discusses related works. Section 3 describes three basic perceptual hashing algorithms and two improved perceptual hashing methods. In Section 4, perceptual hashing-based tracking is presented. Experimental settings and results are given in Section 5. We conclude the paper and point out directions for future work in Section 6.

2   Related Work

Researches have attempted to deal with various kinds of challenges in visual tracking, including occlusion, object pose variation, background distraction, lighting changes and out-of-plane rotation. Kwon et al. [7] presented a robust tracking algorithm to track an object with drastically changed appearance and motion. For efficient design of observation and motion models as well as trackers, the observation model was decomposed into multiple basic models, which are constructed by sparse principal component analysis of a set of feature templates. Karavasilis and Nikou [8] proposed to use the differential Earth Mover's Distance between Gaussian mixtures models and Kalman filtering. This method showed significant improvement in the tracking performance with respect to occlusion. Recently, target templates have been generated via sparse representation and used in the l_1 tracker. This tracker demonstrates promising robustness although the computational complexity is rather high. Considering the high-computational cost problem, Li et al. [9] proposed a real-time visual tracking system by exploiting the signal recovery power of compressive sensing (CS). The l_1 tracker was extended to accelerate the tracking process by using the orthogonal matching pursuit algorithm.

In recent years, perceptual hashing [10] has been well known for its ability to map an image's content to a short digest, regardless of data formats and/or manipulations it suffers from. In computer vision, it is a promising solution for multimedia content authentication [14] and usually used to measure the similarity of two images in image retrieval [10]. In the last decade, a lot of perceptual hashing algorithms have been developed. Yang et al. [10] used the mean of image blocks to obtain a perceptual hash code. Jie [11] proposed a block-DCT and PCA-based image perceptual hashing technique. Laradji and Lahouari [12] presented a new perceptual image hashing approach that exploited the image color information using hypercomplex and QFT representations. Simulation results clearly indicated the superior retrieval performance of the proposed QFT-based technique. Although perceptual hashing has been used in a wide range of applications, there is limited research applying perceptual hashing to visual tracking.

3   Perceptual Hashing Algorithms

In this section, we first describe three basic perceptual hashing algorithms, i.e., average hash (AHash) [13], perceptive hash (pHash) [14], and difference hash (dHash). Afterwards, we introduce the proposed perceptual hashing algorithms in detail, namely the Laplace-based hash (LHash) and Laplace-based difference hash (LDHash). As shown in Figure 1, the common framework of these algorithms consists of the following four stages: preprocessing, feature extraction, quantization, and hash computation, and the key component of perceptual hashing-based tracking is to obtain the perceptual hash codes.

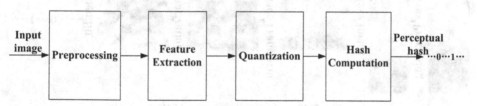

Fig. 1. Flowchart of the perceptual hashing methods.

Average hash (aHash) is one of the simplest perceptual hashes, which mainly uses low-frequency information of an image. By reducing the image size and comparing each pixel value with the mean value of the image, aHash can be simply obtained.

Although aHash is simple and quick, it may be too rigid to use the mean value. Perceptive hash (pHash) is a more robust algorithm which extends the average method to the extreme by using a discrete cosine transform (DCT) to obtain most sensitive information of the human visual system. The DCT transforms an image from the pixel domain to the frequency domain by separating it into a collection of frequencies and scalars, where only the coefficients of a small part of the frequency components are not 0.

Difference hash (dHash) works on the difference between two adjacent pixels. Like aHash, it is very simple to implement but far more accurate in tracking relative gradient directions.

3.1   Proposed Laplace-Based Hash (LHash)

The Laplace transform is widely used in image processing such as image sharpening and edge detection. In visual tracking, some objects in video sequences move very fast. As a result, motion blur is inevitable. Image sharpening is to perform a low-pass filtering operation on images, and therefore an inverse operation, such as a difference operation, can be performed to highlight the image details. The Laplace transform is such a kind of differential operator that can enhance the edge regions of an image while weakening slowly varying gray areas. In this paper, to solve the problem of image blur in order to track objects more accurately, we apply the Laplace transform to generate perceptual hash codes, which is named the Laplace-based Hash (LHash). We find that the Laplace transform is an image sharpening operator which is suitable for edge detection. This is consisted with the simulation results [15] which show that edge detection-based perceptual hash codes have strong robustness and good discriminability in the identification and authentication of multimedia content. This motivated us to apply LHash to object tracking.

Take an image of the "Singer1" video sequences as an example, the flowchart of the proposed LHash is given in Figure 2.

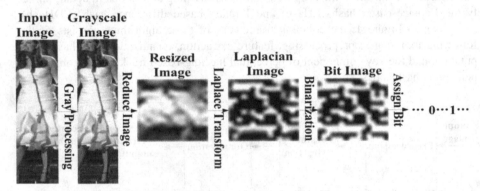

Fig. 2. Flowchart of the proposed LHash.

3.2   Proposed Laplace-Based Difference Hash (LDHash)

Most existing perceptual image hash functions use edge detectors for feature extraction [16,17], where the gradient function can be taken into account to explore abrupt grey-level changes in edge detection. To this end, we propose Laplace-based difference hash (LDHash) for object tracking by combining gradient and Laplacian in edge detection.

For an image $f(x, y)$, a one-dimensional gradient image $f_d(x, y)$ can be obtained by calculating the difference between adjacent pixels. The gradient image represents the relative changes in brightness and intensity of the input image, and it identifies the relative gradient directions. Since Laplacian locates the positions where the zero-crossings of $f_d^{''}(x, y)$ occur and only occur at the edge points of $f_d(x, y)$, we perform the Laplace transform on $f_d(x, y)$ and obtain the edge image $f_{dLap}(x, y)$.

The implementation process of the algorithm is described in Figure 3. Firstly, the input image is converted to grayscale and resized to 17×16. Secondly, we calculate the difference between adjacent pixels. Afterwards, a Laplace transform is performed on the difference image and the hash code is obtained by binarization.

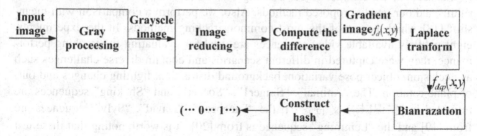

Fig. 3. Flow diagram of the proposed LDHash.

4   Object Tracking Based on Perceptual Hashing

In this section, we apply three basic perceptual hashing methods (i.e., aHash, pHash, dHash) and our proposed algorithms (i.e., LHash and LDHash) to object tracking. This task is achieved by minimizing the similarity between the tracked object region of the current frame and the search window in the next frame. The implementation procedure contains the following steps:

1. Capture the first frame of a video sequence and use a mouse to select the target regions with a rectangle search window of size $A \times B$, where the values of A and B are selected based on the ground truth box;
2. Capture the next frame of the video sequence and select the image area around the target in the previous frame as scanned regions;
3. Search the defined scanned regions with the rectangle scan window and separately calculate the perceptual hash codes H_0 of the target region and H of search windows. After computing the Hamming distances between H_0 and each H, all obtained Hamming distances in this scanned region are compared and the position (x, y) of scan windows with the smallest Hamming distance is selected;
4. Update the object rectangle region position as (x, y) and mark the object in order to adapt to variations of the targets;
5. End the cycle if the video sequence is finished, otherwise go to step (2) to capture the next frame and track objects.

5   Experimental Results

5.1   Experimental Environment

In this section, we conduct object tracking with three basic perceptual hashing algorithms and our newly proposed methods. Also, we perform a comparison with mean-shift [18] which is one of the most common tracking methods. In the experiments, eight publicly available video sequences are used for evaluating the tracking performance, they were captured in different scenarios and contain diverse challenges such as occlusion, object pose variation, background distraction, lighting changes and out-of-plane rotation. The "Animal", "Singer1", "Soccer", and "Shaking" sequences are provided in [错误!未定义书签。]; the "Faceocc2", "David", "Sylv" sequences are from [19] and the "Lemming" sequence is from [20]. It is worth noting that these test sequences are the most challenging sequences from existing works [错误!未定义书签。, 19]. Using a sliding window-search scheme, our proposed object tracking methods perform object localization based on perceptual hashing. Like the template matching method, we perform object tracking by computing the similarity between object regions of the previous frame and the search window in the current frame.

For quantitative performance comparison, we introduce the popular tracking accuracy measurement termed as tracking success rate (TSR) to evaluate the proposed tracking algorithms, compressive tracking (CT) [23] and the mean-shift tracker [42]. Given the tracking window and the ground truth bounding box, TSR is defined as

$$TSR = \frac{area(ROI_t \cap ROI_g)}{area(ROI_g)} \in [0,1] \tag{1}$$

5.2   Qualitative Comparison of All Tracking Algorithms

In this section, we present a qualitative comparison of five perceptual hashing-based tracking methods, the mean-shift algorithm and compressive tracking (CT) on eight video sequences. Based on the qualitative tracking results, we demonstrate the effectiveness and low-computational cost of perceptual hashing-based tracking algorithms, especially our proposed LHash and LDHash achieve the best performance.

The "Singer1" video sequence entails the challenges of illumination variation, complex background and target size changes. Owing to the view and focal length changes of the camera, the point of view and size of the tracked singer change, and illumination variation occurs drastically. Experimental results show that throughout the entire video sequence, our proposed perceptual hashing trackers present outstanding performance compared with the mean-shift tracker, although there is significant illumination change, complex background in the scene and varying size of the target. As we can see, the singer tracked by mean-shift begins to be lost at the 10th frame; as illumination severally changes at the 98th frame, the singer is absolutely lost. Furthermore, by comparing the tracking results of five perceptual hashing-based methods, we find that our proposed LHash and LDHash can track objects more accurately

Fig. 4. Qualitative tracking results of the five perceptual hashing-based tracking algorithms, the mean-shift tracker, and compressive tracking on eight video sequences.

than the other three basic perceptual hashing-based algorithms. For example, at the 160th frame and the 264th frame, the tracking windows of three basic perceptual hash codes appear to shift slightly relative to the target location, but the tracking windows of LHash and LDHash almost locate at the center of the singer.

The "Animal" sequence exhibits background distraction and target motion blur caused by abrupt motion of the object, where either the mean-shift tracker or compressive tracking fails to track the target after the 24th frame. On the contrary, both the newly proposed LHash and LDHash-based trackers can successfully track the target throughout the whole video sequence. Especially, LDHash locates the animal the most accurately.

For the "David" sequence, both the illumination and pose of the target change gradually. All the five perceptual hashing-based methods can track the object better than the mean-shift tracker. At the 161th frame, due to drastic change of object pose, the mean-shift tracking window appears big drift, while there is only little drift with aHash, pHash, dHash and LHash.

Table 1. Average Tracking Success Rate (TSR). Values in boldface indicate the optimal accuracy.

	aHash	dHash	pHash	LHash	LDHash	CT	Mshift
David	0.6541	0.69708	0.70154	0.72285	0.712	**0.7247**	0.34853
Animal	0.7414	0.73374	0.733	0.72781	**0.75153**	0.11	0.10673
Faceocc2	0.7309	0.7624	0.67166	0.74283	**0.80197**	0.73615	0.34607
Lemming	0.3221	0.51565	0.20421	0.38287	**0.59153**	0.46246	0.57449
Shaking	0.3306	0.35613	0.40831	0.4985	**0.52859**	0.35038	0.28934
Singer1	0.5974	0.76088	0.60969	0.70104	**0.76437**	0.4703	0.20903
Soccer	0.6108	**0.75444**	0.61774	0.44589	0.73969	0.2036	0.30566
Sylv	0.467	0.38494	0.55637	**0.72543**	0.2968	0.71574	0.25747

Table 2. Average Time Cost (ms).

	aHash	dHash	pHash	LHash	LDHash	CT	Mshift
David	99.421	317.716	1153.12	219.714	234.302	786.704	15.1174
Animal	109.47	435.844	1437.92	274.056	685.143	760.903	50.8487
Faceocc2	135.98	474.853	1588.01	358.463	723.061	805.017	15.9237
Lemming	64.556	96.711	429.021	185.193	282.765	774.703	68.1336
Shaking	37.032	274.095	199.346	207.067	147.53	728.503	38.6215
Singer1	86.346	273.337	1115.05	169.045	221.017	594.995	46.0655
Soccer	106.06	321.04	1331.23	197.935	402.165	770.368	63.8318
Sylv	39.904	123.37	418.84	88.069	106.37	791.191	16.2994

5.3   Quantitative Comparison of All Tracking Algorithms

We measure the TSR and time cost of seven tracking methods on eight video sequences. As seen from Table 1, we can easily find that all the perceptual hashing-based algorithms achieve better tracking performance than the commonly used mean-shift tracker. Besides, our newly proposed perceptual hashing-based trackers (i.e., LHash and LDHash) outperform the compressive tracking (CT) in most video sequences. As shown in Table 1, values in boldface are the largest TSR indicating the optimal accuracy. Among the eight video sequences, LDHash presents the best tracking results on five video sequences (i.e., "Animal", "Faceocc2", "Lemming", "Shaking", "Singer1"), while LHash shows the best tracking performance on "Sylv" and dHash achieves the best result on "Soccer". By contrast, CT only achieves the best result on "David"; nevertheless, its TSR is very close to those of LHash and LDHash.

5.4 Computational Cost Comparison of Tracking Algorithms

In real-time visual tracking systems, there is no doubt that we demand high accuracy and low-computational cost. What we try to do is to ensure the effectiveness of tracking while reducing the computational cost at the same time. In this paper, we use the average time cost of tracking in each frame of video sequences to measure the computational cost. Table 2 shows the average time cost of seven algorithms on each test video sequence. As we can see, the average time cost of aHash, dHash, LHash and LDHash are all lower than that of CT on all test sequences. Especially, our proposed perceptual hashing-based tracking algorithms outperform the others in terms of accuracy and computational cost. On the contrary, even though the mean-shift tracker requires low computational cost, its tracking accuracy is not satisfactory.

6   Conclusions and Future Works

In this paper, we propose new visual tracking methods based on perceptual hashing. Applying three existing perceptual hashing (i.e., aHash, dHash, and pHash) as an effective image representation, we obtain three robust trackers for visual tracking on challenging video sequences. Compared with the most commonly used mean-shift algorithm and the state-of-the art compressive tracking algorithm, our methods achieve higher accuracy while requiring low-computational cost. To further improve the accuracy of object localization and the robustness of tracking, we propose two new perceptual hashing methods for tracking, which are Laplace-based hash (LHash) and Laplace-based difference hash (LDHash). Compared with three basic perceptual hashing algorithms on eight video sequences, experimental results indicate that the newly proposed methods are able to achieve more accurate and robust tracking, at the same time they demand low-computational cost. Therefore, we believe that LHash and LDHash should have good potential in more complex scenarios, such as face tracking, robot localization and navigation, infrared thermal imaging tracking applications, and so on. For future topics, we intend to combine more information with our algorithms, such as color features and angle features.

References

1. Chen, J.: UAV-guided Navigation for Ground Robot Tele-operation in a Military Reconnaissance Environment. Ergonomics 53(8), 940–950 (2010)
2. Wang, P.K., Torrione, P.A.: Rapid position estimation and tracking for autonomous driving. In: Proc. SPIE 8387. Unmanned Systems Technology XIV (2012)

3. Bulling, A., et al.: Eye Movement Analysis for Activity Recognition Using Electrooculo-graphy. IEEE Trans. Pattern Analysis and Machine Intelligence **33**(4), 741–753 (2011)
4. Zhang, K., Zhang, L., Yang, M.H.: Real-time compressive tracking. In: European Conference on Computer Vision (2012)
5. Zhang, S., Yao, H., Zhou, H., Sun, X., Liu, S.: Robust Visual Tracking based on Online Learning Sparse Representation. Neurocomputing **100**, 31–40 (2013)
6. Kalal, Z., Matas, J., Mikolajczyk, K.: P-n learning: bootstrapping binary classifier by structural constraints. In: Computer Vision and Pattern Recognition, pp. 49–56 (2010)
7. Kwon, J., Lee, K.: Visual tracking decomposition. In: Computer Vision and Pattern Recognition, pp. 1269-1276 (2010)
8. Vasileios, K., Christophoros, N.: Visual Tracking Using the Earth Mover's Distance (EMD) Between Gaussian Mixtures Models and Kalman Filtering. Image and Vision Computing **29**, 295–305 (2011)
9. Li, H., Shen, C., Shi, Q.: Real-time visual tracking using compressive sensing. In: Computer Vision and Pattern Recognition, pp. 1305–1312 (2011)
10. Bian, Y., Fan, G., Xiamu, N.: Block mean value based image perceptual hashing. In: International Conference on Intelligent Information Hiding and Multimedia Signal Processing, pp. 167–172 (2006)
11. Jie, Z.: A Novel Block-DCT and PCA based Image Perceptual Hashing Algorithm. IJCSI International Journal of Computer Science Issues **10**(3) (2013)
12. Laradji, I.H., Lahouari, G.: Perceptual hashing of color images using hypercomplex representations. In: IEEE International Conference on Image Processing (2013)
13. http://hackerfactor.com/blog/index.php?/archives/432-Looks-Like-It.html
14. Monga, V.: Evans Perceptual Image Hashing via Feature Points: Performance Evaluation and Trade Offs. IEEE Transactions on Image Processing **15**(11), 3453–3466 (2006)
15. Weng, L., Preneel, B.: Shape-based features for image hashing. In: Proc. of IEEE International Conference on Multimedia & Expo (2009)
16. Bhattacharjee, S., Kutter, M.: Compression tolerant image authentication. In: Proceeding of the International Conference on Image Processing, vol. 1, pp. 435–439 (1998)
17. Singh, S.: Improved Hash based Approach for Secure Color Image Steganography Using Edge Detection Method. International Journal of Computer Science and Network Security **14**(7) (2014)
18. Comaniciu, D., Ramesh, V., Meer, P.: Real-time tracking of non-rigid objects using mean shift. In: Computer Vision and Pattern Recognition, vol. 2, pp. 142–149 (2000)
19. Babenko, B., Yang, M.H., Belongie, S.: Robust Object Tracking with Online Multiple Instance Learning. IEEE Trans. Pattern Analysis and Machine Intelligence, 1619–1632 (2011)
20. Santner, J., Leistner, C., Saffari, A.: PROST Parallel Robust Online Simple Tracking. In: Computer Vision and Pattern Recognition (2010)

Vision-Based Entity Chinese Chess Playing Robot Design and Realization

Xiangwei Wang and Qijun Chen[⊠]

School of Electronics and Information Engineering,
Tongji University, Shanghai 201804, China
{10kongwei,qjchen}@tongji.edu.cn

Abstract. This paper describes a lowcost method to design and realize a vision-based entity robot system which is able to play Chinese chess. One inexpensive camera is used with vision program to recognise the position, color and role of each chess by camera calibration, coordinate transformation, color segmentation, morphological method and some prior knowledge on Chinese chess. A robot arm with four cheap stears is designed and realised by inverse kinematics, trajectory planning and other method to move the chess. communication system is made to transfer information between control system and robot arm. At last Alpha beta pruning algorithm is realised as the robot's bahavior. At last, the four parts above are combined to realize the whole syetem.

Keywords: Computer vision · Robot arm · Artifical intelligence · Chinese chess

1 Introduction

In recent years, with the developing of the information technology, people have been addicted to the virtual world such as social network and computer games more and more seriously. However as real human being, we should live in the real world, so it is pretty important to make some real entertainment tool for people, expecially for children. Chess recognition, robot arm design and algorithm to play Chinese chess have been nearly mature. However entity chess robot basing on vision and robot arm has not been mature and common because of costing. In this paper, a lowcost method to design and realize a vision-based entity robot system is described which can be programmed to playing Chinese chess and other possible things you like. A solution on how to make the system playing Chinese chess is taken as an example. The system consists four subsystems which are vision system, communication system, robot arm system, control and assistance system.

There is three contributions in my paper. The first contribution is that we design and realize a entity robot in a low cost which contains a vision-based recognition system, a robot arm based actuator. Then we realize a complete set of code including code for recognition, code for communication and code for robot arm control with good code structure. The last but not least, we design

© Springer International Publishing Switzerland 2015
H. Liu et al. (Eds.): ICIRA 2015, Part III, LNAI 9246, pp. 341–351, 2015.
DOI: 10.1007/978-3-319-22873-0_30

a series of software to adjust parameter of vision system and robot arm system and simulate the robot arm.

The paper is organised as follows: Section 2 reviews work related to entity Chinese chess robot. Section 3 presents a brief introduction of the structure of the whole system and vision system is presented in Section 3.1, communication system in Section 3.2 and Section 3.3 introduces how to design and make the robot arm and the arm simulation software. In addition, we introduce how can the robot play Chinese chess briefly in Section 3.4. Finally, our conclusion is given in Section 4.

2 Related Work

This section reviews the related approaches for each subsystem of our system. Section 2.1 reviews the chess recognition. Section 2.2 discusses the whole chess robot.

2.1 Chess Recognition

Chess recognition is the task of recognition of each chess' position, color and role. Various approaches are used in practice. There are three main methods based on machine vision which are recognition by feature description and classification [1],[2],[3],[4],[5], recognition by prior knowledge [7] and the combined method [8]. The first kind of approach can recognize all the information we need at one time. However one drawback of the method is that they requir the camera to capture clearly the character of the chess which means that the camera's optical axis should be perpendicular to chessboard but it is not suitable for a entity chess robot with robot arm. Conversely, for the second method the camera's loaction is more flexible, but the current recognition is based on the last recognition, so if there is something wrong sometimes, there is almost no way for it to recognitize correctly again. And the combined method take the advantage of the two methods and it use the second method in normal case and when there is something wrong it use the first method to reintialize the recognition.

2.2 Chess Robot

Most chess robots available are virtual and they are just a program with chess playing algorithm in computer. The entity chess robot should be with recognition system and actuator. In [9] Du introduce a chess playing robot without recognition which can move the real chess when someone play a computer chess game they design. And its actuactor is based on the rectangular coordinates system. In [10], Huang describes a kind of chess robot use Hall element to get the information of chess. A vision-based robot is needed because it can be used for any kind of chessborad and a robot arm is required to making to be similar to human.

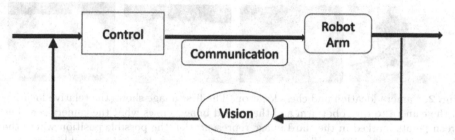

Fig. 1. Structure of Whole System

3   Structure of the Whole System

As shown in Fig. 1, there are four main subsystems. We use a camera with our vision software and program as vision system to obtain visual information of work space. And a robot arm with four degree of freedom is used to be the actor to execute the commands from control system and change something in the workspace. The control system read the information from the vision system and control the robot arm to do what we want to do. And the communication system's responsibility to transfer information between the control system and robot arm system.

3.1   Vision System

The vision system is to obtain information of the Chinese chess game state with one camera. The information during a Chinese chess game we need contains the position, color and role of each chess on the chessboard. We first mount the camera beside and facing to work space where the camere can see the whole work space as shown in Fig .2. Because of the distortion of camera, we calibrate the camera with OpenCV based tools firstly.And then we calculate the coordinate transformation matrix from the real word coordinate to image coordinate by a tiny software I design.

From [13], we can know that for the same point whose coordinate at world coordinates is (X_w, Y_w, Z_w) and position at image (u, v), we can get the equation below.

$$Z_c \begin{pmatrix} u \\ v \\ 1 \end{pmatrix} = L_w \begin{pmatrix} X_w \\ Y_w \\ Z_w \\ 1 \end{pmatrix} \tag{1}$$

Here Z_c is the coordinate in Z direction of Camera coordinates and has no influence on the following calculation.L_w is a 3 by 4 matrix related to the focal length and the relative position between the world coordinates and camera coordinates.

$$L_w = \begin{pmatrix} w_{11} & w_{12} & w_{13} & w_{14} \\ w_{21} & w_{22} & w_{23} & w_{24} \\ w_{31} & w_{32} & w_{33} & w_{34} \end{pmatrix} \tag{2}$$

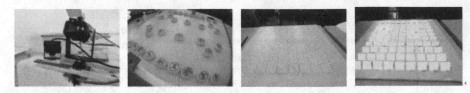

Fig. 2. Camera location and chess location. The first image shows the relative location of the camera to the robot arm and the second image shows what the camera see. The green points marked in the third image representation the possible position where the chess may show and the yellow regions in the last image shows the possible area.

Because w_{34} has no effect on the equation, we set $w_{34} = 1$. Then

$$
\begin{pmatrix} w_{11} & w_{12} & w_{13} \\ uw_{31} & uw_{32} & uw_{33} \end{pmatrix} \begin{pmatrix} X_w \\ Y_w \\ Z_w \end{pmatrix} + w_{14} = u \tag{3}
$$

$$
\begin{pmatrix} w_{21} & w_{22} & w_{23} \\ vw_{31} & vw_{32} & vw_{33} \end{pmatrix} \begin{pmatrix} X_w \\ Y_w \\ Z_w \end{pmatrix} + w_{24} = v \tag{4}
$$

if there are n pairs of corresponding points, we can get following equation

$$
AL = U \tag{5}
$$

where

$$
A = \begin{pmatrix} X_{w1} & Y_{w1} & Z_{w1} & 1 & 0 & 0 & 0 & 0 & -u_1Y_{w1} & -u_1Y_{w1} & -u_1Z_{w1} \\ 0 & 0 & 0 & 0 & X_{w1} & Y_{w1} & Z_{w1} & 1 & -v_1Y_{w1} & -v_1Y_{w1} & -v_1Z_{w1} \\ & & & & \cdots & \cdots & & & & & \\ X_{wn} & Y_{wn} & Z_{wn} & 1 & 0 & 0 & 0 & 0 & -u_nY_{wn} & -u_nY_{wn} & -u_nZ_{wn} \\ 0 & 0 & 0 & 0 & X_{wn} & Y_{wn} & Z_{wn} & 1 & -v_nY_{wn} & -v_nY_{wn} & -v_nZ_{wn} \end{pmatrix} \tag{6}
$$

$$
L = (w_{11}, w_{12}, w_{13}, \ldots\ldots, w_{2,4}, w_{3,1}, w_{3,2}, w_{3,3})^T \tag{7}
$$

$$
U = (u_1, v_1, ..., u_n, v_n)^T \tag{8}
$$

if n is more than 6, we can get the least-square solution

$$
L = (A^T A)^{-1} A^T U \tag{9}
$$

Our tiny software is designed to gather the point pairs and calculate the value of L. If we know the coordinate transformation matrix from the real word coordinate to image coordinate, for a chessboard it is know that the possible position where one chess maybe appear(See Fig. 2). Taking the size of a Chiness chess into account, we got the possible area where one chess maybe appears as shown in Fig. 2. Next we erode the image we got from the camera in order to make the color more distinct.

Fig. 3. Chess position and color recognition. The first one is the origin calibrate image. The next three shows the eroded image, binary image after eroded and labeled image by black and red points. The last image is result image, the possible position marked red exists red hess and position marked blue exists black chess

We check whether there is a chess and the color of the chess on every possible area by two process. To ensure whether there is a chess or not, we change the image into binary image by a suitable threshold value [16](See Fig. 3). And it is easy to get the result by comparing the density of black pixels with a fixed value, because it is very different between the area with a chess and the area without a chess. To judge the color of the chess, we change the RGB image into HSV image and use decision tree to judge each pixel is red, black or white. And we get the color of the chess based on the result of decision tree by calculating the ratio of the number of red pixeds to that of black pixels which represent the possibility of the chess to be red one or black one. Then take the number of chess of each color into account, it is easy to identify the color of chess by comparint the ratio. The Fig. 3 shows the result of recognition the each chess's position and color.

Color label decision tree program

```
Input:   h,s and v value of a pixel;
Outout:  Lable which may be black, red or white;
1: if h<30 Lable = red;
2: else if h>=30 and v>50 Label = red;
3: else if h>=30 and v<50 Label = black;
4: else if v<60 Label= black;
5: else Label=white;
6: Output(Label)
```

The recognition system can not see the character label of chess, so it is impossible to recognise the chess character by the method in [2],[1],[6],[5]. A logic based method is used to judge the role of each chess similar to [7]. As we all know that the initial position of each chess is fixed when someone play chiness chess in normal case and one can only move one chess per step, so if how the chess move at each step is availabel, it is easy to know the role of each chess at the current time. For each step, there are only three options for each possible area. One options is the chess disappears, one is an chess appeears, and the last one is the color of chess is changed. All of the three options is available to be recognitioned if where there is a chess and the color of the chess have been known.

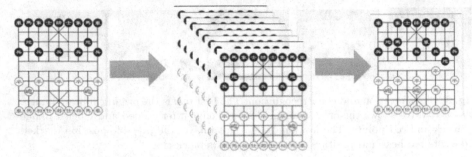

Fig. 4. Recognition movement. We know the initial state(left) and detect the move of chess(middle), we can know the current state(right).

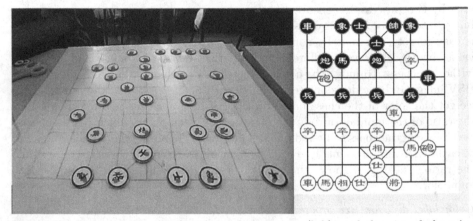

Fig. 5. Whole recognition result. The real chess move(left), and the virtual chess in computer(right) move follow it, so they are the same.

The Fig. 5 shows that when someone move the real chess, the recognition system changes the virtual chess in ourcomputer and the Fig. 4 shows the process.

3.2 Communication System

There is information to be transfered between the computer and single chip microcomputer. The single chip microcomputer takes responsibility for controlling the robot arm by controlling the PWM for each steer of the robot arm and how should the robot arm move is computed by the computer, so there are some information to transfer between the computer and the single chip microcomputer. The information from the computer to microcomputer is the time series of the steers' action and the information from the microcomputer to conputer is state of the microcomputer such as ready to move, moving, done and so on.

In the communication system, a serial port is used as the physical. And Fig.6 shows the communication protocol. There are three flags which are Action_Done representing whether the robot arm have completed the action, Action_Need

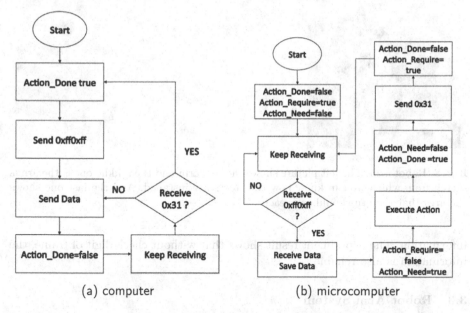

(a) computer   (b) microcomputer

Fig. 6. Communication protocol. The left picture shows how the computer part send and receive message and the right one shows that of microcomputer.

Action First

0xff	0xff	Angle_1a	Angle_1b	Angle_1c	Angle_1d	...	...
Head		The angle of steer 1 of the first action	The angle of steer 2 of the first action	The angle of steer 3 of the first action	The angle of steer 4 of the first action		
Angle_9a	Angle_9b	Angle_9c	Angle_9d	Angle_10a	Angle_10b	Angle_10c	Angle_10d

Action Tenth

Fig. 7. Communication frame. This picture shows the format of the frame send by computer.

representing whether the robot arm have received the action command but not completed the action and Action_Require and it means that the robot arm have complete the last action, have sent a message to the computer and are ready for the next action when Action_Require is ture. The Fig. 7 shows the frame format of our communication system. The first two byte is frame head, and the following 40 byte is time series of the action where every four byte represent an state. Because we use wired connection between the computer and the microcomputer and it is near between the two part, there is no check digit in the information frame and we just use check digit of serial port to ensure the reliability of the

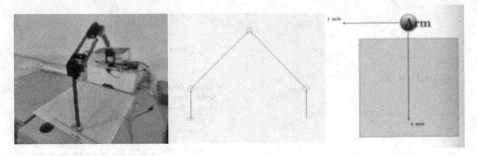

Fig. 8. Robot arm. The left picture shows the real arm and the middle one is the arm's sketch from which we can know how the steers are installed. And the last one shows the area that the arn should can reach.

information. The experiment result shows that without check digit of frame, the information is also reliable.

3.3   Robot Arm System

A kind of four degree of freedom robot arm with four steering engine and a single chip controller is designed and realised(See Fig.8 left). The end of the robot arm is fixed with an electromagent to lift and move something ferromagnetic such as ferromagnetic chess. And Fig.8 middle shows the method how the four steers are installed and all of the them are revolute joints. The first joint moves yaw and the others are pitch axis. Taking the limit of the angle and moment of each steer and the length of each links into account, our robot arm can arrive any place in an area 300mm by 300mm in 50mm away(See Fig. 8 right).

By inverse kinematics, the angle of each joint which make the end of the robot arm reach desired point can be calculated. Here if (x, y, z) is the point we desire to arrive, l_1, l_2, l_3 and l_4 are the length of the robot's links, we calculated the angles as following equations show.

$$\theta_1 = \arctan \frac{y}{x} \tag{10}$$

$$x_c = \sqrt{x^2 + y^2} \tag{11}$$

$$y_c = z + l_4 - l_1 \tag{12}$$

$$r = \sqrt{x_c^2 + y_c^2} \tag{13}$$

$$\alpha_1 = \arccos((r^2 + l_1^2 - l_2^2)/(2rl_1)) \tag{14}$$

$$\alpha_2 = \arccos((r^2 + l_2^2 - l_1^2)/(2rl_2)) \tag{15}$$

$$\delta = \arctan((u_c/x_c) \tag{16}$$

$$\theta_2 = \pi/2 - \alpha_1 - \delta \tag{17}$$

$$\theta_3 = -\alpha_1 - \alpha_2 \tag{18}$$

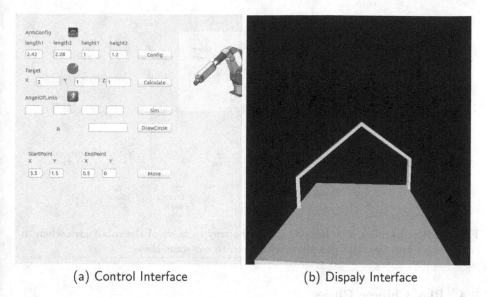

(a) Control Interface   (b) Dispaly Interface

Fig. 9. Robot arm simulation software. The left one is the control interface and the right one is display interface.

$$\theta_4 = -(\pi - \theta_2 - \theta_3) \tag{19}$$

There is also software to simulate our robot arm with OpenGL library. As shown in Fig. 9, there are two interfaces in our simulation software. One is the interface for configuration and controlling(Fig. 9a) and the other is for displaying the simulated result dynamic and in real time(Fig. 9b). The software has four usage: 1) to check whether the end of our robot arm can reach anywhere needed within the limit of the length of links and available angle of joints, 2)to protect the real robot arm by moving before the real robot arm with the parameters calculated by computers.

When the robot arm need move from one point to another point, we ought to plan a path between begin point and the end point [14]. Our path is a semicircle perpendicular to horizontal plane. If the start point is $(x_s, y_s, 0)$ and the end point is $(x_e, y_e, 0)$, the distance (r) between the start point and the end point is $\sqrt{(x_e - x_s)^2 + (y_e - y_s)^2}$ which is the diameter of the path, the center $(x_c, y_c, 0)$ is $(\frac{x_s + x_e}{2}, \frac{y_s + y_e}{2}, 0)$,the angle between the line, which connects the start point and end point, and x-axis is $\beta = \arctan \frac{y_e - y_s}{x_e - x_s}$. Then the point (x_p, y_p, z_p) on the path from the start point to the end point should satify the equations:

$$x_p = r * \cos \gamma * \cos \beta + x_c \tag{20}$$

$$y_p = r * \cos \gamma * \sin \beta + y_c \tag{21}$$

$$z_p = r * \sin \gamma \tag{22}$$

if γ changes from π to 0, (x_p, y_p, z_p) will move from the start point to the end point with a semicircle path.

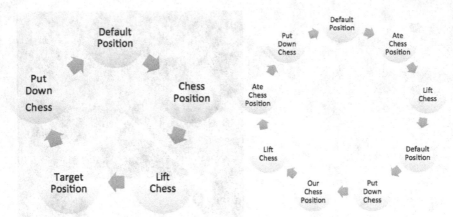

Fig. 10. How to move. The left one shows the way to move of the robot arm when to eat no chess and the right one show that when to eat some chess.

3.4 Play Chinese Chess

There are many kinds of Chinese chess algorithm such as alpha-beta pruning [11] temporal difference learning [17] and minors hash table [15]. We use vision system to recognize the chess and implement alpha-beta pruning [12]algorithm with C++ to play Chinese chess [18]. Then we use our robot arm to move the chess. We should know whether we will eat other's one role at the step, if the answer is yes, the robot arm should move as Fig. 10 left,or as Fig. 10 right.

4   Conclusion

A kind of entity Chinese chess playing robot is designed and realised with vision system to recognize the chess state, robot arm to move chess and communication system to transfer information between controller and robot arm. Under normal circumstance when the luminance changes slightly our system can recognitize accurately. When luminance changes obviously we can use our assistance system to adjust parameters conveniently. The communication system can transfer information fast and accurately, and no error appears for all the test. As for robot arm system, some little reach position error exists because of no feedback but it can be corrected by assitance system.(All software and code are available on github https://github.com/shikongwei/Chinesechess).

Acknowledgments. This work was supported in part by the Key Program for Basic Research of Science and Technology Commission Foundation of Shanghai (No. 12JC1408800).

References

1. Chen, W.: Chinese-chess image recognition by using feature comparison techniques. Appl. Math. Inf. Sci. **8**, 2443–2453 (2014)
2. Hu, P., Luo, Y., Li, C.: Chinese chess recognition based on projection histogram of polar coordinates image and FFT. In: CCPR, pp 1–5 (2009)
3. Du, J., Huang, X.: Design ofchinese chessrobotvision system. Computer Technology and its Application, 133–136 (2007)
4. Feng, Y., Wang, S., Liu, N., Wang, G.: Application of machine vision technology in design of ches playing inteligent robot. Computer Engineering and Design **30**, 3371–3379 (2009)
5. Zhang, H., Ping, Z., Hai, Y., Wang, K.: Chinese chess character recognition with radial harmonic fourier moments. In: International Conference on Document Analysis and Recognition (2011)
6. Fang, J.: A machine vision system for chinese chess-playing robot. In: Zhang, T. (ed.) Mechanical Engineering and Technology. AISC, vol. 125, pp. 379–386. Springer, Heidelberg (2012)
7. Lou, L.Q., Duan, L., Xing, S.H.: Chessman ecognition of chinese chess based on video image understanding. Journal of South-Central University for Nationalities (Nat. Sci. Edition) (2014)
8. Zhai, N.: Improved Chinese chessboard recognition method. Journal of Computer Applocations **30**, 980–981 (2010)
9. Du, G., Chen, W., Wang, Y. : The Design of Chinese Chess Robot System Based Rectangular Coordinates. Robot Technique and application (2013)
10. Huang, S., Chen, L., Shen, X., Jiang, X.: The Manufacture of Chess Robot of Intelligent Chessboard for Man-machine Chess Board. Physical Experiment of College **27**, 42–46 (2014)
11. Moore, R.W., Knuth, D.E.: An analysis of alpha-beta prunning. Artificial Intelligence **6**, 293–326 (1975)
12. Fishburn, J.P., Finkel, R.A.: Improved speedup bounds for parallel alpha-beta search. IEEE Transactions on Pattern Analysis and Machine Intelligence **5**, 89–92 (1983)
13. Gao, H.: Computer Based Binocular Vision. Publishing house of electronics industry (2012)
14. Lin, H.C., Hsiao, S.H., Huang, G.S. Tung, C.K.: Inverse kinematics analysis trajectory planning for a robot arm. In: IEEE Control Conference (2011)
15. Xu, X.H., Wang, J., Li, S.Z.: A minors hash table in chinese-chess programs. ICGA Journal (2009)
16. Otsu, N.: A threshold selection method from gray-level histograms. IEEE Trans. System, Man and Cybernetics **9**, 62–66 (1979)
17. Thong, N.D., Trinh, B., Bashi, A.S.: Temporal difference learning in Chinese Chess. Springer, Heidelberg (1998)
18. Feng, S., Yue, J.: Improvement on alpha-beta search algorithm in chinese chess. Journal of Beijing Normal University(Natural Science) (2009)

Robust 3D Local SIFT Features for 3D Face Recognition

Yue Ming[1]($\boxtimes$) and Yi Jin[2]

[1] School of Electronic Engineering, Beijing University of Posts and
Telecommunications, Beijing 100876, People's Republic of China
`myname358752350126.com`
[2] School of Computer and Information Technology, Beijing Jiaotong University,
Beijing 100044, People's Republic of China

Abstract. In this paper, a robust 3D local SIFT feature is proposed for
3D face recognition. For preprocessing the original 3D face data, facial
regional segmentation is first employed by fusing curvature characteristics and shape band mechanism. Then, we design a new local descriptor for the extracted regions, called 3D local Scale-Invariant Feature
Transform (3D LSIFT). The key point detection based on 3D LSIFT
can effectively reflect the geometric characteristic of 3D facial surface
by encoding the gray and depth information captured by 3D face data.
Then, 3D LSIFT descriptor extends to describe the discrimination on 3D
faces. Experimental results based on the common international 3D face
databases demonstrate the higher-qualified performance of our proposed
algorithm with effectiveness, robustness, and universality.

Keywords: 3D face recognition · 3D Local Scale-Invariant Feature
Transform · Facial region segmentation · Depth information

1 Introduction

Biometrics systems have been presented for several decades with wide applications, such as 3D movies, human-machine interaction, and intelligent monitoring
[1]. Among them, face recognition has a high-level preference for a huge number of researchers and organizations, mainly because of its non-invasiveness and
user-friendliness. However, face recognition developed by 2D images has been
hindered by the obstacles induced by pose, illumination, expressions, and other
varied characteristics in real-world situations. With the rapid development of 3D
digital capturing devices, facial recognition in 3D data has been introduced to
solve the challenging issues using a variety of methods.

A sufficient broad investigation of face recognition has been provided in [2],
specifically on 3D face recognition. Empirical study shows that facial shape has
significant variations in terms of different regions on the facial surface. In order
to effectively encode facial anatomical structure and describe the discriminant
features, we introduce segmenting scheme to address the facial region attributes

© Springer International Publishing Switzerland 2015
H. Liu et al. (Eds.): ICIRA 2015, Part III, LNAI 9246, pp. 352–359, 2015.
DOI: 10.1007/978-3-319-22873-0_31

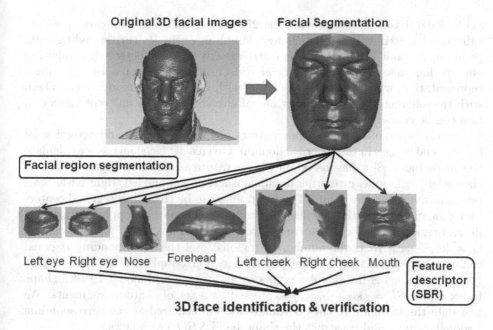

Fig. 1. The framework of our proposed 3D face recognition system

and develop a new recognition framework to 3D facial data. Our framework for 3D face recognition is composed of three parts: facial region segmentation, feature detection and description, and 3D face recognition as shown in Figure 1.

In our framework, a group of facial local regions can be coarsely located by curvature information and shape band [3] algorithm is introduced to refine the localizations which possess more discriminant power. Then, we exploit the original Scale-Invariant Feature Transform [4] to 3D local surface on the subregions of 3D face data, and encode the gray and depth data for the detected key point description. 3D local Scale-Invariant Feature Transform (3D LSIFT) can be extracted to describe the regional characteristics for achieving the 3D face recognition. Our method is more robust to image artifacts, lighting variance, winkles, and occlusions and corruptions and shows the generalization based on the challenging databases.

The organization of the paper is as follows. Facial regional segmentation is described in Section 2. Section 3 proposes 3D local Scale-Invariant Feature Transform (3D LSIFT). Section 4 reports the experimental results based on the challenging 3D face databases and compares the performance of algorithms. Finally, the paper is concluded in Section 5.

2   3D Facial Regional Segmentation

The original 3D facial images in the databases usually contain some non-facial areas, such as ears, necks and shoulders as shown in Figure 1. The common

3D face databases mainly provide the 2D texture image and its corresponding valid point matrix with 3D face images simultaneously. By corresponding valid point on 2D and 3D images, the facial area is coarsely detected. Exploiting our previous research [5][6], the facial registration is calculated by Axis-angle representation, which can align the input with the reference model fixed. Then, with the different values and directions of curvature [7], the different areas of a face can be coarsely detected.

In this paper, our refined region extraction exploits shape band detection [3] on 3D facial images in the spherical domain. Given a 3D facial image, we calculate the shape index [8] values based on the curvature as shown in Figure 2 and we choose left/right inner eye corner points, nasal tip point and left/right nasal basis points as five facial key points. The key points can be treated as regional centers for region matching with reference template. Based on facial cropping parameters [9] and regional shape characteristics, we can obtain a series of regional centers $C = [c_1, \cdots, c_n]$ (n is the number of regions) and the corresponding regional radius $r = [r_1, \cdots, r_n]$. We first match a region template $P = \{p_1, \cdots, p_m\}$ ($p_i(i = 1, \cdots, m)$ is the samples points of an average face model T) to a shape index image $SI = \{si_1, \cdots, si_n\}$ expressed as a set of contour fragments. We translate the template to the coordinate system centered at the corresponding regional centers, and construct the shape band $SB(P)$ as follows,

$$SB(P(c)) = \{ip \in T | \exists p_i \| ip - (p_i + c_i)\|_2 \le r\} \tag{1}$$

where $\|\cdot\|_2$ denotes the l_2 norm.

The detection and matching of the input face model I can be treated as the selection of the optimal central points $C^* = [c_1^*, \cdots, c_n^*] \in I$ and a subset $SI^* = [si_1^*, \cdots, si_n^*]$ as follows,

$$D(P(c^*), SI^*) = \min_{c \in T, SI' \subset SI} D(P(c), SI') \tag{2}$$

where D is a shape band distance defined in literature [3].

Then, Shape Context distance SC is used to the fine process to determine the optimal segmentation. For each shape index segment si_i in SI, its minimum distance can be introduced to select the regions with the preset thresholds Th_i (the value depends on the different databases). If the minimum distance defined in Equation (2) is less than the threshold, then si_i is adjunct to $P(c_i)$, denoted the adjunct segments as $AS(C_i)$,

$$\min Dist(si_i, P(c_i, \varepsilon_i)) = \min_{p_i \in P, q \in Q} \|q - (\varepsilon_i p_i + c_i)\|_2 \tag{3}$$

where Q is the point set in the template segment si_i and ε_i is the scale value of the selected regions. SC can be calculated by the distance between the shape index segments and the template position. We treat the minimal shape distance as the matching segments of the input face model aligned with the regional template.

$$SD(c_i) = \min_{S \subseteq AS(c_i)} SC(US, P(c_i)) \tag{4}$$

The final detected regions of the input face model are the one with the smallest shape distance as shown in Figure 2.

$$region = \arg\min_{c_i \in C} SD(c_i) \tag{5}$$

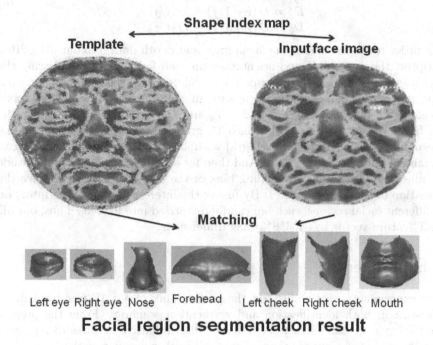

Shape Index map

Template   **Input face image**

Matching

Left eye Right eye Nose   Forehead   Left cheek Right cheek Mouth

Facial region segmentation result

Fig. 2. 3D facial regional segmentation

3   3D Local Scale-Invariant Feature Transform (3D LSIFT)

Inspired by the facial geometric attributes with depth information and scale invariant descriptors for the task of discrimination, a novel descriptor is proposed for a group of facial regions, denoted as 3D Local Scale-Invariant Feature Transform (3D LSIFT). Our feature extraction can be divided into two steps, e.g., the key point detection and description. For the detection, density-invariant Gaussian filters are defined to calculate the filtered mesh sets on the facial surface geometry. We only use the corresponding mesh image of 3D face image and employ with the similar way as SIFT to fix the positions of the key points, which are the local extreme of the Gaussian pyramid.

The second step for 3D LSIFT is to build the descriptor for each key point on the different facial regions. Since the key point detection is the derivative of the original SIFT algorithm, the detected points is with the invariance on scale and rotation. In order to extend the invariance to 3D space, the depth information around the key points is added to describe such points. Suppose the key point is positioned at (i, j), and the gray image is denoted by I, and the depth image as D. For each pixel with in the local facial region, we compute three gradients for it,

$$
\begin{aligned}
I_x &= I(i+1, j) - I(i, j) \\
I_y &= I(i, j+1) - I(i, j) \\
D_z^x &= D(i+1, j) - D(i, j) \\
D_z^y &= D(i, j+1) - D(i, j)
\end{aligned} \tag{6}
$$

In order to describe both the facial gray and depth data into our 3D LSIFT descriptor, the magnitude and orientation on each facial pixel neighboring the key point is calculated in the corresponding scale space of the Gaussian pyramid. For xy, xz, and yz planes, the gradient's magnitude and orientation of each pixel are calculated by I_x and I_y, I_x and D_z^x, I_y and D_z^y, respectively. We divide the local facial region with size 16*16 into 16 grids around the key points in the corresponding scale space. For each grid with 4*4 pixels, the facial rotation can be quantized into eight directions. And then for each orientation, the magnitude is summed up into the corresponding bins of the histogram. We normalize the summation of the histogram to 1. By fusing the interesting point descriptors on the different facial regions, each bin can be quantized into [0, 255]. Thus, our 3D LSIFT feature vector has 4*4*8*3=384 dimensions.

4   Experiments

In the experimental section, we test the performance of our 3D face recognition framework in both identification and verification scenarios. From the preprocessed images, we can extract the different discriminative features to represent the individuals and compare the accuracy with other popular methods based on the same application purpose. The similarity measure is Euclidean distance.

4.1   Comparison Evaluation for Verification Scenarios

In these experiments, we present the comparative evaluations on FRGC v2.0 3D face databases [10] to show the performance of verification scenarios and compare the widely used systems in the literature of 3D face recognition. For the verification scenario, receiver operating characteristics (ROC) curves for three different FRGC masks, namely ROC I, II, and III are shown in Figure 3. To all of masks at an FAR (False Accept Rate) of 0.1%, our method yields the better verification results through all the thresholds, which corresponds to optimal feature subspaces.

In the Table 1, we demonstrate verification results for ROC I, ROC II, and ROC III protocols, which is treated as the standard evaluation on FRGC v2 3D face recognition. For expression issues, some popular methods is based on the different testing sets, including Neutral vs. Neutral (N vs. N), Non-Neutral vs. Neutral (Non-N vs. N), All vs. Neutral (A vs. N). Table 1 showed the verification results with the FAR of 0.1%. We can conclude that the verification rates

Table 1. Verification results (%) with the different 3D face recognition methods on FRGC v2.0 database

Methods	ROC I	ROC II	ROC III	N vs. N	Non-N vs. N	A vs. N
FRGC [10]	-	-	-	-	40	45
Bettetti et al. [11]	-	-	-	97.7	91.4	95.5
Passalis et al. [12]	-	-	-	94.9	79.4	81.5
Cook et al. [13]	93.71	92.91	92.01	-	-	-
Kakadiaris et al. [14]	97.3	97.2	97	-	-	-
Ours	96.97	96.5	96.1	96.4	92.5	96.3

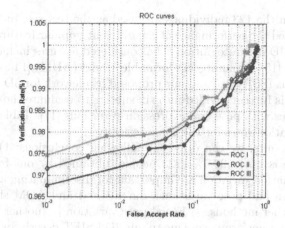

Fig. 3. The ROC curves of FRGC v2.0 3D face database.

of our method are better than other ones, but worse than Kakadiaris et al. [14]. However, Kakadiaris et al. [14] used wavelets and Pyramid transformation to describe the facial scale information and reported the performance was 97% verification at a 0.1% False Acceptance Rate (FAR) in the Face Recognition Grand Challenge. Compared to our algorithm, Kakadiaris et al. [14] utilized a complex algorithm that required a huge computational cost and storage space. For our method, there is a substantial advantage in speed. Our 3D LSIFT algorithm has a much lower computational complexity and is much easier to implement. The performance will be improved when the training data can be significantly increased.

4.2 Performance Evaluation Based on the CASIA 3D Face Database

Here, we employ the CASIA 3D face database [15] and demonstrate the sensor-invariance of our system. This database constitutes challenging variations, including large expressions and poses. The total 4625 facial images can be divided into two subsets, including the training set and the test set. The training set is composed of 615 images, selected 5 facial images for each individual. The rest

Table 2. Rank-1 identification results (%) based on the CASIA 3D face database

Test sets	CO	ADM	SSR	Ours
Illumination Variants	48.21	98.33	96.47	98.5
Expression Variants	45.74	95.73	93.08	95.2
Small Pose Variants	45.27	93.97	92.83	94.86
Large Pose Variants	32.64	56.85	60.99	80.5
Small Pose Variants with Smiling	43.76	90.38	82.15	87.57
Large Pose Variants with Smiling	31.79	52.14	58.43	74.5

of the images from the 123 individuals are used as the test set. In these experiments, we compared the performance of the different popular feature description methods used for 3D face recognition. The considered features include COSMOS shape index (CO) [15], Annotated Deformable Model (ADM) [14], sparse spherical representations (SSR) [16] and our proposed framework for 3D face recognition. The test set is further divided into six subsets to evaluate the performance of different features with pose and expression variations [17]. Table 2 shows the rank-one identification rates for the same test set.

From these results, we can draw the following conclusions: 1) the highest identification rate is up to 98.5% (123 people) obtained by our framework. 2) Shape and illumination variation are important for discriminating an individual. 3) With expression and pose variations, our method and ADM show the better results than other methods. 4) Facial pose variation is another major factor affecting recognition performance. Due to our 3D LSIFT descriptor, our method can capture the shape characteristics of an individual's face and represent this 3D geometric information in an efficient facial regional domain, which shows superior performance relative to other methods that are employed. Thus, our method is robust with the different 3D face database with high generality.

5   Conclusions

A new 3D face recognition method is proposed in this paper by combining a novel facial segmentation method and a novel feature descriptor based on 3D LSIFT. We have utilized a shape bands method for refined region segmentation. Then, we design a unified 3D facial feature descriptor 3D LSIFT fusing appearance and depth information. A huge number of redundant key points can be decreased by adding the facial depth information caused by illumination and pose variations. As a result, our method nicely inherits the invariance of local scale and rotation and increase separability, which overcomes expression and pose variations to some extent. Finally, experimental results demonstrate the improved performance than other popular approaches with a good generalization.

Acknowledgments. The work presented in this paper was supported by the National Natural Science Foundation of China (Grants No. NSFC-61402046 and NSFC-61403024), Fund for the Doctoral Program of Higher Education of China (Grants

No.20120005110002), Beijing Municipal Commission of Education Build Together Project, Principal Fund Project, and President Funding of Beijing University of Posts and Telecommunications.

References

1. Ming, Y., Ruan, Q.: A mandarin edutainment system integrated virtual learning environments. Speech Communication **55**, 71–83 (2013)
2. Bowyer, K.W., Chang, K., Flynn, P.: A survey of approaches and challenges in 3d and multi-modal 3d+2d face recognition. Computer Vision and Image Understanding **101**, 1–15 (2006)
3. Bai, X., Li, Q., Latecki, L.J., Liu, W.: Shape band: a deformable object detection approach. In: CVPR 2010, pp. 1335–1342 (2010)
4. Ming, Y., Ruan, Q., Hauptmann, A.: Activity recognition from kinect with 3d local spatio-temporal features. In: ICME 2012, pp. 344–349 (2012)
5. Ming, Y., Ruan, Q.: Robust sparse bounding sphere for 3d face recognition. Image and Vision Computing **30**, 524–534 (2012)
6. Ming, Y., Ruan, Q., Ni, R.: Learning effective features for 3d face recognition. In: ICIP 2010, pp. 2421–2424 (2010)
7. Moreno, A.B., Sanchez, A., Velez, J.F., Diaz, F.J.: Face recognition using 3d surface-extracted descriptors. In: IMVIP 2003 (2003)
8. Alyuz, N., Gokberk, B., Akarun, L.: Regional registration for expression resistant 3d face recognition. IEEE Trans. Information Forensics and Security **5**, 425–440 (2010)
9. Faltemier, T.C., Bowyer, K.W., Flynn, P.J.: A region ensemble for 3d face recognition. IEEE Trans. Information Forensics and Security **3**, 62–73 (2008)
10. Phillips, P.J., Flynn, P.J., Scruggs, T., Bowyer, K.W., Chang, J., Hoffman, K., Marques, J., Min, J., Worek, W.: Overview of the face recognition grand challenge. In: CVPR 2005, pp. 947–954 (2005)
11. Berretti, S., Bimbo, A.D., Pala, P.: 3d face recognition using isogeodesic stripes. IEEE Trans. Pattern Analysis and Machine Intelligence **32**, 2162–2177 (2010)
12. Passalis, G., Kakadiaris, I.A., Theoharis, T., Toderici, G., Murtuza, N.: Evaluation of 3d face recognition in the presence of facial expressions: an annotated deformable model approach. In: FRG 2005 (2005)
13. Cook, J., McCool, C., Chandran, V., Sridharan, S.: Combined 2d/3d face recognition using log-gabor templates. In: ICVSBS 2006 (2006)
14. Kakadiaris, I.A., Passalis, G., Toderici, G., Murtuza, M.N., Lu, Y., Karampatziakis, N., Theoharis, T.: Three-dimensional face recognition in the presence of facial expressions: an annotated deformable model approach. IEEE Trans. Pattern Analysis and Machine Intelligence **29**, 640–649 (2007)
15. Beumier, C., Acheroy, M.: Automatic 3d face authentication. Image and Vision Computing **18**, 315–321 (2000)
16. Llonch, R.S., Kokiopoulou, E., Tosic, I., Frossard, P.: 3d face recognition with sparse spherical representations. Pattern Recognition **43**, 824–834 (2010)
17. Xu, C., Li, S., Tan, T., Quan, L.: Automatic 3d face recognition from depth and intensity gabor features. Patter recognition **42**, 1895–1905 (2009)

Recognizing Scenes by Simulating Implied Social Interaction Networks

MaryAnne Fields[1], Craig Lennon[1], Christian Lebiere[2], and Michael K. Martin[2(✉)]

[1] Army Research Laboratory, Aberdeen, MD, USA
{mary.a.fields22.civ,craig.t.lennon.civ}@mail.mil
[2] Carnegie Mellon University, Pittsburgh, PA, USA
{cl,mkm}@andrew.cmu.edu

Abstract. Indoor scene recognition remains a challenging problem for autonomous systems. Recognizing public spaces (e.g., libraries, classrooms), which contain collections of commonplace objects (e.g., chairs, tables), is particularly vexing; different furniture arrangements imply different types of social interaction, hence different scene labels. If people arrange rooms to support social interactions of one type or another, then object relationships that reflect the general notion of social immediacy may resolve some of the ambiguity encountered during scene recognition. We thus describe an approach to indoor scene recognition that uses the context provided by inferred social affordances as input to a hybrid cognitive architecture (ACT-R) that can represent, apply and learn knowledge relevant to classifying scenes. To provide common ground, we demonstrate how sub-symbolic learning processes in ACT-R, which plausibly give rise to human cognition, can mimic the performance of a simple, widely used machine learning technique (k-nearest neighbor classification).

Keywords: Indoor scene recognition · ACT-R · K-Nearest neighbor classification · Machine learning · Cognitive robotics · Social networks

1 Introduction

Like humans, autonomous systems often exploit statistical regularities in low-level visual features to recognize objects and scenes. Unlike humans, autonomous systems tend to have difficulties using context to aid in the scene recognition process [1–3]. We believe the creation of a feedback loop between perceptual processes and an established cognitive architecture (ACT-R) will foster the use of context in the classification of scenes, and of the objects within those scenes.

Context is dynamically produced by a subtle interplay of environment, an agent's relevant knowledge and the agent's goals. Some of the mechanisms underlying this interplay form an inherent part of ACT-R. These mechanisms are similar to machine learning techniques employed by computer vision scientists, but are integrated in a unified cognitive framework.

© Springer International Publishing Switzerland 2015
H. Liu et al. (Eds.): ICIRA 2015, Part III, LNAI 9246, pp. 360–371, 2015.
DOI: 10.1007/978-3-319-22873-0_32

The following discussion demonstrates similarities between ACT-R and k-Nearest Neighbor (KNN) when classifying indoor scenes that are better distinguished by contextual features than by object collections. To keep the comparison between KNN and ACT-R straightforward, we generated contextual features with a process external to ACT-R. Our demonstration, therefore, is based on static context rather the dynamic context that will be produced ultimately by fine-grained bidirectional interaction between perceptual processes and the cognitive architecture.

In the next section, we describe the current state of scene classification. We then describe the pattern-matching mechanisms considered: ACT-R and KNN. Subsequently, we describe our methods for constructing scene graphs, feature sets, and noise models, followed by a description of KNN and ACT-R classification performance, a comparison of their mechanisms, and a discussion of future work.

1.1   Autonomous Scene Classification

Scene recognition remains a challenging problem that has engaged the computer vision community for the past decade. Current methods use object or parts recognition, along with the co-occurrence of salient features, to recognize interior scenes. However, labeling objects and activities in scenes may not be sufficient to recognize regions in public spaces such as offices, libraries, or school buildings that contain large collections of identical (or nearly identical) tables, chairs or other furnishings. Humans easily recognize and categorize the purpose of these regions from the content and the layout of the space. An autonomous system also needs this ability to recognize the spatial and contextual relationships between objects, and to use those relationships to infer labels for ensembles of objects.

In this work, we propose a method to classify scenes based on how arrangements of constituent objects might impact social interactions. Based on the position and orientation of objects in a scene, we use simulation to hypothesize possible social networks of humans using these objects. We use two types of relationships between objects, distance and inter-visibility, to build these networks. We chose these relationships because the distance between chairs, and the orientation of chairs relative to one another, are logical choices for objective metrics underlying social immediacy.

In their recent work, [4] use the physical proximity of objects for the purposes of scene classification. The authors assume that the objects relevant for identification of a scene have been appropriately labeled, then link physically adjacent objects into local groups to establish a compressed adjacency graph upon which scene classification can be made. The adjacency graph uses only the physical positions of the objects, and not features such as orientation of the chairs. The wide range of objects that the authors assume to be labeled includes coffeemakers, pillows, switches, bottles, glasses, paintings and candles, making such a method difficult to implement given the current state of semantic perception. However, larger objects, such as chairs, tables, walls, and television screens can be more easily identified and modeled as a complete scene, as done in [5]. This makes it reasonable for us to use such objects, and their orientation, in determining the purpose of a room.

While solving the different problem of how a robot might appropriately place new objects for reasonable human use, Jiang and his colleagues [6] found that a robot can more easily determine appropriate positions for the objects if it knows what position a human will occupy while using those objects. In the absence of knowledge of a human position, one can establish a range of likely positions by hypothesizing human positions based on furniture arrangements [6], and choose object positions that are likely to be appropriate given these hypothesized positions.

Our work builds on the work above by first assuming that chairs and tables can be appropriately labeled and modeled, as in [5]. Inspired by the work of [6], we imagine chairs as being occupied by humans, and use the viewing distance and orientation of the implied human in a chair to link the labeled objects in the room, as well as linking them by physical proximity. By examining simple features of these simulated social networks, we develop a semantic classification of the room.

1.2   k-Nearest Neighbors Classification

KNN is a method of classification based on the assumption that a space of feature values, under a suitable metric, is continuous enough that a point within the space is likely to have the same label as the points near it. In order to apply KNN, one needs a training set with quantitative features and a metric - the Euclidian metric in our case. When a new observation needs to be classified, one measures the distance of the new point from all points in the training set, and classifies it according to the modal label of the K closest training set points. The training set thus partitions the feature space into contiguous regions of one class.

With KNN, new classes can be taught to a system online by adding at least K instances of them to the training set. This method can account for noisy data by increasing the value of K, and can be used in a fashion that incorporates prior knowledge by returning the class with the highest proportion among the K closest points. When applied by a domain expert who can choose an appropriate metric, a suitable value of K, and who can incorporate the results with other prior knowledge, KNN is a versatile tool. For a more detailed description of KNN, see [7].

1.3   ACT-R

ACT-R [8, 9] is a cognitive architecture, i.e., a computational implementation of a unified theory of cognition [10]. The structure of the architecture (see Fig. 1) is composed of a set of modules, including perceptual, motor and memory modules.

Modules are a combination of symbolic information representation and processing, combined with statistical quantities learned through experience. Each module operates asynchronously, with execution requests and their results communicated through buffer(s) associated with them. The associated buffers can each hold a single chunk of information. Interaction among modules is controlled by a central procedural module, implemented as a constrained production system, through a set of limited-capacity buffers. The production system contains simple condition-action rules for testing and modifying the contents of the buffers. Production rule selection is governed by an associated learned quantity called utility that reflects the effectiveness of each rule in achieving its goal.

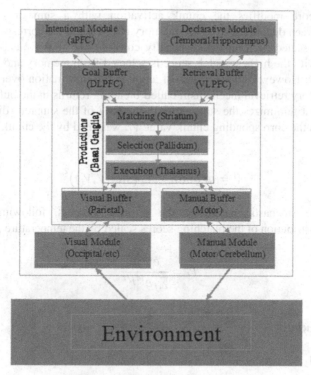

Fig. 1. ACT-R cognitive architecture

In this paper we will focus on the operations of the declarative memory module. This module contains information in the form of chunks, which are structures associating a small set of data items. The retrieval of each chunk i from long-term memory is governed by a quantity A_i called activation, representing the log odds of retrieval, and computed as the sum of factors reflecting the chunk's content and history of use:

$$A_i = \log \sum_k t_k^{-d} + \sum_j W_j S_{ji} + N(0, \sigma) \tag{1}$$

The first term, called the base-level activation, reflects the prior history of the chunk by summing over all past uses k a power law function of the time t_i since that occurrence with a common decay rate d of 0.5, yielding both the power law of practice and power law of forgetting. The second sum, called spreading activation, provides the current context by summing the attentional weight element W_j multiplied by the associative strength, S_{ji} between the ith chunk and all other chunks j currently held in buffers. The third term is a logistic noise function that makes the retrieval process stochastic. Retrieving a chunk from declarative memory involves first specifying a pattern in the associated retrieval buffer. Then the match score M_i of each memory chunk i is computed based on both its activation and the degree to which it matches the pattern:

$$M_i = A_i + MP \sum_d sim(v, d) \tag{2}$$

The match score modifies the chunk activation with a sum of (dis)similarity $sim(v, d)$ between data items from the memory chunk i and the corresponding values in the pattern, scaled by a mismatch penalty constant MP. For typical purposes the single chunk with the highest match score is selected from memory and placed in the retrieval buffer. However, for purposes of aggregating information over a number of chunks, a memory retrieval mechanism called blending returns in the buffer a consensus value V that minimizes the sum over all chunks i of the squared (dis)similarities between V and the corresponding chunk value V_i, weighted by the chunk's probability of retrieval P_i:

$$V = argmin \sum_i P_i \left(sim(V, V_i) \right)^2 \tag{3}$$

P_i reflects the stochastic nature of the retrieval process, following a softmax (Boltzmann) distribution of the matching scores scaled by the temperature parameter t:

$$P_i = \frac{e^{M_i/t}}{\sum_j e^{M_j/t}} \tag{4}$$

2   Methods

2.1   Stimuli

We constructed stimuli to address two issues. The first concerned whether contextual features based on social affordances improved indoor scene recognition compared to recognition based on a collection of objects. To this end, we constructed two feature sets according to the procedure described below. The counts of chairs and tables in a room served as object-based features, and the counts of proximity and visibility links served as affordance-based features. Each feature set was altered with an appropriate type of noise. The object-based feature set was altered to reflect an imperfect recognition of the objects, while the social affordances feature set was altered by changing the orientation and position of the chairs, thus changing the social dynamics of the room. We employed low and high levels of each type of noise.

The second issue concerned the impact of the size of the training set on the classification performance of KNN and ACT-R. We thus created three training set sizes that used 1%, 10%, or 100% of our training stimuli for training, which were all followed by testing on our entire set of testing stimuli.

Room Types. Event and space planners arrange tables, chairs, and other furniture to support desired modes of social interaction. Using [11] as a guide, we considered five types of furniture configurations: instructional, theater, café, conversational, and boardroom. In the instructional layout, participants are seated in rows of tables facing an instructor, podium or screen. A theater has a similar arrangement of chairs, but without tables. These arrangements maximize the number of participants, and have a single focal point, discouraging small group interactions. Café and conversation layouts cluster participants around individual tables (or focus points) to encourage small group discussions. Finally, a boardroom layout clusters all the participants around a central table. Fig. 2 shows examples of the layouts, with chairs represented as semi-circles and tables shown as rectangles.

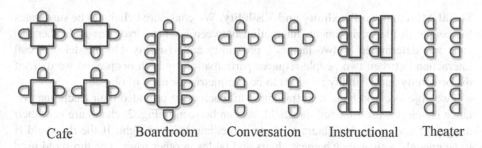

Cafe   Boardroom   Conversation   Instructional   Theater

Fig. 2. Example room layouts

Room Layout. The canonical examples shown in Fig. 2 can be populated with variable numbers of chairs and tables. Boardrooms have one table and an even number of chairs. All other rooms can be divided into rows, which go from top to bottom, and sections, which run from right to left within each row. Each section within a row can have a line of chairs (theater), a group of chairs facing inward (conversation area), a table with chairs facing the center (café), or a line of chairs with a table in front (instructional). For all rooms except boardrooms, we used from 2 to 4 rows and from 2 to 4 sections within each row. Within each section, we allowed from 2 to 6 chairs to be grouped with one table (or focus point).

We then manipulated the degree to which the arrangement of furniture diverged from the canonical arrangements shown in Fig. 2 to simulate two levels of messiness in the room observed (i.e., high and low noise levels). Specifically, we allowed chairs to shift and rotate from their original position. We represented the noise introduced by the movement of chairs as a vector-valued random variable: $\eta = \langle x, y, \theta \rangle$. To lay out a room, chairs were placed at their canonical position, and then shifted along the table edge by up to x inches to the right or left, and away from the table by up to y inches. Movement in the x and y directions was uniform over the ranges of the variables described in Table 1, with right and backward as the positive directions for x and y, and left and forward as the negative directions. Then the chair was rotated by $\theta°$. The distribution of θ was Gaussian, with mean 0 and a standard deviation of s, with clockwise being the positive direction ($\theta > 0$). The components of η were chosen independently of one another, and independently of the noise vectors of all other chairs. Tables were not perturbed from their canonical positions.

Parameters used to create low and high noise in the affordance-based features are shown in Table 1. Noise in the object-based features was created by randomly relabeling tables as chairs and chairs as tables with a low or high probability (Table 1). Thus each room was generated in a fashion that allowed testing of the robustness of classification to various levels of noise in the social dynamics (created by chair movement) and in the object identification (created by mislabeling of objects).

Table 1. Parameters for the random variables composing η

Noise Level	x (left/right)	y (front/back)	s	Labeling Error
Low	[-6, 6] in.	[0, 6] in.	15°	0.05
High	[-12,12] in.	[0, 12] in.	45°	0.20

Social Affordances: Proximity and Visibility. We considered chairs to be surrogates for people in the environment. Interaction between human surrogates and other objects was determined by two factors – proximity and visibility. Our model of social interaction between two people requires participation by both people, so we defined the proximity and visibility relations to be symmetric for pairs of objects.

Two objects a and b are near (proximal to) each other provided their Euclidian distance is below a user-defined threshold. As can be seen in Fig. 2, chairs are near their assigned table, and other chairs on their immediate left and right. If the threshold is large enough, a chair may be near chairs and tables in other rows. The threshold used here was 60 inches.

Visibility measures potential "eye contact" between two chairs. It is a function of both the field-of-view and the distance between the two chairs. Further, eye contact requires the chairs to face each other. Chairs a and b, with orientations O_a and O_b respectively, will be considered visible to each other if they meet 4 conditions based on user defined parameters for view angle (α_{FOV}), visibility distance (d_v), and orientation (ϵ): (1) a is within b's field of view, (2) b is within a's field of view, (3) a and b are close enough, and (4) b is closer to a than any other chair along the same ray.

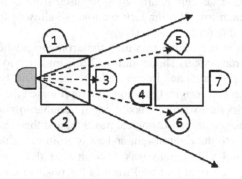

Fig. 3. Visibility links shown as dashed lines

Fig. 3 illustrates the visibility between the gray chair and other chairs in the room. The field of view for the gray chair is the arc between the solid black arrows. In this instance, it is 60°. A visibility link with chair 3 will be present if the view distance is short enough. The link with chairs 5 and 6 also depends on the choice of d_v. None of the other chairs can be linked to the gray chair with this view angle, because chairs 1 and 2 are not within the field-of-view for the gray chair, chair 4 faces the wrong direction, and chair 7 is blocked by chair 3.

The chairs and tables provided the nodes of the room graphs we used in our analyses, with binary proximity and visibility links determined by our choice of distance threshold (60 inches), view angle (190°) and visibility distance (60 inches).

2.2   Procedure

All room types had between 2 and 6 chairs in a section. Boardrooms have only one table (i.e., 1 section, 1 row), while the other rooms had between 2 and 4 rows, and between 2 and 4 sections within each row. For each room type, we created 100 simulated rooms for a total of 18,500 instances at each level of noise (Low, High). For each instance, we constructed two feature sets (Object-based, Affordance-based) for use by our classifiers. To examine the effect of different training-set sizes (1%, 10%, 100%) on classification performance, we first generated a training set and a testing set at each level of noise. In the 1% case, 1% of the training set was used for training, followed by testing using the entire test-set (with results averaged over all 100 partitions of the training set). The 10% and 100% cases were sampled and tested in a similar fashion.

3   Results and Discussion

We trained two classifiers (KNN, ACT-R) to recognize the five room types using two feature sets (Object-based, Affordance-based), two noise levels (High, Low), and three training-set sizes (1%, 10%, 100%). For the KNN classifier, similarity depended on a neighborhood size of $k = 1, 5$, and 10 (for the 1%, 10% and 100% training sets, respectively) as determined by Euclidean distance. For the ACT-R classifier, the model most similar to KNN is an application of instance-based learning (IBL) methodology [12] to the classification problem [13].

In our ACT-R classifier model, each training instance was represented as a single chunk in memory, associating the room category to the representation features (either the number of tables and chairs or the number of proximity and visibility links). Given a new room configuration specified by a set of feature values, those values are placed in the retrieval buffer associated with declarative memory and a blending retrieval process is triggered to produce a consensus category value. The similarity between feature values was defined as a ratio function used in other models:

$$sim(x, y) = \frac{min(x, y)}{max(x, y)} - 1 \tag{5}$$

Thus if $x = y$ the similarity is perfect (0), otherwise it declines to the maximum dissimilarity of -1 as the ratio between the two numbers increases[1].

---

[1]   Other relevant parameters of the model include mismatch penalty MP of 2.0 and temperature t of 0.1.

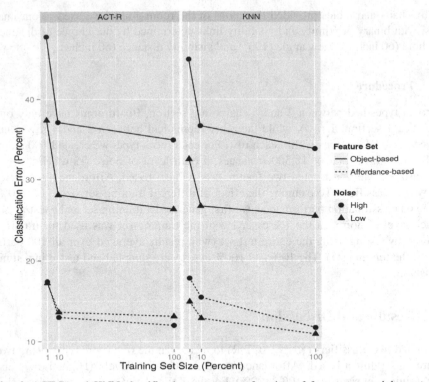

Fig. 4. ACT-R and KNN classification errors as a function of feature set, training set size, and noise

Both classifiers (KNN, ACT-R) recognized rooms more accurately by using affordance-based features rather than object-based features (see Fig. 4). Both classifiers responded similarly to the degree of noise present in the stimuli (High, Low) especially for the object-based features: low noise stimuli tended to reduce classification errors relative to high noise stimuli. However, for affordance-based features high noise improves performance marginally in the ACT-R classifier while still decreasing it slightly for KNN. Both classifiers were robust to decreases in training-set size (1%, 10%, 100%). They performed best with full sampling (i.e., 100%), yet performance at 10% sampling was nearly as good.

Fig. 5 shows room-type confusions for each classifier with each feature-set for what was usually the best training condition (full sampling with low noise). Again, a close similarity in classifier performance can be seen. For the object-based representation, the most common confusion pairs are theater/conversation and instructional/cafe, which results because the former pair doesn't include tables while the latter pair does. Boardroom is not very confusable because of the unique table/chair ratio. For the affordance-based representation, theater and instructional are the most confusable categories because their relation structure is basically identical, differing only in the presence of tables. The same is true to a lesser extent for cafe and conversation. Again, boardroom is not very confusable because of the unique structure of having a single network component regardless of size.

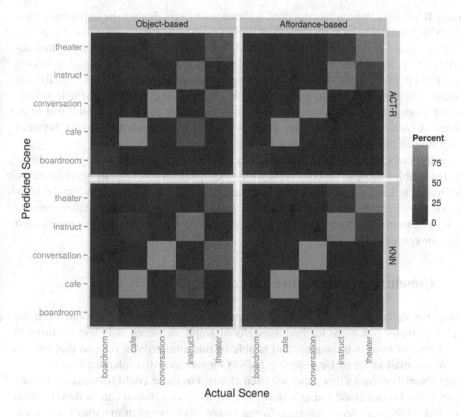

Fig. 5. Percent confusions of room-types for classifiers x feature-set

ACT-R memory retrieval can be conceived as a generalization of KNN. Each chunk in declarative memory corresponds to a training instance. For comparison to the current stimulus, the partial matching mechanism is akin to the distance computation in KNN. There are however two specific differences: (a) while a linear similarity function would correspond to the Euclidean distance, the ratio similarity function used here does not; (b) the addition over the various pattern mismatches of the partial matching equation computes a Manhattan distance rather than a Euclidean distance. More generally, partial matching can be used to compute differences over arbitrary multi-dimensional spaces by defining the relevant similarities between chunk values. Finally, while other components of the activation equation were not used in this model, base-level learning enables the process to be sensitive to the recency and frequency of training instances, while spreading activation enables for semantic priming effects.

The pooling of answers also operates similarly in both approaches. Just as KNN selects the most common answer among the K best candidates, if the values are discrete and do not have any similarity between them, the blending equation will also select the most popular answer among the candidate values. However, there are also a number of aspects in which the blending process is more general than the KNN voting process. First, blending operates over all instances in memory rather than the most similar K of them, which has the dual advantages of broadening the experience base upon which the decision is made, and of removing the need for the modeler to specify a proper value for the K parameter. Second, the memory instances are weighted according to their probability, which partly reflects their degree of match, allowing more similar examples to have a higher impact than more distant ones. Finally, the process of aggregating answers in blending is much more general than simple voting, ranging from averaging over values for which similarity functions are defined (e.g., numerical values), to finding consensus values among symbolic chunks between which similarities have been defined.

Finally, embedding generalizations of algorithms such as KNN in cognitive architectures enables them to be combined with other mechanisms such as heuristic rules, reinforcement learning and Bayesian learning. More generally, it allows for the leverage of knowledge about the semantics of the domain to be naturally taken into account, an essential part of cognitive decision making in complex environments such as indoor spaces [14].

4   Conclusions and Future Directions

This paper demonstrates a method to detect the social purpose of a room from the furniture layout. Since our method uses global graph measures, it is less sensitive to the number of objects in an area, and is able to find similarity in regions that vary in size from small cafes to large dining halls. Our method also tolerates less accurate object identification. In this paper we used chairs, but these could be replaced by any object that a human could use as a chair, so long as an orientation can be derived from it, or indeed the persons themselves. In our future work, we plan to explore the sensitivity of this classification method to perception failures. We are interested in 3 types of failures: missing objects, mislabeled objects, and hallucinated objects. We believe that the social network provides contextual information that could be used to correct inaccurate object labels. Thus a room type derived from a consideration of objects in the room could be used to generate a more informed relabeling of those same objects.

In this paper, we considered networks that represented single purpose layouts. However, many public spaces, such as libraries and offices, have large rooms that serve multiple purposes. A future extension of our methodology will enable robots to recognize the functionality of sub-regions within a larger room. Cognitive architectures can support that goal by providing a framework to integrate mechanisms akin to common machine learning algorithms such as KNN with general purpose knowledge and reasoning capabilities.

Acknowledgements. This research was supported by OSD ASD (R&E) and by the Army Research Laboratory's Robotics Collaborative Technology Alliance.

References

1. Oliva, A., Torralba, A.: The role of context in object recognition. Trends Cogn. Sci. **11**, 520–527 (2007)
2. Quattoni, A., Torralba, A.: Recognizing indoor scenes. IEEE Conf. Comput. Vis. Pattern Recognit. (2009)
3. Wagemans, J., Elder, J.H., Kubovy, M., Palmer, S.E., Peterson, M.A., Singh, M., von der Heydt, R.: A century of Gestalt psychology in visual perception: I. Perceptual grouping and figure–ground organization. Psychol. Bull. **138**, 1172–1217 (2012)
4. Sadovnik, A., Chen, T.: Hierarchical object groups for scene classification. In: 2012 19th IEEE International Conference on Image Processing (ICIP), pp. 1881–1884. IEEE (2012)
5. Shao, T., Xu, W., Zhou, K., Wang, J., Li, D., Guo, B.: An interactive approach to semantic modeling of indoor scenes with an RGBD camera. ACM Trans. Graph. **31**, 1 (2012)
6. Jiang, Y., Zheng, C., Lim, M., Saxena, A.: Learning to place new objects. In: 2012 IEEE International Conference on Robotics and Automation (ICRA), pp. 3088–3095. IEEE (2012)
7. Hastie, T., Tibshirani, R., Friedman, J.: The elements of statistical learning. Springer series in statistics. Springer, Berlin (2009)
8. Anderson, J.R., Lebiere, C.: The atomic components of thought. Lawrence Erlbaum Associates, Mahwah (1998)
9. Anderson, J.R.: How can the human mind occur in the physical universe? Oxford University Press, New York (2007)
10. Newell, A.: Unified Theories of Cognition. Harvard University Press, Cambridge (1990)
11. Collins, P.: Meeting Room Configurations. http://www.jordan-webb.net/downloads/
12. Gonzalez, C., Lerch, J.F., Lebiere, C.: Instance-based learning in dynamic decision making. Cogn. Sci. **27**, 591–635 (2003)
13. Lebiere, C.: Constrained functionality: application of the ACT-R cognitive architecture to the AMBR modeling comparison. In: Gluck, K., Pew, R. (eds.) Modeling Human Behavior with Integrated Cognitive Architectures. Erlbaum, Mahwah (2005)
14. Oltramari, A., Vinokurov, Y., Lebiere, C., Oh, J., Stentz, A.: Ontology-based cognitive system for contextual reasoning in robot architectures. In: 2014 AAAI Spring Symposium Series (2014)

On Modeling and Least Squares Fitting of Cylinders from Single and Multiple Views Using Contour Line Features

Mark Becke[✉]

Regensburg Robotics Research Unit, Faculty of Mechanical Engineering,
Ostbayerische Technische Hochschule Regensburg, Regensburg, Germany
Mark.Becke@oth-regensburg.de

Abstract. In this paper, a new method for a minimum-error pose estimation of cylinder axes based on apparent contour line features from multiple views is presented. Novel model equations for both single and particularly multiple views are derived, and based upon these, an algorithm for least squares fitting the model to imaged cylinder contour line features is introduced. The good performance of the proposed algorithm is shown by solving an exemplary fitting problem.

Keywords: Straight Homogeneous Circular Cylinder · Apparent contour · Computer vision · Multiple view geometry · Least squares fitting

1 Introduction

The reconstruction of the 3D-pose of objects with geometrical shapes based on images generated by perspective projection is a well known and widely discussed topic in computer vision. There is a large amount of literature dealing with model-based pose reconstruction from geometrical features such as points, lines, spheres, or a combination of these features. A survey of these pose estimation problems is given, e.g., by [1]. Compared to that, however, literature regarding the pose estimation of cylindrical objects, in particular those cylinders with constant radius and a straight line as symmetry axis perpendicular to a circular cross section (Straight Homogeneous Circular Cylinder), is somewhat sparse. There is only some fundamental and in recent publications often cited work covering the pose estimation of cylinder axes from a perspective single view, but no efforts on further development of this work in the recent years are reported. There are three major approaches depending each on the evaluation of either elliptical projections of cylinder cross sections onto the image, as in [2,3], the projected apparent contour lines of an (infinite) cylinder, as in [4–7] or the combination of the previous two methods [8,9]. The case of presence of noise, or imaging errors, respectively, in a single view pose estimation is discussed in [5,6], but the only work [7] dealing with pose estimation of cylinder axes based on multiple views and deploying appropriate analytic models considers only the

© Springer International Publishing Switzerland 2015
H. Liu et al. (Eds.): ICIRA 2015, Part III, LNAI 9246, pp. 372–385, 2015.
DOI: 10.1007/978-3-319-22873-0_33

ideal, noise-free case. Although there are sophisticated methods for model-based pose estimation minimizing the overall distances between detected point features from multiple views and the model [10], an adaption or modification of these or similar methods for application on cylinder pose estimation by evaluation of line features is not reported yet.

Due to that, in this paper, a new method for a minimum-error reconstruction of the axis of an (infinite) straight homogeneous cylinder exploiting contour line features is presented for the pose estimation of cylindrical bodies such as, e.g., cylindrical markers, from multiple camera views. The model equations for perspective projection of cylinder contour lines are derived in section 2, and a least squares fitting algorithm is proposed in section 3. In section 4, an exemplary fitting problem is solved and discussed, section 5 concludes this paper.

2   Cylinder Modeling from Single and Multiple Views

2.1   Single View Perspective Projection of a Cylinder

A camera with orientation $^{\mathrm{ref}}R \in SO(3)$ and position $^{\mathrm{ref}}t \in \mathbb{R}^3$, given relative to an arbitrary reference coordinate system, is directed onto a straight homogeneous cylinder with radius $r = \mathrm{const} > 0$ (and infinite length), whose apparent contour in the camera image consists of two straight lines. The origin of the camera coordinate system is located within the focal point, the z-axis is perpendicular to the image plane and directed towards the scene. Setting the camera coordinate system as reference frame, i.e., $^{\mathrm{ref}}R = I$, where I is the identity matrix, and $^{\mathrm{ref}}t = 0$, the image plane is then

$$(0\,0\,1)\,x = f \quad \forall x \in \mathbb{R}^3, \tag{1}$$

where $f > 0$ is the focal length of the camera. The 3×4 projection matrix P maps points from 3D space, where $x^\top = (x, y, z)$ and $^{\mathrm{ref}}x = {}^{\mathrm{ref}}R\,x + {}^{\mathrm{ref}}t$, to the 2D image space, $u^\top = (u, v)$, via perspective projection up to a scale w [11],

$$w \begin{pmatrix} u \\ 1 \end{pmatrix} = \underbrace{K\,^{\mathrm{ref}}R\left(I\big|-{}^{\mathrm{ref}}t\right)}_{P} \begin{pmatrix} ^{\mathrm{ref}}x \\ 1 \end{pmatrix} = Kx, \tag{2}$$

where K is the 3×3 camera calibration matrix. The backprojection of any line $l \in \mathbb{R}^3$ given in Hesse normal form on the image plane with

$$l^\top \begin{pmatrix} u \\ 1 \end{pmatrix} = 0 \quad \forall u \in \mathbb{R}^2 \tag{3}$$

is a plane whose normal vector has the direction $K^\top l$ [11].

Considering the above relations, all rays through the focal point of the camera which are perpendicular to the viewed cylinder with axis direction $k \in S^2$, where $S^2 = \{ x \in \mathbb{R}^3 \big| \|x\| = 1 \}$ is the unit sphere, form a plane through the focal point with normal vector k,

$$k^\top x = 0 \quad \forall x \in \mathbb{R}^3. \tag{4}$$

i.e., the so called normal plane of that particular cylinder viewed with this camera, see Fig. 1(a). The cylinder axis pierces the normal plane in the point p with distance $d = \|p\| > r$ to the focal point, and $m = p/d \in S^2$. The normal vector $n = k \times m \in S^2$ defines the symmetry plane containing both the focal point and the cylinder axis. Then, a coordinate system $R_{\text{cyl}} \in SO(3)$ given w.r.t. the camera coordinate system and located within the focal point can be defined as

$$R_{\text{cyl}} = (\, m\ n\ k\,)\,,\tag{5}$$

which is the canonical image coordinate system of the particular cylinder, see Fig. 1(b). Since $t_{\text{cyl}} = p = dm$ and R_{cyl} both define the pose of the cylinder axis relative to the camera frame, one has the pose relative to any given reference coordinate system in terms of a 4×4 homogeneous transformation matrix

$$\begin{pmatrix} {}^{\text{ref}}R_{\text{cyl}} & {}^{\text{ref}}t_{\text{cyl}} \\ 0 & 1 \end{pmatrix} = \begin{pmatrix} {}^{\text{ref}}R & {}^{\text{ref}}t \\ 0 & 1 \end{pmatrix} \begin{pmatrix} R_{\text{cyl}} & t_{\text{cyl}} \\ 0 & 1 \end{pmatrix}.\tag{6}$$

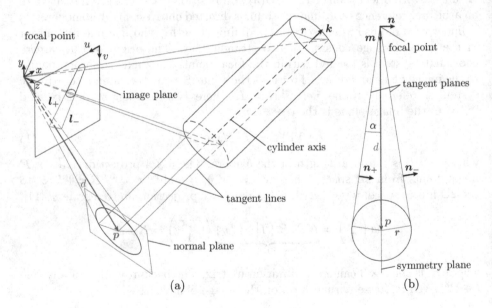

(a)   (b)

Fig. 1. Perspective projection of a cylinder contour (a) and geometric relations on the normal plane containing the focal point (b).

If the contour image of the cylinder consists of two lines on the image plane, l_+ and l_-, the backprojection of these lines are planes tangent to the cylinder with the unit normal vectors

$$n_+ = \frac{K^\top l_+}{\|K^\top l_+\|}, \qquad\qquad n_- = \frac{K^\top l_-}{\|K^\top l_-\|}.\tag{7}$$

The absolute value of the angle between the symmetry plane and the tangent planes is determined by $\sin |\alpha| = r/d$ with $r < d$. This means, that only the contour of that particular cylinder is imaged as two straight lines, whose axis does not pierce or touch the sphere $S_r^2 = \{ \boldsymbol{x} \in \mathbb{R}^3 | \; \|\boldsymbol{x}\| = r \}$, otherwise the focal point could lie within the cylinder. So, the described imaging geometry is only valid if the cylinder has an axis defined by

$$d\boldsymbol{m} + t\boldsymbol{k} \in \{ \boldsymbol{x} \in \mathbb{R}^3 | \; \|\boldsymbol{x}\| > r \} \quad \forall t \in \mathbb{R} . \tag{8}$$

2.2   Cylinder Modeling from Single and Multiple View

Considering now the case where $n \geq 1$ cameras are viewing the same cylinder. Provided that $d_i > r \, \forall i \in \{1, \ldots, n\}$, the contour of the (infinite) cylinder is imaged as straight lines $\boldsymbol{l}_{+,i}$ and $\boldsymbol{l}_{-,i}$ in each camera i, and this multiple view case is equivalent to n separate single view cases as in 2.1. The backprojections of these lines by means of the according camera calibration matrix $\boldsymbol{K}_i$ are planes through the focal points of the according cameras and tangent to the cylinder with the unit normal vectors [11]

$$^i\boldsymbol{n}_{+,i} = \frac{\boldsymbol{K}_i^\top \boldsymbol{l}_{+,i}}{\|\boldsymbol{K}_i^\top \boldsymbol{l}_{+,i}\|}, \qquad ^i\boldsymbol{n}_{-,i} = \frac{\boldsymbol{K}_i^\top \boldsymbol{l}_{-,i}}{\|\boldsymbol{K}_i^\top \boldsymbol{l}_{-,i}\|} . \tag{9}$$

The coordinate system of camera 1 is defined as reference frame and the orientation and position of the i-th camera coordinate system relative to the frame of camera 1 is given with $\boldsymbol{R}_i = {}^1\boldsymbol{R}_i$ and $\boldsymbol{t}_i = {}^1\boldsymbol{t}_i$, see Fig. 2(a). Note that for $\boldsymbol{R}_i = \boldsymbol{I}$ and $\boldsymbol{t}_i = \boldsymbol{0}$, camera i coincides with camera 1.

Since the direction of the cylinder axis is $\boldsymbol{k}_i = \boldsymbol{R}_i {}^i\boldsymbol{k}_i = \boldsymbol{k}_1$, one can determine the points where the cylinder axis pierces the particular normal planes of camera 1 and camera i by

$$\boldsymbol{p}_1 = d_1 \, \boldsymbol{m}_1 , \qquad {}^i\boldsymbol{p}_i = d_i \, {}^i\boldsymbol{m}_i . \tag{10}$$

The unit direction vector ${}^i\boldsymbol{m}_i$ for each camera i can be expressed in coordinates of camera 1 as $\boldsymbol{m}_i$ and depending on $\boldsymbol{n}_{\pm,i} = \boldsymbol{R}_i {}^i\boldsymbol{n}_{\pm,i}$ by

$$\boldsymbol{m}_i = \boldsymbol{R}_i {}^i\boldsymbol{m}_i = \boldsymbol{R}_i \frac{{}^i\boldsymbol{n}_{+,i} - {}^i\boldsymbol{n}_{-,i}}{\|{}^i\boldsymbol{n}_{+,i} - {}^i\boldsymbol{n}_{-,i}\|} = \frac{\boldsymbol{n}_{+,i} - \boldsymbol{n}_{-,i}}{\|\boldsymbol{n}_{+,i} - \boldsymbol{n}_{-,i}\|} , \tag{11a}$$

see Fig. 2(b), and, analogously,

$$\boldsymbol{n}_i = \frac{\boldsymbol{n}_{+,i} + \boldsymbol{n}_{-,i}}{\|\boldsymbol{n}_{+,i} + \boldsymbol{n}_{-,i}\|} , \qquad \boldsymbol{k}_i = \frac{\boldsymbol{n}_{+,i} \times \boldsymbol{n}_{-,i}}{\|\boldsymbol{n}_{+,i} \times \boldsymbol{n}_{-,i}\|} . \tag{11b}$$

Provided that $\boldsymbol{n}_{+,i}^\top \boldsymbol{n}_{-,i} > 0$, and denoting that $\boldsymbol{n}_{-,i}^\top \boldsymbol{m}_i > 0$, or $\boldsymbol{n}_{-,i}^\top \boldsymbol{m}_i < 0$, respectively, the distance between focal point i and the cylinder axis is determined with the angle α_i between the symmetry plane the tangent planes and $\cos 2\alpha_i = \boldsymbol{n}_{+,i}^\top \boldsymbol{n}_{-,i}$ as

$$d_i = r \sqrt{\frac{2}{1 - \boldsymbol{n}_{+,i}^\top \boldsymbol{n}_{-,i}}} > 0 . \tag{12}$$

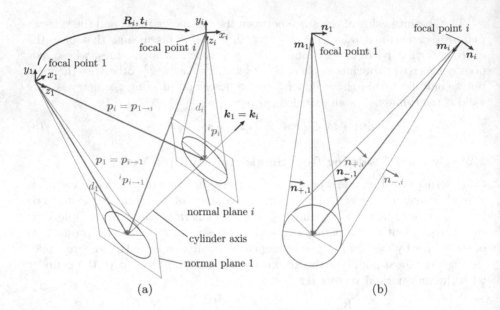

Fig. 2. Relations between the focal points and the normal planes (a) and relations between the canonical image coordinate systems of the cameras and the unit normal vectors of the tangent planes (b) in the multiple view case.

Since the pose of camera i is known, one can determine the point where the cylinder axis pierces the normal plane of camera i in terms of camera 1 with

$$p_{1\to i} = d_1 m_1 + k_1^\top t_i k_1 \tag{13}$$

As Fig. 2(a) shows, this point is identical with ${}^i p_i$ in terms of camera 1, i.e.,

$$p_{1\to i} \overset{!}{=} p_i = R_i {}^i p_i + t_i, \tag{14}$$

and, in return, that point $p_{i\to 1}$, in which the cylinder axis determined by camera i pierces the normal plane of camera 1,

$$p_{i\to 1} = d_i m_i - k_i^\top t_i k_i + t_i, \tag{15}$$

coincides with p_1, i.e., $p_{i\to 1} = p_1$. Hence, d_i can be desribed alternatively as

$$d_i = \|p_{1\to i} - t_i\| . \tag{16}$$

Then, the directions of the axes of the canonical coordinate system $R_{\mathrm{cyl},i} \in$ SO(3) of the cylinder located in t_i, i.e., focal point i, are in terms of camera 1

$$m_i = \frac{p_{1\to i} - t_i}{\|p_{1\to i} - t_i\|} = \frac{p_{1\to i} - t_i}{d_i}, \qquad n_i = k_1 \times m_i, \qquad k_i = k_1. \tag{17}$$

With this, one can express the unit normal vectors of the tangent planes of camera i via the identities (13) to (17) in terms of camera 1 with

$$n_{\pm,i} = \sqrt{1 - \left(\frac{r}{d_i}\right)^2}\, n_i \pm \frac{r}{d_i}\, m_i,\tag{18}$$

where again $d_i > r$ has to be fulfilled, otherwise the cylinder contour is not pictured by two straight lines in camera i. However, $d_i > 0$ is guaranteed for $n_{+,i}^\top n_{-,i} > 0$ and $n_{+,i}^\top m_i > 0$.

2.3   Reconstruction Error

Since the reconstruction of the pose of a cylinder axis is based on line features in the image, the quality of the reconstruction relies on the quality of the detection of these line features. Due to errors in the image induced by, e.g., not completely compensated optical aberration, pixel discretization, or parameter influences of image processing algorithms such as Hough transform, the detected line features $\tilde{l}_{\pm,i}$ are disturbed, i.e., $l_{\pm,i} \neq \tilde{l}_{\pm,i}$, where $l_{\pm,i}$ are the theoretical, undisturbed lines as intersection of the cylinder tangent plane and the image plane. The errors in the determined lines $\tilde{l}_{\pm,i}$ hence affect $\tilde{n}_{\pm,i}$, and by that, the overall reconstruction of the cylinder pose. Denoting in the following measured quantities with a tilde and estimated quantities with a hat, one has $\tilde{n}_{\pm,i} = n_{\pm,i}(a) + e_{\pm,i}$ and $\hat{n}_{\pm,i} = n_{\pm,i}(\hat{a})$, where $n_{\pm,i}(a)$ are the mathematical models, a are the ideal parameters and $\hat{a}$ the estimated parameters, respectively, and $e_{\pm,i}$ are propagated measurement errors which, however, maintain $\|\tilde{n}_{\pm,i}\| = 1$. The vectors $\tilde{n}_{\pm,i}$ lie on the surface of two error cones with $0 < \tilde{n}_{\pm,i}^\top n_{\pm,i} \leq 1$, which affects the reconstruction of the pose of the cylinder axis notably. The distance (12) is sensitive to any disturbances of $n_{\pm,i}$, which holds in particular for $d_i \gg r$.

In the multiple view case where $n > 1$ cameras view the same cylinder, e.g., camera 1 and camera i with $i \neq 1$, the separate evaluation of the apparent cylinder contour lines of each image would lead due to the presence of measurement errors to discrepancies in the n results, and hence to

$$\tilde{p}_1 \neq \tilde{p}_{i\to 1}, \qquad \tilde{p}_{1\to i} \neq \tilde{p}_i, \qquad d_i \neq \|\tilde{p}_{1\to i} - t_i\|, \qquad \tilde{k}_1 \neq \tilde{k}_i.\tag{19}$$

3   Least Squares Fitting of Cylinders from Multiple Views

3.1   Fitting Problem Based on Line Features

The reconstruction of the pose of a straight homogenous cylinder from n views is equivalent to the task of least squares fitting a distinct cylinder model using the identities (13) to (17) to certain features depending on "measured" contour lines $\tilde{l}_{\pm,i}$ by estimation of a set of parameters $\hat{a}$. Then, the task is to search for an optimal set of parameters

$$\hat{a}^* = \arg\min_{\hat{a}} \frac{1}{2}\delta^\top \delta,\tag{20a}$$

with the proposed residuum vector

$$\delta(\widehat{a}) = \underbrace{\begin{pmatrix} \widetilde{n}_{+,1} \\ \widetilde{n}_{-,1} \\ \vdots \\ \widetilde{n}_{+,n} \\ \widetilde{n}_{-,n} \end{pmatrix}}_{const} - \underbrace{\begin{pmatrix} \widehat{n}_{+,1} \\ \widehat{n}_{-,1} \\ \vdots \\ \widehat{n}_{+,n} \\ \widehat{n}_{-,n} \end{pmatrix}}_{f(\widehat{a})} \in \mathbb{R}^{6n} , \tag{20b}$$

and subject to the constraint

$$\widehat{d}_i - r > 0 \quad \forall i \in \{1, \dots, n\} . \tag{20c}$$

The proposed residuum vector (20b) deploys the constant "measured" unit normal vectors on the one side, and the model equations gathered in $f(\widehat{a})$ on the other side. This is equivalent to a fit by combined variation of both the position $\widehat{p}_1$ and orientation $\widehat{k}_1$ of the cylinder axis. Note that all quantities are given w.r.t. to camera 1.

Additionally, the constraint (20c) has to be satsified. Due to the fact, that there is in real camera setups always a minimum distance between the viewed cylinder and the focal point with $d_{\min,i} > r$, one can modify (20c) to $\widehat{d}_i - d_{\min,i} \geq 0$ and handle these constraints, e.g., by introduction of separate slack variables for each d_i in a small neighborhood of $d_{\min,i}$. In the following, however, only the case $d_i \gg d_{\min,i} \, \forall i \in \{1, \dots, n\}$ is considered in particular, so the constraint even can be neglected and the remaining problem can be treated as unconstrained fitting problem.

In any case where on the image planes of each camera i two distinct contour lines of the same cylinder can be evaluated, $d_i > 0 \, \forall i \in \{1, \dots, n\}$ is guaranteed due to $n_{+,i}^\top n_{-,i} > 0$ and $n_{+,i}^\top m_i > 0$. Then, the solution of (20) is unique up to the sign of $k_1 = k_i$, depending only on which contour line is actually assigned to $l_{+,i}$, and which to $l_{-,i}$, respectively. If the cylinder contour is not imaged as two straight lines in a certain camera, this particular camera has to be omitted from evaluation.

3.2   Parameterization

Since the fitting problem (20) is equivalent to a search of an optimal pose of the cylinder with $p_1^* = d_1^* m_1^*$ and k_1^* relative to camera 1, the parameters to choose for (20) have to have influence onto both position and orientation of the model w.r.t. to camera 1. In addition, the deployed model has to be continuous an differentiable w.r.t. the chosen parameters. Therefore, m_1 and k_1 have to be parameterized besides d_1.

Any rotation matrix $R \in SO(3)$ can be parameterized [12] as $R(v) \colon \mathbb{R}^3 \to SO(3)$ deploying Rodrigues' formula

$$R = I + \sin \|v\| \frac{[v]_\times}{\|v\|} + (1 - \cos \|v\|) \frac{[v]_\times [v]_\times}{\|v\|^2} , \tag{21}$$

with the parameter vector $v \in \mathbb{R}^3$, where the length

$$\|v\| = \arccos\left(\frac{\text{trace}(R) - 1}{2}\right) = \theta,$$ (22)

is the rotation angle, and for $\theta \neq 0$ with the entries of R, the unit direction vector of the rotation is

$$\frac{v}{\|v\|} = \frac{1}{2\sin\theta}\begin{pmatrix} R_{32} - R_{23} \\ R_{13} - R_{31} \\ R_{21} - R_{12} \end{pmatrix} = r.$$ (23)

Hence, $v = \theta r$ is the product of (22) and (23). In the special case $\theta = 0$ no rotation occurs, $R = I$, and any arbitrary 3×1 unit vector can be chosen for v.

The skew symmetric matrix of any arbitrary vector $a^\top = (a_1, a_2, a_3)$ is defined that $[a]_\times b = a \times b$ for any $b \in \mathbb{R}^3$ with

$$[a]_\times = \begin{pmatrix} 0 & -a_3 & a_2 \\ a_3 & 0 & -a_1 \\ -a_2 & a_1 & 0 \end{pmatrix}.$$ (24)

Since $R_{\text{cyl},1} = (m_1, n_1, k_1) = R(v_1)$, one can extract the j-th column by multiplying $R_{\text{cyl},1}$ with some 3×1 unit vector e_j, which has a 1 as j-th element and a 0 elsewhere.

The derivatives of $m_1 = R_{\text{cyl},1} e_1$, $n_1 = R_{\text{cyl},1} e_2$, and $k_1 = R_{\text{cyl},1} e_3$ are according to [12]

$$\frac{\partial (R_{\text{cyl},1} e_j)}{\partial v_1} = -R_{\text{cyl},1} [e_j]_\times \frac{v_1 v_1^\top + \left(R_{\text{cyl},1}^\top - I\right)[v_1]_\times}{\|v_1\|^2},$$ (25)

which furthermore simplifies due to the derived model properties and

$$R_{\text{cyl},1} [e_1]_\times = \begin{pmatrix} 0 & k_1 & -n_1 \end{pmatrix},$$ (26)
$$R_{\text{cyl},1} [e_2]_\times = \begin{pmatrix} -k_1 & 0 & m_1 \end{pmatrix},$$ (27)
$$R_{\text{cyl},1} [e_3]_\times = \begin{pmatrix} n_1 & -m_1 & 0 \end{pmatrix}.$$ (28)

The orientation of the cylinder axis can now be parameterized via (21), and the parameter vector for the complete description of the pose of the cylinder axis is proposed as

$$a = \left(v_1^\top \; d_1\right)^\top \in \mathbb{R}^4.$$ (29)

3.3 Algorithm

The task of fitting a cylinder to the images of multiple cameras according to (20) can be solved using the formulas derived in the previous sections by appropriate calculation of the residuum vector (20b). On the one hand, the constant part of

δ consists of the tangent plane normal vectors $\tilde{n}_\pm$ depending on the "measured" line features $\tilde{l}_\pm$. On the other hand, the model part of δ gathered in $f(\hat{a})$ consists of the according tangent plane normal vectors $\hat{n}_\pm = n_\pm(\hat{a})$ calculated by means of the estimated parameters $\hat{a}^\top = \left(\hat{v}_1^\top \; \hat{d}_1 \right)$.

The proposed algorithm to compute $\tilde{n}_\pm$ and the pose of the cylinder axis depending on the evaluated $\tilde{l}_\pm$ imaged in the actually regarded camera relative to camera 1 is depicted in Fig. 3(a), and the algorithm for computing $\hat{n}_\pm$ based on $\hat{v}_1$ and $\hat{d}_1$ is illustrated in Fig. 3(b).

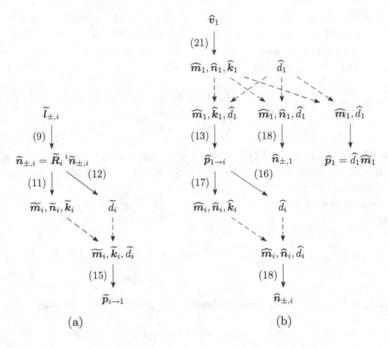

Fig. 3. Proposed algorithms to compute the tangent plane normal vectors and the pose of the cylinder axis relative to camera 1 based on detected apparent contour lines $\tilde{l}_\pm$ (a), and based on estimated parameters $\hat{v}_1$ and $\hat{d}_1$ (b).

Deploying the established Levenberg-Marquardt algorithm (LMA) to solve the fitting problem (20), a local minimum can be found by starting from an initial guess $\hat{a}_0$ and proceeding each step k by updating the estimated parameters $\hat{a}_k$ until convergence [11] with

$$\hat{a}_{k+1} = \hat{a}_k + \left(J_k^\top J_k + \lambda_k D_k \right)^+ J_k^\top \delta_k . \tag{30}$$

$(\bullet)^+$ denotes herein the Moore-Penrose inverse of $\bullet$, J_k is the Jacobian matrix of the model w.r.t. the parameter vector a, i.e., $J_k = J(\hat{a}_k) = \partial f / \partial a|_{a=\hat{a}_k}$, the diagonal matrix D_k contains the main diagonal elements of $J_k^\top J_k$, and $\delta_k = \delta(\hat{a}_k)$, respectively.

For small values of the damping factor $\lambda_k \geq 0$, the LMA does a Gauss-Newton iteration, for large values of λ_k, the LMA does a gradient descent. Depending on the residuum, λ_{k+1} can be decreased by a factor if $\|\delta(\hat{a}_k)\| < \|\delta(\hat{a}_{k-1})\|$, or increased by a factor if $\|\delta(\hat{a}_k)\| \geq \|\delta(\hat{a}_{k-1})\|$.

Feasible initial parameter values can be chosen by evaluating the imaged cylinder contour lines of a single camera according to algorithm in Fig. 3(a). Using (15) with (22) and (23) for evaluation of

$$\tilde{R}_{\text{cyl},i\to 1} = \left(\frac{\tilde{p}_{i\to 1}}{\|\tilde{p}_{i\to 1}\|} \, \tilde{k}_i \times \frac{\tilde{p}_{i\to 1}}{\|\tilde{p}_{i\to 1}\|} \, \tilde{k}_i \right) \to \tilde{v}_{i\to 1}, \tag{31a}$$

$$\tilde{d}_{i\to 1} = \|\tilde{p}_{i\to 1}\|, \tag{31b}$$

one obtains the initial parameters $\hat{a}_0^\top = \left(\tilde{v}_{i\to 1}^\top, \tilde{d}_{i\to 1} \right)$ by evaluation of camera i. Note that these parameters are derived by (not necessarily ideal) cylinder pose reconstruction by means of camera i followed by subsequent ideal transformation to camera 1, so (19) might apply. For $i = 1$, it is obvious that $\tilde{v}_{i\to 1} = \tilde{v}_1$ and $\tilde{d}_{i\to 1} = \tilde{d}_1$, and $\hat{a}_0$ can be derived directly from $\tilde{R}_{\text{cyl},1}$ and $\tilde{p}_1$.

Convergence of the LMA is achieved and the iteration can be stopped if the parameters do not change any longer distinctly, i.e., with a small $\epsilon > 0$

$$\|\hat{a}_{k+1} - \hat{a}_k\| < \epsilon. \tag{32}$$

Since the solution of (20) is unique up to the sign of $k_1 = k_i$ for $n_{+,i}^\top n_{-,i} > 0$ and $n_{+,i}^\top m_i > 0$, the LMA converges to the only minimum for any initial pose implying $d_i > 0 \, \forall i \in \{1, \ldots, n\}$.

Presuming $d_i \gg d_{\min,i} \, \forall i \in \{1, \ldots, n\}$, the constraint (20c) can be neglected, and the Jacobian matrix is then

$$J = \frac{\partial f}{\partial a} = \begin{pmatrix} \frac{\partial n_{+,1}}{\partial a} \\ \frac{\partial n_{-,1}}{\partial a} \\ \vdots \\ \frac{\partial n_{+,n}}{\partial a} \\ \frac{\partial n_{-,n}}{\partial a} \end{pmatrix} = \begin{pmatrix} \frac{\partial n_{+,1}}{\partial v_1} & \frac{\partial n_{+,1}}{\partial d_1} \\ \frac{\partial n_{-,1}}{\partial v_1} & \frac{\partial n_{-,1}}{\partial d_1} \\ \vdots & \vdots \\ \frac{\partial n_{+,n}}{\partial v_1} & \frac{\partial n_{+,n}}{\partial d_1} \\ \frac{\partial n_{-,n}}{\partial v_1} & \frac{\partial n_{-,n}}{\partial d_1} \end{pmatrix} \in \mathbb{R}^{6n \times 4}. \tag{33}$$

Denoting for shortness $s_i = r/d_i$ and $c_i = \sqrt{1 - s_i^2}$ in the following, all entries of the Jacobian matrix and the according derivatives w.r.t. v_1 and d_1 are given as below. Note that all quantities are given w.r.t. the coordinate system of camera 1.

In general, the entries of (33) are the derivatives of (18). The derivatives of m_1, n_1, k_1 w.r.t. v_1 are provided by (25) with (26) to (28), their derivatives w.r.t. d_1 vanish since v_1 and d_1 are independent parameters.

For $i = 1$, the entries of (33) simplify to

$$\frac{\partial n_{\pm,1}}{\partial v_1} = c_1 \frac{\partial n_1}{\partial v_1} \pm s_1 \frac{\partial m_1}{\partial v_1}, \qquad \frac{\partial n_{\pm,1}}{\partial d_1} = \frac{1}{d_1} \left(\frac{s_1^2}{c_1} n_1 \mp s_1 m_1 \right). \tag{34}$$

For $i \neq 1$, the entries of (33) are

$$\frac{\partial n_{\pm,i}}{\partial v_1} = c_i \frac{\partial n_i}{\partial v_1} + n_i \frac{\partial c_i}{\partial v_1} \pm \left(s_i \frac{\partial m_i}{\partial v_1} + m_i \frac{\partial s_i}{\partial v_1} \right), \tag{35a}$$

$$\frac{\partial n_{\pm,i}}{\partial d_1} = c_i \frac{\partial n_i}{\partial d_1} + n_i \frac{\partial c_i}{\partial d_1} \pm \left(s_i \frac{\partial m_i}{\partial d_1} + m_i \frac{\partial s_i}{\partial d_1} \right). \tag{35b}$$

Regarding (17), the derivatives of m_i and n_i are

$$\frac{\partial m_i}{\partial v_1} = \frac{1}{d_i} \left(I - m_i m_i^\top \right) \frac{\partial p_{1 \to i}}{\partial v_1}, \qquad \frac{\partial m_i}{\partial d_1} = \frac{1}{d_i} \left(I - m_i m_i^\top \right) m_1. \tag{36}$$

$$\frac{\partial n_i}{\partial v_1} = [k_1]_\times \frac{\partial m_i}{\partial v_1} - [m_i]_\times \frac{\partial k_1}{\partial v_1}, \qquad \frac{\partial n_i}{\partial d_1} = [k_1]_\times \frac{\partial m_i}{\partial d_1}. \tag{37}$$

Note that $\partial k_i / \partial v_1 = \partial k_1 / \partial v_1$ and $\partial k_i / \partial d_1 = 0$ since $k_i = k_1$. The derivatives of s_i and c_i are given with

$$\frac{\partial s_i}{\partial v_1} = -\frac{1}{d_i} s_i^2 m_i^\top \frac{\partial p_{1 \to i}}{\partial v_1}, \qquad \frac{\partial s_i}{\partial d_1} = -\frac{1}{d_i} s_i^2 m_i^\top m_1, \tag{38}$$

$$\frac{\partial c_i}{\partial v_1} = \frac{1}{d_i} \frac{s_i^2}{c_i} m_i^\top \frac{\partial p_{1 \to i}}{\partial v_1}, \qquad \frac{\partial c_i}{\partial d_1} = \frac{1}{d_i} \frac{s_i^2}{c_i} m_i^\top m_1, \tag{39}$$

and the derivatives of (16) are

$$\frac{\partial d_i}{\partial v_1} = m_i^\top \frac{\partial p_{1 \to i}}{\partial v_1}, \qquad \frac{\partial d_i}{\partial d_1} = m_i^\top m_1. \tag{40}$$

Finally, the derivatives of (13) are obtained through

$$\frac{\partial p_{1 \to i}}{\partial v_1} = d_1 \frac{\partial m_1}{\partial v_1} + \left(k_1 t_i^\top + k_1^\top t_i I \right) \frac{\partial k_1}{\partial v_1}, \qquad \frac{\partial p_{1 \to i}}{\partial d_1} = m_1. \tag{41}$$

4   Results

In order to show the performance of the proposed algorithm exemplarily, the reconstruction of the pose of a cylinder axis in an optimal way according to (20) is discussed in the following.

The task is to estimate the pose of the axis of a cylinder with $r = 10$ based on both nominal and roughly disturbed contour lines, $l_{\pm,i}$ and $\tilde{l}_{\pm,i}$, in the images of two cameras, see Tab. 1. Each projection matrix is $K_i = \mathrm{diag}(f, f, 1)$ with $f = 16$. The pose of camera 2 relative to camera 1 is defined by $R_2(v)$ with $v^\top = (0, 0.2618, 0)$, and $t_2^\top = (-200, -150, 0)$. Regarding the LMA, the damping factor is initialized with $\lambda_0 = 10^{-5}$ and decreased/increased both by factor 10 depending on iteration and actual residuum.

Fig. 4(a) shows the nominal cylinder axis, the reconstructed cylinder axis based on single view (SV) evaluation of disturbed $\tilde{l}_{\pm,1}$ with parameters $\tilde{v}_1$ and $\tilde{d}_1$,

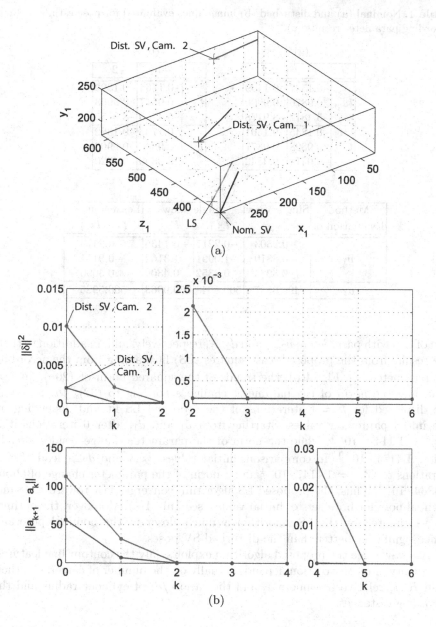

Fig. 4. Pose estimation relative to camera 1 (a) of nominal cylinder axis by means of SV evaluation of $l_{\pm,1}$ or $l_{\pm,2}$ (black), based on disturbed SV evaluation of $\tilde{l}_{\pm,1}$ (blue), and $\tilde{l}_{\pm,2}$ (red), respectively, and from proposed LS fit as result with initial values based on $\tilde{l}_{\pm,1}$ (cyan), and on $\tilde{l}_{\pm,2}$ (magenta), respectively. According trends (b) for each iteration k of LS fit of squared residuum norm and norm of parameter change, starting from disturbed SV pose based on $\tilde{l}_{\pm,1}$ (blue), and $\tilde{l}_{\pm,2}$ (red).

Table 1. Nominal (a) and disturbed (b) image lines evaluated for pose estimation and according parameter results (c).

(a)

i	1	2
$l_{+,i}$	-0.1567	-0.1409
	0.9876	0.9900
	-5.4812	-8.9178
$l_{-,i}$	-0.1845	-0.1561
	0.9828	0.9877
	-6.0418	-9.4842

(b)

i	1	2
$\widetilde{l}_{+,i}$	-0.1564	-0.1392
	0.9877	0.9903
	-5.4780	-8.9155
$\widetilde{l}_{-,i}$	-0.1736	-0.1392
	0.9848	0.9903
	-6.0390	-9.4820

(c)

Method	Single View	Single View		Least Squares
Evaluation of	$l_{\pm,i}$	$\widetilde{l}_{\pm,1}$	$\widetilde{l}_{\pm,2}$	$\widetilde{l}_{\pm,1}, \widetilde{l}_{\pm,2}$
$\widehat{v}_1$	-0.2503	-0.2217	-0.1435	-0.2473
	-0.8819	-1.0991	-1.3162	-0.9151
	0.3084	0.3352	0.4309	0.3090
$\widehat{d}_1$	490.0537	569.5648	650.9963	506.5925

and of $\widetilde{l}_{\pm,2}$ with parameters $\widetilde{v}_{2\to1}$ and $\widetilde{d}_{2\to1}$, respectively, and, in addition to that, the results from the proposed least squares (LS) fit starting from the according SV parameters. In Fig. 4(b), the trends of the squared norm of (20b), $\|\delta\|^2 = \delta^\top \delta$, and the norm of the parameter change according to (32), $\|\widehat{a}_{k+1} - \widehat{a}_k\|$, are displayed for $k = 6$ iterations of the proposed LS fit and depending on the initial parameter values. Starting from $\widetilde{v}_1$ and $\widetilde{d}_1$, after 6 iterations it is $\|\delta\|^2 = 1.1147 \cdot 10^{-4}$, while the norm of the parameter change resides already below $1.4174 \cdot 10^{-9}$. In comparison, initial values $\widetilde{v}_{2\to1}$ and $\widetilde{d}_{2\to1}$ lead after 6 iterations to $\|\delta\|^2 = 1.1147 \cdot 10^{-4}$ and a norm of the parameter change of about $1.8864 \cdot 10^{-7}$. Thus, the proposed LS algorithm converges fast toward the same optimal pose for both sets of initial values, see Tab. 1(c). Moreover, the optimal pose estimated by the LS algorithm is much closer to the nominal pose and hence significant better than the disturbed SV poses.

Anyway, since the proposed algorithm exploits only the contour line features, the quality of reconstruction depends actually on the number of cameras n, their pose R_i, t_i relative to camera 1, and the ratio r/d_i of cylinder radius and the according distances.

5   Conclusions

In this paper, a new method for a minimum-error reconstruction of the axis of an cylinder based on contour line features is presented. Novel model equations for single and especially multiple views are derived, and an algorithm for the least squares fit of the cylinder axis is proposed. The convergence and good performance of the algorithm is shown by an exemplary pose estimation, however, the proposed method estimates the pose of any straight homogeneous cylinder, such as e.g., cylindrical markers for videometric devices, from multiple camera views.

However, the reconstruction quality can be increased by adjusting the number and pose of the cameras, or, on the other hand, by increasing the number of evaluated line features per single view. Considering this, current work of the author focuses on supplementary evaluation of lines applied parallel to the axis on the cylinder body, and preliminary results indicate an increase of estimation quality by a factor of nearly two.

References

1. Doignon, C.: An introduction to model-based pose estimation and 3-D tracking techniques. In: Proceedings of the IEEE 1991 National Aerospace and Electronics Conference, NAECON 1991, pp. 359–382 (2007)
2. Shiu, Y. C., Huang, C.: Locating cylindrical objects from perspective projections. In: 10th IEEE International Symposium on High Performance Distributed Computing, pp. 1070–1076 (1991)
3. Puech, W., Chassery, J.-M., Pitas, I.: Cylindrical surface localization in monocular vision. Pattern Recognition Letters 18(8), 711–722 (1997)
4. Veldhuis, H., Vosselman, G.: The 3D reconstruction of straight and curved pipes using digital line photogrammetry. 5ISPRS6 Journal of Photogrammetry and Remote Sensing 53(1), 6–16 (1998)
5. Hanek, R., Navab, N., Appel, M.: Yet another method for pose estimation: a probabilistic approach using points, lines, and cylinders. In: IEEE Computer Society Conference on Computer Vision and Pattern Recognition, 1999, vol. 2, pp. 544–550 (1999)
6. Doignon, C., de Mathelin, M.: A degenerate conic-based method for a direct fitting and 3-d pose of cylinders with a single perspective view. In: 2007 IEEE International Conference on Robotics and Automation, pp. 4220–4225 (2007)
7. Navab, N., Appel, M.: Canonical representation and multi-view geometry of cylinders. International Journal of Computer Vision 70(2), 133–149 (2006)
8. Shiu, Y. C., Huang, C.: Pose determination of circular cylinders using elliptical and side projections. In: IEEE International Conference on Systems Engineering, 1991, pp. 265–268 (1991)
9. Huang, J.-B., Chen, Z., Chi, T.-L.: Pose determination of a cylinder using reprojection transformation. Pattern Recognition Letters 17(10), 1089–1099 (1996)
10. Sarkis, M., Diepold, K.: Camera-pose estimation via projective newton optimization on the manifold. IEEE Transactions on Image Processing 21(4), 1729–1741 (2012)
11. Hartley, R. I., Zisserman, A.: Multiple View Geometry in Computer Vision, 2nd edn. Cambridge University Press (2004)
12. Gallego, G., Yezzi, A.: A compact formula for the derivative of a 3-D rotation in exponential coordinates. Journal of Mathematical Imaging and Vision (2014)

Iterative Template Matching Strategy for Visual Target Detection by Unmanned Surface Vehicle

Jin Xu[1(✉)] and Ming Xie[1,2]

[1] Nanyang Technological University, 50 Nanyang Avenue,
Singapore City 639798, Singapore
jxu011@e.ntu.edu.sg, mmxie@ntu.edu.sg
[2] Nanjing Tech University, 30 Puzhu South Road,
Nanjing City, People's Republic of China
xieming@njtech.edu.cn

Abstract. The development of USV (Unmanned Surface Vehicle) has boomed around the world for military, research and commercial applications. The full autonomy of the USV is a desirable but challenging task. Though GPS is the main sensing system for the vehicle's positioning and guidance, vision is necessary for tasks such as visual target detection and identification, especially color and shape encoded information. This is well demonstrated in the Maritime RobotX challenge 2014, where all of the five competition tasks require the use of vision to complete. The visual target detection for USV is a challenging task as the platform and target are always moving in the open sea area and the lighting condition varies a lot accordingly to weather and time. For real-time onboard performance, template matching is a good choice for the visual detection. In certain scenarios, the normal template matching method needs to be enhanced for robust performance. One of the example algorithm is the proposed iterative template matching, which provides a fast and robust solution for the vision tasks in the Maritime RobotX challenge 2014. By an additional step of searching for the visual context for the target, the robustness of detection is significantly improved without loss of accuracy.

Keywords: Iterative template matching · Visual target detection · Unmanned surface vehicle

1 Introduction

Unmanned Underwater Vehicle has been used for military and research purpose for many year and Unmanned Surface Vehicle is becoming realistic in recent years. Compared to the underwater environment, the USV needs to handle more dynamic and challenging environment and has attracted many research interest into this area [1]. Maritime RobotX challenge (www.robotx.org) is organized to promote the research in this area. Figure 1 shows the NTU team Leviathan base on the provided double hull platform (WAM-V® USVX HULL).

© Springer International Publishing Switzerland 2015
H. Liu et al. (Eds.): ICIRA 2015, Part III, LNAI 9246, pp. 386–391, 2015.
DOI: 10.1007/978-3-319-22873-0_34

Fig. 1. The USV platform built up on the boat provided by RobotX

Figure 2 shows the satellite image of the competition area in Singapore Marina Bay. Though it is within a bay area, there are still significant current which makes the navigation challenging. The weather is typical tropical weather mixed with strong sunlight and sudden heavy thunderstorm, which requires very robust performance for the visual detection algorithm.

Fig. 2. Satellite image of the competition location

To reach full autonomy on the USV platform, one of the tasks to be completed is the visual target identification. Although Lidar has been used widely for the sensing purpose, vision is still indispensable ability for detection of color and shape coded information, as shown in all the five tasks designed in RobotX challenge 2014 (see www.robotx.org).

To provide vision all around the USV, five cameras are installed around the platform. The cameras are all same IP cameras with waterproof protections. Figure 3 shows the camera chosen for the vision task and Table 1 gives the main specifications of the camera. No stereo is used in the system because the main target of vision is to decode the color and shape encoded information and localization task can be either dedicated to a Lidar system (such as a nodding Lidar) or approximated by 2D localization using a single camera as all the objects of interest or their base are close to water surface. As there significant barrel distortion in the lens, a distortion correction needs to be done before the 2D calibration.

Fig. 3. The IP cameras used as the sensor

Table 1. Specifications of the cameras on SUV

Specifications	figure
Power	Power over Ethernet IEEE 802.3af Class 2 (max. 6.49W) 8 – 28 V DC max 4.7 W
Operating Conditions	-20°to 50°
Connectors	RJ45 for 10BASE-T/100BASE-TX PoE RJ12 for the camera unit
Lens	2.8 mm: 81° view*, F2.0, fixed iris, fixed focus *horizontal angle of view
Light sensitivity	1.0 - 10000 lux
Shutter time	1/6 s to 1/24500 s
Resolutions	1280x720 to 320x180
Frame rate H.264&JPEG	25 fps with 50 Hz 30 fps with 60 Hz
Video compression	H.264 (MPEG-4 Part 10/AVC) Motion JPEG
Image sensor	1/4" progressive scan RGB CMOS

2   Visual Target Detection for USV

Visual target identification is involved in all the five tasks of RobotX 2014. The proposed method is best demonstrated with task 3, which requires the USV to dock into one of the bay depending on the given symbol. The dock consists of three similar docking bays as illustrated in Figure 4. The parking bay is formed using two pontoons established at the port side and at the starboard side and another pontoon at the closing end. They are all fixed to the platform and their GPS location will be published before the match. The difference between them is the geometric shape at the end of each docking bay for the reference usage. A black cruciform, a black triangle and a black circle will be put on the placard located at the closing end of the each docking bay to distinguish the bay. Each day a parking symbol will be designed to serve as visual reference for all teams. The symbol may move among the three docking bays randomly but it will remain the same when the team runs in its own time slot.

The craft should be able to recognize the symbol of the docking bay, move to the right bay, and park into the bay for a specific period of time. Then the vehicle can exit the parking lot and proceed to the next task. Light contact with the pontoon is permitted.

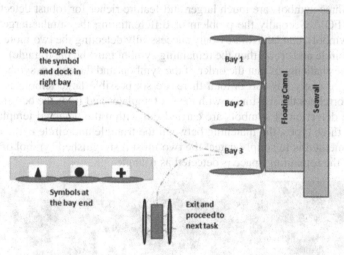

Fig. 4. Layout for the visual target detection in Task 3 of Marine X Challenge

The challenges for this visual target searching task are: 1) each symbol is very small in the camera's FOV at the entering distance and cannot be robustly detected; 2) the difference between triangle and circle is not distinct enough for robust identification using normal template matching method, especially with varying view angles.

3   The Iterative Template Matching Strategy

Template matching is a simple and reliable method for visual target identification inside images or video [2, 3, 4, 5]. The basic principle of template matching is to match a target's template image with all possible sub-images extracted from source images or video. This strategy of searching visual targets works if the source images or video do not contain too much similar visual information. In order to further enhance the performance of template matching, we propose a new strategy of undertaking template matching. The idea is to generate or pre-store a set of composite templates which include a visual target's template as well as the additional information of contexts of a visual target. In this way, the search of a visual target can be achieved by an iterative template matching strategy.

For example, in our case of visual target detection by an autonomous surface vehicle participating in the 2014 RobotX challenge, The implementation of the iterative template matching strategy consists of three stages: 1) identification of the current situation based on GPS information and task manager; 2) detection of the landmark with the composite template: a panel displaying the three symbols for all the three docks; and 3) detection of the target with the visual target's template: respective symbol for each dock.

Our new strategy provides a framework for any specific visual target detection task and solves the two problems mentioned in section 2 in two ways. Firstly, the combined area of the three symbols are much larger and feature richer for robust detection within the camera FOV. Secondly the problem of differentiating the similar targets, triangle and circle symbols, can be avoid by only successfully detecting the two more distinctive symbols of circle and cross (then the remaining symbol must be the triangle).

One of the challenge is that the order of the symbols and the chosen symbol for docking is changed every day. So in total there are six possible large templates. The algorithm first correlates the full image with the six templates and finds the best match.

Then the detection of symbols are carried out within the detected template region. Among the three types, the matching between the triangle and circle is the most similar. So our method is to simply detect the two most distinguished symbol of circle and cross. Then the remaining space is detected as triangle.

Fig. 5. Detection result of the iterative template matching method

Notes should be taken that this iterative template matching method is totally different from pyramid template matching [6], which is a common way of speeding up the matching process through the use of an image pyramid of different resolution in multiple stages. However, the iterative template matching uses template of different level of context for searching.

Once the symbol for the required docking bay is identified, the corresponding dock coordinate is estimated based on a 2D homography which has been estimated via calibration. Figure 5 shows the identification result in one of the frames in real video during the competition. The result is very robust near the entering position of this task.

4   Discussions and Conclusions

Vision is known for its inadequacy of robustness, especially in the natural environment. However, if the vision algorithm can anticipate what to be seen, the robustness can be improved significantly. The proposed iterative template matching method follows this philosophy and formulates a framework based on simple template matching approach. The effectiveness of this method has been verified in the RobotX Maritime Challenge 2014. This general framework can be easily applied into many scenarios for visual target detection and identification.

References

1. Caccia, M., Bibuli, M., Bono, R.: Basic navigation, guidance and control of an unmanned Surface Vehicle. Autonomous Robots **25**, 349–365 (1994). Springer Us
2. Brunelli, R.: Template Matching Techniques in Computer Vision: Theory and Practice. Wiley, ISBN 978-0-470-51706-2 (2009)
3. Li, Y., Jian, L., Jinwen, T., Honbo, X.: A fast rotated template matching based on point feature. In: Proceedings of the SPIE 6043, pp. 453-459 (2005). MIPPR 2005: SAR and Multispectral Image Processing
4. Sirmacek, B., Unsalan, C.: Urban Area and Building Detection Using SIFT Keypoints and Graph Theory. IEEE Transactions on Geoscience and Remote Sensing **47**(4), 1156–1167 (2009)
5. Korman, S., Reichman, D., Tsur, G., Avidan, S.: FAsT-match: fast affine template matching. In: CVPR 2013 (2013)
6. Tamimoto, S.L.: Template Matching in Pyramids. Computer Graphics and Image Processing **16**(4), 356–369 (1981)

Spatial-Temporal Database Based Asynchronous Operation Approach of Fruit-Harvesting Robots

Bin Zhou, Liang Gong, Qianli Chen, Yuanshen Zhao, Xiao Ling, and Chengliang Liu(✉)

Department of Mechanical and Electronic Engineering,
Shanghai Jiao Tong University, Shanghai 200240, China
{zhou_bin,gongliang_mi,chenqianli,lingxiao,chlliu}@sjtu.edu.cn

Abstract. Autonomous fruit-harvesting robots encounter difficulties of low fruit recognition rate and picking efficiency due to the complex unstructured operational environment. To solve this problem, an asynchronous approach has been proposed to discriminate the recognition and manipulation process. The fruit recognition task can be intensified via repetitious inspection or human-robot interaction, meanwhile a spatial-temporal database is constructed to record the recognition information which might facilitate the sequential picking manipulation. In this paper the attributes of a spatial-temporal object are firstly investigated with four elementary constituents attached. Hereby the fruit target is modeled for harvest decision-making. Secondly a three layer database management system is designed as per the modular design principles. Finally, we introduced a picking scheduling application based on this database management system. The picking schedule demonstrates that the Construction of the spatial-temporal database paves the way for the success of paradigm shift from synchronous to asynchronous manipulations of fruit-harvesting robots.

Keywords: Spatial-temporal database · Fruit-harvesting robot · Asynchronous · Intelligence fusion

1 Introduction

Since Prof.kawamura, Kyoto University[1], originated the studies of Autonomous Fruit Picking Machine (AFPM) with a tomato harvesting robot three decades ago, the components of the first fruit harvesting robot were not so quite different from current harvesting robots under development at universities and research institutes. Under the consequence that robots now can be equipped with more powerful computers, cameras with more pixels and sensors with smaller volume, their success rate and operating speed are still far from our satisfactory[2 3]. The reason is mature fruit's recognition, match and location ability, the fruit-harvesting robot's primary ability, has been restricted under complex outdoor environment or greenhouse condition. And exactly, that's what human-beings really good at. As a consequence, a semi-autonomous fruit-harvesting robot

The work is supported by China 863 program under the Grant No. 2013AA102307.

H. Liu et al. (Eds.): ICIRA 2015, Part III, LNAI 9246, pp. 392–400, 2015.
DOI: 10.1007/978-3-319-22873-0_35

system can be a good solution help to limit costs and guarantee a high degree of reliability [4].

Autonomous mobile robot equipped with can collected spatial and temporal data [5 6]. In this system, robot is designed as two arms for multi-task through replacing different executing ends. The head is a binocular Camera which can obtain 3D information of the environment. IPC controls the robot and exchanges information and commands with both mobile devices and server. Recognition and location are synchronous work, which mainly executed by human on mobile devices or PC. The picking action is asynchronous carried out by end-effector, and should be arranged by relative space position of robot and mature tomatoes. In this consequence, the key technology is a data management system, which can handle fruit's temporal and spatial information at the same time after a pretreatment of data gathered by both binocular camera and HRI.

This asynchronous approach mentioned above combines human's advantage in recognition and robot's advantage in executing, trying to asynchronous manipulating between perception, strategy and picking action to improve fruit-harvesting robots' performance in complicated outdoor environment. In this study, we focus on the construction of a spatial-temporal database management system for the asynchronous operation of target recognition and picking.

2   Design of Spatial-Temporal Database

Spatial-temporal databases deal with objects that change their location and/or shape over time by recording their spatial and temporal information [5]. Effective database analyzing applications tracking these changes could be a support for strategy- making [6].

2.1   Spatial-Temporal Object

Spatial-temporal databases are generally used in GIS. It is the key technology of managing subjects that changing over time in space or shape features. Four elementary constituents should be specified in a spatial-temporal object: position, descriptive attributes, spatial attributes and temporal attributes (Figure 1) [7].

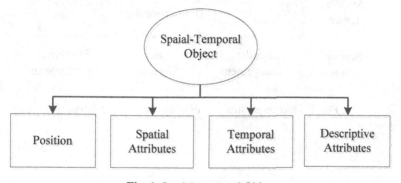

Fig. 1. Spatial-temporal Object

- Position

Position represents spatial location of an object, which can be expressed in form of (x, y, z). Target position is the key attribute and can be recorded through coordinate system transformation.

- Spatial Attributes

Spatial attribute is location-related which inherited property from the spatial position. Enumerated data type (0-pick_enable, 1-pesticides_need, and 2-seed_regeneration) can used to express a spatial attribute, so as to take environment aspects into consideration.

- Temporal Attributes

Temporal attributes are appended to identify inputting time (the first time one object being observed) and observing time (a refresh of the nearest observed time) of the object. Each time attribute can be recorded as the form of bellowing:

$$DD: HH: MM: SS$$

- Descriptive Attributes

Descriptive attributes label other attributes of one object, which is user-defined and extensible. For instance, the maturity is an important attribute of fruit and should be recorded into database.

2.2   Basic Modules of Spatial-Temporal Database

We design a three-layer structure for the whole data management system (Figure 2), which can be broken into 6 modules.

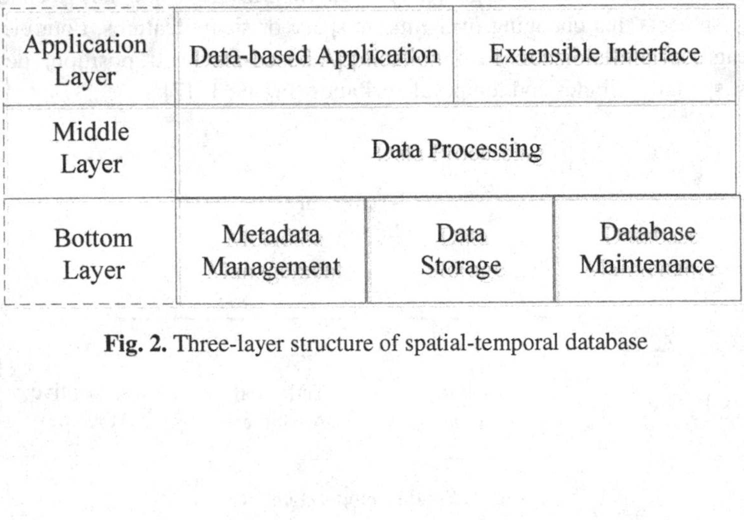

Application Layer	Data-based Application		Extensible Interface	
Middle Layer	Data Processing			
Bottom Layer	Metadata Management	Data Storage	Database Maintenance	

Fig. 2. Three-layer structure of spatial-temporal database

Bottom Layer

- Meta-data Management

We describe data structure, file system and data association by using meta-data standards specific to a particular discipline. Meta-data is data about data. By describing the contents and context of data files, the usefulness of the original data/files is greatly increased.

```xml
<?xml version="1.0" encoding="ISO-8859-1"?>
<! DOCTYPE spatial-temporal DATABASE "std">
<std>
<!--DATA STRUCTURE--!>
<table name="table_target">
  <list id="id">ID</list>
    <list id="x">X</list>
  <list id="y">Y</list>
  <list id="z">Z</list>
  <list id="input_data">INPUT_DATA</list>
  <list id="refresh_data">REFRESH_DATA</list>
  <list id="area">AREA</list>
  <list id="picked">PICKED</list>
  <list id="discribe">DISCRIBE</list>
</table>
<table name="table_picture">
  <list id="id">ID</list>
    <list id="x">X</list>
  <list id="y">Y</list>
  <list id="z">Z</list>
  <list id="input_data">INPUT_DATA</list>
<picture>url</picture>
</table>
<!--DATA ASSOCIATION--!>
<link name="link1" description="" >
<from >......</from>
<to>......</to>
</link>
.....
</std>
```

• Data Storage

We store data with two dimensions: temporal dimension and type dimension (Figure 3).

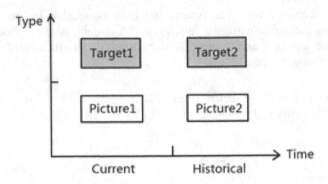

Fig. 3. Storage Dimensions

The temporal dimension consists of historical data, which is data of tomatoes that has been picked, and current data. Robot can learn from past data to generate new levels of intelligence with the help of machine learning algorithms.

Type dimension expresses two different data types stored in database: picture (Table 1) and target (Table 2).

Table 1. Picture Data Type

ID	X	Y	Z	DIRECTION	DATE	DATA
picture1	..	..	..	..	..	..

Picture data is constituted with location data where robot takes it, direction data of the robot, time data when robot takes it and image url. It is the primary data we get from the binocular camera.

Table 2. Target Data Type

ID	X	Y	Z	DATE1	DATE2	DISCRIBE	PICKED
target1	..	..	..	..	..	..	..

As a spatial-temporal object, target data structure involved all four elements as shown in Table 2.

• Database Management

Spatial-temporal database's function of management mainly includes following aspects:

1. Database security management and user management;
2. Copy or Delete operation of spatial-temporal database;
3. Data Import and Export.

Middle Layer

This layer contains all data processing procedures just exampled as below:

- Data Preprocessing

Data preprocessing generate target data according to picture data and human-robot interaction results. The first step is extracting the pixels of tomato, which can be determined by human click or touch, and calculate average depth of these pixels to get the relative position of robot and the target (Figure 4). Then we can get the position of target in the greenhouse space through coordinate system transformation.

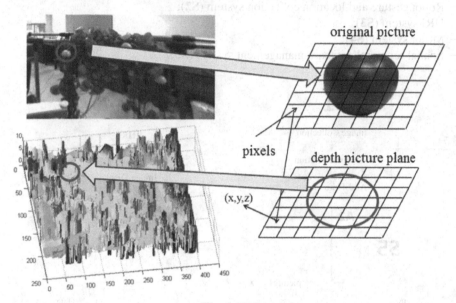

Fig. 4. HRI

- Data Compression

Image data need to be compressed before stored into database as the data size is 1M each.

- Data scrubbing

Data scrubbing maintains the cleanliness of database. When a target has been picked, the data of this target should be transferred to historical database.

Application Layer

- Application

Based on the spatial-temporal database platform and data processing algorithms, plenty of applications can be developed according to customer's requirement. These applications are modular designed and aiming at specific mission.

• Extensible Interfaces

Except applications, we will provide enough interfaces for customers who don't want to make sense of how the data is generated and processed to develop an application.

3   Picking Scheduling Based on Spatial-Temporal Database

The whole system is makes up of five independent small systems (Figure 5):

— Visual information collection system (**S1**);
— Robot gesture and location collection system (**S2**);
— HRI system (**S3**);
— Motor system (**S4**);
— Spatial-temporal database management system (**S5**).

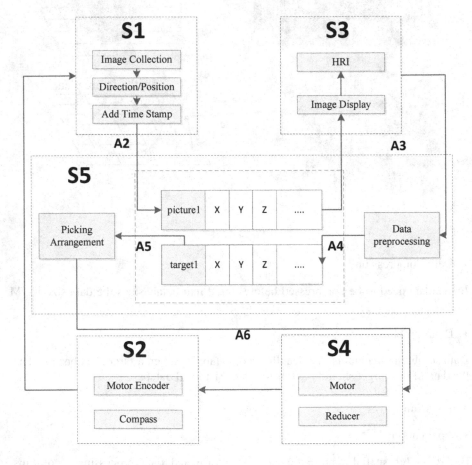

Fig. 5. The diagram of picking arrangement

To applicate a picking scheduling task, five systems cooperate as following steps:

- **A1:** When robot reaches one fixed position, S2 records gesture and location of robot at this moment and then sends this information together with a time stamp to S1.
- **A2:** S1 receives the time stamp as an initiator for binocular camera to capture a picture contains depth information. Then a package together with picture, position, time and direction will be send to S5. S5 unpacks the package, compresses the picture and stores all the data to Table1.
- **A3:** S5 arranges image displaying in terminals (PC or mobile devices) in regard with terminal state and robot location. S3 sends plane coordinate values that generate by human click or touch to S5.
- **A4:** S5 preprocesses the data send by S3 to get target position. After a contrast with existing data, S5 decides whether create a new target or refresh the existing data.
- **A5:** If qualified data reach a given scale, S5 will make a picking schedule in regard with robot location and target position in a scheduling time.
- **A6:** S4 receives the arrangement and start a picking action. Between robot movements, fixed position being stroked one by another, then we return back to step A1 and forms a closed loop.

4 Conclusion and Future Work

4.1 Conclusion

In this study, we proposed a method for human-robot cooperation fruit harvesting, which is asynchronous both in spatial and temporal. In this method, a spatial- temporal database management system is designed to support picking strategy formulating, task scheduling and other appropriate applications of data, and as we can see in the picking scheduling application, this database management system act as the brain of the whole system. The final system is under test, which has not yet achieved a level of productivity capable of replacing human pickers. Further mechanical modifications and more robust and adaptive algorithms are needed to achieve a stronger robot system.

4.2 Future Work

More software Tools should be developed to help robot mining effective information from these data. An intelligence robot can extract information from its environment and make decisions without the help of humans. To achieve this goal, an ideal effect we want to realize is that robot can get the knowledge of how human recognize tomato, which can be integrated with feature extraction of images. In this consequence, besides improving the mechanical condition, we will imply more machine learning and pattern recognition algorithms to this study.

References

1. Kawamura, N., Namikawa, K., Fujiura, T., Ura, M.: Study on agricultural robot (Part 1). Journal of the JSAM **46**(3), 353–358 (1984)
2. Kondo, N., Monta, M., Noguchi, N.: Agri-Robots (II) –Mechanisms and Practice. Corona Publishing Co. Ltd., Tokyo (2006)
3. Baeten, J., Donné, K., Boedrij, S., et al.: Autonomous fruit picking machine: A robotic apple harvester. Field and Service Robotics, 531–539 (2008)
4. Liu, Y., Nejat, G., Doroodgar, B.: Learning based semi-autonomous control for robots in urban search and rescue. Safety, Security, and Rescue Robotics (SSRR), 1–6 (2012)
5. Meguro, J., Ishikawa, K., Amano, Y., et al.: Creating spatial temporal database by autonomous mobile surveillance system (a study of mobile robot surveillance system using spatial temporal GIS part 1). Safety, Security and Rescue Robotics, 143–150 (2005)
6. Ishikawa, K., Meguro, J., Amano, Y., et al.: Parking-vehicles recognition using spatial temporal data (a study of mobile robot surveillance system using spatial temporal gis part 2). Safety, Security and Rescue Robotics, 151–157 (2005)
7. Paton, N.W., Fernandes, A.A.A., Griffiths, T.: Spatio-temporal databases: contentions, components and consolidation. Database and Expert Systems Applications, 851–855 (2000)
8. Zhang, J., Li, X.: A novel index method based on convex hull property of function. Transportation, Mechanical, and Electrical Engineering, 355–358 (2011)
9. Langran, G., Chrisman, N.R.: A framework for temporal geographic information. Cartographica: The International Journal for Geographic Information and Geovisualization **25**(3), 1–14 (1988)

Performance Analysis in Advanced Tele-operation System Based on Introduction of Danger-Avoidance View

Junjie Yang[1(✉)], Mitsuhiro Kamezaki[2], Ryuya Sato[1], Hiroyasu Iwata[1], and Shigeki Sugano[1]

[1] Department of Modern Mechanical Engineering, Waseda University, Tokyo, Japan
yang@sugano.mech.waseda.ac.jp
[2] Research Institute for Science and Engineering, Waseda University, Tokyo, Japan

Abstract. Unmanned construction machines are used after disasters. Compared with manned construction, time efficiency is lower because of incomplete visual information, communication delay, and lack of tactile experience. We have developed an autonomous camera control system to supply appropriate visual information. In order to let operator easy to avoid unexpected collision, we first introduce the frame for each viewport according to its contents. And then, we exchange the danger view with an overlook view. The purpose of this study is to attract operators' attention to the danger area when there is a probable collision if operator won't stop or change their action. Besides, we also want to reveal the relationship between gaze habit and performance. The experimental results conducted using our virtual reality simulator confirms that operator tends to watch danger view when it appears. In addition, work strategy may influence the danger avoidance and work time efficiency in opposite way.

Keywords: Virtual reality · Attention attraction · Tele-operation · Construction machine

1 Introduction

Natural disasters such as earthquakes, tsunamis, and volcanic eruptions destroy living environments and infrastructures and endanger living things. After the disaster, our preferential task is to save more injured people and recovering the living environment as soon as possible. In order to avoid the danger of secondary disaster, we use remote-control construction machines to remove debris and accomplish recovery tasks [1-3]. However, it is represented that the low efficiency is the main demerit and the insufficient visibility is the major cause for lowering the time efficiency [4-5], as shown in Fig. 2 (a). In order to overcome it, we have developed a virtual reality (VR) simulation (Fig. 1) to analyze operators' observation behaviors [6-9]. In order to reduce the camera control time and supply an appropriate real-time vision, we proposed autonomous camera control system [10], as shown in Fig. 2 (b). The control system successfully raised accuracy and decreased the total task time.

However, we found operators can rarely observe the condition around machine itself when there is a probable collision, which make it hard for them avoid collision

© Springer International Publishing Switzerland 2015
H. Liu et al. (Eds.): ICIRA 2015, Part III, LNAI 9246, pp. 401–412, 2015.
DOI: 10.1007/978-3-319-22873-0_36

between machine and other objects. In addition, the posture of manipulator and obstacles as well as their position relationship are not always clear. For the above
reasons, operator is hard to avoid such collision because of the lack of danger information. In many cases, such collision is forbidden which may leads to the failure of
task. To avoid such operation mistake, we need more information around potential
dangerous area first.

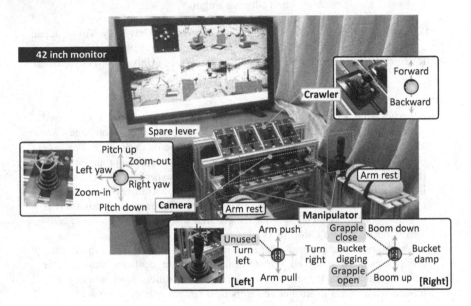

Fig. 1. Developed virtual reality simulator.

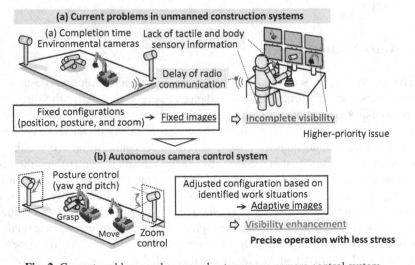

Fig. 2. Current problems and proposed autonomous camera control system.

In this study, we focus on introducing view about dangerous scene if necessary and attract operators' attention to such view first. Through the new view, we try to discover the relationship of gaze habit and work performance. To achieve the target with our proposed autonomous camera control system, we add a new role for cameras to observe the potential dangerous area between manipulator and objects. Cameras fixed on the rear of machine itself for observing potential probable collision is used as well. Either of views from these cameras will exchange an overlook view if the condition is judged dangerous.

In Section 2, the composition of our proposed VR simulator is introduced. In Section 3, new dangerous views and improved autonomous system will be introduced. In section 4, we clarify our experiment and the result will be given in Section 5. In section 6, vision movement is discussed when danger view occurs. In the last section, Section 7, we conclude our research and decide the future works.

2   Autonomous System for Camera Control

In this section, we clarify modules in our VR simulation system and the former autonomous camera control system [10] first and modify it later.

2.1   VR Simulation System

Software Component

We use openGL as the render engine and Open Dynamics Engine (ODE) as the dynamics calculation engine in the VR simulation system. Through ODE, we can set mass, joints for all parts of machine and objects. Thus, we can get real-time forces, moments or posture matrix of each defined objects. In the simulation system, we use a common six two-dimensional (2-D) viewports array (2 x 3) to represent the machine

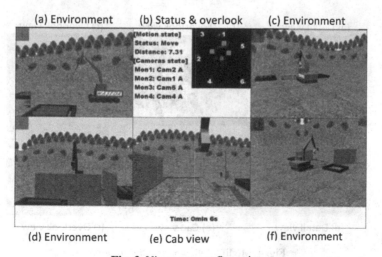

Fig. 3. Viewports configuration

information. Except a viewport with machine status and a 2D overlook view, we set a cab (cockpit) view and four monitors corresponding to environment cameras in order to describe machine itself and acquire the similar view as manned control. Aspect ratio of each monitor was set to 4:3. Besides these views, a time bar is set on the bottom of the screen. The configuration of the monitor is shown as Fig. 3.

Camera Module
In this system, each camera can rotate in yaw and pitch directions or zoom in or out. We should use them to take overlook views and feature views with little distortion. So, we limit the maximum angle of view as 70 and minimum angle of view as 10 considering the properties of wide-angle lens and long focus lenses. The pitch angle is also limited because of the limitation of the mounting brackets of cameras (-70–70).

2.2   Camera Control System

In former camera control system, four camera roles were defined considering effectiveness, independence, and practicality.

However, because of the strict conditions, it is impossible for a few cameras to get posture view or trajectory view of manipulator all the time. When it occurs, operator can hardly conduct operation with high accuracy. Thus we think that compared with discontinuous highly demanded view, a continuous side view with lower demand is better for operators' observation. So, we use two cameras (one from the front and another from the side on the same height level of end-effector) to shoot end-effector (Figs. 4 (a) and (d)). We also use two cameras (one from the front or back and another one from the side of basis of machine at high places) to shoot overlook views (Figs. 4 (c) and (f)). In order to understand the role of each view, we use frames with different colors. Green frame stands for detail view; blue frame stands for overlook view and yellow frame stands for cab view. Considering the amount of time cost and importance

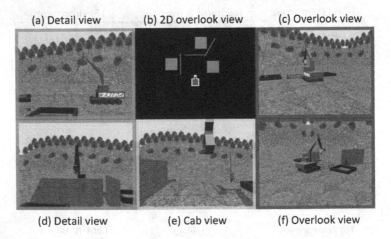

(a) Detail view   (b) 2D overlook view   (c) Overlook view

(d) Detail view   (e) Cab view   (f) Overlook view

Fig. 4. Modified viewports configuration

of manipulation, two end-effector views have higher priority than overlook views. Because we need to balance the information of front view and side view and the cab view is a front-rear view, the only two side views have higher priority than three front-rear views. The assignment system independently assigns camera roles to cameras on the basis of the defined priority [10]. Moreover, considering the effectiveness of sight movement, the status of camera connection will not be displayed because operators rarely watched it in our previous experiment.

2.3   Improved Camera Control System

We find operators prefer to observe enlarged end-effector views in most of the time, because it makes them easy to conduct the detail operation near the end-effector with high accuracy. However, it easily leads to the ignorance of the position relationship between manipulator and obstacles as well as the relationship between machine itself and objects because of the narrow vision, which may cause unexpected collision.

Considering the less importance of the two overlook views, one overview view (Fig. 3 (f)) will switch with a potential danger view when manipulator or the main body of machine is near an obstacle. The design of danger view and modification of task assignment system will be introduced in next section.

3   Introduction of Danger Avoidance View

Because of the insufficient information of potential probable forbidden collision, we introduce two kinds of danger view instead of one of the two overlook views. And the camera assignment system is modified according to the new roles.

3.1   Danger Avoidance Views

Danger between Machine and Objects
We import the view to describe the danger between machine and objects when rotating the main body of machine to possess a better close overlook view, which can make operator easy to avoid the collision between main body of machine and objects.

(a) Area between manipulator and objects   (b) Area between main body and objects

Fig. 5. Introduced danger views

In many cases, environmental cameras cannot play such role because of its fixed position and limited height position. The 2D overlook view cannot do it as well because machine and objects cannot be described exactly which may cause misleading. So we use a camera positioned over its main body and on the axis of rotation to play such role. In addition, the upper direction will not change all the time (Fig. 5 (a)).

Danger between Manipulator and Objects
The view between manipulator and objects should describe each posture and the position relationship between them. In order to describe as accurately as possible, the orientation of camera and the orientation of distance should be as vertical as possible. And the center of the distance always keeps at the center of view. So it is shoot by a proper environmental camera (Fig. 5 (b)).

3.2   Modified Camera Assignment System

Considering handling the global situation of work site, at least one viewport should display overlook view. Thus, at most one of the two danger views will be shown in the right bottom viewport instead of the side overlook view. Meanwhile, considering the importance of the danger view, we endow it with the highest priority. The priority of each view is shown in Table 1. According to the priority of each view, we assign them as our current system.

Table 1. Role priority in two assignment system

Priority	Role (Without danger view)	Role (With danger view)
1st	Side end-effector	Danger view
2nd	Front end-effector	Side end-effector
3rd	Side machine	Front end-effector
4th	Front-rare machine	Side machine

4   Experiment

We conducted a transport experiment by using the VR simulator. To derive the effectiveness of danger view in remote control conducted in VR environment, a vision measurement device called eye tracker is used to record the orbit of vision movement.

4.1   Experimental Setting

Work Site Configuration
We have four environments used for training and another three for experiments (Fig. 6). Each environment includes obstacles, uneven terrain, mound and debris to simply simulate the disaster scene. In training environments, we use such four different environments to simulate the different scene after disaster. In each scene, two debris cylinders are placed in one box or separately placed in two boxes which are

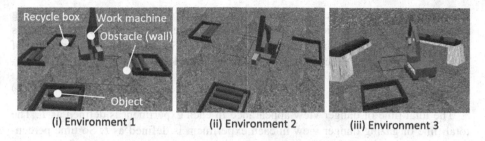

| (i) Environment 1 | (ii) Environment 2 | (iii) Environment 3 |

Fig. 6. Work site environment used for comparison experiment

placed on the ground or on a mound. In the environments used for comparison experiment, three scenes with higher difficulty are used (Fig. 6). Six debris cylinders are placed in different debris boxes which are placed on the ground or on a mound.

Camera Configuration

In order to make operators have a good sense of the work site, we place six environmental cameras as the current system. Four are set on the height of about 4 meters to have a clear view of debris boxes and recycle box. Another two cameras are set on a high position (8 meters), which enable them take aerial views as well. All camera control are controlled autonomously.

Experimental Task

In training environment, operators were asked to control the machine to transport all two debris cylindrical sticks one by one from debris box to a recycle box. In the comparison experiment, six cylindrical sticks are asked to transport. In both tasks, operators should avoid unexpected contact with walls or boxes and grasp the center of each stick to keep it balanced. And, they are expected to conduct as fast as they can.

4.2 Experimental Conditions

Operational skills are trained and measured in the four training environments by using a simple learning curve. And, we then conducted experiments in the other three environments to analyze whether danger view is effective. Twelve operators attended our experiment. As shown in Fig. 6, we use three different environments to conduct experiment. Operators are asked to conduct transport task in each environment 4 times. We recorded the total task time, number of collision in the experiment and vision movement orbits by using an eye-tracking system.

5   Results

We analyze the frequency of vision movement for each group in each day to measure the effectiveness of danger view when it appears. It is evaluated by using average data. In each experiment, the total number of appearance times of danger view is defined as n. During the period between appearance of danger view and the following disappearance of danger view, the total number of gaze times o plus one if operator's vision move to the right bottom viewport (Fig. 4 (f)). So the frequency percentage of gazing danger view f in each experiment can be defined as following equation (1)

$$f = o/n \times 100\%$$ (1)

The total time of danger view appearance in each experiment is defined as T. The total time of gazing danger view in each experiment is defined as t. So time percentage of gazing danger view f_t in each experiment can be written as equation (2)

$$f_t = t/T \times 100\%$$ (2)

According to the calculation, the average frequency percentage of gazing danger view f in each experiment is 42.32% and the standard derivation is 27.82 %; the average time percentage of gazing danger view f_t in each experiment is 17.37% and its standard derivation is 15.43 %.

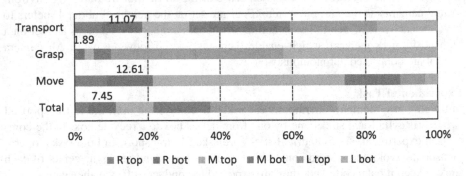

Fig. 7. Average time percentage of gazing each viewport in each state

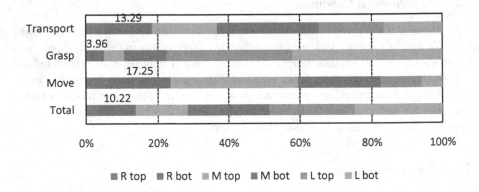

Fig. 8. Average frequency percentage of gazing each viewport in each state

In Figs. 7 and 8, R means the right while M means middle and L means left. According to Fig. 4, we can know the time percentage and frequency percentage of gazing each view. It is easy to find the time percentage of gazing right bottom viewport is 7.45% on average, which is about 42.89% of that when danger view appears; the frequency percentage of gazing right bottom viewport is 10.22% which is about 24.15% of that when danger view appears. It demonstrates that the danger view is successful to attract operators' attention as expected.

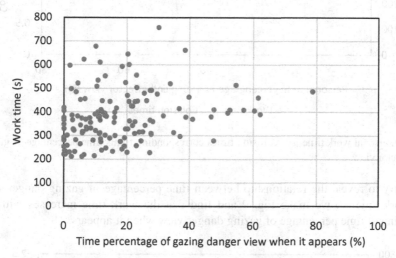

Fig. 9. Work time corresponding to time percentage of gazing danger view when it appears

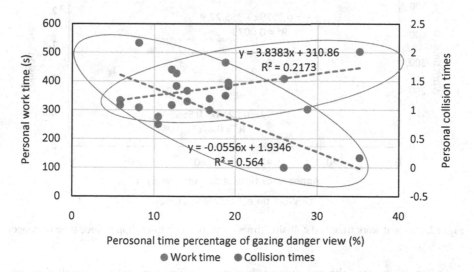

Fig. 10. Personal work time and collision times corresponding to his time percentage of gazing danger view when it appears.

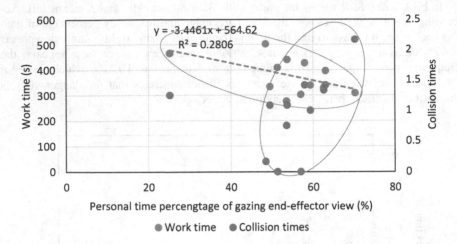

Fig. 11. Personal work time and collision times corresponding to his time percentage of gazing end-effector view

We try to reveal the relationship between time percentage of gazing danger view and work time, so we make Fig. 9 and find that the work time increases with the increasing of time percentage of gazing danger view when it appears.

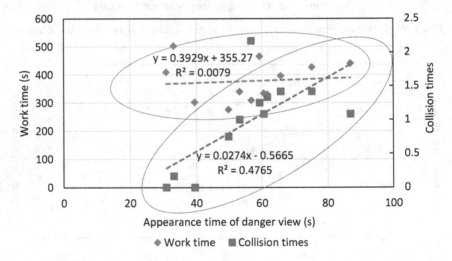

Fig. 12. Personal work time and collision times corresponding to his appearance time of danger view.

In Fig. 10, we find the operator has better time efficiency and worse collision performance with the decrease of time percentage of gazing danger view. Contrarily, in Fig. 11, the better time efficiency and worse collision performance with the increase of time percentage of gazing end-effector view. The higher attention to end-effector views means emphasizing the higher time efficiency to finish task; the higher attention to danger view means higher priority of danger avoidance. Operator can only gaze one viewport one time, so it implies that the less unexpected collision occurs with the sacrifice of work time frequency. Consequently, the balance between collision avoidance and work time frequency depends on his strategy.

In Fig. 12, appearance time of danger view is proportional to work time and collision times. From Figs. 10 and 12, we find operator who did better in work efficiency prefer to gaze danger view with less time percentage because of the less appearance time of danger view. We think, an operator with good skill should trigger less appearance time of danger view because of the less emergency, which makes him less necessary to gaze it.

6   Discussion

We find the gaze frequency f is far more than gaze time frequency f_t, so we think operators prefer to glance at the danger view compared with gazing it when potential danger occurs. In addition, operators with wide field of view may handle the concept of danger information without move their vision. It is observed that, in most cases, operator try to avoid vertical manipulator danger (Fig. 13) by watching left two enlarged end-effector views to know the manipulator information and objects information near dangerous area, because there will be no main body rotation which may lead to horizontal collision. On the other hand, when they try to avoid main body collision or horizontal manipulator danger, they have to ensure no more collision of each other. So operators usually raise manipulator first to make sure it has no chance to collide any objects in horizontal direction. Consequently, in such situation, two end-effector views are also preferred at first as well. In order to make the danger view more effective and vision move less, we think maybe a sub-viewport inside enlarged view is more suitable to avoid danger [11]. The comparison experiment will be conducted in the further research.

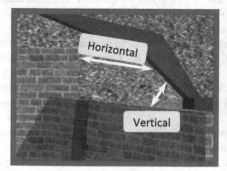

Fig. 13. Potential collision in vertical direction and that in horizontal direction

7   Conclusion

In this paper, we introduce two kinds of danger view which can attract operators' attention to avoid potential probable danger. Either of one will appear on the right bottom viewport if needed. The experiment indicates that our method is successful to make operators pay more attention to danger area. In addition, we discover the relationship between work performance and gaze frequency. In the next step, we will conduct comparison experiments to find some better danger view arrangement which can reduce the vision movement cost.

Acknowledgments. This research was supported in part by JSPS KAKENHI Grant Numbers 26870656 and 25220005 and in part by the Research Institute for Science and Engineering, Waseda Univ.

References

1. Nishigaki, S., Saibara, K., Kitahara, S., Iwasaki, H., Yamada, K., Satoh, H.: ICT-based work management to support un-manned construction for post-disaster restoration. In: Proc. Int. Symp. Automation and Robotics in Construction, pp. 508–513 (2011)
2. Yokoi, K., Nakashima, K., Kobayashi, M., Mihune, H., Hasunuma, H., Yanagihara, Y., Ueno, T,. Gokyuu, T., Endou, K.: A tele-operated humanoid robot drives a backhoe in the open air. In: Proc. IEEE/RSJ Int. Conf. Intelligent Robots and Systems, pp. 1117–1122 (2003)
3. Ban, Y.: Unmanned construction system: present status and challenges. In: Proc. Int. Symp. Automation and Robotics in Const., pp. 241–246 (2002)
4. Honma, M., Ikura, N., Takai, K.: A study on video technologies for unmanned construction: back-hoe operations. In: Proc. Symp. Japan Const. Machinery and Const. Association, pp. 61–64 (2007) (in Japanese)
5. Tang, X., Zhao, D., Yamada, H., Ni, T.: Haptic interaction in tele-operation control system of construction robot based on virtual reality. In: Proc. Int. Conf. Mechatronics and Automation, pp. 78–83 (2009)
6. Yang, J., Kamezaki, M., Iwata, H., Sugano, S.: Analysis of effective environmental-camera images using virtual environment for advanced unmanned construction. In: Proc. IEEE/ASME Int. Conf. Advanced Intelligent Mechatronics, pp. 664–669 (2014)
7. Lee, S., Jung, H., Song, H., Park, S.: Development of immersive augmented reality interface system for construction robotic system. In: Proc. IEEE Int. Conf. Control Automation and Systems, pp. 2309–2314 (2010)
8. Moon, J., Son, Y., Park, S., Kim, J.: Development of immersive augmented reality interface for construction robotic system. In: Proc. IEEE Int. Conf. Control Automation and Systems, pp. 1192–1197 (2007)
9. Moteki, M., Fujino, K., Nishiyama, A.: Research on operator's mastery of unmanned construction. In: Proc. Int. Symp. Automation and Robotics in Construction, pp. 540–547 (2013)
10. Kamezaki, M., Yang, J., Iwata, H., Sugano, S.: An autonomous multi-camera control system using situation-based role assignment for tele-operated work machines. In: Proc. IEEE Int. Conf. Robotics and Automation, pp. 5971–5976 (2014)
11. Hirose, K., Ogawa, T., Kiyokawa, K., Takemura, H.: Interactive reconfiguration techniques of reference frame hierarchy in the multi-viewport interface. In: Proc. IEEE Symp. 3D User Interface, pp. 73–80 (2006)

Active Control in Tunneling Boring Machine

Error Compensation for Inclinometer in TBM Attitude Measurement System

Long Wang, Chuncao Zhang, and Guoli Zhu[⊠]

School of Mechanical Science and Engineering,
Huazhong University of Science and Technology, Wuhan 430074, China
longwmail@126.com, chuncaozhang@163.com, glzhu@mail.hust.edu.cn

Abstract. Inclinometer is used to measure roll and pitch angles in laser target system of TBM. A synthetic method based on wavelet analysis is proposed to decrease the measurement error of inclinometer under great vibration, improving the accuracy of TBM attitude angles. It is applied to reduce edge effect based on signal extension and orthogonal polynomials extension of approximate coefficients. Different denoising parameters are chosen to get minimum error according to the characteristic of the signal. Experiment shows that the proposed method is more effective and versatile than average calculation method used in the field and traditional wavelet denoising method, and the error is compensated less than $1mrad$.

Keywords: Wavelet denoising · Edge effect · Orthogonal polynomials · Edge extension

1 Introduction

Inclinometer, which has the advantages of high resolution, small volume, easy to integrate and wide working temperature range [1], is widely used in attitude angles measurement in laser target system of TBM [2]. TBM suffers great vibration and impact during tunneling. In this case, inclinometer's output has obvious fluctuation, and the measurement result can't be used in subsequent processing [3]. Average calculation method has been used in the field to improve measurement accuracy, which averages the data in the last second. However the result cannot meet the accuracy requirement of construction.

Among the recent methods to process the output of inclinometer in vibration, wavelet analysis has been recognized as an effective and robust method due to its capability to deal with non-stationary signals. But edge effect is a significant bottleneck in wavelet application [4,5], especially for our case that the value we measure is the endpoint. A commonly used way to decrease edge effect is edge extension when discrete wavelet transform (DWT) is realized [6]. Mainly used edge extension methods, including zero extension, period extension and symmetric extension, are not well used for data processing of inclinometer. Here we put forward an extension method based on orthogonal polynomials fitting, which preserves

H. Liu et al. (Eds.): ICIRA 2015, Part III, LNAI 9246, pp. 415–425, 2015.
DOI: 10.1007/978-3-319-22873-0_37

continuity at the boundaries up to a predetermined order. It means to extend the approximate coefficients with the fitted orthogonal polynomials by the boundaries. It's proved to be much better than other methods.

It's known that the closer to edge, the bigger edge effect it occurs [7]. So, besides approximate coefficients extension method, another solution to this problem is to extend the signal on the right, making the endpoint in the middle [8]. These two methods are combined to decrease edge effect of wavelet denoising for inclinometer signal, which could result in smaller error for attitude angles.

In this paper, a synthetical method based on signal extension and approximate coefficients extension is proposed to reduce edge effect of wavelet denoising and compensate the error of inclinometer. In simulation, three kinds of field-simulated signals are extended first and then processed with wavelet denoisng based on different approximate coefficients extension methods. Comparing the simulation results, the parameters of the synthetic method are determined. In experiment, the field condition is simulated and the inclinometer output is processed with the proposed method, the traditional wavelet denoising method and the field method. The results show that the proposed method is the best and the error is compensated less than $1mrad$.

2   Theory

2.1   Wavelet Threshold Denoising Method

Wavelet threshold denoising is one of the most commonly used methods in signal processing, which contains three steps [9]:

1. Wavelet decomposition for digital signal;
2. Threshold processing for detail coefficients;
3. Wavelet reconstruction with thresholding coefficients.

As for step 1 and 3, Mallat algorithm is used for wavelet decomposition and reconstruction. Let $h(k), g(k), \tilde{h}(k)$ and $\tilde{g}(k)$, $k=0,1,2..M-1$,be the two-dimension sequences of functions $\phi(t), \psi(t), \tilde{\phi}(t)$ and $\tilde{\psi}(t)$.Then the multiresolution decomposition equations are given by

$$c_{m-1}(n) = \sum_k c_m(k)\tilde{h}(k - 2n) \tag{1}$$

$$d_{m-1}(n) = \sum_k c_m(k)\tilde{g}(k - 2n) \tag{2}$$

And the multiresolution reconstruction equation is

$$c_m(n) = \sum_k c_{m-1}(k)h(n - 2k) + \sum_k d_{m-1}(k)g(n - 2k) \tag{3}$$

where $c_{m-1}(k)$ is approximate coefficient and $d_{m-1}(k)$ is wavelet coefficient. We use sequences $\tilde{h}(k)$ and $\tilde{g}(k)$ to decompose for step 1 and $h(k)$ and $g(k)$ to reconstruct for step 3 with biorthogonal wavelet. If it is orthogonal wavelet, $\tilde{h}(k)$ and $h(k)$ are the same, and $\tilde{g}(k)$ and $g(k)$ are also the same [10]. As for step 2, in this paper we choose *Birge-Massart* strategy to calculate the thresholds instead of global thresholds [11].

2.2 Edge Extension in Mallat Algorithm

Equations (1) to (3) are applied to infinite sequence in theory. However the actual sequence is finite and undesirable edge effect will occur if these equations are applied without extension [12]. A commonly used method to decrease edge effect is to extend the sequence $c_m(n)$ at the boundaries, such as symmetric extension and orthogonal polynomials extension. For decomposition, we can extend $c_m(n)$ for M-1 on the left and right each, and then doing convolution and two-extraction to get $c_{m-1}(n)$ and $d_{m-1}(n)$. For reconstruction we can do zero-insertion, convolution and summation first, and then choose the former L values to accomplish reconstruction. The whole flow of this algorithm is shown in Fig.1.

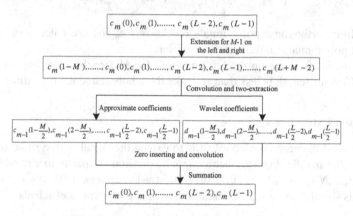

Fig. 1. Edge extension in Mallat algorithm

Hence the most important matter is how we extend the coefficients. In this paper, we only talk about symmetric extension and orthogonal polynomials extension. Symmetric extension means that the extended coefficients are symmetric with the values at the endpoints. The coefficients $c_m(k)$ are extended to

$$c_m(M-2), \dots, c_m(0), c_m(0), \dots, c_m(L-1), c_m(L-1), \dots c_m(L-M+1).$$

Orthogonal polynomials extension will be stated below.

2.3 Orthogonal Polynomials Extension of Approximate Coefficients

Orthogonal polynomials extension of approximate coefficients is to extend the approximate coefficients with fitted orthogonal polynomials according to the boundaries.

It means to calculate the fitted orthogonal polynomials in order M_1 with N_1 points near the endpoints and extend the coefficients for M-1 on the left and right each. Two important extension parameters are fitting order and fitting length. If the fitting length is too big, boundary state can't be reflected. And if the fitting length is too small, randomness has an outstanding impact on orthogonal polynomials [13]. As for fitting order, if it is too big, the normal equations are morbid, and undesirable fitting effect will occur. So the fitting order is controlled less than two. Here we take second-order orthogonal polynomials extension as an example. The extension method is below:

1. Take the left N_1 points $c_m(0), c_m(1), \ldots\ldots, c_m(N_1 - 2), c_m(N_1 - 1)$ to fit the orthogonal polynomials $c'_m(n) = \alpha_0 T_0(n) + \alpha_1 T_1(n) + \alpha_2 T_2(n)$;
2. The orthogonal polynomials on the left are:

$$T_0(n) = 1, \quad T_1(n) = n - \frac{1}{N_1} \sum_{i=0}^{N_1-1} i,$$

$$T_2(n) = n^2 - \left[\sum_{i=0}^{N_1-1} \frac{i^2 T_1(i)}{\sum_{i=0}^{N_1-1} T_1^2(i)} \right] T_1(n) - \frac{1}{N_1} \sum_{i=0}^{N_1-1} i^2.$$

The orthogonal polynomials coefficients on the left are:

$$\alpha_0 = \frac{\sum_{i=0}^{N_1-1} T_0(i) c_m(i)}{\sum_{i=0}^{N_1-1} T_0^2(i)}, \quad \alpha_1 = \frac{\sum_{i=0}^{N_1-1} T_1(i) c_m(i)}{\sum_{i=0}^{N_1-1} T_1^2(i)}, \quad \alpha_2 = \frac{\sum_{i=0}^{N_1-1} T_2(i) c_m(i)}{\sum_{i=0}^{N_1-1} T_2^2(i)}.$$

Then the fitted orthogonal polynomials on the left $c'_m(n)$ are calculated. The fitted orthogonal polynomials on the right are similar.

3. Extend the coefficients on the left and right for M-1 each, according to $c'_m(n)$. If the coefficients length is less than N_1, take the whole sequence for fitting.

2.4   Signal Extension

Another method to decrease edge effect is to extend the signal on the right, making the endpoint in the middle. For symmetric signal extension, it means to extend the signal $x(n)$, $n=0,1,\ldots,N$-1, to $x(0)$, $x(1),\ldots,$ $x(N$-1), $x(N$-1),$\ldots,$ $x(1)$, $x(0)$. As for orthogonal polynomials signal extension, it's similar to that of approximate coefficients.

3   Simulation

The angles of TBM change slowly in the range of $\pm 10°$ during tunneling. Three kinds of ideal tracks such as constant signal, ramp signal and sinusoidal signal are assumed due to unknown actual track of TBM. The non-stationary random noise is added into the track signal to describe the great and random fluctuation of inclinometer's output, which means the variance of the noise changes as time goes by while the mean keeps zero.

In simulation, we choose constant signal $5°$, ramp signals with different slopes less than $0.2°$/s and sinusoidal signals of amplitude $5°$ with different frequencies less than 0.01Hz. The processed window length is set 256. The wavelet we use is db4 as is the most widely used in signal processing. The denoising effect is evaluated by root mean-square errors (RMSE) and max errors (MAX) of the endpoints.

3.1 Parameters Optimization of Orthogonal Polynomials Extension of Approximate Coefficients

We choose fitting length every 20 points ranging from 40 to 120 and fitting order less than 2. The denoising results of orthogonal polynomials extension of approximate coefficients with different fitting length are below:

1. First-order algorithm. The results of constant signal, ramp signal and low frequency signal are similar: when fitting length is over 60, denoising result has little change, as is seen in Fig.2. So fitting length is set 60. For high frequency signal whose frequency is over 0.004Hz, fitting length should decrease to 40.
2. Second-order algorithm. When fitting length is over 60, denoising result has little change for all the signals, as is seen in Fig.3. So fitting length is set 60.
3. By contrast, we find that first-order algorithm is better than second-order algorithm for constant signal, ramp signal and low frequency signal. If the frequency is over 0.03Hz, second-order algorithm is better, as is seen in Fig 4.

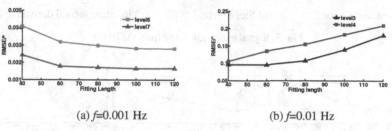

(a) f=0.001 Hz   (b) f=0.01 Hz

Fig. 2. Denoising RMSE of first-order algorithm with different fitting length

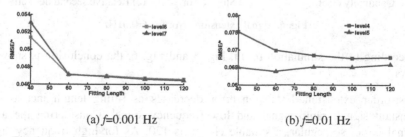

(a) f=0.001 Hz   (b) f=0.01 Hz

Fig. 3. Denoising RMSE of second-order algorithm with different fitting length

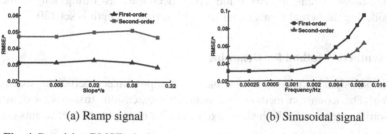

(a) Ramp signal   (b) Sinusoidal signal

Fig. 4. Denoising RMSE of orthogonal algorithm with optimized fitting length

3.2   Parameters Optimization of Orthogonal Polynomials Signal Extension

Fitting order and fitting length are needed with orthogonal polynomials signal extension. We choose fitting length every 20 points ranging from 40 to 180 and fitting order 1 and 2. Continuity and smoothness on the edge are two factors for signal extension. We choose to count the continuity error and slope error for first-order extension, and continuity error, slope error and relative second derivative error for second-order extension. As the results of constant signal, ramp signal and low frequency signal are similar, we just show the results of low frequency signal. The simulation results are shown in Fig. 5 and Fig.6:

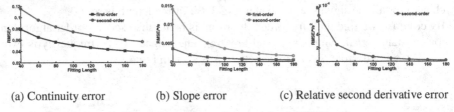

    (a) Continuity error   (b) Slope error   (c) Relative second derivative error

Fig. 5. Signal extension error for f=0.001Hz

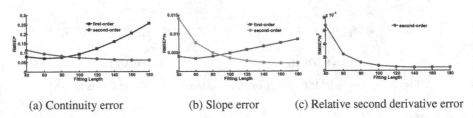

    (a) Continuity error   (b) Slope error   (c) Relative second derivative error

Fig. 6. Signal extension error for f=0.01Hz

According to the simulation result, Fig. 5 and Fig. 6, the conclusion is shown as follows:

1. First-order extension. Extension error decreases as fitting length increases for constant signal, ramp signal and low frequency signal. Considering the actual signal is not so regular, a suitable choice is 120. As for high frequency signal, fitting length should be reduced.
2. Second-order extension. Extension error decreases as fitting length increases. Considering the actual signal is not so regular, fitting length is set 120.

3.3   Synthetic Method for Denoising

In this paper, signal extension is combined with approximate coefficients extension for denoising. The extension methods are symmetric extension, first-order extension and second-order extension for both signal extension and approximate coefficients extension.

Firstly we determine the optimal approximate coefficients extension methods for each signal extension approach. Then comparing these combined methods, we get the preliminary optional methods regardless of decomposition levels. Finally we simulate with these optional methods in differemt levels and get the proposed method through comparison.

The denoisng results of optional methods for ramp signals are shown in Fig. 7. In this Fig, legend '*symmetric first-order*' means symmetric signal extension with first-order approximate coefficients extension, and legend '*first-order*' means first-order approximate coefficients extension without signal extension. As denoising max error has complex randomness, we look on more as RMSE. According to Fig 7, we come to conclusion for ramp signals:

1. If $s<0.00224°/s$, symmetric signal extension with first-order approximate coefficients extension for level 9 is suitable;
2. If $0.00224°/s<s<0.00419°/s$, symmetric signal extension with first-order approximate coefficients extension for level 8 is suitable;
3. If $0.00419°/s$ /s$<s<0.2°/s$, first-order approximate coefficients extension without signal extension for level 7 is suitable.

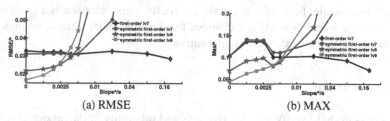

(a) RMSE   (b) MAX

Fig. 7. Denoisng results of optional methods for ramp signals

The denoisng results of optional methods for sinusoidal signals are shown in Fig 8. The legends are similar to those of ramp signals.

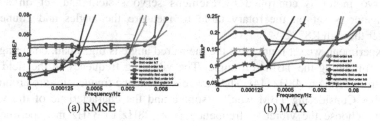

(a) RMSE   (b) MAX

Fig. 8. Denoisng results of optional methods for sinusoidal signals

According to Fig 8, we come to conclusion for sinusoidal signals, which is shown in table 1.

Table 1. Denoising method for sinusoidal signals

frequency/Hz	method
0<f<0.0001	symmetric first-order lv 9
0.0001<f<0.00019	symmetric first-order lv 8
0.00019<f<0.00155	first-order lv 7
0.00155<f<0.00248	first-order lv 6
0.00248<f<0.00354	first-order first-order lv 6
0.00354<f<0.00656	second-order lv 6
0.00625<f<0.01	second-order lv 5

The variables we have chosen are slope and frequency in simulation. However we can only measure the changing speed in the field. In fact they are equal. For example, s=0.2°/s and f=0.01Hz are equal to the average changing speed 0.2°/s. In the field, if the inclinometer output changes linearly, we regard it as ramp signal. Otherwise we regard it as sinusoidal signal. After calculating the changing speed, we can choose which method to use for denoising.

In detail, we can see that when signal changes slowly, symmetric signal extension is a good choice for denoising. And when signal changes fast, signal extension has no big effect. Meanwhile, first-order approximate coefficients extension is a good choice to decrease edge effect. But when it changes fast for sinusoidal signals, second-order approximate coefficients extension is alternative.

4   Experiment

Experiment is made to verify the effectiveness and reliability of the proposed method, comparing with the field method and traditional wavelet denoising method. The hardware system is shown in Fig. 9, simulating TBM in the field. A vibrating table is used to generate vibration, simulating the working condition of TBM. And a rotary table is designed to simulate the motion of TBM, which is joint with a servo motor. The servo motor is controlled by Siemens servo system and set on a holder. Inclinometer is set on the rotary table to measure the angles and connected to computer through RS232.

In experiment, we set four cases: the measured angle is big, small, changing slowly or fast, simulating the TBM roll angle. The vibrating frequency in the field mainly ranges from 4.4Hz to 48Hz [14]. The device's resonance frequencies are around 10Hz and 20Hz. Considering TBM won't resonate and the characteristic of the vibrating table, we choose the vibrating frequency from 28Hz to 33Hz in experiment. The vibrating amplitude is set 80 relatively (max=100). The inclinometer type is SST-260, whose accuracy is 0.01° in static environment. We find that the tracking error of the motor is less than 0.02° in vibration, which means that the predetermined track could be the reference angles. Thus the errors are calculated by the estimated values and reference values.

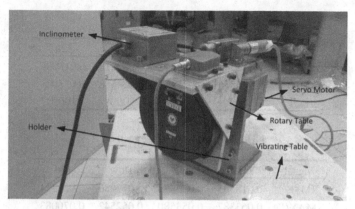

Fig. 9. Experiment Device

As for signal processing, the root mean-squared errors and max errors are adopted to evaluate these methods: the proposed method, period average method and traditional wavelet denoisng method. Traditional wavelet denoisng method means symmetric approximate coefficients extension without signal extension. Meanwhile the percentage of the absolute errors less than $1mrad$ is counted. It is the accuracy needed in the field. The denoising results of these methods are shown in Fig 10. The errors are counted in tables 2, 3 and 4.

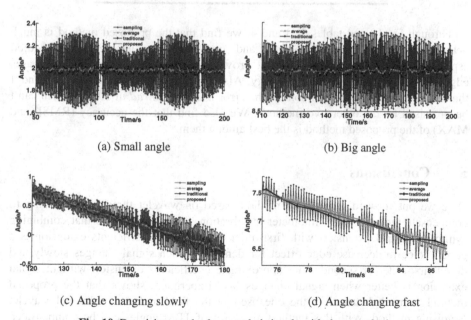

(a) Small angle

(b) Big angle

(c) Angle changing slowly

(d) Angle changing fast

Fig. 10. Denoising results for sampled signals with three methods

Table 2. Proposed method

signal	small	big	slow	fast
RMSE/°	0.012302	0.010845	0.009218	0.018026
MAX/°	0.025516	0.021983	0.024688	0.033026
PCT	100%	100%	100%	100%

Table 3. Period average method

signal	small	big	slow	fast
RMSE/°	0.018553	0.019779	0.017086	0.029324
MAX/°	0.063387	0.053580	0.052542	0.070083
PCT	99.27%	100%	100%	97.01%

Table 4. Traditional wavelet denoising method

signal	small	big	slow	fast
RMSE/°	0.029397	0.034393	0.048877	0.045539
MAX/°	0.076198	0.078493	0.125174	0.116782
PCT	95.24%	94.85%	73.29%	75.08%

Through Fig.10 and tables 2, 3 and 4, we find that the proposed method is much better than period average method and traditional wavelet denoising method especially when angle changes fast. It proves that the proposed method decreases edge effect of wavelet denoising largely. Also it can be seen from tables 2, 3 and 4 that the proposed method can control the error less than $1mrad$ in 100%, which can't be promised by the other two methods. We also find that the accuracy (RMSE and MAX) of the proposed method is the best among them.

5   Conclusions

Here we put forward a synthetic method based on wavelet threshold denoising to compensate the error of inclinometer in vibration. Simulation shows that combining symmetric signal extension with first-order approximate coefficients extension is a good choice to decrease edge effect for denoising when signal changes slowly and direct first-order or second-order approximate coefficients extension without signal extension is better when signal changes fast. Experiment shows that the proposed method is much better than the one used in the field and the traditional wavelet denoising method. With this method, the error of TBM angles can be compensated satisfying the accuracy demand.

References

1. Zhu, G.L., Wen, X.W., Pan, M.H.: Application of curve fitting to compensation of measurement errors from an inclinator (in Chinese). J. Huazhong Univ. of Sci. & Tech. (Nat. Sci. Ed.) **38**(5), 83–85 (2010)
2. Pan, M.H., Zhu, G.L.: Study of Measure Methods of the Automatic Guiding System of Shield Machine (in Chinese). Constr. Tech. **34**(6), 34–36 (2005)
3. Li, H.X., Pan, M.H., Wang, L., Zhu, G.L.: Research of vibration error compensation for inclinometers. In: 2012 International Conference on Mechatronic Systems and Materials Application, vol. 590, pp. 377–384. Elsevier, Qingdao (2012)
4. Mallat, S.: A Wavelet Tour on Signal Processing, 3rd edn. Academic Press, New York (2008)
5. Montanari, L., Basu, B., Spagnoli, A., Broderick, B.M.: A padding method to reduce edge effects for enhanced damage identification using wavelet analysis. Mech. Syst. Signal Proc. **52–53**, 264–277 (2015)
6. Williams, J.R., Amaratunga, K.: A Discrete Wavelet Transform without Edge Effects Using Wavelet Extrapolation. J. Fourier Anal. Appl. **3**(4), 435–449 (1997)
7. Zheng, L.Y., Xiong, J.T., Li, L.C., Yang, J.Y.: Wavelet Border Processing and Real-Time Denoising (in Chinese). Radar Sci. Tech. **5**(4), 300–303 (2007)
8. Liu, Z.C., Cheng, X.G., Li, Y.F., Li, B.: Real time wavelet filtering method for sensor output time series (in Chinese). J. B. Univ. Chem. Tech. **3**(1), 71–75 (2007)
9. Donoho, D.L.: Denoising by soft-thresholding. IEEE Trans. Inform. Theory **41**(3), 613–627 (1995)
10. Mallat, S.: A theory for multiresolution signal decomposition: the wavelet representation. IEEE Trans. Pattern Anal. Machine Intell. **11**(7), 674–693 (1989)
11. Varun, P.G., Pavithran, M., Nishanth, T., Balaji, S., Rajavelu, V., Palanisamy, P.: A novel wavelet based denoising algorithm using level dependent thresholding. In: 2014 International Conference on Electronics and Communication Systems, pp. 1–6. IEEE Press, Coimbatore (2014)
12. Karsson, G., Vetterl, I.M.: Extension of finite length signals for sub-band coding. IEEE Trans. Signal Proc. **17**, 161–168 (1989)
13. Kong, C., Fang, Y.H., Lan, T.G., Xiong, W., Dong, D.M., Li, D.C.: A boundary prolongation method based on orthogonal polynomial fitting in wavelet transform (in Chinese). Chin. J. Quantum Electron. **25**(1), 25–28 (2008)
14. Tian, H.W.: Discussion on the double-layer viaduct bridge pier design of Luotang river (in Chinese). Shanxi Architecture **38**(34), 192–194 (2012)

Synchronization Control of Torque Cylinders for Hard Rock Tunnel Boring Machine

Zhen Zhang, Guofang Gong$^{(\boxtimes)}$, Huayong Yang, Weiqiang Wu, and Tong Liu

State Key Laboratory of Fluid Power Transmission and Control,
Zhejiang University, Hangzhou 310027, China
qlzhangzhen@163.com, gfgong@zju.edu.cn

Abstract. A synchronization control system(SCS) is proposed to reduce synchronization deviation of left and right torque cylinders in vertical steering of hard rock tunnel boring machine(TBM). Hydraulic system based on proportional valves is designed after the introduction of attitude mechanism. Establishing the load model of torque cylinders, this paper carried out simulation comparison between SCS and traditional displacement control system(DCS). The results show that the deviation in SCS is 1.5 mm and 6 mm in DCS. Synchronization control system has a significant effect on reducing displacement deviation of left and right torque cylinders.

Keywords: TBM · Vertical steering · Torque cylinders · Synchronization control · Displacement deviation

1 Introduction

Full face hard rock tunnel boring machine(TBM) is a high integrative product of machinery, hydraulic pressure, electricity and computer, and widely used in hard rock tunneling due to its high excavation rate and circular tunnel profile[1-3]. Fig. 1 shows the typical structure of an open-type TBM. It consists of cutterhead system, gripper and propel system, attitude control system and support system. This machine cuts the rock by the rotation of the cutterhead, generally equipped with disk cutters. When tunneling with a TBM, the gripper shoes connected with the barrels of the propel cylinders push against the excavated tunnel wall, and four propel cylinders are provided to generate the hydraulic thrust force necessary to propel the machine.

Attitue adjustment system plays an important role in curved tunnel and rectifying deviation. Four torque cylinders are the critical components in the vertical steering. It needs to adjust attitude of the machine in vertical plane by extending or retracting the piston rod. As for the attitude adjustment system, precision is focused generally. But one premise is to keep the machine horizontal especially in the process of vetical steering. That's to say the torque cylinders on each side of the main beam must be kept at the same length. Otherwise, machine has the danger of capsizing, and propel system and cutterhead work in harmful condition. In order to stay high-efficiency,

H. Liu et al. (Eds.): ICIRA 2015, Part III, LNAI 9246, pp. 426–435, 2015.
DOI: 10.1007/978-3-319-22873-0_38

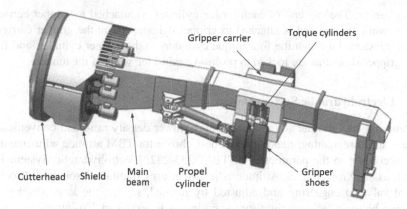

Fig. 1. Three-dimensional model of an open-type TBM

TBM usually keeps moving forward during attitude adjustment. As a result, enormous loads with opposite directions between left and right sides of main beam will act on the torque cylinders when the cylinders are extended or retracted. So it is important to put forward a strategy to make torque cylinders on each side of main beam have the same displacement in process of attitude adjustment.

Some scholars have made research for synchronization control. Han Bo proposed the control system with feedback of synchronous error has a better static and dynamic state characteristic[4]. Xiongbin Peng applied master-slave synchronization in lifting hydraulic cylinders for segment erector[5], and acquire a good effect. But there is no method that has been put into displacement control of torque cylinders for TBM.

In this paper, we design an electrohydraulic control system based on parameters of TBM 253-282, an open type TBM of Robbins, and propose displacement synchronization control strategy. Cutterhead torque and machine gravity are taken into consideration to determine the force acting on left and right torque cylinders. A detailed mathematiacl model of the position process has been developed to design the synchronization control system. Based on the actual attitude adjustment in vetical plane, simulations are carried out to verify the validity of the synchronization control system.

2 Vetical Steering System Design

2.1 Vetical Steering Mechanism

Vetical steering mechanism for the machine is shown in Fig. 2. Four torque cylinders are provided, which can be divided into two groups named as left or right torque cylinders according to its location on left or right side of the main beam, to provide hydraulic force to move the rear of the machine vertically. When the gripper shoes are locked into the bore of the tunnel, extension of all torque cylinders simultaneously moves the rear of the machine upward, resulting in a downward correction at the cutterhead. Conversely, when all torque cylinders are retracted, the rear of the machine moves down, resulting in an upward correction at the cutterhead.

When tunneling with a TBM, cutterhead is driven by variable frequency motors. The torque created from the cutterhead is transmitted through the main beam, and then to the

gripper carrier. The cap end of each torque cylinders is attached to gripper carrier assembly, while the rod end is attached to gripper cylinder. From the gripper carrier, the
load is transferred through the four torque cylinders to the gripper cylinder, and finally
to the gripper shoes that are locked in position against the walls of the tunnel.

2.2   Electrohydraulic System Design

Hydraulic system has the advantages of high power density ratio and convenience of
stepless speed regulating, making it the best choice for TBM attitude adjustment system. According to the parameters of TBM 253-282, Electrohydraulic System is designed, as shown in Fig. 3. Attitude adjustment uses traditional solenoid directional
control valve in engeering and adjusted by manual, so that the large displacement
deviation between left and right torque cylinders is produced. To introduce synchronization control strategy, proportional directional valves are employed. The parameters of hydraulic system are listed in Table 1.

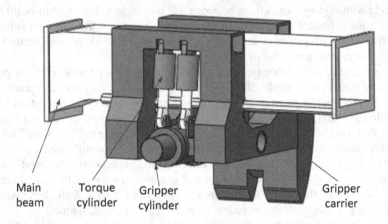

Main   Torque   Gripper   Gripper
beam   cylinder   cylinder   carrier

Fig. 2. Three-dimensional model of vetical steering mechanism

Table 1. Main parameters of hydraulic system

Parameters	values
Size of torque cylinders	Φ380/250×203 mm
Frequency of proportional valve	25 Hz
Flow rate of oil source	120 L/min
Balance valve opening pressure	130 bar
Signal range of proportional valve	±10 V

Two left torque cylinders are connected in parallel[6]. The pressure reduce valve 3
keeps work pressure less than setting value. The proportional directional valve 1 controls the flow rate and the oil flow direction of the hydraulic system thereby the speed

and the flow direction of the cylinders[7]. The counterbalance valve 2 is to balance the force generated by torsional loading when the cylinders are extended and retracted. And two counterbalance valves' control chamber is communicated with each other. This can lock the barrels of the cylinders and keep it in the expected position precisely when the machine attitude need not adjust.

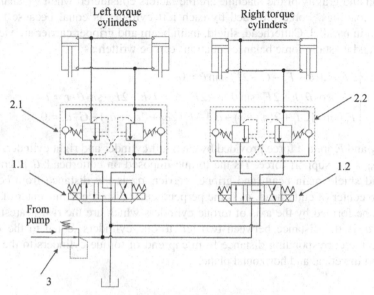

Fig. 3. Hydraulic system of torque cylinders

3   Mathmatical Modeling and Control System Design

3.1   Load Model of Hydraulic Cylinders

As mentioned before, when all torque cylinders are retracted and extended, angle between cutterhead's axis and horizontal plane called pitch angle changes. The pitch angle varies from -0.3 to 0.3 degree depending on the maximum tunnel slope. So the pitch angle is controlled by torque cylinders displacement and the distance from the center of cutterhead to cylinders. The relation of parameters can be expressed as:

$$\begin{cases} \sin\gamma = \dfrac{d_0 + d}{D + s} \\ d_0 = l_0 \cdot \sin\theta \\ d = (l_0 + l) \cdot \sin\theta \end{cases} \tag{1}$$

Where γ is the pitch angle, l_0 is the initial length of torque cylinders when $\gamma = 0$, l is the target displacement, d_0 and d are the cylinders corresponding length or displacement of l_0 and l in vetical plane. What's more, the cylinders are not perpendicular to horizontal plane, and the angle is θ.

Besides the torque from cutterhead and gravity of machine, resistance force is also transmitted to torque cylinders as surrounding rocks of shield are compressed when pitch angle changes. But the amount of rock compression is small because the angle is limited from 0 to 0.3 degree. Also due to leverage effect, the rock resistance force acting on torque cylinders is neglected relative to torsional loading. So torque from cutterhead and gravity of the machine are the factors considered when we analyze the cylinders loadings. Force provided by each left cylinder is equal because they are connected in parallel. Cutterhead, shield, main beam and gripper carrier are viewed as a whole, and its static force balance equations can be written as:

$$\begin{cases} 2F_1 \cdot \sin\theta + F_n - G - 2F_r \cdot \sin\theta = 0 \\ 2F_1 \cdot \cos\theta \cdot h + 2F_1 \cdot \sin\theta \cdot r - 2F_r \cdot \cos\theta \cdot h - 2F_r \cdot \sin\theta \cdot r = T \\ F_1 \cdot \sin\theta[2a_1 + 2(a_2 + x) + a_3] - F_r(2a_1 + 2a_2 + a_3) - G \cdot a_1 = 0 \end{cases} \quad (2)$$

Where F_1 and F_r are the force provided by each left cylinder and right cylinder, respectively, F_n is the support force, T is the torque imposed on cutterhead, G is gravity of cutterhead shield main beam and gripper carrier, a_1 is the distance from center of gravity to center of cutterhead, a_2 is the perpendicular distance from center of gravity to the plane formed by the axis of torque cylinders which are the two nearest to cutterhead, a_3 is the distance between two left torque cylinders ,equal to the right, r and h are the corresponding distance from cap end of torque cylinders to the axis of cutterhead in vetical and horizontal plane.

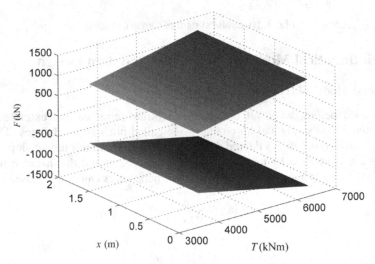

Fig. 4. Loads of left and right torque cylinders

F_1' and F_r', the force acting on left and right torque cylinders, is reactive force of F_1 and F_r. F_1' and F_1' are obatained by equation(2) and θ is approximately equal to 90 degree, and they are expressed as

$$F_1' = \frac{T}{4r} + \frac{1}{2} \frac{G \cdot a_1}{2a_1 + 2(a_2 + x) + a_3}$$

$$F_r' = -\frac{T}{4r} + \frac{1}{2} \frac{G \cdot a_1}{2a_1 + 2(a_2 + x) + a_3}$$

(3)

Using parameters of TBM 253-282, we can get the size and orientation of force intuitively, as shown in Fig.4. The force orientation of left and right is opposite and the size is not equal.

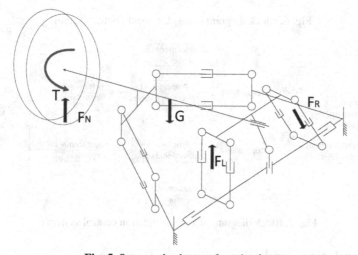

Fig. 5. Structural scheme of mechanism

3.2   Control System Design

To ensure left and right torque cylinders have the same displacement in vertical steering, two kinds of strategies are proposed. Displacement control system (DCS) is shown in Fig.6, and synchronization control system (SCS) is in Fig.7. Generally, the target displacement is a ramp signal in the process of trajectory tracking and rectifying deviation. But the maximum velocity of torque cylinders is limited by the setting flow rate of oil source which is determined by the setting advance speed and limited tunnel slope. Conventional PIDs are introduced to accomplish the closed-loop control in the two systems.

In displacement control system, the same displacement signals are given to left and right controllers. The displacement is measured and fed back to the control unit. Two sides cylinders are controlled by their PID controllers separately.

In synchronization control system, we choose the displacement of left cylinders as reference parameter. The displacement diference of two sides' cylinders will be transmitted to synchronization controller, And the controller output will be added to target displacement signal. The rest of synchronization control system is just the same as the displacement control system.

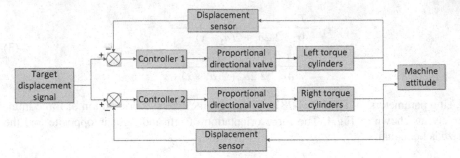

Fig. 6. Block diagram of displacement control system

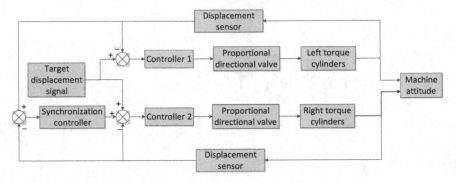

Fig. 7. Block diagram of synchronization control system

4   Simulations and Discussions

To compare the performance of two control systems in terms of synchronization precision. Simulation models are created in AMESim software. We can specify the loads
of torque cylinders according to equation (3). The load in simulation is shown in
Fig.8. There are two step changes in 4th and 5th second to simulate fluctuating load.
The same ramp displacement signals and loads are given to both of the systems.

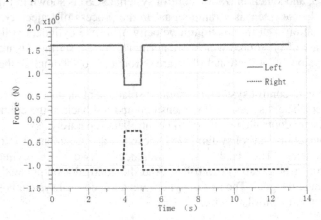

Fig. 8. Loads of torque cylinders

Fig.9 shows the displacement difference between left and right cylinders. We can easily find the difference of left and right cylinders is close to 6 mm in DCS, and 1.5 mm in SCS. The SCS has an obvious inhibiting effect on displacement difference which is reduced to 25% of DCS. But the deviation in both of the control systems will vary when the loads change.

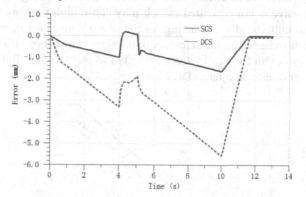

Fig. 9. Displacement difference under DCS and SCS

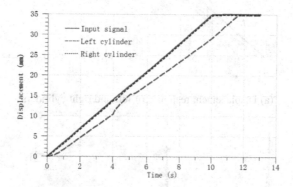

(a) Displacement response of left and right cylinders

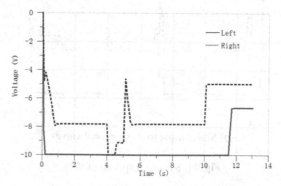

(b) Signal input to proportional valves

Fig. 10. System response of DCS

Fig.10 and Fig.11 show the system response of DCS and SCS. As shown in Fig.10(a), the left cylinders can follow the displacement signal accurately, and the right is hysteretic. But from Fig.10(b), we can see that the signal input to proportional valve of left cylinders has reached its maximum. That's because the effect of counterbalance valves on balancing the loads is limited. Too large setting value of counterbalance valve is not permitted ,which may cause high pressure to damage the cylinders due to cutterhead's fluctuating torque. As a consequence, we choose displacement of left cylinder as the reference parameter. From Fig.11(a), we can find both left and right cylinders are hysteretic relative to input signal. But the total time spending on extensions is similar to DCS.

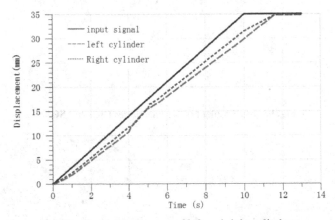

(a) Displacement response of left and right cylinders

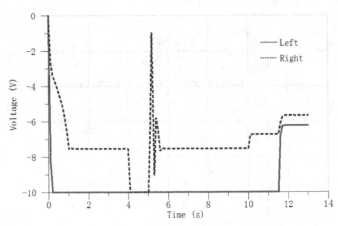

(b) Signal input to proportional valves

Fig. 11. System response of SCS

5   Conclusions

Synchronization control system has a significant effect on reducing displacement deviation of two-side torque cylinders and overturning danger of TBM. The maximum deviation is 1.5 mm in SCS, 25% of DCS. SCS has almost the same ramp response time as DCS, and will not reduce the efficiency of vetical steering. Also, SCS has a good position control precision, resulting a precision vertical steering. And SCS has the advantages of less computation and easier engineering application than other complicated strategies. This research will be helpful to develop an automatic attitude adjustment system, and a more efficient synchronization controller will be the focus of research in the future.

Acknowledgments. This work is supported by National Basic Research Program (973 Program) of China (No. 2013CB035400) and National High-tech Research and Development Program (863 Program) of China (No. 2012AA041803). Their support is gratefully appreciated. The authors also wish to acknowledge the help of the engineers in China Railway Construction Heavy Industry Co.,Ltd.

References

1. Yazdani-Chamzini, A., Yakhchali, S.H.: Tunnel Boring Machine (TBM) selection using fuzzy multicriteria decision making methods. Tunnelling and Underground Space Technology **30**, 194–204 (2012)
2. Mengshu, W.: Construction technique of open TBM for long railway tunnels in very hard or soft rock strata. China Civil Engineering Journal **38**, 54–58 (2005)
3. Saiqun, Z.: Study on drive system of the full-face rock tunnel boring machine. Zhejiang University, Hangzhou (2005)
4. Han, B., Qingfeng, W., Yongxiang, L.: Investigation into control structure of electrohydraulic proportional synchronized control system. Machine Tool and Hydraulics **145**, 7–10 (1997)
5. Xiongbin, P., Guofang, G., Chen, K., Lintao, W.: Synchronization fuzzy PID control of lifting hydraulic cylinders for segment erector. Journal of Zhejiang University (Engineering Science) **18**, 2002–2008 (2014)
6. Lintao, W., Guofang, G., Shi, H., Dianqing, H.: Position precision and impact force control of segment erector for shield tunneling machine. In: IEEE International Conference on Digital Manufacturing & Automation, pp. 612–617, Guilin, China (2012)
7. Shi, H., Guofang, G., Huayong, Y., Lintao, W.: Positioning speed and precision control of a segment erector for a shield tunneling machine. In: IEEE/ASME International Conference on Advanced Intelligent Mechatronics, pp. 1076–1080, Montréal, Canada (2010)

Mechanical Analysis and Prediction
for the Total Thrust on TBMs

Siyang Zhou[1], Yilan Kang[1], Cuixia Su[2], and Qian Zhang[1(✉)]

[1] Tianjin Key Laboratory of Modern Engineering Mechanics, Tianjin, China
zhangqian@tju.edu.cn
[2] China Railway Construction Heavy Industry Group Limited, Changsha, China

Abstract. This paper proposed a new model to predict the total thrust acting on Tunnel Boring Machines (TBMs) based on mechanical analysis. Key parameters involved in the prediction are mainly rock property, operating status and structural feature. Several force components which the total thrust consists of were analyzed. Considering failure criterion, mechanical analysis of the interaction between the rock and the cutterhead was carried out. Furthermore, the comparison was performed between the predicted value and the in-situ measured data of the thrust in a TBM engineering project. The results show that this approximate predictive model is able to reflect the global trend of the acting thrust due to considering geological, operating and structural feature.

Keywords: TBM · Total thrust · Predictive model · Mechanical analysis

1 Introduction

Tunnel boring machines are a kind of hard rock excavating equipment. It takes advantages of high penetration rate and small disturbance on the surrounding environment, thus has been widely used in underground constructions. Due to its complex structure, heavy power, and particularly under complicated geological conditions, there are high requirements on TBMs' parameters control during design and operating process. The total thrust acting on TBMs that reflects the interaction between the machine and the rock is one of the essential parameters. Predictions of the total thrust for TBMs are indispensible for optimizing the design and control during construction.

Recently, there is some relevant research on loads in TBM tunneling progress. In regard to theoretical approaches, Colorado School of Mines (CSM) prediction model developed an estimation of forces based on a uniform distribution of contact pressure between a cutter and the rock[1]; Roxborough et al.[2] analyzed cutting forces of a wedge-shaped cutter by taking into account the contact area and the compressive strength of rock. By using the stress distribution in the Hertz's elastic contact, Sun et al. [3]provided a predictive model on cutting forces, power and energy consumption were also discussed on the basis of cutter load analysis. Wu et al.[4] applied the theory of dimension to calculate normal forces of cutter, and he used the

© Springer International Publishing Switzerland 2015
H. Liu et al. (Eds.): ICIRA 2015, Part III, LNAI 9246, pp. 436–444, 2015.
DOI: 10.1007/978-3-319-22873-0_39

laboratory cutting tests and engineering data to obtain empirical equation.Considering the geometric feature of the contact area between cutters and the rock, Wijk [5]developed predictive formulas to estimate the thrust and torque on a wedge-shaped cutter. Based on mechanical analysis of the disc cutters, prediction of disc cutter wear and specific energy(SE) was investigated by Wang et al.[6].

In respect of experimental work and numerical simulation, Snowdon et al.[7] used laboratory experiments to analyze tool cutters forces and specific energy under four kinds of rock at various spacing and penetration. Rostami et al.[1] studied the form of contact pressure distribution, a regression analysis between measured cutting forces and cutting parameters was utilized to develop an estimate of the pressure. Gong et al.[8] discussed the influence of frictional force by field test and friction experiment, and a rock mass boreability analysis was given. Ma et al.[9] carried out a numerical study on the effect of confining stress on rock fragmentation. To provide a better estimate of the influence on total thrust force, Su et al.[10] analyzed several parameters such as the cutterhead opening ratio and the chamber pressure.

As for engineering data analysis, Yagiz et al.[11]. suggested improvements should be made to conventional CSM model where they conducted investigations on the effect of rock mass fracture and brittleness. Zhang et al.[12]proposed a method for inverse analysis and modeling based on mass on-site measured data, in which dimensional analysis and data mining techniques were combined. In addition, an identification and optimization method for the energy consumption of a shield tunnel machine was presented[13]. In the light of cutter force analysis, a multi-objective genetic algorithm (MOGA) was applied to improve the quality of the disc cutters' plane layout by Huo et al.[14] Zhao et al.[15] studied various geological parameters' effects on rock fragmentation, they concluded that UCS is the most crucial factor, and a statistical prediction model is set up by performing a nonlinear regression analysis. Hassanpour et al.[16] made use of the available data and found strong relationships between geological parameters and field penetration index (FPI). Based on statistical approaches, Torabi et al.[17] ranked the effectiveness of four selected geological parameters on penetration rate in a descending order as follows: UCS, friction angle, Poisson's ratio, and cohesion.

In conclusion, research based on rock experiment, numerical simulation and data analysis did provided valuable results of the relationship between thrust force requirements and various parameters. Moreover, theoretical analysis is essential to study the intrinsic relationship between various crucial factors. However, the majority of related works with regard to TBM load are focused on the cutter force, global mechanical analysis of equipment during excavation is currently lacking, which can give a fully understand of thrust force requirement for TBM with different geological and operational conditions.

This paper carried out a global mechanical analysis of TBMs during excavation. Several force components of the total thrust were analyzed. Considering specific failure criterion, mechanical analysis of the interaction between the rock and the cutterhead was carried out. A model was established to predict the total thrust acting on TBMs with different geological and operational conditions. The comparison was performed between the predicted value and the in-situ measured data of the thrust in a TBM engineering project.

2   Mechanical Analysis and Modeling of the Total Thrust on TBMs

This paper studied the total thrust of the whole equipment, and mechanical analysis along advancing direction (Fig.1) was discussed. According to the mechanical equilibrium along advancing direction during excavation, the total thrust is supposed to equal to the sum of resistance forces. These resistance forces mainly include: the force acting on cutterhead by the rock F_r (kN), friction force of shield F_f (kN), friction force of subsequent equipments F_b (kN).Therefore, the total thrust F (kN) on a TBM can be expressed as:

$$F = F_r + F_f + F_b \tag{1}$$

When advancing, the cutterhead is in contact with the tunneling interface. Resistance forces are determined by the interaction between cutters and the rock. At present, constant cross section (CCS) cutters are commonly adopted, so the paper mainly discussed CCS cutters. From the perspective of mechanics, the excavating progress of TBM is essentially a continuous interaction between tunneling interface of rock and cutters, which is under the confining pressure of surrounding rock. The primary resistance force is the force acting on the cutterhead by the rock. Results of full-scale laboratory cutting tests indicated that a uniform pressure distribution along the cutter penetration edge can be used in most cases for CCS cutters[1]. Taking a random infinitesimal body of rock in contact with the arc surface of cutter to analyze the stress state (Fig.2), the infinitesimal body is under the confining pressure and cutters' pressure simultaneously.

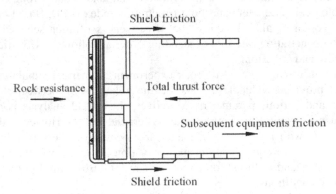

Fig. 1. Forces at TBM along advancing direction.

If the disc is free rolling and neglecting friction[2], the disc cutter is only under the load of radial pressure σ_1 (pointing towards the center) from the rock(ignoring the shearing stress). Based on shearing stress theorem, the side face of the infinitesimal body is under confining pressure σ_3 perpendicular to side face.

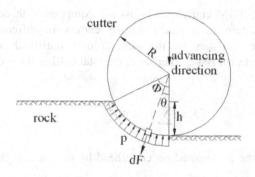

Fig. 2. Mechanical analysis of a cutter.

In the process of excavation, rock is broken by the crushing of cutter. Shear failure occurs under compression status[18]. On the basis of Mohr-Coulomb criteria, the limiting shearing stress of infinitesimal body depends on the friction angle and cohesion of rock[19].

$$\tau = c + \sigma \tan \varphi \tag{2}$$

Where τ (MPa) is shearing stress on failure surface; c (MPa) is rock cohesion; σ (MPa) is normal stress on failure surface; φ (rad) is rock friction angle.

According to the Mohr's stress circle , the limiting radial pressure can be obtained:

$$\sigma_1 = \frac{1+\sin\varphi}{1-\sin\varphi}\sigma_3 + \sigma_c \tag{3}$$

Where σ_c is the uniaxial compressive strength, namely the limiting stress without confining pressure[20,21].

Because of the pressure distribution along the cutter penetration edge is nearly uniform, normal forces F_n can be estimated by:

$$F_n = \int_0^\phi TR\sigma_1 d\theta \cos\theta \tag{4}$$

Where F_n is normal forces; T (mm) is cutter tip thickness; R (mm) is cutter radius; θ (rad) is the radian between radical pressure and advancing direction; ϕ (rad) is the radian of arc surface, which can be obtained by geometrical relationship:
$\phi = \arccos(\dfrac{R-h}{R})$, h is the penetration.

Putting (3) in (4) and making integral calculation, thus the normal force is obtained:

$$F_n = TR(\frac{1+\sin\varphi}{1-\sin\varphi}\sigma_3 + \sigma_c)\sin\phi \tag{5}$$

As it is known to all, TBM cutterhead is usually equipped with dozens of cutters. Despite there has certain discrepancy among cutters in different places, it is appropriate to assume that they are the same from a statistical perspective when analyzing global mechanical issue. Therefore, the total cutting force of cutterhead can be nearly expressed as:

$$F_r = \sum_{i=1}^{N} F_i \approx NF_n \qquad (6)$$

Where F_r (kN) is the force acted on cutterhead by the rock; F_i (kN)is the normal force of single cutter

In order to reduce the vibration and to prevent jamming accidents, TBM should adjust shield position to make the shield contact with the tunnel wall under smaller pressure[22]. As a consequence, there exist frictional forces between shield and rock, and it is determined by contact stress, friction coefficient and shield size:

$$F_f = \mu_1 \pi Dl \frac{\Theta}{360^0} P \qquad (7)$$

Where F_f is the friction force between shield and tunnel wall; μ_1 is the friction coefficient between shield and tunnel wall; D (m) is cutterhead diameter; $l(m)$ is the length of shied; $\Theta(^\circ)$ is the extent of contact area between shield and tunnel wall; P (MPa) is the contact pressure determined by contact conditions.

In order to reduce the vibration and to prevent jamming accidents, TBM should adjust shield position to make the shield contact with the tunnel wall under smaller pressure[23]. As a consequence, there exist frictional forces between shield and rock, and it is determined by contact stress, friction coefficient and shield size:.

$$F_d = \mu_2 mg \qquad (8)$$

Where μ_2 is friction coefficient between subsequent equipments and track; m (t) is the weight of subsequent equipments; g (m/s^2) is the acceleration of gravity.

By Summing up force components mentioned above, an approximate predictive model of thrust force is established (equation(9)). Moreover, this proposed model involves the relationships between the thrust and the geological, operating, and structural parameters, and it is a reasonable prediction from the perspective of mechanical analysis.

$$F = F_r + F_f + F_b = NTR(\frac{1+\sin\varphi}{1-\sin\varphi}\sigma_3 + \sigma_c)\sin\phi + \mu_1 \pi Dl \frac{\theta}{360^0} P + \mu_2 mg \qquad (9)$$

3   Case Study

Data collected from one hard rock TBM tunnel (the Dahuofang Water Conveyance Tunnel) was utilized to verify the predictive capability of presented model. Influences of geological and operating parameters on the thrust were investigated. In this literature, a section of the tunnel being about 4.5 km long was excavated with open-type TBM.

Several main rock types have been identified in the study area as follows: tuffaceous siltstone, volcanic breccia, migmatite, etc, with uniaxial compressive strength ranges from 30~120MPa. Geological parameters for calculation were taken from the report of engineering geological exploration.

In this case study, cutterhead diameter is 8.03m, the length of the shield is 3m and the weight of subsequent equipments is 356t.The main structure parameters of cutter were demonstrated as follows: cutter radius is 241.5 mm; cutter tip thickness is 25mm; and there are 51 cutters on the cutterhead.

In addition, operating data derived from the control system of the machine were recorded in real-time: the rate of penetration ranges from 29~64mm/min, while the range for penetration is 4~13mm/rev.

The curves of predicted and measured thrust force were shown in Fig.3. As is illustrated in the graph, the red line stands for the actual measured data while the blue line represents the calculation result based on the predictive model. Besides, both of these two lines fluctuate between 5000kN to 20000kN.

The results revealed that the predicted one shows a good agreement with measured curve and fluctuates around it. For instance, when the advancing distance is between 700-1000m, the trend of the measured curve declines, and the predicted one also dropped. It could be concluded that the proposed model reflects the overall trend of thrust force, since it allows for the mechanical system consists of ground condition and equipment.

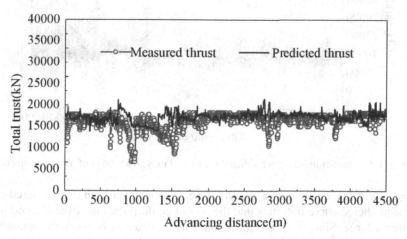

Fig. 3. Comparation between predicted thrust and measured thrust of Dahuofang project.

The second case study is about the measured data from Tunnel no.9 of Yintao Project, which is located in the Dingxi, Guansu Province. Different from the Dahuofang Water Conveyance Tunnel, this project was mainly accomplished by a double-shield TBM.

Complicated geological condition was encountered during excavation. The rock types not only include soft rock like argillaceous siltstone and conglomeratic sandstone, but also hard rock such as marble and granitic gneiss. Furthermore, the maximum uniaxial compressive strength is up to 95MPa while the minimum is only 2.5 MPa. Geological parameters for calculation were taken from the report of engineering geological exploration.

In this case study, cutterhead diameter is 5.75m, the length of the shield is 6m and the weight of subsequent equipments is 523t.The main structure parameters of cutter were demonstrated as follows: cutter radius is 216 mm; cutter tip thickness is 20mm; and there are 42 cutters on the cutterhead.

In addition, operating data derived from the control system of the machine were recorded in real-time: the rate of penetration ranges from 40~140mm/min, while the range for penetration is 5~25mm/rev.

The curves of predicted and measured total thrust were illustrated in Figure4. The results revealed that the total advancing distance could be divided into two parts. The measured thrust of the first part (from 0~6500m) is low due to the soft rock. However, the total thrust of the second part(from 6500~10000m) is relatively high because of the high uniaxial compressive strength of the rock. The predicted one also has a two-stage trend and shows good agreement of the measured curve.

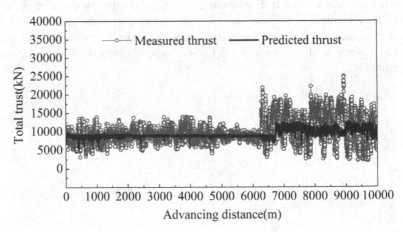

Fig. 4. Comparation between predicted thrust and measured thrust of Yintao project.

Two case studies were presented, and the comparison between the measured curve and the predicted curve indicates that the model could reflect the overall trend of the total thrust force. Since it takes into account the interaction between the ground and the equipment, this prediction is capable of reflecting the influences of geological and operating parameters.

However, the predicted curve was more moderate in contrast with the measured one, this is due to the existence of shear zones ,joints, faults, and other local weakness zones. The excavating equipment will have unstable performances when facing such complicated ground condition. Therefore, more detailed geological information is required for further research of the influences of weakness zones.

4   Conclusions

This paper conducted mechanical analysis and modeling on the total thrust acting on TBMs, and then discussed the relationships between the thrust and the geological, operating, and structural parameters.

Based on the mechanical equilibrium during excavation, several force components acting on the cutterhead, the shield and the subsequent equipment, were fully analyzed. Especially, the force acting on the cutterhead was mainly analyzed. During the progress of excavation, a TBM is affected by the confining pressure as well as geological resistances on the tunneling interface. The paper used infinitesimal body of rock in contact with the arc surface of a cutter to discuss the stress state. According to Mohr-Coulomb criteria, maximum principal stress was calculated, which is the limiting radial stress during rock breaking. Subsequently, mechanical analysis on the interaction between the rock and a cutter was performed. The proposed mechanical formula of the entire cutterhead could reflect influences of geological, operating, and structural parameters. On this basis, a predictive model of the total thrust was set up.

Furthermore, an engineering project was involved to verify and analyze the proposed model. The predicted one shows a good agreement with measured curve, and fluctuates around it. The results indicate that this approximate prediction model takes into account the effects of geological and operating conditions, hence it is able to reflect the overall trend of the total thrust force.

Acknowledgements. This work was supported by National Basic Research Program of China (973 Program, Grant No. 2013CB035402) and National Natural Science Foundation of China (NSFC, Grant Nos. 11127202 and 11302146).

References

1. Ozdemir, L., Rostami, J.: A new model for performance prediction of hard rock TBMs. In: Proceedings of the Rapid Excavation and Tunneling Conference. Society For Mining, Metallogy & Exploration, Inc., pp. 793–793 (1993)
2. Roxborough, F.F., Phillips, H.R.: Rock excavation by disc cutter. International Journal of Rock Mechanics and Mining Sciences & Geomechanics Abstracts, 361–366 (1975)
3. Sun, H., Chen, J., Chen, G.: Study on the cutter force and the calculated load of disc cutters on boring machines. J. Construction Machinery **08**, 1–7 (1980). (in Chinese)
4. Wu, Q.: Mechanical analysis of rock breaking by disc cutter during shield tunneling in mixed face ground conditions. D. Jinan University. (in Chinese)
5. Wijk, G.: A model of tunnel boring machine performance. J. Geotechnical and Geological Engineering **10**, 19–40 (1992)

6. Wang, L., Kang, Y., Cai, Z., Zhang, Q., Zhao, Y., Zhao, H., Su, P.: The energy method to predict disc cutter wear extent for hard rock TBMs. J. Tunnelling and Underground Space Technology **28**, 183–191 (2012)

7. Snowdon, R.A., Ryley, M.D., Temporal, J.: A study of disc cutting in selected British rocks. International Journal of Rock Mechanics and Mining Sciences & Geomechanics Abstracts, 107–121 (1982)

8. Gong, Q.M., Zhao, J., Jiang, Y.S.: In situ TBM penetration tests and rock mass boreability analysis in hard rock tunnels. J. Tunnelling and Underground Space Technology **22**, 303–316 (2007)

9. Ma, H., Yin, L., Ji, H.: Numerical study of the effect of confining stress on rock fragmentation by TBM cutters. International Journal of Rock Mechanics and Mining Sciences **48**, 1021–1033 (2011)

10. Su, J., Gong, G., Yang, H.: Total Thrust Calculation and Test Research for Soil Pressure Balancing Shield. J. Tunneling Construction Machinery and Equipment **39**(1), 13–16 (2008). (in Chinese)

11. Yagi, S.: A model for the prediction of tunnel boring machine performance. In: Proceedings of 10th IAEG Congress (2006)

12. Zhang, Q., Kang, Y., Zheng, Z., Wang, L.: Inverse Analysis and Modeling for Tunneling Thrust on Shield Machine. J. Mathematical Problems in Engineering (2013)

13. Zhang, Q., Qu, C., Kang, Y., Huang, G., Cai, Z., Zhao, Y., Zhao, H., Su, P.: Identification and optimization of energy consumption by shield tunnel machines using a combined mechanical and regression analysis. J. Tunnelling and Underground Space Technology **28**, 350–354 (2012)

14. Huo, J., Sun, W., Chen, J., Su, P., Deng, L.: Optimal disc cutters plane layout design of the full-face rock tunnel boring machine (tbm) based on a multi-objective genetic algorithm. Journal of Mechanical Science and Technology **24**, 521–528 (2010)

15. Gong, Q.M., Zhao, J.: Development of a rock mass characteristics model for TBM penetration rate prediction. International Journal of Rock Mechanics and Mining Sciences **46**, 8–18 (2009)

16. Hassanpour, J., Rostami, J., Zhao, J.: A new hard rock TBM performance prediction model for project planning. J. Tunnelling and Underground Space Technology **26**, 595–603 (2011)

17. Torabi, S., Shirazi, H., Hajali, H., Monjezi, M.: Study of the influence of geotechnical parameters on the TBM performance in Tehran-Shomal highway project using ANN and SPSS. Arabian Journal of Geosciences **6**, 1215–1227 (2013)

18. You, M.: Mechanical Properties of Rocks. Geological Publishing House, Beijing (2007). (in Chinese)

19. Cai, H., Chen, Z.: Rock Mechanics. Science Press, Beijing (2004). (in Chinese)

20. Cai, M.: Rock Mechanics and Engineering. Science Press, Beijing (2002). (in Chinese)

21. Gao, W.: Rock Mechanics. Peking University Medical Press, Beijing (2010). (in Chinese)

22. Wang, M., Li, D., Zhang, J., et al.: Rock tunnel boring machine (TBM) construction and engineering examples. China Railway Press, Beijing (2004). (in Chinese)

23. Du, Y., Du, L.: Full Face Hard Rock Tunnel Boring Machine-System Principles and Integrated Design. Huazhong university of science and technology press, Wuhan (2011). (in Chinese)

Tunneling Simulation and Strength Analysis
of Cutterhead System of TBM

Meidong Han[1], Zongxi Cai[1], Chuanyong Qu[1(✉)], and Kui Chen[2]

[1] School of Mechanical Engineering, Tianjin University, Tianjin, China
qu_chuanyong@tju.edu.cn
[2] State Key Laboratory of Shield Machine and Boring Technology, Zhengzhou, China

Abstract. Cutterhead, cutterhead support and main beam are three main stressed components of TBM in tunneling process. To obtain the loadings on these components, a three-dimensional finite element model was established to simulate the excavating process, in which the progressive damage and failure model was introduced to simulate the formation and separation of cutting scraps. Then the stress and deformation distribution on these three components were analyzed based on the established model. Results reveal that the dangerous point appears in the intersection of corbel and cutterhead panel, with the max stress 237.7MPa which is less than the material yield limit 345MPa. The maximum deformation of the system is 1.96mm, which is a minute quantity comparing to structure size. So both the strength and stiffness of the components meet the requirements. This method can be utilized to get more useful information in complex conditions for the optimization of topological structure of the key components of TBM.

Keywords: Cutterhead · Cutterhead support · Main beam · Tunneling loads · Mechanical properties

1   Introduction

With the acceleration of the urbanization and the development of the construction of infrastructure facilities on a large scale, people's need for space has changed from the surface to the subterranean. A large number of underground tunnels are gradually under construction. With the advantage of safety, environmental protection, economy and efficiency [1,2], TBM is now becoming an effective tool for tunnels of various sizes with lengths over 1.5–2 km [3]. As one main type of TBM, open-style TBM is widely used in tunnels with good rock stability and less soft rock. The main set, which is the core structure of the open-style TBM, is composed mainly of cutterhead, cutterhead support, main beam, gripper system, traction motors, push cylinders and rear support, as shown in Fig.1.

In the tunneling process, gripper system holds the rock tightly, then main beam and cutterhead move forwards under the promotion of push cylinders. At the same time, cutterhead rotates with the help of traction motors installed on the cutterhead support. As the direct tunneling tool and load-transfer components, cutterhead, cutterhead

© Springer International Publishing Switzerland 2015
H. Liu et al. (Eds.): ICIRA 2015, Part III, LNAI 9246, pp. 445–455, 2015.
DOI: 10.1007/978-3-319-22873-0_40

support and main beam are under complex stress state in tunneling. Some researchers have done mechanical analysis on these components by finite element method. Xia et al. [4] calculated the stress and deformation of cutterhead through structure analysis based on ANSYS. Tao et al. [5] established a three-dimensional model of main beam to analyze its strength in five different working conditions. Also, the mechanical properties of cutterhead support is analyzed in ANSYS Workbench by Qin [6]. From the view of stiffness and strength, these researches provides some valuable references for the design and optimization of key components of TBM. Whereas the load employed in these simulation are estimated according to the design parameters which can't reflect the influences of geological, operating and structural parameters in tunneling process. Also, all these studies focus on the mechanical properties of single structure without considering the influence of the interaction between connecting parts.

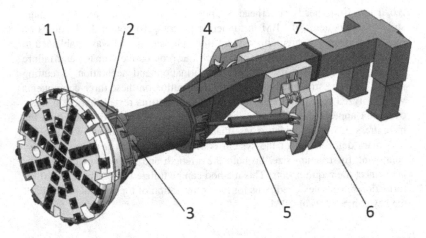

Fig. 1. The overall structure of open-style TBM (1. Cutterhead. 2. Cutterhead support 3. Traction motor 4. Main beam 5. Push cylinder 6. Gripper system 7. Rear support)

This paper initially established a three-dimensional finite element model for dynamic simulation of cutterhead-rock interaction, through which the actual trust and torque in TBM tunneling were calculated. Then we analyzed the deformation and stress state of the system composed of cutterhead, cutterhead support in working conditions. The results could provide some helpful and useful information for the optimization of topological structure of the key components of TBM.

2   Models for Simulation

2.1   Finite Element Modeling of the key Components

Cutterhead, cutterhead support and main beam are the three key components of TBM. Considering the structural complexity, the ANSYS Parametric Design Language was employed for modeling. According to the design drawings of the components, we deleted the bolt holes, the injection ports, the chamfers and the corners which have

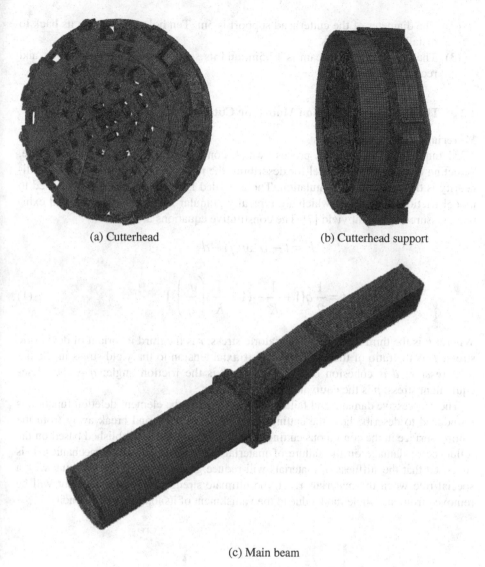

(a) Cutterhead   (b) Cutterhead support

(c) Main beam

Fig. 2. Finite element models of the key components.

little influence on the structural strength. To ensure the quality of meshing, appropriate Boolean operations have been done on the geometric models. The final established finite element models is shown in Fig.2.

(1) The diameter of the cutterhead is 8m. Fifty-one disc cutters, including nine front cutters, eight center cutters and eight gage cutters are installed at different radius on cutterhead.

(2) The diameter of the cutterhead support is 5m. Ten holes are dug on its back to install traction motors.

(3) The length of main beam is 17.5m, and its section is circular in forepart and rectangular in the tail.

2.2   The Dynamic Simulation Model for Cutterhead Tunneling

Material Model for Rock

TBM tunneling is a complex process which contains material damage and failure. Selecting an appropriate model for describing the rock concerning its material nonlinearity is the key step for simulation. The extended Drucker–Prager model is used to model frictional materials, which are typically granular-like soils and rock, and exhibits pressure-dependent yield [7]. The constitutive equations are:

$$F = t - p \tan \beta - d$$

$$t = \frac{1}{2} q[1 + \frac{1}{K} - (1 - \frac{1}{K})\left(\frac{r}{q}\right)^3] \tag{1}$$

Where: F is the third invariant of deviatoric stress; r is the third invariant of deviatoric stress; K is the ratio of the yield stress in triaxial tension to the yield stress in triaxial compression; d is cohesion of the material; β is the friction angle; q is the Mises equivalent stress; p is the equivalent pressure stress.

The progressive damage and failure model including the element-deletion function is introduced to describe how the cutting scrap of rock failure and break away from the cutting surface in the continuous cutting process. This model is established based on the influence of damage on the failure of materials described in fracture mechanics. It is supposed that the stiffness of materials will reduce gradually to zero complying with a special rule when the materials reach the ultimate strength. Then the element will be removed from the whole model due to the vanishment of its load-bearing capacity.

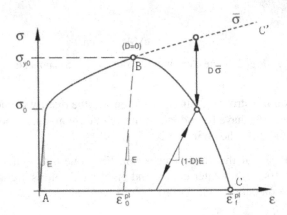

Fig. 3. Stress–strain relation of the materials' progressive damage degradation.

As shown in Fig.3, the stress-strain response is consist of three parts: the material response is initially linear elastic, followed by plastic yielding controlled by The extended Drucker–Prager model; beyond point B there is a marked reduction of load-carrying capacity until rupture, B-C. Point B identifies the material state at the onset of damage, which is referred to as the damage initiation criterion. Once the damage appears in the materials, the characteristic stress–strain behavior is managed by Eq. (2):

$$\sigma = (1 - D)\bar{\sigma} \tag{2}$$

where D is the damage variable which denotes the damage level of the materials and its evolution equation can be described as Eq. (3):

$$\dot{D} = \frac{L\dot{\bar{\varepsilon}}^{pl}}{\bar{u}_f^{pl}} = \frac{\dot{\bar{u}}^{pl}}{\bar{u}_f^{pl}} \tag{3}$$

where L is the characteristic length of the element, $\dot{\bar{\varepsilon}}^{pl}$ is the rate of strain, $\dot{\bar{u}}^{pl}$ is the rate of equivalent plastic displacement, $\bar{u}_f^{pl}$ is the equivalent plastic displacement when the materials are totally damaged.

Cutterhead-rock Interaction Model
In front of the cutterhead, we added the rock model with the shape of cylinder to form the integral finite element model, as shown in Fig.4. The cutterhead, whose stiffness is large enough, was defined as a rigid body with a reference node. This method enables the calculation of the resultant reaction loads acting on the entire cutterhead at this single reference node [8].

Fig. 4. Cutterhead-rock interaction model.

The contact pair algorithm was employed to simulate the cutterhead-rock interaction. For this algorithm, ABAQUS/Explicit enforces contact constraints using a penalty contact method, which searches for slave node penetrations in the current configuration, including node-into-face, node-into-analytical rigid surface, and edge-into-edge penetrations. The penalty stiffness that relates the contact force to the penetration distance is chosen automatically so that the effect on the time increment is minimal yet the penetration is not significant [9].

2.3   The Model for Structural Analysis of key Components

Referring to the relative position shown in Fig.1, we assembly all the key components in ABAQUS/CAE module, as shown in Fig.5.

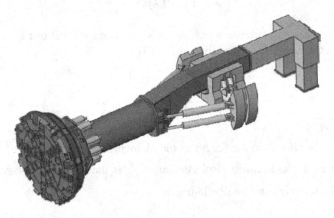

Fig. 5. The assembled components

As the cutterhead, cutterhead support and the main beam are the main stressed components in tunneling, we set them as the deformable body. In order to improve the efficiency of solving, the other components, marked in gray in Fig.5, are set as the display body which specify that a part instance should be used for display purposes only and should not take part in the analysis.

The coupling constraint was employed to describe the connection between cutterhead and cutterhead support. The kinematic method, which constrains the motion of the coupling nodes to the rigid body motion of the reference node was chosen as the coupling type. Once any combination of displacement degrees of freedom at a coupling node is constrained, degrees of freedom at the coupling nodes will be eliminated. As cutterhead and cutterhead support are connected by the high strength bolts and no relative movement is allowed, we constraint all degrees of freedom of the coupled nodes. The method for the connection of cutterhead support and main beam is the same to the above, relative description is omitted.

At a certain moment of tunneling, all degrees of freedom of disc cutter ring are fixed due to its penetration into the rock; the trust and torque for rock breaking are respectively provided by push cylinders and traction motors. Therefore, the loads and boundaries applied on these components in structural analysis are as follows :

(1) Constrain all degrees of freedom of every cutter ring;

(2) Apply the moment in tunneling direction to the installing holes of traction motors on cutterhead support;

(3) Impose the force in tunneling direction on the junction of push cylinders and main beam.

3   Results and Discussion

3.1   Results of Tunneling Simulation

Based on the model introduced in Section 2.2, we simulated the dynamic process when TBM tunneled through granite with the penetration of 12mm, some dynamic results have been obtained from the simulation.

Morphology on Tunneling Face

At the initial stage of excavation, cutters start to contact with the tunneling face, local damage firstly occurs on the rock under cutters, as shown in Fig.6 (a). With the cutterhead moving forward, the enhanced interaction between cutters and rock leads to the damage propagation which is controlled by Eq.3. Fracture is then allowed to occur when D=1.0 and the concerned elements are deleted from the mesh according to the separation criteria. With tunneling forwards, a series of concentric grooves matching with the rolling path of the cutters occur on the tunneling face, as shown in Fig.6 (b). This morphology is consistent with the phenomenon described by Song [10]. After the boring depth reaches 9.7 mm, the damaged area extends from local to the whole tunneling face, as shown in Fig.6 (c), and the overall loading-carrying capacity of rock on tunneling face reduce to none.

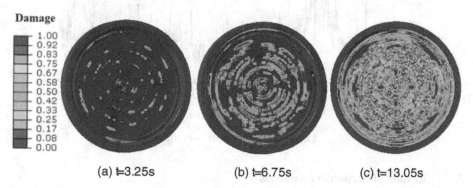

(a) t=3.25s   (b) t=6.75s   (c) t=13.05s

Fig. 6. Damage contour of rock at different time.

Trust and Torque

The excavation loads are the most complicated component which provide a real-time response to geological conditions and operating parameters [11]. Figs.7 and 8 respectively show the corresponding time series curves for trust and torque in working conditions.

During the whole tunneling process, the loads fluctuate severely due to the failure mechanism of quasi-brittle material. Both the trust and toque gradually increase at the beginning due to the accelerated motion. Fifteen seconds later, loads come into a relatively steady stage in which the effective value of the thrust and torque are respectively 6223kN and 6163kN·m. Based on the design parameters of equipment, the ultimate thrust is 15435kN and the rated moment is 6713kN·m. Comparing the simulation results with the design value, the calculated thrust account for about 40.3% of the ultimate force and the calculated torque account for about 91.8% of the rated moment in this case.

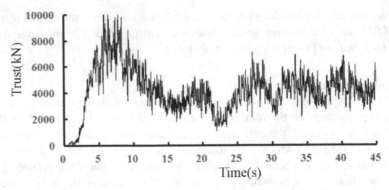

Fig. 7. Curve of trust with time.

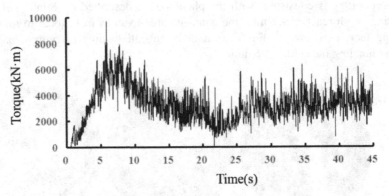

Fig. 8. Curve of torque with time.

3.2   Results of Structural Analysis

Apply the trust and torque obtained from tunneling simulation to the model introduced in Section 2.3, then we can get the mechanical properties of the equipment system consists of the cutterhead, cutterhead support and main beam through structural analysis.

Stiffness, depending on materials, structures and size, is of great importance to the structural to resist the deformation when TBM bores into the rock. In this simulation, the stiffness of the structure is demonstrated by the displacement obtained when TBM works in the steady state. Fig.9 shows the displacement distribution on the whole system, from which we can see that main beam has the relatively large deformation and the max displacement occurs at the junction of push cylinders and main beam, with the max value of 1.96mm, which is a minute quantity comparing to length of the equipment. So the structural has enough stiffness to tunnel the rock.

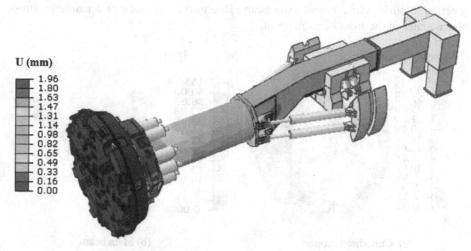

Fig. 9. Displacement distribution of the structural.

To so large welded steel structure, the stress distributions of the key components are analyzed in this simulation to find where is most vulnerable to stress failure.

Fig.10 shows the Mises stress distribution on the cutterhead. Affected by the layout of disc cutters, the stress on the cutterhead panel distributes in the radial direction; the high stress appears in the intersection of corbel and cutter head panel, with the max stress 237.74MPa.

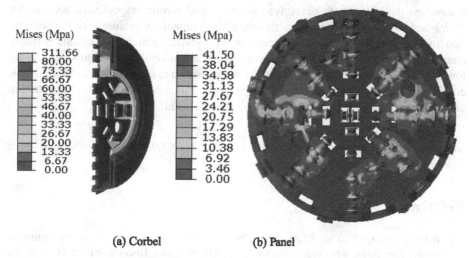

(a) Corbel   (b) Panel

Fig. 10. Mises stress distribution of cutterhead

Figs.11 shows the stress state of cutterhead support and main beam. Due to the moment provided by the traction motors, the stress on the cutterhead support distributes in the circumferential direction. As to the main beam, high stress occurs at the junction of push cylinders and main beam. Due to the existence of a manhole, stress concentrates in the front of main beam.

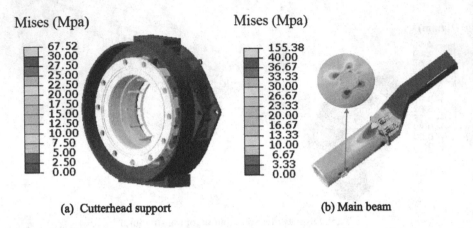

(a) Cutterhead support (b) Main beam

Fig. 11. Mises stress distribution of cutterhead support and main beam

4 Conclusion

Dynamic simulation of cutterhead tunneling is performed to calculate the excavation loads and structural analysis is made to investigate the mechanical properties of the key components when TBM works in the normal condition. The results show that excavation loads fluctuate severely; the thrust and torque respectively account for 40.3% and 91.8% of the design value when TBM tunnels in granite with the penetration of 12mm. The maximum deformation is 1.96mm, which is a minute quantity comparing to the length of the structural. All the stress on the key components is far lower than the ultimate strength. Consider fatigue damage caused by alternating loads, corner and high welded strength should be applied to reduce the stress concentration and further reinforce the structure in the high stress point.

Acknowledgment. This work is supported by grant 2013CB035402 of the National Basic Research Program of China.

References

1. Wang, L., Kang, Y., Cai, Z., Zhang, Q., Zhao, Y., Zhao, H., Su, P.: The energy method to predict disc cutter wear extent for hard rock TBMs. Tunn. Undergr. Space Technol. **28**, 183–191 (2012)
2. Ramoni, M., Anagnostou, G.: Tunnel boring machines under squeezing conditions. Tunn. Undergr. Space Technol. **25**, 139–157 (2010)
3. Hassanpour, J., Rostami, J., Zhao, J.: A new hard rock TBM performance prediction model for project planning. Tunn. Undergr. Space Technol. **26**, 595–603 (2011)

4. Xia, Y., Wu, Y., Wu, F., Liu, W., Lin, L.: Study on simulation and mechanical properties of tunnel shield cutterhead. Computer Engineering and Applications **49**, 248–251 (2013). (in Chinese)
5. Tao, L., Ning, X., He, F.: Finite element analysis for main beam of open-style TBM. Mechanical Engineering and Automation, 43–47 (2013). (in Chinese)
6. Qin, S., Li, R., Wang, D., Zhang, Y.: Finite element analysis of TBM ctterhead support based on ANSYS Workbench. Coal Mine Machinery **36**, 238–240 (2015). (in Chinese)
7. Naderi-Boldaji, M., Alimardani, R., Hemmat, A., Sharifi, A., Keyhani, A., Tekeste, M.Z., Keller, T.: 3D finite element simulation of a single-tip horizontal penetrometer–soil interaction. Part I: Development of the Model and Evaluation of the Model Parameters Soil Tillage Res. **134**, 153–162 (2013)
8. Abo-Elnor, M., Hamilton, R., Boyle, J.T.: 3D Dynamic analysis of soil–tool interaction using the finite element method. J. Terramech. **40**, 51–62 (2003)
9. Hibbitt, Karlsson & Sorensen, Inc., ABAQUS Analysis User's Manual Help Online, ABAQUS Documentation
10. Song, K., Wang, B.: Operation principle analysis of disc cutter on TBM. Construction Machinery, 71–74 (2007). (in Chinese)
11. Zhang, Q., Qu, C., Cai, Z., Kang, Y., Huang, T.: Modeling of the thrust and torque acting on shield machines during tunneling. Autom Constr. **40**, 60–67 (2014)

Research on the Soft-Starting Characteristics of Wet Clutches in TBM Cutter-Head Driving System

Haibo Xie, Huasheng Gong[✉], and Huayong Yang

The State Key Laboratory of Fluid Power Transmission and Control,
Zhejiang University, Hangzhou, Zhejiang, China
{hbxie,ghs,yhy}@zju.edu.cn

Abstract. This article proposes to apply wet clutches to the cutter-head driving system of tunneling boring machine (TBM) and also investigates the soft-starting transmission characteristics of the wet clutch under three different strategies. The strategies discussed here refer to the engagement rules of plates in wet clutches, and they are: positive parabola, straight line, negative parabola. The numerical simulations corresponding to those strategies are carried out in the FLUENT, and the result indicates that the numerical method is valid for the good agreement with the analytical one. After comparing the simulation results, we conclude that the positive parabolic engagement strategy can be preferred during the soft-starting of TBM cutter-head if the invalid starting time, starting smoothness and heat dissipation are regarded as the evaluation standard.

Keywords: TBM · Wet clutches · Soft-starting strategy · FLUENT simulation

1 Introduction

The tunnel boring machine (TBM) is a kind of large complex equipment integrating mechanics and intelligent control. It is usually employed in many engineering fields, such as railway tunnel, highway tunnel, water conservancy and so on. However, some problems should also be paid more attention, especially for the lack of ability to escape when the cutter-head is trapped by the collapsed rock. If this happens, larger driving torque will be needed. What's more, more time will be wasted and economic losses are considerable. So how to help the TBM out becomes an engineering challenge.

According to the above problem, we propose a novel design for the cutter-head driving system. As we know, the cutter-head is justly driven by several variable-frequency motors in conventional driving system, so its maximum output torque cannot be changed once the manufacture of TBM is finished. Thus it will not help at all if larger torque is needed. However, the new design in Fig. 1 is definitely different. The wet clutches are applied to replace parts of variable-frequency motors in the original mechanical structure. When the TBM cutter-head runs normally, the wet clutches can work with the variable-frequency motors (Motor 2 in Fig. 1) synchronously. If the cutter-head is trapped, more separator and friction plates can be added to the wet

H. Liu et al. (Eds.): ICIRA 2015, Part III, LNAI 9246, pp. 456–468, 2015.
DOI: 10.1007/978-3-319-22873-0_41

clutches so that larger output torque will be obtained. So our research focuses on the working characteristics of wet clutches during the starting of head-cutter.

Wet clutches work through shear stresses of lubricant film and have a broad field in engineering application, such as belt conveyor, water pump, fans, cars, etc. As is shown in Fig. 1, when the pressure of control oil acting on the piston is increased, the separator plates will be pushed to the friction plates by the piston, and the thickness of the oil film between the plates will be smaller. As a result, the output torque of wet clutches will become larger based on the theory of hydro-viscous transmission, then the cutter-head will obtain larger power to help the TBM to continue working. This process is usually called soft-starting for wet clutches, and it can help cutter-head to start smoothly.

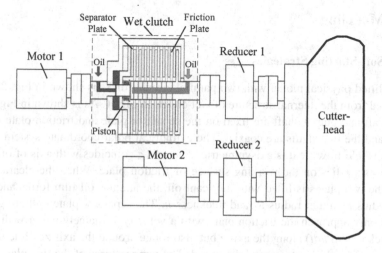

Fig. 1. The driving system of TBM cutter-head

In order to achieve a better understanding of the engagement process in wet clutches, many studies are launched. Andew M. Smith[1] focuses on the shearing mechanism of lubricant film in wet clutches. N.B. Naduvinamani et al.[2] devote themselves into the non-Newtonian fluid, and they not only establish mathematical models of squeeze Rabinowitsch fluid in circular stepped plates but also investigate the load-carrying capacity and response time. Mikael Holgerson[3] weakens the torque fluctuation and reduces the temperature rise of lubricant oil film by changing the normal force acting on the piston in wet clutches of cars. The squeezed lubricant film is studied through a numerical method by Khalid Zarbane[4], where some behaviors of squeezed film are observed and the results show that squeezing frequency and average film thickness can have an influence on the load-carrying capacity. M. Mahbubur Razzaque and Takahisa Kato[5] investigate many factors affecting the squeezing process under the isothermal and no slip boundary conditions. It is found that the angular orientation can significantly affect the squeezing process. Xie FW et al.[6] investigate the distribution and variation of the oil film temperature in hydro-viscous drive (HVD) by the means of simulation and experiment. It is showed that the temperature field will be affected by the grooves on the plates and parallelism of the working surfaces.

The rest of the paper is organized as follows. Section 2 introduces the axisymmetric physical model and three kinds of soft-starting strategies, and then the governing equations and their boundary conditions are described. The numerical simulations in the FLUENT are carried out in section 3, including physical parameter settings and the verification of numerical methods. And also the variations of film temperature and the output torques are analyzed and compared in this section. Finally some conclusions are drawn and future works are introduced.

2   Modeling

2.1   Soft-Starting Strategies

A simplified physical model with two parallel circular disks is shown in Fig. 2, and it is derived from the internal structure of wet clutch in Fig. 1. As is shown in Fig. 2, the input shaft and output shaft are fixed on the separator plate and friction plate respectively, and the two shafts are coaxial. The cylindrical polar coordinate system is used and the whole flow field is axisymmetric. The axis z coincides with axis of one shaft and the axis r lies on the working surface of friction plate. When the clearance between the two plates is filled with lubricant oil, the annular oil film forms and it has inner radius R_1, outer radius R_2 and thickness h. The separator plate without grooves can not only approach the friction plate with a velocity V (namely the growth rate of film thickness $-dh/dt$) along the axis z but also rotate around the axis z, while the friction plate can only rotate around the axis z. The temperatures of the two plates are T_f and T_s, respectively. The inlet temperature and inlet pressure of oil film are T_1 and p_1, and the outlet ones are T_2 and p_2. If the separator plate starts to rotate at an angular speed ω_1, it can drive the friction plate to rotate at an angular speed ω_2 relying on viscous shear stresses of oil film.

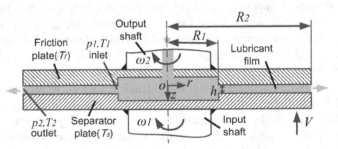

Fig. 2. Schematic drawing of a simplified wet clutch model

In order to find a better way to improve the dynamic working characteristics of cut-ter-head driving system during the soft-starting, the oil film thickness between two plates must be controlled well. The present study adopts three strategies for the varia-tions of oil film thickness: positive parabolic strategy, linear strategy, negative para-bolic strategy. They can be expressed as Eq. 1~Eq. 3. By the way, the 'positive' means that the coefficient of quadratic term is positive, and it is the same to 'nega-tive'.

$$h_{positive}(t) = a_1 t^2 + b_1 t + c_1, \quad 0 \le t \le t_0 \tag{1}$$

$$h_{linear}(t) = a_2 t + b_2, \quad\quad 0 \le t \le t_0 \tag{2}$$

$$h_{negative}(t) = a_3 t^2 + b_3 t + c_3, \quad 0 \le t \le t_0 \tag{3}$$

where $a_1 > 0, a_2 < 0, a_3 < 0$. In our research, the total soft-starting time t_0 is set to 0.5s and the oil film thickness h decreases gradually from 0.2mm to 0.05mm during the time. Thus the unknown coefficients in Eq. 1~Eq. 3 can be solved with above settings and the equations are illustrated in Fig. 3. It is noteworthy that when the three strate-gies are tested respectively, their working conditions are kept the same, such as total starting time, initial and boundary conditions of oil film.

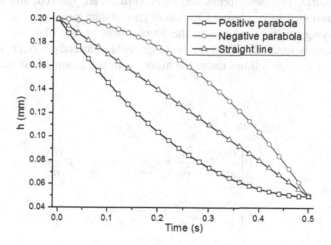

Fig. 3. Three kinds of variations of film thickness

In fact, what really should be concerned about is the growth rate of film thickness, namely the relative approaching velocity V (-dh/dt) of the two plates shown in Fig. 4. And we can conclude that the growth rate of film thickness is constant for the linear strategy, while the others' vary linearly.

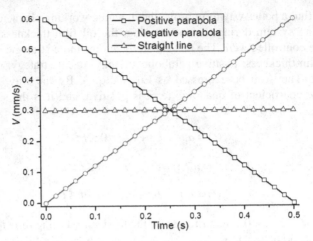

Fig. 4. Relative approaching velocity of the two plates

2.2 Governing Equations

In this paper, the lubricant flow between the plates is treated as Newtonian, laminar and axisymmetric. The mass forces and inertia effects are ignored, and the dynamic viscosity is assumed to be constant for small temperature rise. The film thickness is factually varying all the time during the engagement of wet clutches, so the soft-starting process is unsteady. Then the incompressible unsteady Navier-Stokes equations can be reduced as follows under the aforementioned assumptions and three ones in [7]:

$$\frac{\partial v_r}{\partial r} + \frac{v_r}{r} + \frac{\partial v_z}{\partial z} = 0 \tag{4}$$

$$\rho \left[\frac{\partial v_r}{\partial t} + v_r \frac{\partial v_r}{\partial r} + v_z \frac{\partial v_r}{\partial z} \right] = -\frac{\partial p}{\partial r} + \mu \frac{\partial^2 v_r}{\partial z^2} \tag{5}$$

$$\rho \left[\frac{\partial v_\theta}{\partial t} + v_r \frac{\partial v_\theta}{\partial r} + v_z \frac{\partial v_\theta}{\partial z} \right] = \mu \frac{\partial^2 v_\theta}{\partial z^2} \tag{6}$$

$$\frac{\partial p}{\partial z} = 0 \tag{7}$$

which includes density ρ, dynamic viscosity μ and velocity components (v_r, v_θ, v_z) in the r, θ and z directions. For the sake of simplicity, the angular speed of the friction plate is set to zero, namely $\omega_2 = 0$. Then the boundary conditions for velocity are concluded as follows:

$$v_r = 0, v_\theta = 0, v_z = 0 \text{ at } z = 0;$$
$$v_r = 0, v_\theta = r\omega_1, v_z = V \text{ at } z = h(t).$$

Because the outlet of the oil film contacts the atmosphere directly, the outlet pressure is zero, while the inlet pressure is kept constant p_0 by regulating the output pressure of pressure relief valve. Then the pressure boundary conditions can be expressed like that:

$$p = p_1 = p_0 \text{ at } r = R_1 ;$$
$$p = p_2 = 0 \text{ at } r = R_2 .$$

During the soft-starting process of TBM cutter-head, plenty of useless heat will generate due to the viscous shear stresses in lubricant oil. So the energy equation of the axisymmetric flow must be solved so that we can get a further knowledgement of the temperature variation of oil film. And the energy equation can be described by the following form based on the above assumptions:

$$\frac{\partial T}{\partial t} + v_r \frac{\partial T}{\partial r} + v_z \frac{\partial T}{\partial z} = k \frac{\partial^2 T}{\partial z^2} + \mu \left[\left(\frac{\partial v_\theta}{\partial z} \right)^2 + \left(\frac{\partial v_r}{\partial z} \right)^2 \right] \tag{8}$$

where k is thermal conductivity and the second term on the right side represents the viscous dissipation. It has been verified that the condition ($T_s = T_f = T_l =$ constant) can make the simulation results closer to the experimental results in [8], and that is still used in our research. So the temperature boundary conditions in axial and radial direction are:

$$T_s = T_{s0} \text{ at } z = h(t), \qquad T_f = T_{f0} \text{ at } z = 0 ;$$
$$T_1 = T_{10} \text{ at } r = R_1, \qquad T_2 = T_{20} \text{ at } r = R_2 .$$

where $T_{s0} = T_{f0} = T_{10} =$ constant and T_{20} is the atmosphere temperature, and they are all kept invariant.

3   Numerical Simulations in FLUENT

Before the soft-starting of cutter-head, the film thickness between plates in wet clutches will be kept at an appropriate value (namely initial film thickness h_0) under which the clutch cannot work at all. Then the input shaft will be driven by the motor at a desired speed ω_1. The engagement of plates will happen after the oil film is steady. So a steady simulation must be carried out firstly with invariant film thickness h_0 and the results will be regarded as the initial conditions of the unsteady process. After that, the three strategies can be employed respectively under the same initial conditions.

3.1   Methodology and Parameters

The 3D oil film model is meshed in the ICEM and then imported into the FLUENT where the numerical simulation is performed. The momentum equations in Section 2.2

are discretized by the finite volume method, while the energy equation is discretized through the second order upwind. The SIMPLEC algorithm is used to solve the discretized equations and the axial motion of friction plate is realized by programming user-defined functions (UDFs). Here are the physical parameters listed in Table 1.

Table 1. Physical parameters

Parameter	Value	Unit
R_1	0.06	m
R_2	0.1	m
T_{10}	303	K
T_{20}	298	K
T_{f0}	303	K
T_{s0}	303	K
ρ	872	Kg/m^3
M	0.0439	Pa $\cdot$ s
ω_1	1500	r/min
ω_2	0	r/min
h_0	0.2	mm

3.2   Verification

In order to verify the rationality of numerical method, the numerical solution of pressure will be compared with the analytical one. The simplified pressure equation can be derived from the generated Reynolds' equation[9], and it is as follows:

$$h^3 \frac{1}{r} \frac{\partial}{\partial r}\left(r \frac{\partial p}{\partial r} \right) = \frac{3}{5} \rho h^3 \omega_1^2 + 12 \mu \frac{dh}{dt} \tag{9}$$

Then the analytical equation of pressure will be obtained by integrating Eq. 9 with respect to r from R_1 to r twice, and the result can be gotten if taking the pressure boundary conditions into account:

$$p = \frac{1}{4} K(r^2 - R_2^2) + C_1 \ln \frac{r}{R_2} \tag{10}$$

where $C_1 = [p_1 - \frac{1}{4} K(R_1^2 - R_2^2)] / \ln \frac{R_1}{R_2}$, $K = \frac{3}{5} \rho \omega_1^2 + \frac{12 \mu}{h^3} \frac{dh}{dt}$.

The numerical and analytical pressure solutions under different strategies, at $r = 76$mm for example, are shown in Fig. 5. The results indicate that the numerical solutions coincide with the analytical one well, which confirms the numerical method is right. Furthermore, we can find that the film pressure can vary with the change of oil film thickness, which will induce different load-carrying capacity of wet clutches.

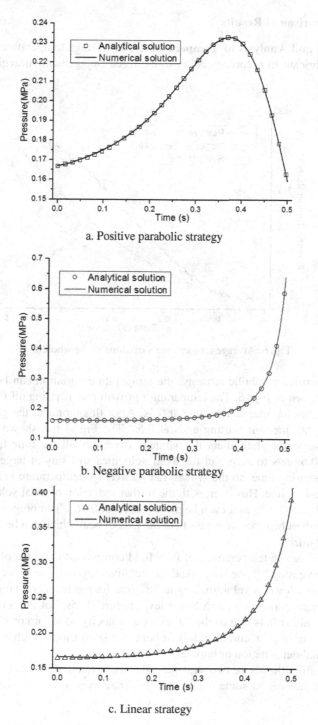

a. Positive parabolic strategy

b. Negative parabolic strategy

c. Linear strategy

Fig. 5. Numerical and analytical solutions of film pressure (r =76mm)

3.3   Comparison of Results

Comparison and Analysis of Temperature. The average temperature variations of the whole lubricant film corresponding to the three engagement strategies are shown in Fig. 6.

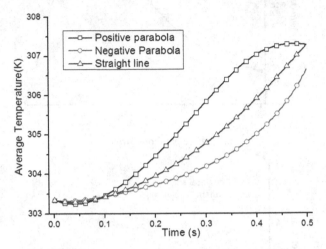

Fig. 6. Average temperature variations of the whole film

For the positive parabolic strategy, the temperature variation can be divided into two stages shown in Fig. 6. The temperature growth rate is rising all the time before 0.3s and reaches the maximum value at 0.3s. After this moment, the growth rate begins to fall until the soft-starting ends at 0.5s. The Fig. 4 has shown that the axial velocity of separator plate about this strategy decreases all the time from the maximum value 0.6mm/s to zero, but the axial velocities still stay at large values in the first part of starting time, so the growth rate of average temperature keeps rising during this period of time. However, with the further reduction of axial velocity, the temperature will rise slowly and even begin to fall from 0.46s. The proper explanation is that the growth of heat becomes less than the loss of heat which can be brought out by the lubricant oil.

While the average temperatures of flow field corresponding to the other two strategies keep increasing all the time. And the relative approaching velocity of the two plates stays at a lower level at the beginning time for the negative parabolic strategy, so its temperature also stays at a lower level before 0.25s, but the temperature rises rapidly from about 0.4s due to the larger axial velocity of separator plate. However, this process can often accumulate lots of heat in a short time, which can result in the plates' thermal deformation or burnout.

Different from the above two strategies, the temperature rise under the linear strategy is much steadier to some extent and this may own to its invariant engagement velocity.

Furthermore, the Fig. 7 illustrates the temperature variations of the annular oil film under positive parabolic strategy at different moments. We can find that the trend of temperature variation on one section of oil film is similar to the average trend, and the temperature rise will become larger with the increase of the radius.

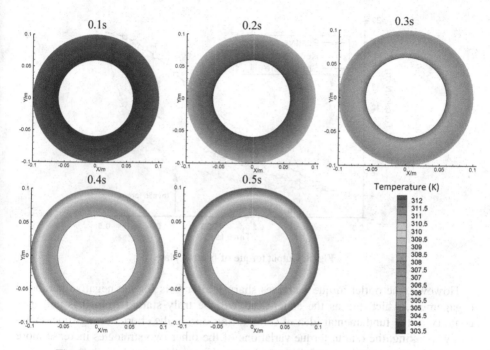

Fig. 7. Temperature variations of positive parabolic engagement at z=0.03mm

Comparing the temperature variations of three kinds of engagements, the positive parabolic strategy or linear one will be preferred. The results of linear strategy (which has constant engagement velocity) have indicated that the temperature of the oil film will rise with the reduction of film thickness, and more heat will generate during this period. So it is significant for the separator plate to slow down or keep a slow speed when the oil film becomes smaller and smaller, such that more heat will be brought out without damaging the plates.

Comparison and Analysis of Torque. The variations of viscous torque acting on the friction plate are illustrated in Fig. 8 and their growth rates are in Fig. 9.

In the practical application of wet clutches, the starting time during which the output torque is less than 30% of the required maximum value is defined as the invalid starting time. So the shorter the invalid starting time is, the higher soft-starting efficiency the TBM cutter-head driving system will obtain. We can find that the three strategies almost have the same maximum torques 18N·m in Fig. 8, which may result from the same oil film thickness at 0.5s. The invalid output torque can be calculated as about 6N·m based on the above definition. Thus for negative parabolic strategy, the invalid starting time is the longest lasting for approximately 0.3s, while the positive parabolic one has the shortest lasting for 0.12s and the linear strategy has 0.18s. Obviously, the positive parabolic strategy can promote the soft-starting transmission efficiency.

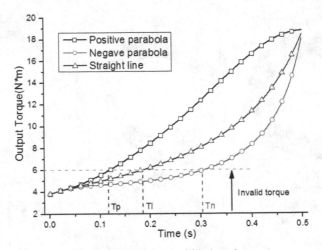

Fig. 8. Output torque of friction plate

However, the outlet torque increases sharply after 0.4s for the negative parabolic engagement, which means the cutter-head will be truly started from 0.4s and it is contrary to the fundamental idea of soft-starting due to the short starting time. Relatively speaking, the output torque variations of the other two strategies increase more steadily than the negative parabolic one during soft-starting.

The growth rate of output torque of the positive parabolic strategy in Fig. 9 can be divided into two stages: growing to the maximum value 44.5N·m/s and then falling to 2.0N·m/s gradually. That is to say, the output torque of wet clutches rises rapidly before 0.35s and rises slowly from 0.35s to the end. Moreover, we can also find that its growth rate of output torque stays at the highest level among the three during the time of 0~0.38s. However, the growth rates of output torques for both negative parabola and straight line are growing with time. The one of negative parabolic strategy rises most slowly before 0.4s and the rises sharply from about 0.42s, while the linear strategy grows smoothly by contrast.

For the soft-starting of TBM cutter-head driving system, it is preferred that the output torque grows quickly in the beginning time and then grows slowly before the ending time. This process can help to reduce invalid starting time and avoid wasting more power, moreover it can also help to keep the output torque vary smoothly before the soft-starting finishes. In conclusion, the positive parabolic engagement strategy is the perfect one absolutely at this point, and the mechanical driving system of TBM can start smoothly and suffer less from rigid impact with it.

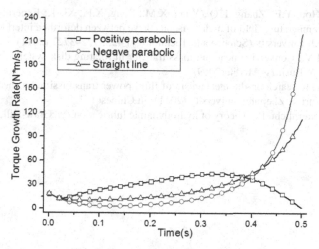

Fig. 9. Growth rate of output torque

4   Conclusions

The oil film behaviors and soft-starting transmission characteristics of wet clutches applied in TBM cutter-head driving system are studied under different soft-starting controlling strategies. On one hand, the positive parabolic engagement strategy has the highest soft-starting efficiency due to its shortest invalid starting time, and it can also help the cutter-head start smoothly. On the other hand, the negative parabolic engagement can produce a large temperature growth rate in a short time, which may burnout the plates due to heat accumulation, while the other two strategies have smaller temperature rises. In summary, the positive parabolic engagement strategy is the excellent way for the soft-starting of TBM cutter-head driving system. In our future work, the stabilities of transmission during the soft-starting of cutter-head will be treated as the key point, such as whether the turbulent flow can occur in oil film and how to control it.

References

1. Smith, A.M.: Hydrodynamic lubrication theory in rotating disk clutches. University of Notre Dame (1997)
2. Naduvinamani, N., Rajashekar, M., Kadadi, A.: Squeeze film lubrication between circular stepped plates: Rabinowitsch fluid model. Tribology International **73**, 78–82 (2014)
3. Zarbane, K., Zeghloul, T., Hajjam, M.: Optimizing the smoothness and temperatures of a wet clutch engagement through control of the normal force and drive torque. Journal of tribology **122**, 119–123 (2000)
4. Holgerson, M.: A numerical study of lubricant film behaviour subject to periodic loading. Tribology International **44**, 1659–1667 (2011)
5. Razzaque, M.M., Kato, T.: Squeezing of a Porous Faced Rotating Annular Disk Over a Grooved Annular Disk. Tribology Transactions **44**, 97–103 (2001)

6. Xie, F.W., Hou, Y.F., Zhang, L.Q., Yuan, X.M., Song, X.F., Xi, T.: Experimental research on oil film temperature field of hydro-viscous drive between deformed interfaces. Journal of Central South University (Science and Technology) **42**, 3722–3727 (2011)
7. Shevchuk, I.V.: Convective heat and mass transfer in rotating disk systems, vol. 45. Springer Science & Business Media (2009)
8. Huang, J.H.: Research on the mechanism of fluid power transmission by shear stress in hydro-viscous drive. Zhejiang University (2011) (in Chinese)
9. Pinkus, O., Sternlicht, B.: Theory of hydrodynamic lubrication. McGraw-Hill, New York

Sensitivity Analysis of Major Equipment Based on Radial Basis Function Metamodel

Xin Ding, Wei Sun, Lintao Wang, Junzhou Huo, Qingchao Sun,
and Xueguan Song[⊠]

School of Mechanical Engineering, Dalian University of Technology, Dalian 116024, China
wwwdingxin@163.com, {weisun,sxg}@dlut.edu.cn

Abstract. A major equipment is generally an extraordinarily large and complex machine containing lots of parameters, understanding and assessment the performance of a major equipment is a great challenge as each of these parameters has uncertainty. In this work, a metamodel-based global sensitivity analysis (MBGSA) method is proposed for understanding the influence of the parameters on the different outputs. The MBGSA consists of global sensitivity analysis (GSA) method to identify the impact of each parameters on each of the outputs, moderate-fidelity computational models to mimic the physical model, and metamodeling technique that constructs the mapping between input and output with a limited sampling data. An example on applying the MBGSA in the dynamics analysis of a tunnel boring machine's driving system is presented, where hierarchical dynamics model, radial basis function (RBF) metamodel and Sobol's GSA are bonded together to achieve the aim. The results indicate that the proposed MBGSA is available and efficient for the analysis of TBM and thus can be of great help for other large and complex major equipments at the early stage of design.

Keywords: Major equipment · Global sensitivity analysis · Metamodel · Tunnel boring machine · Dynamics performance

1   Introduction

Major equipment is generally an extraordinarily large and complex machines consisting of many sub-systems and lots of components, the demand for major equipment like tunnel boring machine (TBM) has been rapidly increasing during the past decade. For a major equipment, experimental analysis of a full model or main subsystem/component under real condition is generally not practicable, which results in a lack of understanding and assessment of its realistic performance. Recently, due to the advances in computer hardware and numerical methods, computational simulation has been becoming a popular and promising method/tool in the early stage of a major equipment design to assess its various performances such as structural strength, fluid characteristics, dynamics performance and control characteristics. Much work has been done in an effort to understand the complex behavior of various major equipment and considerable information of value has been obtained [1-4]. However, most

H. Liu et al. (Eds.): ICIRA 2015, Part III, LNAI 9246, pp. 469–480, 2015.
DOI: 10.1007/978-3-319-22873-0_42

of previous work falls into the category of conventional deterministic analysis, in which the effect of statistical variation has not been taken into account. On the contrary, there are usually many parameters involved in even one sub-system or one component, and most of them are associated with variation due to the uncertainties of material properties, manufacturing errors and assembling tolerance. Therefore, it is inadequate to investigate a major equipment in a small and deterministic design space, a method to deal with these uncertainties involved and to identify the most important uncertain parameters is necessary for better understanding and further optimization of a major equipment.

Global sensitivity analysis (GSA) is a promising approach to achieve it as it is capable of identifying the effect of input parameter's uncertainty/variation on output response [5, 6]. Among these GSA methods [7-10], Sobol's method has been widely used recently in a variety of fields such as geologic, chemical and mechanical engineering [11-13]. The availability and reliability of Sobol's GSA have been verified through these research. Nevertheless, the efficiency of Sobol's method has also been found not adequately high as Sobol's method necessitates a large number of calculations, which becomes a major barrier for Sobol's GSA to apply in computationally expensive simulation such as the dynamics simulation of TBMs.

Metamodeling is an alternative to overcome the difficulty mentioned. Metamodeling is able to approximate the implicit relationship between input and output with a limited number of computational simulation, and thus efficient to infer output values at yet-unseen input points. To date, there has been a wide variety of metamodels developed and applied in various engineering problems [14-18]. Metamodels such as polynomial response surface (PRS) model, Kriging model and Radial Basis Function (RBF) model have been proven reliable and efficient in many application. The present study is focusing on a RBF model for constructing the mapping between the input and output, as long as its errors or metrics meet the accuracy requirements.

Therefore, the aim of this paper is to develop a novel kind of metamodel-based GSA (MBGSA) methodology for uncertainty analysis of major equipments. It demonstrates how a large and complex major equipment can be investigated through the proposed process. The rest of the paper is organized as follows. Section 2 describes the MBGSA process which consists of GSA and a typical metamodel, RBF model. Section 3 gives an example of applying MBGSA in dynamics analysis of a practical TBM. Section 4 draws some conclusions.

2   Metamodel-Based Global Sensitivity Analysis (MBGSA)

2.1   Sobol' Global Sensitivity Analysis

A number of GSA methods have been developed, the Sobol's method is adopted in this paper. The Sobol' method is a variance-based Monte Carlo method that allows the computation of both the sensitivity indices of individual parameters, Si, and the interactions between these parameters through the ratio of each sensitivity index to the corresponding total sensitivity index, S_i^{tot} [10,19]. Specifically, considering an integrable function, $f(\mathbf{x})$, which can be written in the form:

$$f(\mathbf{x}) = f_0 + \sum_{s=1}^{n} \sum_{i_1 < \cdots < i_s}^{n} f_{i_1 \cdots i_s}\left(x_{i_1}, \ldots, x_{i_s}\right) \tag{1}$$

where $1 \le i_1 < \cdots < i_s \le n$, $f_{i_1 \cdots i_s}\left(x_{i_1}, \ldots, x_{i_s}\right)$ is a function of a unique subset of variables from $\mathbf{x}$. Assume now that $f(\mathbf{x})$ is square integrable, then the following equation can be obtained,

$$D = \int f^2(\mathbf{x})d\mathbf{x} - f_0^2 = \sum_{s=1}^{n} \sum_{i_1 < \cdots < i_s}^{n} \int f^2_{i_1 \cdots i_s} dx_{i_1} \, dx_{i_s} = \sum_{s=1}^{n} \sum_{i_1 < \cdots < i_s}^{n} \int D_{i_1 \cdots i_s} \tag{2}$$

where $D_{i_1 \cdots i_s}$ is the partial variance in the model response brought about by simultaneous changes in factors i1 to is and D is the total variance of $f(x)$. Now, Let y denote the set of m variables ($1 \le m \le n-1$), and Z denote the set of n-m complementary variables, thus $x = (y, z)$, let $K = (k_1, \cdots, k_m)$. The variance corresponding to the subset y can be defined as

$$D_y = \sum_{s=1}^{m} \sum_{(i_1 < \cdots < i_s) \in K}^{m} \int D_{i_1 \cdots i_s} \tag{3}$$

Therefore the individual sensitivity indices (or main sensitivity indices) and the total sensitivity indices can be calculated as follows,

$$S_y = \frac{D_y}{D} \tag{4}$$

$$S_y^{tot} = \frac{D_y^{tot}}{D} \tag{5}$$

2.2   Radial Basis Function Metamodel

The radial basis function model was originally developed to approximate multivariate functions based on scattered data [20]. For a data set consisting of input variables and responses at n sampling points, the typical RBF model takes the following form:

$$y(\mathbf{x}) = \sum_{s=1}^{S} \lambda_s \phi(\|\mathbf{x} - \mathbf{x}_s\|) + \sum_{k=1}^{K} C_k p_k(\mathbf{x}) \tag{6}$$

where S is the number of sampling points, $\|\mathbf{x} - \mathbf{x}_s\|$ is the Euclidean norm between design variables $\mathbf{x}$ and the sth sampling point $\mathbf{x}_s$, ϕ is a basis function, λ_s is the unknown weighting factor positioned at the sth sampling point, $p_k(\mathbf{x})$ are polynomial terms, K is the number of polynomial terms (usually $K < S$), and $C_k (k = 1, 2, \cdots, K)$

is the coefficient for $p_k(\mathbf{x})$. Therefore, an RBF model is actually a linear combination of S RBF and K polynomial terms with the weighted coefficients.

2.3   Implementation of the MBGSA Method

The proposed MBGSA approach consists of the following five steps as shown in Fig. 1. STEP 1: The GSA problem including the input parameters, the ranges of input parameters and the output response is firstly defined. STEP 2: A sampling strategy is determined and a series of training sampling points are generated in the design space, computational simulations are carried out at these sampling points and corresponding response values are collected. STEP 3: Constructing the metamodels based on the training sample data, and checking the accuracy of the metamodels, if the accuracy satisfies the required criterion, then it goes to the next step, otherwise, more sampling data should be added in terms of sequential sampling strategies and new metamodels with more accuracy are constructed. STEP 4: Carrying out GSA based on the meta-models to investigate the effects of input parameters on the output responses.

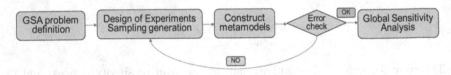

Fig. 1. Flowchat of the metamodel-based GSA methodology

3   Case Study: Dynamics Analysis of TBM's Driving System

A tunnel boring machine is a classical large and complex major equipment, which has been widely used in excavation of tunnels under the ground. TBM based excavation is a very complex and rigorous multidisciplinary and multi-objective engineering problem including mechanical, civil, control, hydraulic and planning problems, whose security has to been guaranteed as the associated risk can be extremely high. It has

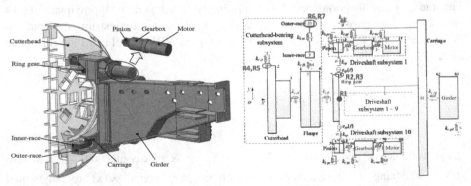

Fig. 2. Structure (left) and dynamics model (right) of the driving system of TBM

been found that the dynamics performance of TBM is mainly determined by the dynamics performance of its drive system, which is a pivotal transmission system shown in Fig. 2 (left). During the operation, the cutterhead is driven by multiple drive units with a uniform distribution on it, each drive unit is composed of a pinion, a planetary reducer and an inverter motor. The main bearing and carriage provide continuous and reliable support for the driving system.

3.1   Hierarchical Dynamics Modelling of the Driving System

The driving system of TBM is composed of lots of components, so high-fidelity modelling containing all information is neither practical nor necessary, a simplified dynamics modelling would be more appropriate considering both the computational burden and accuracy. However, traditional modeling method is difficult to establish the fully coupled model of the cutter head driving system in tunneling boring machine (TBM) owing to the structural complexity. In this paper, a moderate-fidelity model of the driving system is established in terms of hierarchical dynamics theory, which is cited the finite element modeling method. The modeling process is as follows: The driving system is divided into one cutterhead-bearing subsystem and several identical driveshaft subsystems firstly. The main components such as the cutterhead, the flange, the pinion and the motor in each subsystem are equivalent to Timoshenko beam elements with the same mass and stiffness characteristics. The dynamic model of planetary gears is established using lumped mass method [4]. The supporting between the cutterhead and rock and the connections such as that between the cutterhead and flange can be equivalent to the spring-damper elements. Coupling the FE model of cuntterhead-bearing subsystem with particular number of driveshaft subsystems by gear mesh elements, the dynamics model of the whole driving system can be established as shown in Fig. 2 (right) be solved based on the classical dynamics equation:

$$M\ddot{q}(t) + C\dot{q}(t) + Kq(t) = F(t) \tag{7}$$

where, M is the mass matrix; C is the damping matrix; K is the stiffness matrix; F is the force vector including the external additional load; q is the displacement vector of the all nodes.

3.2   Dynamics Results

The dynamic response of the example driving system (Table 1) can be obtained by solving Eq. 7 with the precise time-integration method. Fig. 3 illustrates the response of angular acceleration at the node of ring gear when applying a cyclic torque load of 6000 (kNm) on the first node of cutter head. It can be found that the angular acceleration varies periodically.

Table 1. Main parameters of the example TBM cutterhead driving system

Parameters	Value
Diameter of cutterhead (m)	8.53
Rated speed of cutterhead (r/min)	5.5
Rated torque of cutterhead (kNm)	5695
Tooth number of ring gear	198
Tooth number of pinion	19
Drive ratio of gearbox	29.34
Number of pinions	10

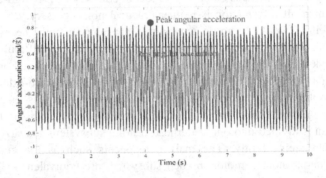

Fig. 3. Angular acceleration at the node of ring gear

3.3   MBGSA of the TBM Driving System

STEP 1: Definition of the GSA Problem

To comparatively assess the dynamics performance of the driving system, it is essential to define the input parameters and output responses based on the dynamics results. Generally the load and acceleration are used to evaluate the dynamics characteristics. In this study, totally seven indicators at two different positions are used as the output responses, they are listed in Table 2.

Table 2. Seven outputs for the dynamics assessment

R1	the peak meshing load of the multiple pinions driving
R2	the peak angular acceleration at the node of ring gear
R3	the RMS angular acceleration at the node of ring gear
R4	the peak angular acceleration at the node of cutterhead
R5	the RMS angular acceleration at the node of cutterhead
R6	the peak angular acceleration at the node of bearing outer-race
R7	the RMS angular acceleration at the node of bearing outer-race

Totally seven variables are selected and defined as the input parameters, they are the average mesh stiffness (x_1), the mesh error (x_2), the gear backlash (x_3), the damping ratio of the gear mesh (x_4), the moment of inertia of the ring gear (x_5), the speed of the cutterhead (x_6), and the torque of the cutterhead (x_7). It's very certain that each of these sevem parameters has uncertainty due to the variation of material property and/or manufacturing errors. However, mathematically representing the exact uncertainty is difficult for this large and complex major equipment as empirical probability distributions requires a large amount of measurements, which is impractical in this case. As a result, an alternative approach is to assume a uniform probability distribution based on expert knowledge. The seven input parameters with the lower and upper bounds of the uncertainties are given in Table 3.

Table 3. Input parameters and their uncertainty

Name	Lower bound	Upper bound
x_1, Average mesh stiffness $\bar{k}_{rp}$ (N/m)	10^8	10^{10}
x_2, Mesh error e_{rp} (m)	10^{-6}	10^{-4}
x_3, Gear backlash b (μm)	100	500
x_4, Damping ratio of gear mesh ζ	0.005	0.03
x_5, Moment of inertia of ring gear ($kg \cdot m^2$)	800	1500
x_6, Speed of cutterhead	5	6
x_7, Torque of cutterhead	4000	8000

STEP 2: Sampling Generation

Having defined the GSA problems, the seven outputs can be calculated by using the dynamics model. The training sample number, n_s should be defined considering both the computational cost and the metamodel accuracy. As there are totally seven input parameters, 100 sample points are generated by using Latin hypercube sampling (LHS) method to explore the domain. Fig. 4 illustrates the distribution of the 100 samples over the uncertainty domain. It could be found that the samples points based on LHS method distribute evenly over the whole domain, which is quite important for the domain exploration. Each dynamics model takes about three minutes to run, and thus this 100 samples takes about five hours to run. Sampling data are then collected for the metamodel construction.

STEP 3: Metamodel Construction and Error Check

As mentioned above, there are many kinds of basic functions in RBF metamodel, and no one will know which one is the best for a particular output, so five widely used basis functions are comparatively used to construct the RBF metamodels for the seven objectives. The accuracy of the metamodels has to be guaranteed so that the subsequent analysis will be not misleading. There are generally two kinds of error checking methods, one is to use the existing initial training sampling data to validate the metamodel, it does not generate extra samples and is thus simple and fast. Of many different methods, cross validation (CV) is a very popular and reliable one. The other one is to use extra test sampling data to validate the constructed metamodel, it will be quite lengthy if each computer run is costly, but it provides more credible and reliable validation information for the metamodel. As each computer run in this case is not very costly, so extra 30 test samples are generated based on LHS method to validate the metamodels. Similarly, an even distribution is good at domain exploration and thus the validation of the global accuracy of the metamodels. For each metamodel, the coefficient R^2 is used to validate the accuracy, the equations is given in Eq. (8).

$$R^2 = \frac{SSR}{SST} = \frac{\sum_{i=1}^{n_t}(\hat{y}(x_i)-\bar{y})^2}{\sum_{i=1}^{n_t}(y_i(x_i)-\bar{y})^2} \tag{8}$$

where n_t is the number of extra sample points for validation. A higher R^2 that close to 1 indicates that the metamodel is accurate to fit the mapping between input and output. A threshold of 50% is used in this case to qualify the predictive quality as satisfactory. Fig. 5 compares the R^2 values using different basis functions, where it can be identified easily that the Thin-plate spline (TPS) function performs best for the first three outputs (R1, R2, and R3), and Gaussian function performs best for the rest four outputs (R4, R5, R6 and R7). It also can be found that RBF with TPS basis function generate satisfactory approximation for the three outputs.

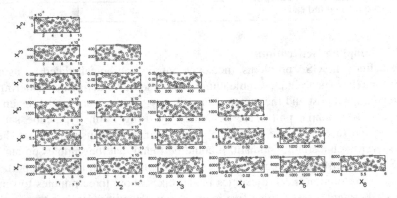

Fig. 4. Initial 100 training sample points over the uncertainty domain

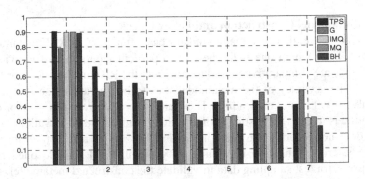

Fig. 5. R^2 comparison with different basis functions

STEP 4: Global Sensitivity Analysis

GSA generally requires a large number of Monte Carlo (MC) calculations to ensure its accuracy. In this case, 20,000 MC calculation are conducted, corresponding to about 35mintus of computation time on a workstation with 3.5GHz CPUs. The same GSA would require 41 days if it is based on the direct dynamics modeling.

Fig. 6 to Fig. 9 provide a guideline to the effects of seven input parameters on the seven outputs by varying them over the uncertainty space, respectively. Specially, it can be found in Fig. 6 that the average mesh stiffness (x_1) is the most significant parameters with an individual effect of about 90% for the peak meshing load of the multiple pinions driving (R1), while the other 6 parameters has little effect on the output. This is due to the fact that the load is obtained mainly in terms of mesh stiffness and displacement along the path of contact.

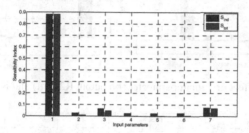

Fig. 6. Effect of 7 input parameters on output R1

For the peak angular acceleration at the node of ring gear (R2) as shown in Fig. 7 (left), the average mesh stiffness (x_1) remains to be the first most important with an individual effect of over 40% and a total effect of over 70%, and the speed of cutterhead (x_6) becomes the second most important factor with an individual effect of over 20% and a total effect of over 40%. The high difference between the individual and total effect means that there is a strong interaction between the average mesh stiffness and the speed of the cutterhead. This is similar to the GSA result for the RMS angular acceleration at the node of ring gear as shown in Fig. 7 (right), where the mesh stiffness and the speed of the cutterhead are the first and second most important parameters, respectively. This is because that increasing the speed of cutterhead leads to the increment of contact excitation frequency between the ring gear and pinions. It is also observed that the total effect of the mesh error (x_3) increases to about 10%, which indicates that the mesh error becomes a non-negligible parameter.

For the peak and RMS angular acceleration at the node of the cutterhead as shown in Fig. 8 respectively, the speed of the cutterhead increases to the first most important parameter, and the mesh stiffness and mesh error become the second and third most important parameters, respectively. The similar phenomenon can be found for the peak and RMS angular acceleration at the node of the bearing outer-race as shown in Fig. 9, the speed of the cutterhead, the mesh stiffness and the mesh error rank the top three most important parameters. Meanwhile, the torque of the cutterhead is found to be influential on the output R6 and R7.

In summary, the mess stiffness, the speed of the cutterhead and the mesh error are the most important parameters, the sum of the individual effect of these three parameters account for more than 70% for the dynamics performance of the TBM's driving system. That is, to achieve a better performance with optimization method, more attentions have to be paid on these three parameters as a small variation of them will lead to a significant variation of the outputs, and the other four parameters can be assumed to be constants while doing further investigation.

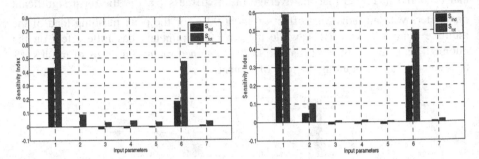

Fig. 7. Effect of 7 input parameters on output R2 (left) and R3 (right)

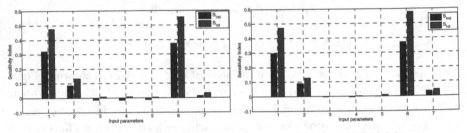

Fig. 8. Effect of 7 input parameters on output R4 (left) and R5 (right)

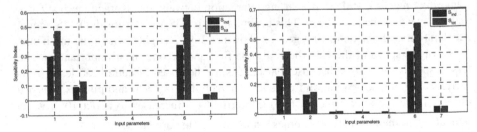

Fig. 9. Effect of 7 input parameters on output R6 (left) and R7 (right)

4   Conclusions

Major equipment involves a large number of input parameters with uncertainty. Their influence and importance have to be taken into account for better understanding and assessing the performance of major equipment when computational simulation is carried out at the early stage of design. A GSA method is a feasible approach to do it, but it usually requires a huge number of direct simulations, which make it less useful and even unfeasible for practical engineering problem. For this reason, a metamodel-based GSA method is proposed to achieve rapid sensitivity analysis of major equipment. The proposed MBGSA combines Sobol's GSA method, metamodeling and computational simulation of major equipment, where metamodeling is used to construct mapping between input parameters and output responses with a limited number of computational simulation, and then GSA can be carried out based on the metamodels within an affordable computational and time cost.

A case study has been given by using the proposed MBGSA, in which the dynamics analysis of TBM's driving system is investigated. LHS method is used to generate 100 training and 30 test samples, the whole driving system is modelled by using hierarchical dynamics method, and Sobol's method is used for the final sensitivity analysis. Totally seven objectives related to the dynamics performance of the TBM's driving system are investigated. The results show that MBGSA is capable of identifying the most important parameters among many different parameters and can help to omit parameters with less effect for the further investigation.

Despite of this, much work such as the definition/assumption of the uncertainty range and selection of better metamodels should be further addressed as expert knowledge on it might be misleading for some cases.

Acknowledgments. The research is supported by National Basic Research Program of China (973 Program, Grant No. 2013CB035402), the Fundamental Research Funds for Central Universities (Grant No. DUT14RC(3) 133 and Grand No. DUT12JN04) and Open Foundation of the State Key Laboratory of Fluid Power Transmission and Control of Zhejiang University of China (Grant No. GZKF- 201414).

References

1. Delisio, A., Zhao, J., Einstein, H.H.: Analysis and prediction of TBM performance in blocky rock conditions at the Lötschberg Base Tunnel. Tunn. Undergr. Sp. Tech. **33**, 131–142 (2013)
2. Zhang, K.Z., Yu, H.D., Liu, Z.P., et al.: Dynamic characteristic analysis of TBM tunneling in mixed-face conditions. Simul. Model. Pract. Th. **18**(7), 1019–1031 (2010)
3. Wei, J., Sun, Q.C., Sun, W., et al.: Load-sharing characteristic of multiple pinions driving in tunneling boring machine. Chin. J. Mech. Eng-En. **26**(3), 532–540 (2013)
4. Sun, W., Ding, X., Wei, J., et al.: An analyzing method of coupled modes in multi-stage planetary gear system. Int. J. Precis. Eng. Man. **15**(11), 2357–2366 (2014)
5. Karkee, M., Steward, B.L.: Local and global sensitivity analysis of a tractor and single axle grain cart dynamic system model. Biosyst. Eng. **106**(4), 352–366 (2010)
6. Saltelli, A., Andres, T., Homma, T.: Sensitivity analysis of model output. Performance of the iterated fractional factorial design method. Comput. Stat. Data. Anal. **20**(4), 387–407 (1995)
7. Saltelli, A., Tarantola, S., Chan, K.: A quantitative model-independent method for global sensitivity analysis of model output. Technometrics **41**(1), 39–56 (1999)

8. Sobol, I.M.: Sensitivity estimates for nonlinear mathematical models. Matematicheskoe Modelirovanie **2**(1), 407–414 (1990)

9. McRae, G.J., Tilden, J.W., Seinfeld, J.H.: Global sensitivity analysis—a computational implementation of the Fourier amplitude sensitivity test (FAST). Comput. Chem. Eng. **6**(1), 15–25 (1982)

10. Sobol, I.M.: Global sensitivity indices for nonlinear mathematical models and their Monte Carlo estimates. Math. Comput. Simulat. **55**(1–3), 271–280 (2001)

11. Rohmer, J., Foerster, E.: Global sensitivity analysis of large-scale numerical landslide models based on Gaussian-Process meta-modeling. Comput. Geost-UK **37**(7), 917–927 (2011)

12. Song, X., Zhang, J., Kang, S., et al.: Surrogate-based analysis and optimization for the design of heat sinks with jet impingement. IEEE Trans. Compon., Packag., Manuf. Technol. **4**(3), 429–437 (2014)

13. Xing, L., Song, X., Scott, K., et al.: Multi-variable optimisation of PEMFC cathodes based on surrogate modelling. Int. J. Hydrogen. Energ. **38**(33), 14295–14313 (2013)

14. Jin, R., Chen, W., Simpson, T.W.: Comparative studies of metamodeling techniques under multiple modeling criteria. Struct. Multidiscip. O. **23**(1), 1–13 (2001)

15. Simpson, T.W., Poplinski, J.D., Koch, P.N., Allen, J.K.: Metamodels for computer-based engineering design: survey and recommendations. Eng. Comput-Germany. **17**(2), 129–150 (2001)

16. Mohammed, F.H., Russel, R.B., Sanjay, B.J.: Metamodeling: Radial basis functions, versus polynomials, computing. Eur. J. Oper. Res. **138**(1), 142–154 (2002)

17. Lee, K.H., Kang, D.H.: Structural optimization of an automotive door using the kriging interpolation method. P. I. Mech. Eng. D-J. Aut. **221**(12), 1525–1534 (2007)

18. Marzbanrad, J., Ebrahimi, M.R.: Multi-objective optimization of aluminum hollow tubes for vehicle crash energy absorption using a genetic algorithm and neural networks. Thin. Wall. Struct. **49**(12), 1605–1615 (2011)

19. Sobol's, I.M., Kucherenko, S.S.: On global sensitivity analysis of quasi-Monte Carlo algorithms. Monte Carlo Methods and Applications **11**(1), 1–9 (2005)

20. Hardy, R.L.: Multiquadric equations of topography and other irregular surfaces. J. Geophys. Res. **76**(8), 1905–1915 (1971)

Finite Element Modeling of Gripping-Thrusting-Regripping Mechanism of TBM Based on SAMCEF

Xiaxin Gao and Yuhu Yang[✉]

Department of Mechanical Engineering, Tianjin University, Tianjin 300072, China
gaoxxtju2013@163.com, yangyuhu@tju.edu.cn

Abstract. Took GTR (Gripping-Thrusting-Regripping) Mechanism of TBM (Tunneling Boring Machine) as research object, a FEM (Finite Element Modeling) method of the mechanism was developed in SAMCEF^{TM}. Based on system analysis of the constitution and operating principle, aimed at different working conditions, the finite element model of host machine was established. Since TBM is a rigid-flexible coupling system, the model took surrounding rock constraints, hydraulic oil stiffness and the key components' elasticity into account. Through simulation, the paper analyzed the stress and deformation of GTR Mechanism. Finally, the model was verified to be valid through engineering examples.

Keywords: GTR Mechanism · FEM · SAMCEF^{TM} software · Rigid-flexible coupling · Simulation

1 Introduction

The GTR Mechanism is a lower DOF spatial parallel mechanism, which has a major impact on the continuous and precise tunneling, rapid and flexible steering of the TBM. Different from normal parallel mechanism, it is incompletely restrained.

Currently, TBM abroad have already been mature, such as the products of American Robbins Company and Germany Wirth Company. However, the research of TBM has just started in China. Some researchers focused on the cutter head of TBM, used ADAMS to build model, simulated and analyzed the installation radius and angle of the cutter, the results provided method for structural design of the cutter head [1-2]. Another part of scholars chose to study on the main beam of TBM, and the simulation results guided design of the main beam [3-4].

Most of the domestic scholars just studied on a single component, while the research of GTR Mechanism was rare. Therefore, this paper chose a single support open TBM as example, based on SAMCEF^{TM}, established the finite element model of GTR Mechanism.

2 Simulation Platform and Research Object

SAMCEF^{TM} is finite element analysis software which can be used to simulate rigid-flexible coupling system. The solving process of SAMCEF^{TM} can be divided into three stages of pre-treatment, solving and post-processing [5]. The flow chart of TBM simulation based on SAMCEF^{TM} is shown in Fig. 1.

© Springer International Publishing Switzerland 2015
H. Liu et al. (Eds.): ICIRA 2015, Part III, LNAI 9246, pp. 481–488, 2015.
DOI: 10.1007/978-3-319-22873-0_43

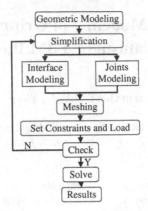

Fig. 1. The flow chart of TBM simulation

A typical GTR Mechanism contains main beam, saddletree, propulsion cylinders, torque cylinders, support cylinders, grippers and rear legs, etc. The structural composition of TBM is shown in Fig. 2.

During tunneling, the support cylinders push the grippers to hold rock tightly, and then the propulsion cylinders drive the cutter breaking rock until finish a tunneling stroke, finally the rear legs land onto ground, the propulsion cylinders and support cylinders retreat quickly and reset the saddletree. This three-steps above cycles and makes continuous tunneling come true. Through manipulating the torque cylinders and support cylinders, the GTR Mechanism could rectify deviation of trajectory.

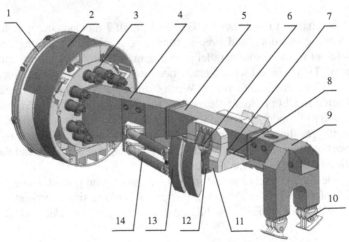

1. Cutter Head 2. Shield3. Reduction Gearbox4. Main Beam I 5. Main Beam II 6. Torque Cylinder 7. Slider 8. Guide9. Main Beam III 10. Rear Leg11. Saddletree12. Support Cylinder13. Gripper14. Propulsion Cylinder

Fig. 2. The structural composition of TBM

3   Finite Element Modeling

3.1   Geometric Model

Delete the components which have little impact on the results, like motors, etc; erase all the technologic holes, chamfering and rounded corners; add surrounding rock to ensure its determined motion. The simplified model is shown in Fig. 3.

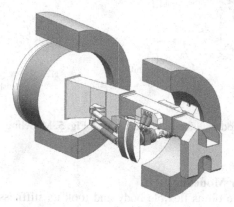

Fig. 3. Geometric model of TBM

3.2   Virtual Assembly

During tunneling process, the surrounding rock support and constrain the host machine. Compared with the host machine, the stiffness of surrounding rock is higher and its elasticity could be ignored. Thus, the host machine is regarded as multiflexible body system, while the surrounding rock is taken as rigid body.

Mechanism Modeling
GTR Mechanism contains prismatic joint, universal joint, spherical joint, cylindric joint and revolute joint [6]. The spherical joint is treated as three orthogonal revolute joint, and the universal joint is regarded as two axes intersecting revolute joint.

The guide mechanism can be deemed as a prismatic joint. The finite element modeling method of guide and slider is presented in Fig. 4. The mechanism is represented by beam element (guide), two rigid nodes and shell element (slider). Two rigid nodes are connected by a bushing unit, which is used to simulate the contact stiffness under the direction of the 3 axes of Cartesian coordinate system. The rigid nodes are united respectively with the shell element and the beam element by the mean unit, which can ensure the force and the moments be applied uniformly on the slider and guide.

The hinge joint is a kinematic pair that connects two rigid nodes of bodies N1 and N2 by constraining them to rotate about a common axis. As shown in Fig.5, the constraints between the three translational DOF and two rotational DOF around y and z axis of nodes are imposed by setting the corresponding stiffness in the bushing unit to a large value. Therefore, the rotational DOF of x axis can be released.

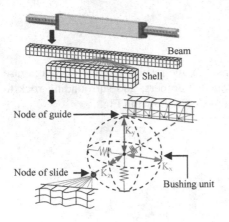

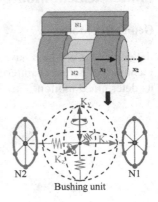

Fig. 4. Modeling method of guide and slider

Fig. 5. Modeling method of hinge joint

Hydraulic Cylinder Modeling

Regard the hydraulic oil as flexible body and took its stiffness [7] into account. The hydraulic cylinder is simulated by two rigid nodes, prop surface and cylinder surface, the modeling approach of hydraulic cylinder is shown in Fig. 6. Two rigid nodes are united by a bushing unit whose reference frame is cylinder coordinate system, so a cylindric joint can be simulated through setting the axial damping. Use mean unit to connect the rigid nodes with the surface of prop and cylinder respectively.

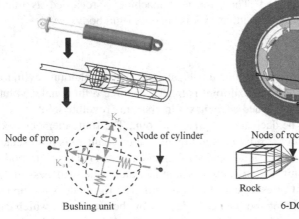

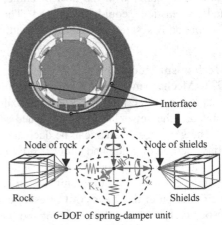

Fig. 6. Modeling method of cylinder

Fig. 7. Modeling method of interface

The Interface Between Rock and Host Machine Modeling

The contact between surrounding rock and shields is deemed as resilient and it affects the force condition of host machine. As shown in Fig. 7, it is similar to the method described above. The bushing unit can simulate a spring-damping element with 6 DOFs.

Grippers hold tightly with the surrounding rock, so set glue unit to unite grippers and surrounding rock.

3.3   Mesh Generating

The whole body of the finite element model is set to continuum element, and the mesh type is set to tetrahedron. The finite element model with 34145 elements and 14295 nodes is shown in Fig. 8. The parameters of the mesh for the key components are given in Table 1.

Table 1. The parameters of the finite mesh

Component	Mesh type	Mesh size(cm)
Cross pin	Tetrahedron	10
Guide	Beam	10
Cylinder	Tetrahedron	20
Saddletree	Tetrahedron	40

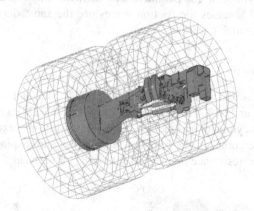

Fig. 8. Finite element model of TBM

4   Example

4.1   Parameters Setting

According to the modeling method above, a single support open TBM with a diameter of 8040mm has been fully established in the SAMCEF^{TM}. Take advancing and steering conditions as examples to simulate the force state of the GTR Mechanism. The material properties are given in Table 2.

Table 2. Material properties

	Material	Elastic Modulus (GPa)	Poisson's ratio	Density (kg/m³)	Bulk Density (N/m³)
TBM	Q235	206	0.33	7800	—
Rock	Hard rock	67.5	0.14	—	27.3

Table 3. The parameters of each working conditions

	Action point	Value (Advancing condition)	Value (Steering condition)
P	Center of the cutter head	17.6e6N	—
M	Center of the cutter head	5.4e6N·m	—
F_{N1}	Center of the left Shield	3e6N	—
F_{N2}	Center of the right Shield	3e6N	3e6N[8]
F_{N3}	Center of the left Gripper	2e7N	3e7N
F_{N4}	Center of the right Gripper	2e7N	1e7N
$T_{1(2,3,4)}$	Cylinder sectional center	—	5e6N

The loads and boundary constraints are as follows: set an axial load P and resistance moment M on the centre of the cutter head, the pressure $F_{N1\,(2)}$ and $F_{N3\,(4)}$ are applied on the shields and grippers respectively, and the pushing force $T_{1\,(2,\,3,\,4)}$ are applied on torque cylinders. The parameters of each working conditions are shown in Table 3. The model assumes the friction forces are the same during simulation. Set the constraints of the rock to clamp.

4.2   Results Analysis

The equivalent stress and nodal displacement of GTR Mechanism during advancing condition is shown as Fig.9. The maximum stress is 246.72MPa and it occurs in the prop of the support cylinders and the spherical joint, while the extreme displacement is 5.4mm and in the corner of the grippers. Both stress and displacement is not symmetry because of the resistance moment M set on the cutter head.

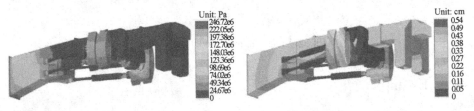

Fig. 9. Equivalent stress (left) and nodal displacement (right) of GTR Mechanism under advancing condition

The equivalent stress and nodal displacement of the main beam during two working conditions is shown as Fig. 10 and Fig. 11 respectively. Ignore the main beam Ⅲ since its little effects on the whole mechanism.

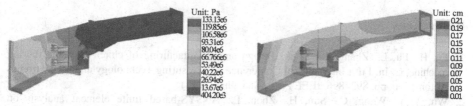

Fig. 10. Equivalent stress (left) and nodal displacement (right) of main beam under advancing condition

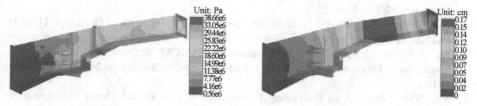

Fig. 11. Equivalent stress (left) and nodal displacement (right) of main beam under steering condition

Under advancing condition, because of the force from the cutter head and propulsion cylinders, stress concentration occurs in the front of the main beam Ⅰ. Under steering condition, stress mainly concentrates on the main beam Ⅱ, since the force of torque cylinders and support cylinders is transmitted through the guide to the main beam. The details are shown in Table 4.

Table 4. The maximum stress and displacement under each working condition

Working condition	The maximum stress(MPa)	The maximum displacement(mm)
Advancing	133.13	2.10
Steering	36.66	1.69

5   Conclusion

In this paper, a finite element modeling method for single support open style TBM is studied. The establishment of surrounding rock and shield-rock combined surface are used as the model constraints. Through simulation, the stress and displacement of GTR Mechanism under two working conditions are got. The results showed that the model is valid.

Acknowledgments. The research is supported by National Basic Research Program of China (973 Pro-gram, Grant No. 2013CB035403), and National High-tech Research Program of China (863 Pro-gram, Grant No. 2012AA04).

References

1. Fu, H., Fu, L., Zhang, G., et al.: Design and manufacturing of clutch in tunnel boring machine, C. In: Fifth International Conference on Measuring Technology and Mechatronics Automation, pp. 893–896. IEEE Press, Shanghai (2011)
2. Wu, Y.-h., Wang, C., Sun, H., Zhou, P.: ANSYS-based finite element analysis on cutter-plate of rock excavators. J. Chinses Journerl of Construction Machinery **10**(2), 171–176 (2012)
3. Zhang, G.-f., Li, Z.-b.: Finite element analysis for TBM's host machine with CATIA. J. Machinery **34**(7), 39–41 (2007)
4. Tao, L., Ning, X.-k., He, F.: Finite Element Analysis for Main Beam of Open-type TBM. J. Mechanical Engineering & Automation **5**, 43–44 (2013)
5. Du, Z.-c., Yu, Y.-q., Su, X.: Applications of SAMCEF Software for Modeling and Simulation of Flexible Parallel Robots. J. Journal of System Simulation **20**(11), 2999–3011 (2008)
6. Niu, W.-w.: Analysis on Excavating Performance of Gripper and Thrust Mechanism of TBM. D. Tianjin University: Tianjin University (2014)
7. Dai, Y.-f.: Calculation of Hydraulic Oil Stiffness. J. Nonferrous Metal Design **1**, 61–63 (1999)
8. David T. Cass, Seattle, Wash.: Earth boring machine and method: US, 3861748. P. 01-21 (1975)

The Vibration Analysis of TBM Tunnelling Parameters Based on Dynamic Model

Hanyang Wu, Junzhou Huo[✉], Xueguan Song, Lintao Wang, and Wei Sun

School of Mechanical Engineering, Dalian University of Technology, Dalian 116024, China
huojunzhou@dlut.edu.cn

Abstract. In TBM tunneling process, the excessive vibration often leads to serious damage in some of TBM main components. Appropriate tunneling parameters such as cutterhead rotation speed, advancing speed, etc have a significant influence on TBM vibration. This study is based on a TBM muti-degree of freedom nonlinear dynamic model. The total force in three direction and torque on cutterhead are calculated according to the real data. The subsequent calculation shows that keeping the rotation speed in the range of 5.88–6.2 rpm can make the cutterhead vibration at a low level. And the effect of different TBM advancing speed is analyzed under this appropriate rotation speed. It shows that there is no significant change in the axial and vertical vibration condition of cutterhead and main frame with the advancing speed changing from 2–3.2 m/h but the horizontal vibration obviously increases. The deviation analysis of TBM motor's input torque is also carried out. The vibration situation of each pinion is analyzed which determines No 7,8,9 pinions are particularly affected.

Keywords: TBM · Coupled dynamics · Rotation speed · Advancing speed · Input torque deviation

1 Introduction

In hard rock condition, the construction process with big torque, thrust and large impact load. The excessive vibration of TBM will cause non-normal damage in critical component and shorten the life of TBM [1]-[2]. In the actual tunneling process, the driver often selected tunneling parameters based on experience and this may lead to accident. Therefore, theoretical analysis of the tunneling parameters that affect TBM vibration is of great important. It can determine a reliable range for the driver and deeper the diver's understanding about these tunneling parameters.

This paper analyzed TBM vibrations situation from the angle of dynamics. The excessive vibration will not only affect the excavation efficiency but shorten the fatigue life of structural component and lead to the accident [3]. Therefore, many scholars started studying the vibration dynamical characteristics of TBM. K.Z.Zhang [4]-[6] et al established a coupling dynamical model of shield machine considering redundant drive system, hydraulic propulsion system, geological conditions, etc, and the dynamical characteristics of the rotary system was studied based on the dynamical

© Springer International Publishing Switzerland 2015
H. Liu et al. (Eds.): ICIRA 2015, Part III, LNAI 9246, pp. 489–500, 2015.
DOI: 10.1007/978-3-319-22873-0_44

model. J.X.Lin and W.Sun [7] established a nonlinear dynamical model of cutterhead system, and analyzed the dynamical characteristics of cutterhead.

Many kinds of TBM dynamic models have been established by domestic and foreign scholars. This paper choose a dynamic model that comprehensively [8] considers time-varying impact load, a multi-component complex relationship between each component from cutters to gripper shoe. This paper analyzes some tunneling parameters such as advancing speed and cutterhead rotation speed that affect TBM vibration which provides theoretical basis for the actual construction.

2   Dynamic Model of TBM

The dynamic calculation in this paper is based on this dynamic model below. Firstly, the major components of TBM are show in fig. 1. TBM mainly includes cutterhead system and propel system. The cutterhead system includes cutterhead, bull gear, pinion, coupling, variable frequency motor, planetary reducer, etc. The propel system includes main frame, support cylinder and gripper shoe. The X, Y, Z represent horizontal, vertical and axial direction as shown in Figure 1.

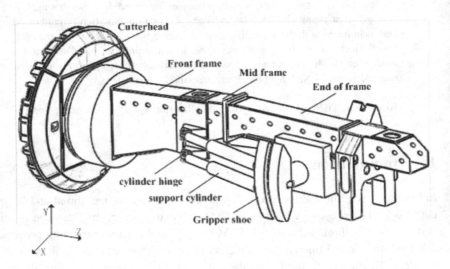

Fig. 1. The component diagram of TBM

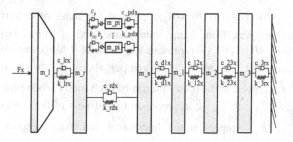

(a) The horizontal degree of freedom

(b) The vertical degree of freedom

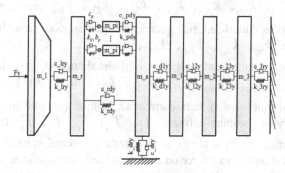

(c) The axial degree of freedom

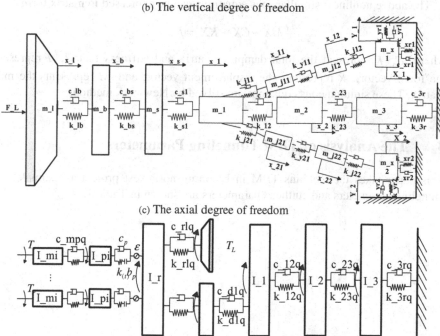

(d) The torsional degree of freedom

Fig. 2. The dynamic model of TBM

where $m_l, m_b, m_s, m_1, m_2, m_3, m_r, m_{pi}$ represent the mass of cutterhead, cutterhead bearing, cutterhead support, front frame, mid frame, end of frame, bull gear and each pinion. $m_{j11}, m_{j12}, m_{j21}, m_{j22}$ represent the mass of 4 hinges as shown in Figure 3 (c). m_{x1}, m_{x2} represent the mass of 2 gripper shoes as shown in Figure 3 (c). I_{mi}, I_{pi}, I_r, I_l represent the rotary inertia of each motor, each pinion, bull gear and cutterhead. $k_{j11}, k_{j12}, k_{j21}, k_{j22}$ represent 4 stiffness of the corresponding hinges. $k_{xr1}, k_{yr1}, k_{xr2}, k_{yr2}$ represent the horizontal and axial support stiffness of the corresponding gripper shoes. $k_{lb}, k_{bs}, k_{s1}, k_{12}, k_{23}, k_{3r}$ represent the structural stiffness of cutterhead, cutterhead bearing, cutterhead support, front frame, mid frame, end of frame. $k_{lrx}, k_{rdx}, k_{pdx}, k_{d1x}, k_{12x}, k_{23x}, k_{3rx}$ represent the horizontal structural stiffness of cutterhead, bull gear, cutterhead support, front frame, mid frame and end of frame. $k_{lry}, k_{rdy}, k_{pdy}, k_{d1y}, k_{12y}, k_{23y}, k_{3ry}$ represent the vertical structural stiffness of cutterhead, bull gear, cutterhead support, front frame, mid frame and end of frame. $k_{mpq}, k_{rlq}, k_{d1q}, k_{12q}, k_{23q}, k_{3rq}, k(t)$ represent the torsional stiffness of transmission shaft, cutterhead, front frame, mid frame, end of frame and the time-varying damping stiffness. $T_L, T_{pi}, F_x, F_y, F_L$ represent the load torque on cutterhead, the input torque of motor, the horizontal unbalanced force on cutterhead, the vertical unbalanced force on cutterhead and the axial force on cutterhead.

The above nonlinear simultaneous equations can be expressed in matrix form:

$$M\ddot{X} + C\dot{X} + KX = F$$

where C, K represent the total damping matrix and stiffness matrix, F represents the force vector, X represents the displacement vector, and M represents the mass matrix. These simultaneous equations are solved by Newmark method.

3   The Analysis of TBM Tunneling Parameters

This paper takes the Robbins TBM in Liaoning northwest project as example. The tunneling parameters and cutthead parameters are shown in Table 1.

Table 1. The tunneling parameters and cutthead parameters

Tunneling parameters		Cutthead parameters	
Advance speed	2.4m/h	Diameter of cutterhead	8.53m
Penetration	7.8mm/rev	Mass of cutterhead	152t
Rotating speed	5.18RPM	Teeth number of bull gear	174
Rock mechanical parameters		Teeth number of pinion	14
Compressive strength	93.6MPa	Number of center cutter	8
Tunnel depth	130-1000m	Number of inner cutter	40
Confining pressure	6-30MPa	Number of gauge cutter	12
		Cutter spacing	75mm

In the process of actual excavation, the computer in master control room will record the TBM tunneling parameters such as the total thrust, the motor output torque, the pressure of side support cylinder, etc. The axial force, horizontal force, vertical force and torque on cutterhead which are used as external incentives when calculating the dynamic model are estimated based on the output parameters above. The loads under different advancing speed and rotate speed are intercepted as the input parameters and the more suitable tunneling parameters are determined comparing the vibration condition of TBM main parts in three direction. Due to the complicated situation in actual excavation, the actual advancing speed always fluctuate from 2 to 3m/h, and the rotate speed change from 2.5 to 6.7 rpm. It's hard to obtain the corresponding load with only one parameter changing. In analysis, the TBM advance speed is considered relatively constant, and the appropriate cutterhead rotate speed is found out firstly. Then effect of different advancing speed on TBM vibration is studied according to the optimal rotate speed.

3.1   The Optimization of Cutterhead Rotate Speed

The advancing speed fluctuates in the range from 2-3m/h in the field data segment, and different rotation speed corresponding data segments are intercepted. These data segments are taken as input load for calculating, and the vibration results of main components are obtained. Due to the random characteristic of these data segments, the data is intercepted in a non-equidistant way. Take the corresponding load when cutterhead rotation speed is 5.5 rpm as example. The axial, horizontal, vertical force and torque on cutterhead in 20 seconds are shown in Fig. 3.

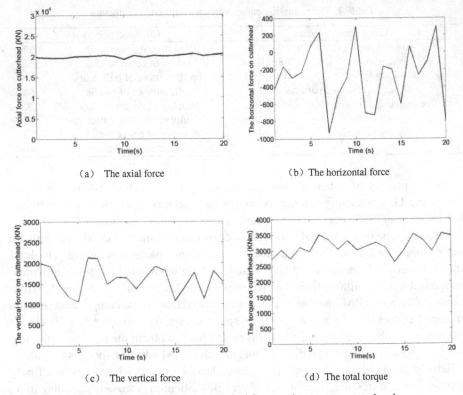

(a) The axial force

(b) The horizontal force

(c) The vertical force

(d) The total torque

Fig. 3. The axial, horizontal, vertical force and torque on cutterhead

The non-equidistant rotation speed sequence is 2.5,3.5,4.2,4.9,5.5,5.7,5.9,6.2,6.4, 6.7rpm. The dynamic calculation is carried out according to the corresponding loads. The results are shown in Fig. 4.

It can be seen that the mean of cutterhead axial vibration increases slowly when the rotation speed is less than 6.2 rpm, and when the rotation, speed is larger than 6.2 rpm, the average axial vibration of cutterhead increases quickly. Thus for cutterhead axial vibration, keeping the rotation speed less than 6.2 rpm can effectively reduce the vibration. The cutterhead horizontal vibration does not change significantly with the increase of rotation speed. The cutthead vertical vibration obviously declines with the rotation speed increasing and drops to 60% of the original value when the rotation speed reaches to 5.88rpm. It's worth noting that the vertical vibration increases after the speed reaches to around 6.5rpm.

In conclusion, when the TBM advancing speed floats around 2-3m/h, keeping the parameter on the range of 5.88-6.2 rpm can make the cutterhead vibration at a low level. This may effectively reduce the TBM machine damage.

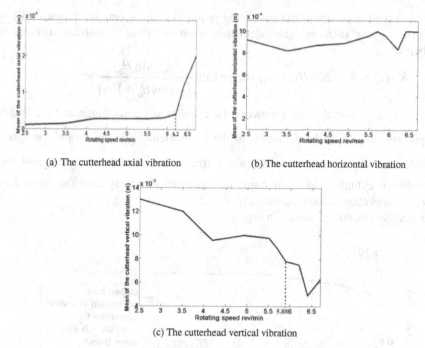

(a) The cutterhead axial vibration   (b) The cutterhead horizontal vibration

(c) The cutterhead vertical vibration

Fig. 4. The calculation results

3.2   The Analysis of Advancing Speed

Set the cutterhead rotation speed in the range of 5.88-6.2rpm, the data segment of advancing speed is 2, 2.2, 2.4, 2.6, 2.8, 3, 3.2m/hour, and each data segment length is 30s. The data is ascending ordered, the respectively cutterhead torque, axial, horizontal and vertical unbalanced force is shown in Fig. 5.

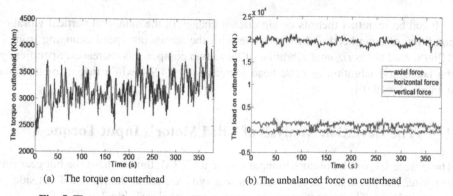

(a)   The torque on cutterhead   (b) The unbalanced force on cutterhead

Fig. 5. The cutterhead torque, axial, horizontal and vertical unbalanced force

In the process of excavation, the advancing speed directly affects the propulsion cylinder time-varying stiffness. The calculation equation of the propulsion stiffness is listed below:

$$K_y(t) = K_y \cdot \cos\theta(t) \quad \theta(t) = ac\tan(\frac{l_0 \cdot \sin\theta_0}{l_0 \cdot \cos\theta_0 + V_t \cdot t})$$

where θ_0, l_0 represent the angle between the cylinder and main frame and the length of propulsion cylinder at the beginning of the excavation. V_t is the propulsion speed of cylinder and K_y is the cylinder support stiffness. With the advancing progress, the angle between cylinder and main frame decreases continuously, and the equivalent support stiffness changes correspondingly.

The analysis results are shown in Fig. 6.

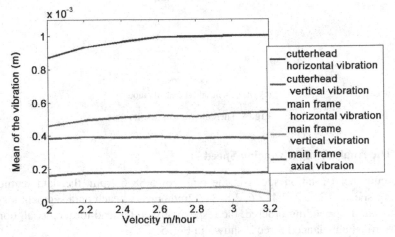

Fig. 6. The analysis results

It can be seen that there is no significant change in the axial and vertical vibration condition of cutterhead and main frame with the advancing speed changing from 2-3.2m/h. And the horizontal vibration of these two components increases. Specifically, the horizontal vibration of cutterhead increases by 14.9% while that of main frame increases by 16.9%.

4   The Deviation Analysis of TBM Motor's Input Torque

There were 10 parallel motors to input power for TBM, thus to drive the big gear ring to rotate. The phase layout of 10 motors was symmetrical in the right and left side, as shown in Fig 7. The respective phase angles were listed in Table 2.

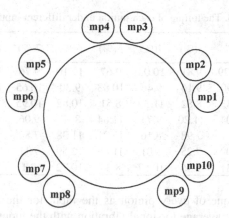

Fig. 7. The phase layout of 10 motors

Table 2. The respective phase angles

No.	1	2	3	4	5	6	7	8	9	10
Phase angle (°)	11.25	33.75	78.75	101.25	146.25	168.75	213.75	236.25	303.75	326.25

In actual project, due to the reasons such as gear assembly error, there would be certain error among the input torque of the motors. This kind of error would cause the change of the TBM vibration. Assume the total torque of the motors was T_{total}, the torque of every pinion was $10\% T_{total}$ when the input torque of every motor was the same. Under actual circumstances, assume the input torque of the gear which has the maximum torque was $(\Delta_{max}\% + 10\%) \cdot T_{total}$, the input torque of the gear which has the minimum torque was $(\Delta 10\% - \Delta_{min}\%) \cdot T_{total}$. The input error of the gear torque input error was defined as $w\%$.

$$w\% = \frac{(\Delta_{max}\% + 10\%) \cdot T_{total} - (\Delta 10\% - \Delta_{min}\%) \cdot T_{total}}{10\% T_{total}}$$

Take $w\% = 10\%$ as example, $\Delta_{max}\% - \Delta_{min}\% = 1\%$ from the equation above. The data set that the error was from 10% to 70% was studied. For each set, 10 torque between $(\Delta 10\% - \Delta_{min}\%) \cdot T_{total}$ and $(\Delta_{max}\% + 10\%) \cdot T_{total}$ were generated randomly. Meanwhile, the sum of the 10 pinion s' input torque was equal to the total torque, as shown in Table 3.

Table 3. The torque of each pinion under different input error

Error(%)	1(%)	2(%)	3(%)	4(%)	5(%)	6(%)	7(%)	8(%)	9(%)	10(%)
10	9.67	10.29	9.81	10.03	9.67	10.10	9.76	10.15	10.19	10.33
20	10.50	9.90	9.16	9.46	10.83	9.30	10.65	10.08	10.99	9.13
30	8.73	9.83	8.82	11.39	8.51	10.82	10.95	11.11	8.75	11.09
40	9.60	9.04	11.20	9.73	11.64	8.73	9.06	8.58	11.48	10.94
50	11.40	8.71	7.98	8.16	12.21	11.38	7.80	8.67	12.11	11.58
60	11.92	7.09	7.25	8.01	11.68	12.59	7.11	12.88	12.40	9.07
70	7.28	8.31	9.36	10.67	8.34	10.72	12.11	13.12	6.67	13.42

Take the input torque of every pinion as the input for the dynamic model. The changes of 10 pinions' average torsional vibration with the input error increases were shown in Fig. The changes of cutterhead and main frame's average horizontal-vertical vibration with the input error increases were shown in Fig. 8.

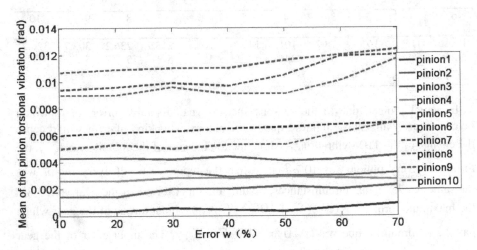

Fig. 8. The calculation results

Form the Fig above,

1. The No.7, No.8 and No.9 pinion has bigger torsional vibration which is in relatively tough vibration situation. This kind of situation may be related to the TBM's special pinion phase.
2. The vibration of No.1-No.5 pinion was small. Meanwhile, the change of the average vibration was not obvious as the input torque error increases.
3. The vibration situation of No.6-No.10 was relatively tough. The average vibration of No.6, No.8 and No.10 gear increased by 15.0%, 15.9%, 40.0% as the input error increases from 10% to 70%, and the process was approximately linear. The average vibration of No.7 gear increased by 27.7% as the input error increases from 10% to 70%. In particular, the average vibration increased by 22.7% as the input error

increases from 40% to 60%. The average vibration of No.9 gear increased by30.0% as the input error increases from 10% to 70%. Especially when the input error increases from 40% to 60%, the average vibration increased by 28.5% respectively.

It can be seem from the Fig. 9 that the pinion's input torque error affects cutterhead's average horizontal-vertical vibration significantly, but not obvious with main frame. The horizontal vibration of cutter increased by 7.9% as the input error increased from 10% to 70%. The increase of vertical vibration of cutterhead was 12.9%.

To sum up, the input torque error of the pinion had a significant effect on the overall vibration of the TBM. No.5-No.10 pinion's vibration increased with the growth of the torque error significantly.No.7 and No.9 pinion would have obvious vibration increment with small input error. For a major part of the TBM, especially for cutterhead, the growth of the pinion input torque error increased the horizontal-vertical vibration of the cutter evidently.

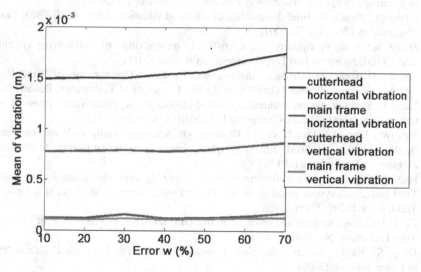

Fig. 9. The calculation results

5   Conclusion

In this paper, first hand real tunneling parameters are output and taken as input incentive for calculating a dynamic model. The suitable rotation speed is determined based on a TBM multi-degree dynamic model. The analysis results show that the rotation speed in the range of 5.88-6.2 rpm can make the cutterhead vibration at a low level. And the effect of different TBM advancing speed is analyzed under the appropriate rotation speed. It shows that there is no significant change in the axial and vertical vibration condition of cutterhead and main frame with the advancing speed changing from 2-3.2m/h but the horizontal vibration obviously increases. The horizontal vibration of cutterhead increases by 14.9% while that of main frame increases by 16.9%. In addition, the deviation analysis of TBM motor's input torque is also carried out. The analysis results show that the pinion vibration intensifies as the input error increases,especially for the pinion numbered 7,8,9. Meanwhile,the pinion's input torque error affects cutterhead's average horizontal-vertical vibration significantly which increases by 7.9% in horizontal direction and increases by 12.9% in vertical direction as the input error increased from 10% to 70%.

References

1. Promotion Center for Science & Technology Achievements of Ministry of Water Resources: Full face rock tunnel boring machine (TBM). Petroleum Industry Press, Beijing (2005)
2. The Book Compilation Committee: Rock tunnel boring machine (TBM) construction and engineering examples. China railway publishing house, Beijing (2004)
3. Zhang, H., Zhang, N.: Brief discussion on cutterhead vibration of type 803E TBM. Tunnel Construction 27(6), 76–78 (2007)
4. Zhang, K.: Study on dynamic characteristics of redundantly driven revolving system of shield TBM. Shanghai Jiao Tong University, Shanghai (2011)
5. Cai, J.: Dynamic Performance Analysis and Parameter Optimization of Wind Power Speed-up Machine Planetary Gear Trains. Dalian University of Technology, Dalian (2012)
6. Yang, J.: Study on dynamics characteristics of planetary gear transmission system of wind turbine under varying loads. Chongqing University, Chongqing (2012)
7. Sun, W., Ling, J., Huo, J., et al.: Dynamic characteristics study with multidegree-of-freedom coupling in TBM cutterhead system based on complex factors. Mathematical Problems in Engineering (2013)
8. Huo, J., Wu, H., et al.: Multi-directional coupling dynamic characteristics analysis of TBM cutterhead system based on tunneling field test. Journal of Mechanical Science and Technology, 8(29) (2015)
9. Fu, J.: Dynamics analysis and research on the TBM tunneling mechanism. East China Jiao Tong University, Nanchang (2009)
10. Deng, S., Jia, Q.: Design principles of rolling bearings, pp. 111–116. Standards Press of China, Beijjing (2008)

Pressure Balance Control System for Slurry Shield Based on Predictive Function Control

Xiuliang Li[1(✉)], Hepei Zhang[2], Yifan Xue[3], and Chengjun Shao[4]

[1] State Key Laboratory of Shield Machine and Boring Technology, State Key Laboratory of Industrial Control Technology, Zhejiang University, Zhejiang 310027, China
xiuliangli@csc.zju.edu.cn
[2] State Key Laboratory of Shield Machine and Boring Technology, Zhengzhou 450001, China
[3] Control Department, Zhejiang University, Zhejiang 310027, China
[4] Institute of Cyber-System and Control, Zhejiang University, Zhejiang 310027, China

Abstract. The most important part of the excavation for slurry shield machine is keeping the earth pressure and slurry circulation system pressure in balance. In this paper, an excavating face pressure balance principle for direct type slurry shield machine is analyzed and the pressure balance dynamic model is introduced. Then, a controller is designed based on predictive function control method. Finally, the controller initialization method is proposed to deal with the problem of controller switching from manual to automatic mode. Simulation results show the improved performance of proposed method.

Keywords: Slurry shield · Slurry circulation system · Predictive function control

1 Introduction

The slurry shield is an important branch of the modern shield method, which is widely applied around the world. Slurry shield method is the most commonly used method, especially underwater tunnel in soft soil in the river and sea [1-3]. The control of pressure for excavating face is an important part in the process of tunneling across the river. Once the control is improper, it will cause the working face collapse, the river flow backward and a series of safety problems.

According to the slurry circulation pressure control methods, the slurry shield can be divided into two basic types, indirect control type (German style with bubble chamber) and direct control type as Fig. 1 (Japanese style without bubble chamber) [1]. Unlike German style shield, direct control shield does not have bubble chamber part, the pressure of chamber is controlled by the flow of feed or discharge slurry pump. Slurry in chamber of direct control shield will be easier to be discharged than German style shield in the absence of bubble chamber. So the slurry in chamber will

This work was supported part by National Basic Research Program of China (973 Program 2013CB035406), The National High Technology Research and Development Program of China (863 Program 2012AA041702).

H. Liu et al. (Eds.): ICIRA 2015, Part III, LNAI 9246, pp. 501–510, 2015.
DOI: 10.1007/978-3-319-22873-0_45

be less able to precipitate. However, the pressure in slurry will be under the big and fast disturbance by input/output feed pump flow and the driving speed. As the shield does not have bubble chamber part to buffer the impact of disturbance, the pressure balance control is more difficult compared to the German style shield. Yunpu Song has designed a dynamic model and controller for German style shield. But the design of pressure balance controller of direct control shield is rarely mentioned [4].

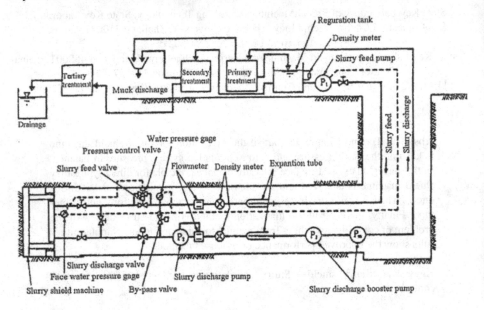

Fig. 1. Direct control type slurry shield: Slurry circulation system

This paper analyzes the composition and principle of circulation system, then establishes the pressure balance dynamic model for circulation system. A controller is designed based on predictive function control method which is easier to be implemented in program logic controller (PLC).

2   The Dynamic Model of Slurry Circulation System in Continuous Excavation Process

In order to simplify the problem, this paper only take the pressure balance problem in continuous excavation mode into account, ignoring the shutdown mode and by-pass mode. In continuous excavation mode, the pressure of chamber is controlled by the flowrate of slurry feed pump P1, and the slurry is discharged by the flowrate of slurry discharge pump P2. The schematic diagram of slurry circulation system is Fig. 2.

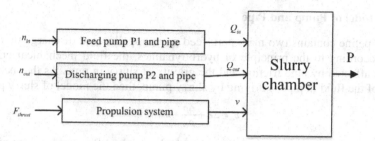

Fig. 2. Schematic diagram of slurry circulation system

2.1 Pressure Model of Slurry Chamber

This paper assumes the chamber is in a sealed environment and the pressure gradient distribution in chamber is ignored. In continuous excavation mode, the amount of slurry in chamber will be changed by the excavation soil, the slurry flowrate of feed pump and discharge pump, the elastic modulus of slurry in chamber can be expressed as [5]

$$K = -\frac{dP}{dV/V_0} \tag{1}$$

which K is modulus of elasticity. P is pressure of slurry, V_0 is the volume of chamber.

The volume variation of slurry in chamber is

$$dV = Q_{out} - Q_{in} - v \times A \tag{2}$$

where v is shield advanced rate, A is excavation area of shield, Q_{out} is flowrate of feed pump P1, Q_{in} is discharge flowrate of pump P2.

Combine equation 1 and equation 2

$$dP = -\frac{Q_{out} - Q_{in} - v\pi r^2}{\pi r^2 d} \tag{3}$$

where r is radius of excavation face, d is the length of chamber.

At normal state, the chamber is not full of slurry and there is some air in the top of chamber), which will affect transient process of pressure. Density of slurry will change by excavating. The transfer function model of pressure in chamber can be described as below

$$P(s) = \frac{K_{in}}{s(T_{in} \times s + 1)} Q_{in}(s) + \frac{K_v}{s(T_v \times s + 1)} v(s) - \frac{K_{out}}{s(T_{out} \times s + 1)} Q_{out}(s) \tag{4}$$

2.2   Model of Pump and Pipe

Slurry pipeline contains two main part: feed pump pipeline and discharge pump pipelines. According to the principle of hydrodynamics, the fluid mechanical energy of slurry is affected by pipe friction and the gravity of slurry in pipe. As the mechanical energy of the fluid in pipe is driving by slurry pump, then the model of slurry pipe is

$$h_e = \Delta Z + \frac{\Delta p}{\rho g} + h_f \tag{5}$$

which h_e is the pressure head of pipe, ΔZ is the height difference between pipe inlet and outlet, Δp is the static pressure difference between pipe inlet and outlet. ρ is the density of slurry, g is acceleration of gravity. h_f is the pressure head caused by friction loss.

h_f is include two part, first part is the friction loss h_{f1} which caused by the friction between liquid flow and pipe

$$h_{f1} = \lambda \frac{l}{d} \frac{u^2}{2g} \tag{6}$$

where u is fluid flowrate, l is the length of pipe, d is the diameter of pipe, λ is friction coefficient, $\lambda = 64 \frac{1}{\text{Re}} = 64 \frac{du\rho}{\mu}$ as the slurry flow in pipe is laminar flow. The other part is the pressure loss h_{f2} caused by turn valves and sudden changes in fluid which can be calculated by local resistance coefficient method.

$$f_{f2} = \xi \frac{u^2}{2g} \tag{7}$$

where ξ is constant for a certain type of pipes or valves.

Since slurry is transported by the centrifugal pump, the centrifugal pump characteristic curve equation can be used to describe the relationship between pump work point and fluid velocity in pipe.

$$K_{motor} n^3 = \frac{HQ\rho g}{\eta} \tag{8}$$

where K_{motor} is the power conversion coefficient from motor to bearing, n is rotational speed of motor, H is the lift head of pump, Q is flowrate of pump, ρ is liquid density, η is the efficiency of pump.

2.3   Model of Propulsion System

Since the pressure of excavation face and thrust pressure must keep in balance, according to Newton's law of motion, the speed of shield is

$$m\dot{v} = F_{thrust} - PA + cv + f \tag{9}$$

where c is damping coefficient, f is sliding friction.

3   Controller Design for Slurry Circulation System

In the design of the controller for slurry shield excavating face balance system, PID control method is typically used in project. However, equation 9 shows that the excavating face pressure is affected not only by the advance rate, but also by the feed pump flowrate and the discharge pump flowrate, even by slurry density. To prevent the mud sedimentation, the flowrate of discharge pump should not be below critical velocity. In order to provide adequate pressure head, the feed pump should work in a small work point region. Conventional PID control method cannot take these measured disturbance and constraints into account, overshoot and fluctuations will appear in the excavation pressure control. Based on predictive function control (PFC) method which can deal with above factors, the controller can be designed as Fig. 3.

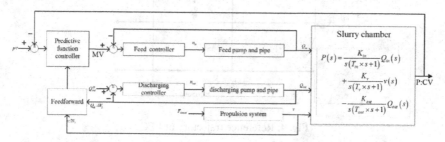

Fig. 3. Controller design for slurry circulation system

The flowrate of feed pump and discharge pump are controlled by traditional controller, such as PID. The pressure balance system is controlled by PFC with discharge flowrate and advanced velocity as feedforward [6].

3.1   Predictive Function Control

The desired increment Δp in Fig. 4 at $n+h$ will be given by

$$\Delta p = \varepsilon(n+h) = \varepsilon(n), \text{ where } \Delta p = \varepsilon(n)\left(1-\lambda^h\right), \lambda = e^{\frac{T_s}{T}}$$
$$\text{and } \Delta p = \left(\text{Setpoint} - \text{Process output}(n)\right)\left(1-\lambda^h\right) \qquad (10)$$

where T is the time constant of system, T_s is sample period of controller.

The model output increment Δm may be easily calculated because it is evaluated in the mathematical domain where all relevant information is known. The increment will be composed of the free output $S_L(n+h)$ in conjunction with the forced output $S_F(n+h)$ that contains the projection of the required MV:

$$\left(\text{Setpoint} - \text{Process output}(n)\right)\left(1-\lambda^h\right) = \text{forced output}(n+h) + \text{free output}(n+h) - \text{model output}(n) \qquad (11)$$

$$MV(n) = \frac{\text{Desired increment-Free output increment}}{\text{Unit forced output}} \qquad (12)$$

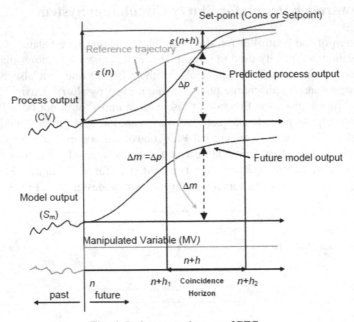

Fig. 4. Reference trajectory of PFC

3.2   PFC of an Integrator Process

From equation 12, the process between CV and MV can be considered as an integrator process system with a first order transfer function.

$$CV(s) = MV(s)\frac{K_p}{s(1+sT)} = MV(s)M_0(s) \tag{13}$$

As the process contains integral action, the model should be decomposed the unstable model into two stable processes as Fig. 5.

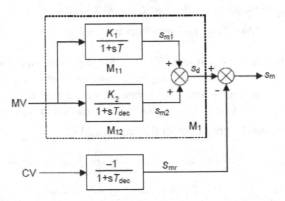

Fig. 5. Decomposition of a first-order process with integrator

The control equation is given by:

$$MV(n) = \frac{(\text{Setpoint-CV}(n))l_h + s_{m1}(n)b_{mh} + s_{m2}b_{sh} - SS}{K_1 b_{mh} + K_2 b_{sh}}$$

where $l_h = 1 - e^{\frac{3hT_s}{TRBF}}, b_{mh} = 1 - a_m^h, b_{sh} = 1 - a_s^h, SS = s_{mr}(n)b_{sh} - CV(n)b_{sh}$

$$s_{m1}(n) = s_{m1}(n-1)a_m + b_m K_1 e(n-1)$$
$$s_{m2}(n) = s_{m2}(n-1)a_s + b_s K_2 e(n-1) \tag{14}$$
$$s_{mr}(n) = s_{mr}(n-1)a_s + CV(n-1)$$
$$s_m(n) = s_{m1}(n) + s_{m2}(n) + s_{mr}(n)$$

$$K_2 = \frac{K_p T_{dec}^2}{T_{dec} - T}, K_1 = K_p T_{dec} - K_2, a_s = e^{-\frac{T_s}{T_{dec}}}, b_s = 1 - a_s$$

3.3   Feedforward Compensation

Measured disturbance compensation is a procedure of great practical benefit, which should be incorporated as often as possible into the regulator design process. The concept is simple: to counteract the effects of a disturbance before it appears. The disturbance at instant $n+h$ creates a control increment Δ_{pert} that depends on the free and forced outputs of the process. The past measured disturbance produces an output $s_{m_{pert}}(n)$ in response to the known disturbance transfer function. Under these conditions, the free output $s_{L_{pert}}(n+h)$ which depends only on the past measured disturbance is known. On the other hand, the forced output $s_{F_{pert}}(n+i)$ is unknown and therefore a prediction of the disturbance must be made as $\text{Pert}(n+i) = \text{Pert}(n), 0 < i < h-1$. This results in a step characteristic response of the process multiplied by the local value of the measured disturbance

$$S_{F_{pert}}(n+h) = G_0(h)\text{Pert}(n), \text{ where } G_0(h) \text{ is a gain function of } h \tag{15}$$

The term $\Delta \text{pert}(n+h)$ is added to the control equation as feedforward, given.

$$MV(n) = \frac{(\text{Setpoint-CV}(n))l_h + s_m(n) - \left(s_m(n)a_m^k + s_{L_{pert}}(n+h) + s_{F_{pert}}(n+h) - s_{m_{pert}}(n)\right)}{K_1 b_{mh} + K_2 b_{sh}} \tag{16}$$

3.4   Controller Initialization When Switch from Manual to Automatic

It is a relatively simple exercise to transfer from an automatic to manual mode of regulation. But, switching from a manual to automatic mode is always a little agonizing for operators, particularly when the system contains an integrator. We need to provide a smooth transition between the MV's at the point of switching. The PFC controller has been installed for a long period of time and the operator wishes to transfer from manual to automatic mode for some reason. Any installed PFC, working offline, is permanently computing its MV, which is not applied. This particular PFC mode is defined by two characteristics: 1).Tracking mode: The set-point is set equal to the CV; 2).Internal model: The applied MV, generated by another controller or from manual mode, is the input to the internal model.

4   Simulation Results

Model parameters are divided into two categories, one is the mechanism parameters such as friction coefficient in pipe, fluid density, sliding friction coefficient and so on which is remained unchanged during the construction process, the other type is construction data such as advanced speed, pressure, flowrate and so on. These parameters must be determined before simulation conclusion

The elastic modulus of slurry mixed with soil is approximately equal to water $K = 10^9 \text{Pa}$.

Assume the length of chamber is $4m$ and diameter of chamber is $8m$. Then, $V_0 = 201.0619 m^3$.

Assume the mass of head shield is $m = 100000 kg$, sliding friction in practice can be set to $f = 0.03 F_{thrust}$, advanced speed is $v = 6.67 \times 10^{-4} m/s \, (2.4 m/h)$, the pressure of excavation face is $P = 6 \times 10^5 \text{Pa}$. $Q_{out} = 0.0556 m^3/s \, (200 m^3/h)$.

The parameters above are original data needed to determine the model. Under the condition of steady state operating point of the system, the other parameter can be calculated $F_{thrust} = 3.11 \times 10^7 N, Q_{in} = 0.0221 m^3/s$.

According to pump and pipe model, the steady state rotational speed of feed and discharge pump motor is $n_{in} = 7.3579 r/s, n_{out} = 2.7289 r/s.$

With these parameter, the model simulation result can be seen in Fig. 6

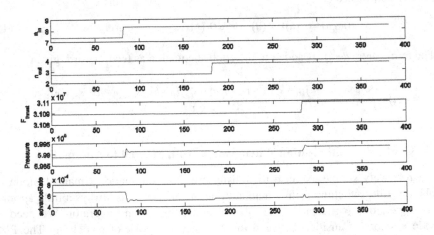

Fig. 6. Simulation result for slurry circulation system model

In order to simplify the system analysis and simulation, the equation 16 is dimensionless. The simulation result can be seen in Fig. 7. Here, the controlled variable is drifting at the beginning of the simulation. The INIT period of the controller lasts 50seconds. During that time span, the controller is estimating the slope of the controlled variable, in order to properly initialize the model. A manual phase is considered from 270 to 300s. From simulation result, we can see that the controller could track pressure setpoint changes even under the disturbance, and the controller could smoothly switch from manual to automatic mode.

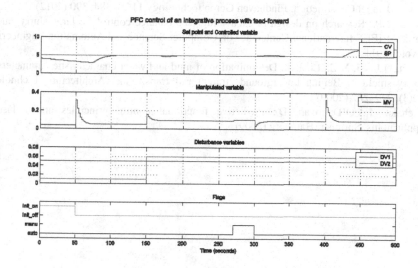

Fig. 7. PFC controller simulation result

5   Conclusion

The most important part of the excavation for slurry shield machine is keeping balance between the earth pressure and slurry circulation system pressure, especially for direct control type shield without air chamber. In this paper, an excavating face pressure balance principle for of direct type slurry shield machine is analyzed and its pressure balance dynamic model is introduced. Then, a controller is designed based on predictive function control method. Finally, the controller initialization method is proposed to deal with the problem of controller switching from manual to automatic mode. Simulation result shows the model and controller have achieved the desired design goals.

References

1. Xiang, Z.C., Lou, R.Y., Fu, D.M.: New technology information for slurry shield machines. Shanghai tunnel Engineering Institute of construction technology of joint-stock company scientific and technical information (2001)
2. Huayong, Y., Hu, S., Guofang, G., Guoliang, H.: Electro-hydraulic Proportional Control of Thrust System for Shield Tunneling Machine. Automation in Construction **18**, 950–956 (2009)
3. Shen, X., Lu, M., Fernando, S., AbouRizk, S.: Tunnel Boring Machine Positioning Automation in Tunnel Construction. Eindhoven Geron technology **11**(2), 384–390 (2012)
4. Song, Y.P.: Research on design of excavating face balance control for large slurry shield. In: 2011 IEEE International Conference on Computer Science and Automation Engineering, Kyoto (2011)
5. Yuan, D.J., Li, X.G., Li, J.H.: Determination of mud and water characteristic parameters of slurry shield of Beijing underground straight rail transit line. Architecture Technology **40**(3), 279–282 (2009)
6. Richalet, J., O'Donovan, D.: Predictive functional control principles and industrial applicaiotns. Springer, Hamburg (2009)

Fuzzy-PID Based Induction Motor Control and Its Application to TBM Cutter Head Systems

Yaqi Zhao[1,2], Jingcheng Wang[1,2] (✉), Langwen Zhang[1,2], and Tao Hu[1,2]

[1] Department of Automation, Shanghai Jiao Tong University, Shanghai, China
zhaoyq2013@126.com
[2] Key Laboratory of System Control and Information Processing,
Ministry of Education of China, Shanghai 200240, China

Abstract. Induction motor is used to drive cutter head system of tunnel boring machine (TBM). Due to the complex working environment, the conventional PID based vector control method cannot meet the varying motor parameters and load disturbances. In this paper, the fuzzy theory and PID control are investigated and applied to the speed control and electromagnetic torque control of induction motor for TBM cutter head systems. Based on the fuzzy-PID control strategy of each induction motor, a multi-motor synchronization control using master-slave control strategy is studied and analyzed. Simulation results show the effectiveness and robustness of the proposed method.

Keywords: Induction motor · Fuzzy-PID · TBM · Synchronization control

1 Introduction

Nowadays, the world is witnessing an increasing need for tunnels because of their unique characteristics and potential applications. Tunnels are artificial underground space in order to provide a capacity for particular goals such as water transfer, road tunnels, and mine. TBM is commonly used for tunneling due to its high safety, rapid excavation, and low manual labor [1] . The cutter-head undertakes the task of excavating rocks and soil. The cutter-head driving system plays an important role in TBM. Traditionally, the cutter is driven by hydraulic system, but in recent years, three-phase AC asynchronous motor controlled by transducer is applied in cutter head driving system, as shown in Fig. 1. Compared with hydraulic drive system, motor drive system is simple in mechanical design, installation and maintenance.

Fig. 1. TBM cutter-head driving motor

© Springer International Publishing Switzerland 2015
H. Liu et al. (Eds.): ICIRA 2015, Part III, LNAI 9246, pp. 511–522, 2015.
DOI: 10.1007/978-3-319-22873-0_46

TBM cutter system is driven by multi-motor, if one of the motor speed is lower or higher than the other motor speed, the motor will heat and probably get damaged if this lasts for a long time. To achieve stability of cutter-head driving, speed control is of crucial importance. However, induction motors are much more difficult to control and not suitable for high dynamic performance applications because of their complex nonlinear dynamics. Vector control provides decoupled control of the flux magnitude and the torque producing current, which is commonly used in asynchronous motor. Classical PI controller is a simple method used in the control of induction motor drive. However, the drawbacks of PI controller are the sensitivity of performance to the system-parameter variations and inadequate rejection of external disturbances and load changes [2]. Considering the complex underground tunneling environment and the changes of motor parameters, we proposed a self-tuning fuzzy-PID controller is proposed for a vector control based induction motor drive.

The rest of this paper are organized as follows. In Section 2, mathematical model of induction motor is built. Section 3 presents the vector control theory. In Section 4, we investigate the fuzzy PID based vector control of induction motor. Section 5 gives the simulation results and Section 6 concludes this paper.

2   Mathematical Model of Induction Motor

A three-phase induction motor is used in TBM cutter-head drive system. The fundamentals of assuming that the three-phase AC voltages are balanced and the stator windings are uniformly distributed. The mathematical model [3] of induction motor can be expressed as follows:

$$
\begin{cases}
\dfrac{d\omega}{dt} = \dfrac{n_p^2 L_m}{J L_r}\left(i_{s\beta}\Psi_{r\alpha} - i_{s\alpha}\Psi_{r\beta}\right) - \dfrac{n_p}{J}T_L \\[2mm]
\dfrac{d\Psi_{r\alpha}}{dt} = -\dfrac{1}{T_r}\Psi_{r\alpha} - \omega\Psi_{r\beta} + \dfrac{L_m}{T_r}i_{s\alpha} \\[2mm]
\dfrac{d\Psi_{r\beta}}{dt} = -\dfrac{1}{T_r}\Psi_{r\beta} + \omega\Psi_{r\alpha} + \dfrac{L_m}{T_r}i_{s\beta} \\[2mm]
\dfrac{di_{s\alpha}}{dt} = \dfrac{L_m}{\sigma L_s L_r T_r}\Psi_{r\alpha} + \dfrac{L_m}{\sigma L_s L_r}\omega\Psi_{r\beta} - \dfrac{R_s L_r^2 + R_r L_m^2}{\sigma L_s L_r^2}i_{s\alpha} + \dfrac{u_{s\alpha}}{\sigma L_s} \\[2mm]
\dfrac{di_{s\beta}}{dt} = \dfrac{L_m}{\sigma L_s L_r T_r}\Psi_{r\beta} - \dfrac{L_m}{\sigma L_s L_r}\omega\Psi_{r\beta} - \dfrac{R_s L_r^2 + R_r L_m^2}{\sigma L_s L_r^2}i_{s\beta} + \dfrac{u_{s\beta}}{\sigma L_s}
\end{cases}
\tag{1}
$$

where $T_r = \dfrac{L_r}{R_r}$, $\sigma = 1 - \dfrac{L_m^2}{L_s L_r}$.

Moreover, electromagnetic torque equation can be represented as:

$$T_e = \frac{n_p L_m}{L_r}(i_{s\beta}\Psi_{r\alpha} - i_{s\alpha}\Psi_{r\beta}) \tag{2}$$

In (1), the subscripts s and r refer to the stator and rotor, and subscript α and β denote the mathematical model in a synchronous rotating reference frame for a three phase induction motor. The features of parameters in dynamic model of induction motor are shown in Table 1.

Table 1. Parameter of dynamic model of induction motor

Parameters	Feature
ω	Rotor speed
n_p	The number of pole pairs
L_m	Mutual inductance
L_r, L_s	Stator and rotor inductances
R_s, R_r	Stator and rotor resistence
J	Moment of inertia
$i_{s\beta}, i_{s\alpha}$	Stator current components
$\Psi_{r\alpha}, \Psi_{r\beta}$	Rotor flux components
T_L	Load torque

3 Vector Control Theory

In the induction motor dynamic model, there is a direct coupling between the parameters of rotor and stator. The speed of induction motor is difficult to control. Vector control is a control method where the stator currents of a three-phase induction motor are identified as two orthogonal components that can be visualized with a vector. The vector control leads to control torque and the flux independently likes in the case of DC motors using d-q rotating reference frame [4]. If synchronously rotating d-q-0 frame was selected, which d-axes is precisely adjusted with the rotor field, the q component of the rotor flux would be zero, that is:

$$\Psi_{rm} = \Psi_{rd} = \Psi_r \tag{3}$$

$$\Psi_{rt} = \Psi_{rq} = 0 \tag{4}$$

Hence the equations (1) can be written as

$$
\begin{cases}
\dfrac{d\omega}{dt} = \dfrac{n_p^2 L_m}{JL_r}\left(i_{s\beta}\psi_r\right) - \dfrac{n_p}{J}T_L \\[2mm]
\dfrac{d\psi_r}{dt} = -\dfrac{1}{T_r}\psi_r + \dfrac{L_m}{T_r}i_{sm} \\[2mm]
\dfrac{di_{sm}}{dt} = \dfrac{L_m}{\sigma L_s L_r T_r}\psi_r - \dfrac{R_s L_r^2 + R_r L_m^2}{\sigma L_s L_r^2}i_{sm} + \dfrac{u_{sm}}{\sigma L_s} \\[2mm]
\dfrac{di_{st}}{dt} = \dfrac{L_m}{\sigma L_s L_r T_r}\psi_r - \dfrac{R_s L_r^2 + R_r L_m^2}{\sigma L_s L_r^2}i_{s\beta} + \dfrac{u_{st}}{\sigma L_s}
\end{cases}
\tag{5}
$$

The motor phase currents i_a, i_b, i_c are converted to i_d and i_q in the stationary reference frame. These are then converted to the synchronously rotating reference frame d-q currents, i_{ds} and i_{qs}. Using these conditions, the induction motor can be controlled as a DC motor.

A block diagram of a vector control method using SVPWM voltage-fed inverter is shown as Fig. 2. The rotor speed ω is compared with the reference speed ω^* and adjusted by PI controller. The output is the quadratic current i_q^*. There are three control loops including the speed control, torque control and flux control. Rotor speed and flux is controlled in a closed-loop form.

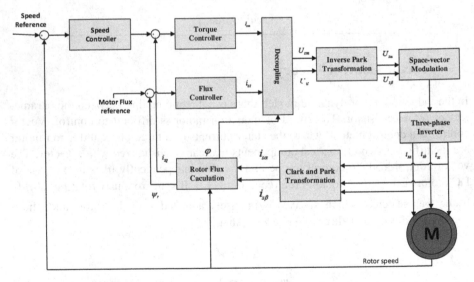

Fig. 2. Vector control block diagram with rotor flux orientation

4   Fuzzy Vector Control of Induction Motor

The conventional PID controller has advantages of simple structure and simply implemented [5]. However, it cannot adjust its coefficient adaptively to the change of motor parameters and system disturbance. Fuzzy PID control, combining the advantage of PID control and fuzzy logic, demonstrates the significant performance improvement over the conventional PID control [6]. The fuzzy PID control structure is shown in Fig. 3.

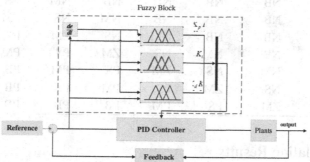

Fig. 3. Self-tuning fuzzy PID control structure

According to Fig. 3, the three coefficients K_p, K_i and K_d is tuned by fuzzy tuners using the following equation:

$$K_a = K_{a0} + U_a \Delta K_a, U_a \in [0,1] \quad a \text{ is } p, i \text{ or } d \tag{6}$$

where U_a is the parameter obtained from the output of the fuzzy controllers, ΔK_a are the correction coefficients. Three coefficients K_p, K_i and K_d are tuned by using the three independent fuzzy tuners. Consequently, the three separate fuzzy controllers are combined to form the fuzzy PID controller.

The fuzzy inference is based on fuzzy set theory [7]. Fig. 4 shows the membership functions for the controller inputs on the common interval [-1, 1] (i.e., normalized error E and normalized change of error EC), and output of fuzzy controller is on common interval [0, 1] [8]. Table 2 shows the proposed two dimensional rule base.

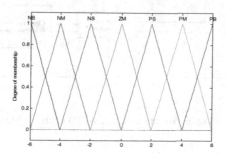

Fig. 4. Member function of fuzzy inference

Table 2. Fuzzy rule

		E						
		NB	NM	NS	ZM	PS	PM	PB
EC	NB	NB	NB	NB	NB	NM	NS	ZM
	NM	NB	NB	NB	NM	NS	ZM	PS
	NS	NB	NB	NM	NS	ZM	PS	PM
	ZM	NB	NM	NS	ZM	PS	PM	PB
	PS	NM	NS	ZM	PS	PM	PB	PB
	PM	NS	ZM	PS	PM	PB	PB	PB
	PB	ZM	PS	PM	PB	PB	PB	PB

5   Simulation Results

5.1   Fuzzy-PID Based Induction Motor Control

The simulation of the proposed controller of induction motor are carried out in MATLB/SIMULINK. The block diagram of implementation of fuzzy PID vector control model of induction motor is shown in Fig. 5. Table 3 shows the motor parameters which is used in TBM cutter head driving system [9]. Sampling time in this simulation is 3s.

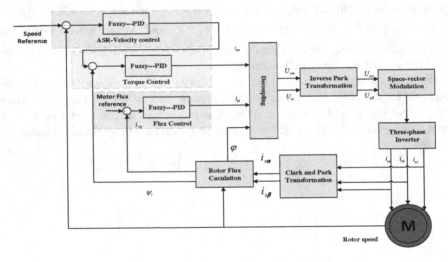

Fig. 5. Structure of fuzzy-PID control system

Table 3. Parameters value of induction motor

Mutual inductance	Rotor inductance	Stator inductance	Stator resistance	Rotor resistance
34.7mH	0.8mH	0.8mH	0.087 Ω	0.087 Ω

Inertia	Pole pairs	Frequency	Power
1.662kg. m^2	2	50Hz	50HP

Set the reference speed as 120 rad/s, and the load torque as 0. The results of rotor speed, flux and magnetic torque response of fuzzy PID control are shown in Fig. 6. The results show that the speed can reach the reference value quickly without over-shoot. Meanwhile, the rotor flux and the magnetic torque can reach steady state fast.

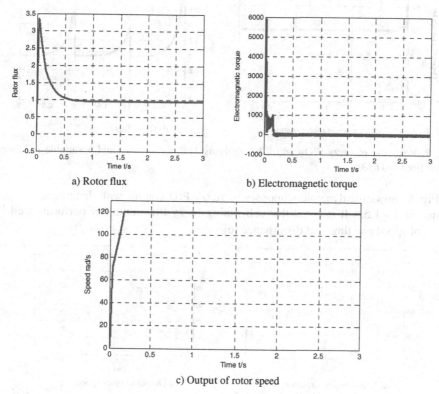

a) Rotor flux

b) Electromagnetic torque

c) Output of rotor speed

Fig. 6. Response of fuzzy-PID control by applying load torque T=0 Nm and reference speed as 120rad/s between t=0 to 3s

Moreover, the control of time-varying speed references are given in Fig. 7. Fig. 7 (a) gives the reference speed.

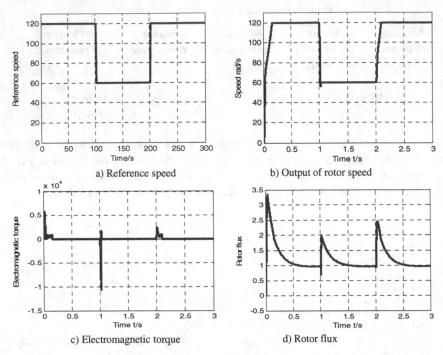

a) Reference speed

b) Output of rotor speed

c) Electromagnetic torque

d) Rotor flux

Fig. 7. Response of fuzzy-PID control by applying load torque T=0 and a variable reference speed from t=0 to 3s.

Fig. 8 shows the dynamic responses of fuzzy PID control with disturbance on the torque at $t = 1.5s$. It is noted that self-tuning fuzzy PID controller performs well in terms of speed rise time and disturbance rejection.

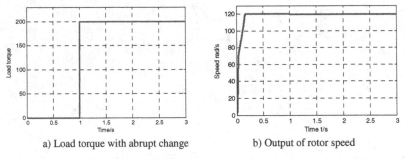

a) Load torque with abrupt change

b) Output of rotor speed

Fig. 8. Response of Fuzzy-PID controller by applying reference speed as 120 rad/s and load torque with abrupt change at t=1.5 from 0 to 200Nm

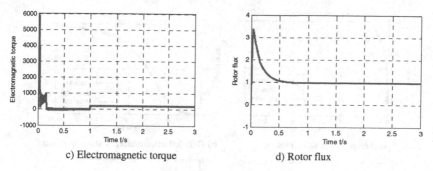

c) Electromagnetic torque   d) Rotor flux

Fig. 8. (*Continued*)

5.2   Application to Synchronization Control of TBM Cutter Head Systems

To verify effectiveness of the proposed fuzzy-PID controller of induction motor used in TBM cutter head driving system, a multiple motor system including three induction motors is established and simulated in MATLAB based on master-slave synchronization techniques [10]. The master-slave control strategy is widely adopted in cutter head rotation electric systems. Master-slave configuration for a three-motor system is shown in Fig. 9. The output speed of the master motor serves as the speed reference of the other two slave motors.

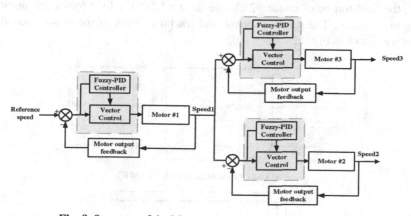

Fig. 9. Structure of the Master-slave for a three motor system

In actual construction, even if we select the same motors, the parameters cannot be exactly the same because of environmental disturbs and some other reasons [11]. In this simulation, we set the three motors with different parameters, where the rotor resistance value of motor #1 is half of the other two motors, and the inductance value of motor #2 is half of motor #1 and motor#3, the mutual inductance value of motor #3 is half of motor #1 and motor #2. The simulation results are shown in Fig. 10.

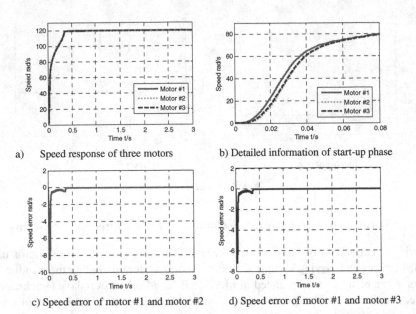

a)   Speed response of three motors

b) Detailed information of start-up phase

c) Speed error of motor #1 and motor #2

d) Speed error of motor #1 and motor #3

Fig. 10. Response of speed master-slave control of three motors with different parameters

Then we consider the load change case. The load torque of motor #1 change at $t = 1s$, the load torque of motor #2 change at $t = 1.5s$, the load torque of motor #3 change at $t = 2.5s$. The speed response and tracking errors of the master and slave motors are shown in Fig. 11.

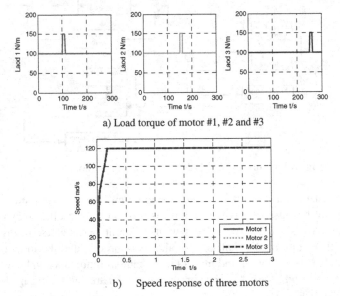

a) Load torque of motor #1, #2 and #3

b)   Speed response of three motors

Fig. 11. Response of speed master-salve control of three motors with random load change

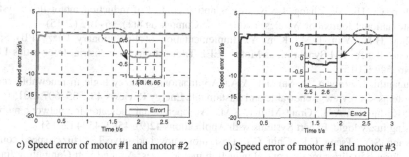

c) Speed error of motor #1 and motor #2   d) Speed error of motor #1 and motor #3

Fig.11. (*Continued*)

From the above results we can conclude that multi-motor system with fuzzy-PID based induction motor can achieve fast track performance and high synchronization accuracy. It can effectively eliminate the impact of motor parameter variations and abrupt load change of different motors.

6   Conclusion

In this paper, a Fuzzy-PID controllers of a field-oriented induction motor is proposed and implemented. The design of the controller is achieved using a simple fuzzy logic adaptation mechanism. Through simulation analysis, it can be concluded that the designed adaptive fuzzy-PID controller achieves a good dynamic behavior with various reference speed and a rapid setting time without overshoot. Meanwhile, its application to TBM cutter head driving system using master-slave synchronization techniques is accuracy which verify effectiveness of the proposed fuzzy-PID controller.

Acknowledgements. This work was supported by National 973 Program of China (No. 2013CB035406), National Natural Science Foundation of China (No. 61174059, 61233004, 61433002).

References

1. Talebi, K., et al.: Modeling of soil movement in the screw conveyor of the earth pressure balance machines (EPBM) using computational fluid dynamics. Tunnelling and Underground Space Technology **47**, 136–142 (2015)
2. Ling, J., et al.: Study of TBM cutterhead fatigue crack propagation life based on multi-degree of freedom coupling system dynamics. Computers and Industrial Engineering **83**, 1–14 (2015)
3. Zhou, D., Zhao, J., Liu, Y.: Predictive torque control scheme for three-phase four-switch inverter-fed induction motor drives with DC-link voltages offset suppression. IEEE Transactions on Power Electronics **30**(6), 3309–3318 (2015)
4. Wang, S.Y., et al.: An adaptive supervisory sliding fuzzy cerebellar model articulation controller for sensorless vector-controlled induction motor drive systems. Sensors (Switzerland) **15**(4), 7323–7348 (2015)

5. Chen, C., Hu, G., Zhang, L.: Neuron adaptive control of induction motor drive system. International Journal of Wireless and Mobile Computing **8**(1), 95–102 (2015)
6. Choi, H.H., Yun, H.M., Kim, Y.: Implementation of Evolutionary Fuzzy PID Speed Controller for PM Synchronous Motor. IEEE Transactions on Industrial Informatics **11**(2), 540–547 (2015)
7. Suganthi, L., Iniyan, S., Samuel, A.A.: Applications of fuzzy logic in renewable energy systems - A review. Renewable and Sustainable Energy Reviews **48**, 585–607 (2015)
8. El-Rashidy, R.A.H., Grant-Muller, S.M.: An operational indicator for network mobility using fuzzy logic. Expert Systems with Applications **42**(9), 4582–4594 (2015)
9. Wang, S.Y., Tseng, C.L., Chiu, C.J.: Design of a novel adaptive TSK-fuzzy speed controller for use in direct torque control induction motor drives. Applied Soft Computing Journal **31**, 396–404 (2015)
10. Perez-Pinal, F.J., et al.: Comparison of multi-motor synchronization techniques. In: 30th Annual Conference of IEEE on Industrial Electronics Society. IECON 2004 (2004)
11. Jeftenic, et al.: Controlled multi-motor drives. In: International Symposium on Power Electronics, Electrical Drives, Automation and Motion. SPEEDAM 2006 (2006)

Industrial Robot and Its Applications

Dynamics Modeling and Simulation
of Robot Manipulator

Lixin Yang and Xianming Zhang[✉]

Guangdong Provincial Key Laboratory of Precision Equipment and Manufacturing
Technology, School of Mechanical and Automotive Engineering,
South China University of Technology, Guangzhou, China
{melxyang,zhangxm}@scut.edu.cn

Abstract. Robot manipulators are extensively used in industry and research operations. To increase system stability and precision and to obtain small amplitudes of vibration, it is important to describe the relationship between force and motion. In this paper the dynamic model of 6 DOF robot manipulator is developed by Newton–Euler equations and its dynamical behavior was analyzed. Furthermore, the result is compared with mechanical system dynamics analysis software, ADAMS. It is presented as groundwork for future work.

Keywords: Dynamic model · Robot manipulator · ADAMS

1 Introduction

The modeling and simulation of 6 DOF robot manipulators have received great consideration in recent years in order to increase system stability and precision [1-4]. Due to the field's importance and applicability, such as to industry and research operations, 6 DOF robot manipulator dynamics and control is considered to be a challenging research problem.

The dynamic response of a robot manipulator is affected by wear, material deformation or joint clearance of the parts [4-7], which results in a high level of vibration and a low ability to perform accurate and safe operations. To better understand and improve their mechanical response, accurate mathematical modeling and simulation of such systems have been considered by researchers worldwide [8-11]. Robot manipulator is a spatial multibody system, the problem of the dynamic behavior of planar multibody systems are not valid for spatial multibody systems, where the system motion is not limited to be planar. The main purpose of this paper is to present effective methodologies for spatial multibody systems' dynamical behavior analysis. Due to their relevance for this paper, some aspects of the multibody formulation for spatial systems, based on the Cartesian coordinates, are reviewed here to introduce the basic aspects on the dynamic modeling of spatial multibody systems. In addition, dynamic simulations are performed by the dynamic simulation software ADAMS. The computational results of the mechanical model are in agreement with the simulation results of ADAMS, proving that the mechanical model is reasonable.

© Springer International Publishing Switzerland 2015
H. Liu et al. (Eds.): ICIRA 2015, Part III, LNAI 9246, pp. 525–535, 2015.
DOI: 10.1007/978-3-319-22873-0_47

2   Dynamics Modeling of Robot Manipulator

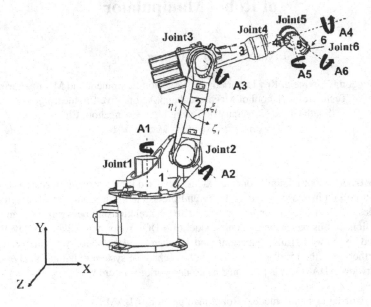

Fig. 1. 6 DOF Robot Manipulator

2.1   6D Spatial Vector

Spatial dynamics analysis requires powerful mathematical techniques, particularly for describing the angular orientation of a body in a global coordinate system. Therefore, this section is mostly devoted to developing the techniques concentrate on a set of orientation coordinates known as Euler parameters, which may drastically simplify the mathematical formulations for large-scale computer programs.

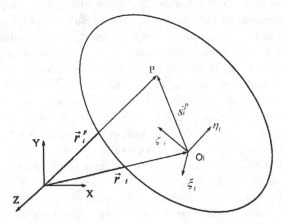

Fig. 2. Cartesian coordinates for a rigid body

As shown in Figure 2, an unconstrained rigid body i in space requires six indepen-dent coordinates to determine its three-coordinate configurations to specify translation and three specify rotation. The six coordinates define the location of a Cartesian coor-dinate system $\xi_i\eta_i\zeta_i$ that is attached at its center of mass relative to the global coordi-nate system XYZ, the position and orientation of the rigid body are defined by a set of translational and rotational coordinates.

The position of the body with respect to global coordinate system XYZ is defined by the coordinate vector

$$r_i = \begin{bmatrix} x\ y\ z \end{bmatrix}_i^T, i = 1 \sim 6 \qquad (1)$$

That represents the location of the local reference frame $\xi_i\eta_i\zeta_i$.

The orientation of the body is described by the rotational coordinate's vector

$$p_i = \begin{bmatrix} e_0\ e_1\ e_2\ e_3 \end{bmatrix}_i^T, i = 1 \sim 6 \qquad (2)$$

Which includes the Euler parameters e_0, e_1, e_2, e_3 for the rigid body [12].

Therefore, the vector of coordinates that describes completely the rigid body i is,

$$q_i = \begin{bmatrix} r_i^T\ p_i^T \end{bmatrix}^T, i = 1 \sim 6 \qquad (3)$$

$$q = \begin{bmatrix} q_1^T, q_2^T, ..., q_{nb}^T \end{bmatrix}^T, \ nb = 6 \qquad (4)$$

Where, nb is the Robot Manipulator's member.

2.2   Kinematic Analysis of Multibody Systems

The kinematic analysis consists in the study of the system's motion independently of the forces that produce it. In particular, it analysis involves the determination of posi-tion, velocity and acceleration of the system components, and only the interaction between the geometry and the motions of the system is obtained. Since the interaction between the forces and the system's motion is not considered, the motion of the sys-tem need to be specified to some extent, that is, the kinematic characteristics of some driving elements need to be prescribed, while the kinematic motion characteristics of the remaining elements are obtained using the kinematic constraint equations, which describe the topology of the system.

Point P on body i can be defined by position vector s_i^P, which represents the loca-tion of point P with respect to the body-fixed reference frame $\xi_i\eta_i\zeta_i$, and by the global position vector r_i, that is,

$$r_i^P = r_i + s_i^P = r_i + A_i s_i^{'P} \qquad (5)$$

Where A_i is the transformation matrix for body i that defines the orientation of the referential $\xi_i \eta_i \zeta_i$ with respect to the referential frame XYZ. The transformation matrix is expressed as function of the four Euler parameters as [12],

$$
A_i = 2 \begin{bmatrix} e_0^2 + e_1^2 - \dfrac{1}{2} & e_1 e_2 - e_0 e_3 & e_1 e_3 + e_0 e_2 \\[2mm] e_1 e_2 + e_0 e_3 & e_0^2 + e_2^2 - \dfrac{1}{2} & e_2 e_3 - e_0 e_1 \\[2mm] e_1 e_3 - e_0 e_2 & e_2 e_3 + e_0 e_1 & e_0^2 + e_3^2 - \dfrac{1}{2} \end{bmatrix}_i
\tag{6}
$$

Notice that the vector s_i^P is expressed in global coordinates whereas the vector $s_i'^P$ is defined in the body i fixed coordinate system. Throughout the formulation presented in this work, the quantities with (.)' means that (.) is expressed in local system coordinates.

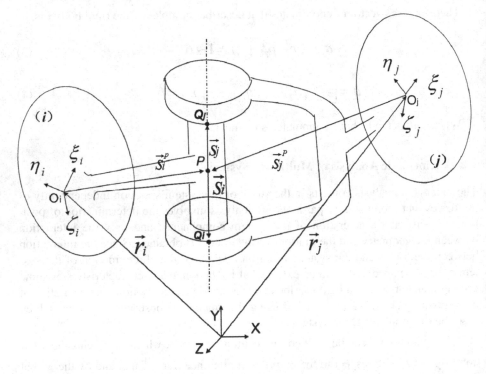

Fig. 3. Spatial revolute joint

A revolute joint between bodies i and j is shown in Fig. 3. An arbitrary point P point on the revolute-joint axis has constant coordinates in both local coordinate systems. Two other points, Q_i on body i and Q_j on body j, are also chosen arbitrarily on the joint axis. It is clear that vectors S_i and S_j must remain parallel. Therefore, there are five constraint equations for a revolute joint:

$$\phi \equiv \phi(q) \equiv \phi(r_1, p_1, ..., r_b, p_b) = 0 \tag{7}$$

$$\phi^{(r,5)} \equiv \begin{cases} \phi^{(s,3)} \\ \phi^{(p_1,2)} \end{cases} = \begin{cases} r_i + A_i s_i'^P - r_j - A_j s_j'^P = 0 \\ \tilde{s}_i s_j = 0 \end{cases} \tag{8}$$

There is only one relative degree of freedom between two bodies connected by a revolute joint.

The first time derivative with respect to time of Eq. (7) provides the velocity constraint equations

$$\dot{\phi} \equiv \phi_q \dot{q} = \begin{bmatrix} \phi_{r1}, \phi_{p1}, ..., \phi_{rb}, \phi_{pb} \end{bmatrix} \begin{bmatrix} \dot{r}_1 \\ \dot{p}_1 \\ \cdot \\ \cdot \\ \cdot \\ \dot{r}_b \\ \dot{p}_b \end{bmatrix} = \begin{bmatrix} \phi_{r1}, \frac{1}{2}\phi_{p1}L_1^T, ..., \phi_{rb}, \frac{1}{2}\phi_{pb}L_b^T \end{bmatrix} \begin{bmatrix} \dot{r}_1 \\ \omega_1' \\ \cdot \\ \cdot \\ \cdot \\ \dot{r}_b \\ \omega_b' \end{bmatrix} = -\phi_t \equiv v \tag{9}$$

where ϕ_q is Jacobian Matrix of the constraint equations, that is, $\phi_q = \dfrac{\partial \phi}{\partial q}$, and

similarly, $\phi_t = \dfrac{\partial \phi}{\partial t}$, $\phi_r = \dfrac{\partial \phi}{\partial r}$, $\phi_p = \dfrac{\partial \phi}{\partial p}$.

A second differentiation of Eq. (7) with respect to time leads to the acceleration constraint equations

$$\phi_q \ddot{q} = \begin{bmatrix} \phi_{r1}, \phi_{p1}, ..., \phi_{rb}, \phi_{Pb} \end{bmatrix} \begin{bmatrix} \ddot{r}_i \\ \ddot{p}_1 \\ \cdot \\ \cdot \\ \cdot \\ \ddot{r}_b \\ \ddot{p}_b \end{bmatrix} = \gamma \tag{10}$$

The velocities and accelerations of body i use the angular velocities ω_i' and accelerations $\dot{\omega}_i'$ instead of the time derivatives of the Euler parameters, which simplifies the mathematical formulation and do not have critical singular cases. When Euler parameters are employed as rotational coordinates, the relation between their time derivatives, $\dot{P}_i$, and the angular velocities is expressed by [12],

$$\dot{P}_i = \frac{1}{2} L^T \omega_i' \tag{11}$$

Where the auxiliary matrix L is the function of Euler parameters [12],

$$L_i = \begin{bmatrix} -e_1 & e_0 & e_3 & -e_2 \\ -e_2 & -e_3 & e_0 & e_1 \\ -e_3 & e_2 & -e_1 & e_0 \end{bmatrix}_i \tag{12}$$

Then,

$$\ddot{P}_i = \frac{1}{2} L_1^T \dot{\omega}_i' - \frac{1}{4} \omega_i'^2 P_i \tag{13}$$

Then,

$$\phi_{pi} \ddot{P}_i = \phi_{pi} \left(\frac{1}{2} L_1^T \dot{\omega}_i' - \frac{1}{4} \omega_i'^2 P_i \right) = \frac{1}{2} \phi_{pi} L_1^T \dot{\omega}' - \frac{1}{4} \omega_i'^2 \phi_{pi} P_i \tag{14}$$

Then,

$$\left[\phi_{r1}, \frac{1}{2} \phi_{p1} L_1^T, \dots, \phi_{rb}, \frac{1}{2} \phi_{pb} L_b^T \right] \left[\ddot{r}_1 \quad \dot{\omega}_1' \quad \cdots \quad \ddot{r}_b \quad \dot{\omega}_b' \right]^T = \gamma^\# \tag{15}$$

Where, when $\phi^{(s,3)} = r_i + A_i s_i'^P - r_j - A_j s_j'^P$, $\gamma^{\#(s,3)} = -\tilde{\omega}_i \dot{s}_i^P + \tilde{\omega}_j \dot{s}_j^P$, when $\phi^{(P_1,2)} = \tilde{s}_i s_j$, $\gamma^{\#(P_1,2)} = -2\tilde{s}_i \dot{s}_j + \tilde{s}_j \tilde{\omega}_i \dot{s}_i - \tilde{s}_i \tilde{\omega}_j \dot{s}_j$.

Eq. (8) – (10) are the system kinematic constraint equations.

2.3   Dynamic Analysis of Multibody Systems

The dynamic analysis of multibody systems aims to understanding the relationship between the motion of the system parts and the causes that produce the motion including external applied forces and moments. The Newton–Euler equations of the system in Cartesian coordinates are written as

$$M\dot{h}+b=g \tag{16}$$

$$M\dot{h}+b=g+g^{c} \tag{17}$$

$$g^{c}=B^{T}\lambda \tag{18}$$

$$M\dot{h}+b-B^{T}\lambda=g \tag{19}$$

Where M is the system mass matrix,

$$M=\begin{bmatrix} N_{1} & & & & \\ & J_{1}' & & 0 & \\ & & \cdot & & \\ & & & \cdot & \\ & 0 & & \cdot & \\ & & & & N_{b} \\ & & & & & J_{b}' \end{bmatrix} \tag{20}$$

λ is the vector of Lagrange multipliers,

$$\lambda=\left[\lambda_{1},\lambda_{2},...,\lambda_{m}\right]^{T} \tag{21}$$

g is the generalized force vector,

$$g=\begin{bmatrix} f_{1} & n_{1}' & \cdot & \cdot & \cdot & f_{b} & n_{b}' \end{bmatrix}^{T} \tag{22}$$

$$B=\left[\phi_{r1},\frac{1}{2}\phi_{p1}L_{1}^{T},...,\phi_{rb},\frac{1}{2}\phi_{pb}L_{b}^{T}\right] \tag{23}$$

$$\dot{h}=\begin{bmatrix} \ddot{r}_{1} & \dot{\omega}_{1}' & \cdot & \cdot & \cdot & \ddot{r}_{b} & \dot{\omega}_{b}' \end{bmatrix}^{T} \tag{24}$$

$$b=\begin{bmatrix} 0 & \tilde{\omega}_{1}'J_{1}'\omega_{1}' & \cdot & \cdot & \cdot & 0 & \tilde{\omega}_{b}'J_{b}'\omega_{b}' \end{bmatrix}^{T} \tag{25}$$

Eq. (19) is appended to Eq. (15), yielding a system of differential algebraic equations, written as

$$\begin{bmatrix} M & B^{T} \\ B & 0 \end{bmatrix}\begin{bmatrix} \dot{h} \\ -\lambda \end{bmatrix}+\begin{bmatrix} b \\ 0 \end{bmatrix}=\begin{bmatrix} g \\ \gamma^{\#} \end{bmatrix} \tag{26}$$

2.4   Computational Strategy to Solve the Equations of Motion

As shown before, for a multibody system of kinematic constrained rigid bodies, the equations of motion are given by Eq. (26) and solved for $\dot{h}$ and λ, with initial conditions for q^0 and h^0. The constraint violations result from accumulated numerical integration errors, and become more apparent with stiff systems, that is, when the natural frequencies of the system are widely spread. The stiffness can be produced by the physical characteristics of the multibody system, such as components with large differences in their masses, stiffness and/or damping [13]. Even though the initial conditions guarantee the non-violation of constraint equations on position and velocity level, during the course of numerical integration the numerical errors do not satisfy the constraint equations. The effect of these errors increases with time. Consequently, the constant distances cease to be constant and the points of the same element progressively move closer to or further away from their original position. So Eq. (26) represents an unstable system. The Baumgarte stabilization method is used to allow constraints to be slightly violated before corrective actions can take place, in order to force the violation to vanish [14]. The equations of motion for dynamic system utilizing Baumgarte's approach are stated in the form Eq. (27).

$$\begin{bmatrix} M & B^T \\ B & 0 \end{bmatrix} \begin{bmatrix} \dot{h} \\ -\lambda \end{bmatrix} + \begin{bmatrix} b \\ 0 \end{bmatrix} = \begin{bmatrix} g \\ \gamma^{\#} - 2\alpha\dot{\phi} - \beta^2\phi \end{bmatrix} \tag{27}$$

Where α, β are the Baumgarte parameters, which are positive constants.

2.5   Dynamic Simulation and Results

To simulate the motion of a manipulator we must make use of a model of the dynamics (Eq. (27)), which we have just developed. Simulation requires solving the dynamic equation for acceleration, then we apply Matlab to integrate the acceleration to positions and velocities. Manipulator system is a highly nonlinear system. In order to understand the properties of the manipulator to lay a foundation for the controller design and investigate its dynamical response to the load, the system is released from this initial configuration only subjected to gravity. The relevant parameters and simplified notations for simulation are given in table 1. The simulation results are shown from Fig. 4.

Table 1. Parameters for simulation

Arms	Length[mm]	Mass[kg]	Inertia characteristics[kgm^2]		
			$I_{\xi\xi}$	$I_{\eta\eta}$	$I_{\zeta\zeta}$
Arm 1	372.55	136.54	3.38	4.87	4.20
Arm 2	682.47	105.46	4.65	4.64	0.46
Arm 3	536.62	83.94	0.94	1.72	2.14
Arm 4	151.78	29.99	0.11	0.26	0.29
Arm 5	133.82	13.93	0.08	0.06	0.07
Arm 6	39.90	0.35	0.0001	0.0001	0.0002

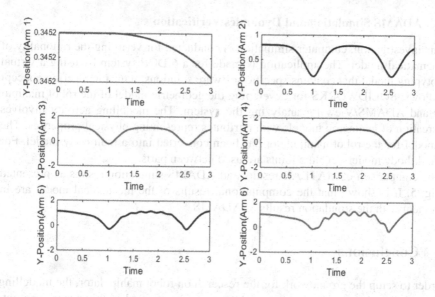

Fig. 4. The Y-Position laws for center mass of arms are 1, 2, 3, 4, 5 and 6

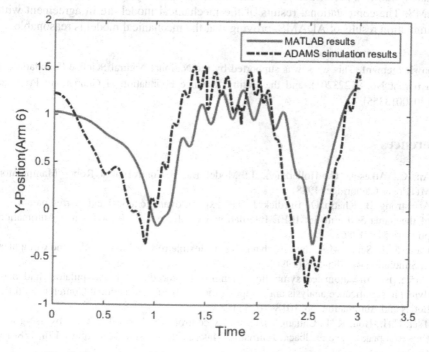

Fig. 5. Comparison of MATLAB results and ADAMS simulation results

2.6   ADAMS Simulation and Dynamics Verification

In this investigation, computer simulation is conducted for verifying the rationality of mathematical model. The application is made for a 6 DOF system by using a virtual prototyping model that contains specific software solutions in the engineering concept as follows: SOLIDWORKS for developing the geometric model of the robot manipulator and ADAMS/View for analyzing the system. The modelling activity involves the creation of geometric models with attributes representing physical properties. The 3D model for the robot manipulator has been converted into a multibody model. For the multibody model, revolute joints are used between parts.

The comparison of MATLAB results and ADAMS simulation results as presented in Fig. 5. It is shown that the computational results of the mechanical model are in agreement with the simulation results of ADAMS.

3   Conclusions

In order to setup the groundwork for the research on robot manipulator, the modelling and dynamic response of robot manipulator was investigated in this paper. The outcome of this research investigation is that, mainly dynamics analysis is successfully simulated, and it is performed by the mechanical system dynamics analysis software, ADAMS. The computational results of the mechanical model are in agreement with the simulation results of ADAMS, proving that the mechanical model is reasonable.

Acknowledgement. This work was supported by the National Natural Science Foundation of China (Grant No. 91223201), and the Natural Science Foundation of Guangdong Province (S2013030013355).

References

1. An, C., Atkeson, C., Hollerbach, J.: Model Based Control of a Robot Manipulator. MITPress, Cambridge (1988)
2. Armstrong, B., Khatib, O., Burdick, J.: The explicit dynamic model and inertial parameters of the puma 560 arm. In: IEEE International Conference on Robotics and Automation, pp. 510–518 (1986)
3. Carrera, E., Serna, M.A.: Inverse dynamics of flexible robots. Mathematics and Computers in Simulation **41**, 485–508 (1996)
4. Corke, P.: An automated symbolic and numeric procedure for manipulator rigid body dynamic significance analysis and simplication. In: IEEE International Conference on Robotics and Automation, pp. 1018–1023 (1986)
5. Tsai, C.H., Hou, K.H., Chuang, H.T.: Fuzzy control of pulsed GTA welds by using real-time root bead image feedback. Journal of Materials Processing Technology **176**, 158–167 (2006)
6. Bauchau, O.A., Rodrigez, J.: Modelling of joints with clearance in flexible multibody systems. International Journal of Solids and Structures **39**, 41–63 (2002)

7. Flores, P., Ambrosio, J., Claro, J.P.: Dynamic analysis for planar multibody mechanical systems with lubricated joints. Multibody System Dynamics 12, 47–74 (2004)
8. Flores, P., Ambrosio, J., Claro, J.C.P., Lankarani, H.M.: Dynamic behavior of planar rigid multibody systems including revolute joints with clearance. Proceedings of the Institution of Mechanical Engineers Part K: Journal of Multi-body Dynamics 221, 161–174 (2007)
9. Flores, P., Ambrosio, J., Claro, H.C.P., Lankarani, H.M., Koshy, C.S.: A study on dynamics of mechanical systems including joints with clearance and lubrication. Mechanism and Machine Theory 41, 247–261 (2006)
10. Flores, P.: Modeling and simulation of wear in revolute clearance joints in multibody systems. Mechanism and Machine Theory 44, 1211–1222 (2009)
11. Chang, C.O., Nikravesh, P.E.: An adaptive constraint violation stabilization method for dynamic analysis of mechanical systems. Journal of Mechanisms, Transmissions, and Automation in Design 107, 488–492 (1985)
12. Nikravesh, P.E.: Computer-Aided Analysis of Mechanical Systems. Prentice Hall, Englewood Cliffs (1988)
13. Flores, P., Machado, M., Seabra, E., Sliva, M.T.: A parametric study on the Baumgarte stabilization method for forward dynamics of constrained multibody systems. ASME J. Comput. Nonlinear Dyn. 6(1), 1–9 (2011)
14. Garcia, D.E., Jalon, J., Bayo, E.: Kinematic and dynamic simulations of multibody systems. Springer, New York (1994)

A Vision Location System Design
of Glue Dispensing Robot

Nianfeng Wang[✉], Jinghui Liu, Shuai Wei, and Xianmin Zhang[✉]

School of Mechanical and Automobile Engineering,
South China University of Technology, Guangzhou, China
{menfwang,zhangxm}@scut.edu.cn

Abstract. The location accuracy of the workpiece is the key problem in the glue dispensing operation of the automobile engine cover. A vision location system is designed for the glue dispensing robot to locate the position of the feature points in the engine cover. The system correct the original teaching trajectory in real time to maintain the accuracy of the gluing operation. The system is based on a digital signal processor (DSP) dedicated to image processing for its real-time performance. Image preprocessing operations including gray value linear transformation and median filtering are applied to modify the image quality for the subsequent feature-point detection procedure. Experiments are conducted and the result shows that the system is feasible after the calibration of the camera.

Keywords: Vision location · Robot gluing · DSP · Feature point detection

1 Introduction

In conventional industrial robot control system, online teaching and offline programming method are commonly used to produce job tasks. In those ways, the robot can only perform a specified task with fixed location, which leads to strict requirements for the position and orientation of the workpiece. Once the errors of the positioning of the workpiece or the geometric error of workpiece is too large to be ignored, there will be a great impact on the quality of production, even leading to annoying stop of the production line. In view of this situation, an industrial robot combined with the vision system is necessary. In use of the locating information from the vision system, the robot can be guided. It is capable of effectively improving the operating efficiency and production quality [1-5].

The visual function existing in robot systems can be divided into two categories: the closed visual systems and the open visual systems. There are many closed systems in

This work is supported by National Natural Science Foundation of China (Grant Nos. 51205134, 91223201), Science and Technology Program of Guangzhou (Grant No. 2014Y2-00217), Research Project of State Key Laboratory of Mechanical System and Vibration (MSV201405), the Fundamental Research Funds for the Central University (Fund No. 2015ZZ007) and Natural Science Foundation of Guangdong Province (S2013030013355).

H. Liu et al. (Eds.): ICIRA 2015, Part III, LNAI 9246, pp. 536–551, 2015.
DOI: 10.1007/978-3-319-22873-0_48

the market, such as ABB's True View [6], Fanuc's iRvision [7] as well as Epson's Vision Guide CV1. This type of vision systems can only be specified in conjunction with the specific robot controller. Thus does not have the versatility. Open robot vision system is not limited to a particular robot controller, such as Cognex VisonPro [8], Keyence's CV-X200. As long as the robot controller is open, it can be easily integrated with the open robot vision system. Since the industrial robot control system developed in this paper is open-ended, the vision system can be designed as open from so as to embedding it into the existing system as a functional module. It will not only be able to reduce the room of equipment, but also perform unified management through Human Machine Interface (HMI) module of the robot controller.

2   Location System Layout Based on DSP

2.1   Error Sources Analysis of Glue Dispensing Trajectory

In the production line, the robot just needed to be taught for one time in case of the workpiece of the same kind. The operation is repeated after the first call to teaching program, so that the path relative to the base operating system of the robot is fixed. But in practice, the position of the workpiece without precision fixtures will change randomly, making positioning error. In the automobile engine cover's rubberized platform, the error sources of rubberized track include the following two aspects:

(1) Tooling positioning error

The cover of automobile engine is moving from the origin station to the station on the glue position through the holder. Since the piece is heavy, and there are no obvious positioning features, it may cause certain positioning error. Also, due to the large size of the workpiece, small positioning errors are likely to cause a large tracking error.

(2) Error due to the cylinder

In order to facilitate the movement of the workpiece to the next station, gluing table controlled by the piston of the cylinder is installed. The vibration due to the cylinder during the movement is severe, and table locations may change during each gumming cycle, leading to offset compared to the teaching trajectory.

After the experiments, it's found that the glue track error caused by the above two factors is large and cannot be ignored. Re-teaching on every workpiece gumming circle is unwise taking into account the operational efficiency. Therefore, it's needed to develop vision system to handle the errors.

2.2   Visual Positioning Scheme of Glue Dispensing Based on DSP

For the gluing trajectory error problem, this paper presents a visual system, as shown in Figure 1. First, glue trajectory of workpiece is determined via teaching operation. After each gluing operation cycle, the feature points of the workpiece are detected via visual positioning system after the workpiece is in place, after which the actual position of the workpiece is obtained. After compared with the teaching trajectory point, the relative

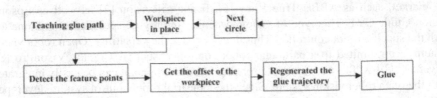

Fig. 1. Visual system design of glue operation

displacement of the workpiece can be calculated. Then the trajectory can be adjusted accordingly based on the offset value, namely, the overall translation or rotation operation. A new trajectory corresponding to the current workpiece can be generated.

The current implementations of image processing are mainly based on industrial computer (IPC) or DSP chip. The operation based on IPC is most commonly used for its simplicity, which has rich resources. However, because of its high CPU consumption, the processing speed is relatively slow, leading to bad real time performance. Therefore, for robotic vision positioning applications with the high real-time requirements, DSP-based implementation has more advantages.

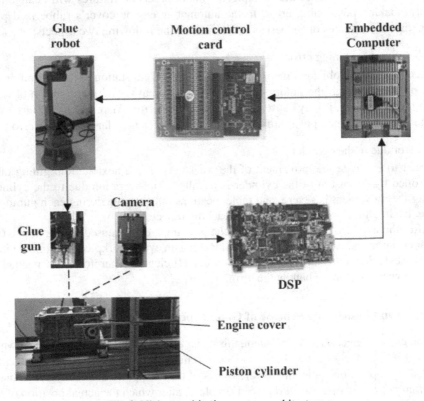

Fig. 2. Vision positioning system architecture

The proposed vision positioning system architecture is shown in Figure 2, where the camera is mounted on the end of the robot body. DSP-based image capture, processing and output units are integrated into the robot controller as embedded subsystems. The function of the CCD camera is to convert the optical signal of the workpiece and its surroundings into an electric signal, namely the image. The image sequences of analog video signal are output into the VPORT of DSP. Since VPORT can be seamlessly connected with the general image A/D circuit, the image signal captured by the CCD sensor is transferred into the DSP memory for calling of the image processing program.

The DSP is integrated with hardware multiplier and pipelined instruction set for handling high-speed and parallel calculation, leading great efficient processing of image data. The DSP is also integrated on-chip peripherals (EMAC), it's easily to directly connect seamlessly with the embedded computer inside the robot controller via the network port expansion chip PHY.

The workflow of the system is as follows: when the workpiece is in place, the robot controller sends a directive signal to the visual positioning system via Ethernet; After receiving the relevant instruction via the network transmission program, VPORT interrupt service of the DSP will be enabled, and image processing programs are performed; the image processing program gets the image data calling the VPORT from memory, and the associated algorithm calculations are performed, obtaining the final position information; The network transmission program transmits the position data into the robot controller in the form of a string obtained from the image processing and closes VPORT interrupt service, waiting for the next instruction; the robot controller receives the results of visual positioning system, modifies the location information of the corresponding point, and then executes the control program.

The software was developed in the Code Composer Studio 3.3 (CCS 3.3) with the C language and the software architecture is shown in Figure 3.

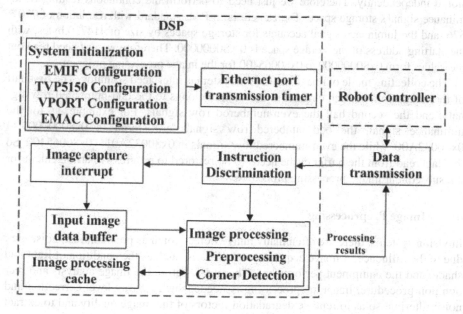

Fig. 3. Software System Architecture

After powering up the DSP, the system will be initialized, including the configuration of the External Memory Interface (EMIF), Video decoder (TVP5150), VPORT and Ethernet Media Access Controller (EMAC). TMS320DM642 links external FLASH and SDRAM by the EMIF. The interface timing of these two different storages can be configured through the initialization, so the program can easily manage the data in memory. TVP5150 is a high performance video decoder which can turn the NTSC, PAL video signals into digital color difference signals (YUV 4: 2: 2) and the output format is ITU-R BT.656. Since TMS320DM642 supports BT.656 video data stream format, so will be able to get the video stream seamlessly by TVP5150. TMS32 0DM642 has three video ports (VP0, VP1 and VP2), and the system requires only one video port (VP1). DSP needs to receive instructions from the robot controller and simultaneously transmit the results of image processing back to the robot controller, and this process is accomplished via Ethernet communication, and the EMAC need to be configured for establishing the connection between the two communications.

3   Image Processing Procedure

3.1   Image Acquisition

After the image decoding through TVP5150, the image data format is YUV 4:2:2, containing the brightness signal - Y, red difference signal - RY (U) and blue difference signals - B Y (V). Because the system applies the Harris corner detection algorithm, the information of the gray level image is required, namely the luminance signal Y. In this system, image data is stored into $\overline{CE0}$ space of the EMIF, and Y, U and V signals are continuously stored independently. Therefore we just need to perform the continuous reading of luminance signal's storage space. The camera is in PAL standard with resolution of 720 * 576, and the luminance signal accounts for storage spaces by size of 414720 bytes, with the starting address of the storage space is 0 x80000000. Therefore we just need read out the values form 0x80000000 to 0x80065400 for the image processing program.

The collecting mode of image data of the system is interlaced scanning. The first half of the image data of each channel in the cache belongs to the odd-numbered row signals, and the second half the even-numbered row signals. For example, as for the luminance signal, the odd-numbered row signals are stored in 0x80000000 ~ 0x80032A00, while the even-numbered row signals in 0x80032A01 ~ 0x80065400. So the data read from the buffer cache needs to be restored to progressive arrangement for the subsequent image processing procedure.

3.2   Image Preprocessing

In vision systems, there are often many image defects such as poor contrast, noise etc., due to the influence of image acquisition conditions, such as the condition of light and shade, and the equipment performance. Therefore, before the image analysis and recognition procedure, image preprocessing is needed such as gray level correction and noise filtering, so as to remove degradation factors of the image quality and to extract useful information easily.

Gray Value Linear Transformation

In the case of under- or over-exposure, the image may be confined to the gray in a smaller range. After performing the gray value linear transformation, the dynamic range of the image is increased and the contrast is extended, making it easier to distinguish the workpiece from the background with a more distinct features.

The image gray value linear transformation is transformed the gray value of all the points in accordance with gray value linear transform function. The transform equation is given by:

$$g(x, y) = f_A f(x, y) + f_B$$

$f(x, y)$ is the gray value of the pixel point (x, y) in the input image, and the $g(x, y)$ is the gray value after the transformation of this pixel point. To extend the image contrast, the value of f_A is set to 2 and f_B is set to -128. The gray value level is from 0 to 255. In order to ensure the consistency of gray level, the actual transform equation is given by:

$$g(x, y) = \begin{cases} 0 & f(x, y) \le 64 \\ 2f(x, y) - 128 & 64 < f(x, y) < 192 \\ 255 & f(x, y) \ge 192 \end{cases}$$

Figure 4 shows the gray value linear transform process. The gray value of the original image is magnified two times in the range of 64 to 192, while others in both ends were intercepted with minimum and maximum gray value. The process is equivalent to threshold segmentation, which can extend the contrast and make the features more distinct.

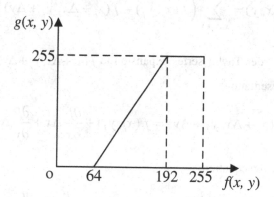

Fig. 4. Gray value linear transformation

Median Filtering

Median filtering is a nonlinear filtering method. The basic principle is to replace the value of an image pixel with the median value of its neighbor region pixels. It can overcome the ambiguity problem of image detail brought from the linear filter such as minimum mean square filter and mean filtering, which is particularly effective in filter out pulse noise and salt and pepper noise. Different shapes and sizes of windows have great influence on the filtering result. Thus the selection of mask is greatly depended on the image quality and different application requirements. Commonly used two-dimension median filtering mask have the shape of a square, circular, cross-shaped and circular, etc. For images with smooth contours, a square or circular mask is adopted, and for images with sharp corners, a cross window is better. The mask size should not exceed the size of the smallest valid object in the image. Employed herein is the cross-shaped mask with the size of 5.

The gray value of the median filtering mask w is $f(x, y)$, and the result is given by:

$$g(x, y) = \underset{w}{Med}\{f(x, y)\}$$

The bubble sort algorithm is applied to compute the $g(x, y)$. The gray value of each pixel in the mask is compared one by one from small to large order and the median value is selected as the result.

3.3   Feature Point Detection

The basic principle of Harris algorithm [9] is selecting a mask with the center coincide with the target pixel and calculating the changes of the gray value after moving the mask in all directions, so as to detect the feature points.

The mask (w) move a small displacement $(\Delta x, \Delta y)$ in any direction and the sum of squares of the change of gray value in w is given by:

$$G(x, y) = \sum_{x_i, y_j \in w} \left(f(x_i, y_j) - f(x_i + \Delta x, y_j + \Delta y) \right)^2$$

Perform the first-order Taylor series expansion to $f(x_i + \Delta x, y_j + \Delta y)$, and we get an approximate representation:

$$f(x_i + \Delta x, y_j + \Delta y) = f(x_i, y_j) + \frac{\partial f}{\partial x} \Delta x + \frac{\partial f}{\partial y} \Delta y$$

We can get the follow equation:

$$G(x, y) = \sum_{x_i, y_j \in w} \left(f(x_i, y_j) - f(x_i, y_j) - \frac{\partial f}{\partial x} \Delta x - \frac{\partial f}{\partial y} \Delta y \right)^2$$

$$= [\Delta x, \Delta y] M' \begin{bmatrix} \Delta x \\ \Delta y \end{bmatrix}$$

M' is the second derivative at the origin, which can be given by:

$$M' = \begin{bmatrix} \sum\limits_{x_i, y_j \in w} \dfrac{\partial^2 f}{\partial x^2} & \sum\limits_{x_i, y_j \in w} \dfrac{\partial f}{\partial x} \dfrac{\partial f}{\partial y} \\ \sum\limits_{x_i, y_j \in w} \dfrac{\partial f}{\partial x} \dfrac{\partial f}{\partial y} & \sum\limits_{x_i, y_j \in w} \dfrac{\partial^2 f}{\partial y^2} \end{bmatrix}$$

In order to reduce the noise, Gaussian filter mask is commonly used, and the window function is given by:

$$h(u, v) = \frac{1}{2\pi\sigma^2} e^{-\frac{u^2 + v^2}{2\sigma^2}}$$

The u and v represent the size of a Gaussian mask, and σ is standard deviation. Local autocorrelation function of gradation image can be obtained:

$$E(x, y) = \sum\limits_{x_i, y_j \in w} h(u, v)\left(f(x_i, y_j) - f(x_i + \Delta x, y_j + \Delta y) \right)^2$$

$$= [\Delta x, \Delta y] M \begin{bmatrix} \Delta x \\ \Delta y \end{bmatrix}$$

M is the autocorrelation matrix, which has the form:

$$M = \begin{bmatrix} A & C \\ C & B \end{bmatrix}$$

In which:

$$\begin{cases} A = \dfrac{\partial^2 f}{\partial x^2} \otimes h \\ B = \dfrac{\partial^2 f}{\partial y^2} \otimes h \\ C = \dfrac{\partial f}{\partial x} \dfrac{\partial f}{\partial y} \otimes h \end{cases}$$

$\otimes$ denotes convolution procedure.

Let $\lambda 1$, $\lambda 2$ the two eigenvalues of the matrix M, and the $\lambda 1$, $\lambda 2$ represents the curvature of the local autocorrelation function E. Through the analysis of the eigenvalues, we can draw the following three conditions:

(1) If the two eigenvalues are very small, the curvature of the partial autocorrelation function is flat, which means a small change function along any direction. The gray area of the mask is approximately constant.

(2) If one eigenvalue is large and the other is very small, the local autocorrelation function is ridged. When there is very little changes along the ridge direction, there will be great variation in the vertical direction, which means the mask is located in an edge region.

(3) If the two eigenvalues are large, the curvature of the partial autocorrelation function will change a lot when the mask moving in any direction. The image region of the mask is like winger.

When getting the third case, it can be determined whether the related point a corner or not through further analysis of the local area.

In the visual system, we implement the algorithm in the following specific steps:

(1) Choose the moving masks of x and y direction respectively:

$$
x_{msk} = \begin{bmatrix} -1 & 0 & 1 \\ -1 & 0 & 1 \\ -1 & 0 & 1 \end{bmatrix} \qquad y_{msk} = \begin{bmatrix} -1 & -1 & -1 \\ 0 & 0 & 0 \\ 1 & 1 & 1 \end{bmatrix}
$$

The gray gradient of the pixel point $f(x, y)$ of the two directions is given by:

$$
\begin{cases}
\dfrac{\partial f}{\partial x} = f(x-1, y+1) - f(x-1, y-1) + f(x, y+1) - f(x, y-1) \\
\qquad + f(x+1, y+1) - f(x+1, y-1) \\
\dfrac{\partial f}{\partial y} = f(x+1, y-1) - f(x-1, y-1) + f(x+1, y) - f(x-1, y) \\
\qquad + f(x+1, y+1) - f(x-1, y+1)
\end{cases}
$$

(2) The size of the Gaussian smoothing mask selected is 5, namely u = 5 and v = 5. The standard deviation σ is 0.8. The calculated Gaussian mask is given by:

$$
h = \frac{1}{32768} \begin{bmatrix}
15 & 163 & 358 & 163 & 15 \\
163 & 1708 & 3730 & 1708 & 163 \\
358 & 3730 & 8148 & 3730 & 358 \\
163 & 1708 & 3730 & 1708 & 163 \\
15 & 163 & 358 & 163 & 15
\end{bmatrix}
$$

(3) After performing the convolution of M' and h, the M is given by:

$$\begin{cases} A = \sum_{i=-2}^{2}\sum_{j=-2}^{2}\left(\frac{\partial^2 f_{i,j}}{\partial x^2}h_{i+2,j+2}\right) \\ B = \sum_{i=-2}^{2}\sum_{j=-2}^{2}\left(\frac{\partial^2 f_{i,j}}{\partial y^2}h_{i+2,j+2}\right) \\ C = \sum_{i=-2}^{2}\sum_{j=-2}^{2}\left(\frac{\partial f_{i,j}}{\partial x}\frac{\partial f_{i,j}}{\partial y}h_{i+2,j+2}\right) \end{cases}$$

(4) In order to avoid solving the eigenvalues, define the response function R of calculating corner:

$$R = Det(M) - kTrace^2(M)$$

Wherein, $Det(M) = \lambda_1\lambda_2 = AB - C2$, $Trace(M) = \lambda_1 + \lambda_2 = A + B$. The value of k is set to 0.04 based on experience. For the corner points, the R value is greater than 0.

(5) In order to remove the false corners, a suitable threshold value T is selected for the response function R to filter out those corner with a smaller value. Then within the rest corners, those with a local maximum value are selected in appropriate selected masks as result corners. The local mask's size selected is 10.

3.4 Monocular Positioning

When the target points needed to locate on the workpiece are in the same plane and have the same distant to robot end-effector, the workpiece can be located by single camera.

After the calibration procedure of the camera [10], the camera parameters can be obtained, including its internal reference matrix A, external reference matrix $[R\ t]$ and the radial distortion coefficient, namely k_1 and k_2. The actual position of the workpiece's point (X, Y) corresponding to the image coordinates (u, v) has the following correspondence:

$$Z_c\begin{bmatrix} u \\ v \\ 1 \end{bmatrix} = A\begin{bmatrix} R & t \end{bmatrix}\begin{bmatrix} 1+k_1R^2 & 0 & 0 \\ 0 & 1+k_2R^2 & 0 \\ 0 & 0 & 1 \end{bmatrix}\begin{bmatrix} X \\ Y \\ 1 \end{bmatrix} = N\begin{bmatrix} X \\ Y \\ 1 \end{bmatrix}$$

Z_c represents the distance from the workpiece to the optical center of the lens. R represents the radial distance of the image point. The sensor's pixel size in the horizontal direction and the vertical direction are respectively P_x and P_y and R can be obtained by the following formula:

$$R = \sqrt{(uP_x)^2 + (vP_y)^2}$$

After the camera is installed and the calibration is completed, Z_c is a fixed value. The pixel coordinates (u, v) of the feature point of the workpiece can be obtained by image acquisition, the actual position can be evaluated according to equation:

$$\begin{bmatrix} X \\ Y \\ 1 \end{bmatrix} = Z_c N^{-1} \begin{bmatrix} u \\ v \\ 1 \end{bmatrix}$$

4   Experiment and Analysis

The visual experiment system is based on an experimental box (SEED-DTK-VP M642), as shown in Figure 5. The platform includes emulators, camera, DSP (TMS320DM642), monitor and other components. With the help of the emulators, the program can be debugged in real time, and the DSP program can be easily written and burn into the DSP; the image processing results can be displayed on the monitor in real time. After DSP captured an image by the camera, the feature point position obtained through image processing will be transmitted through the Ethernet to the robot controller, and the robot will be able to create a new job or perform trajectory correction of existing jobs in accordance with the obtained position information of the trajectory.

Fig. 5. The visual experiment platform

4.1   Preprocessing and Corner Detection Experiment

In order to verify the performance of image processing algorithms in different environments, two checkerboard images with different background are selected to perform this experiments, shown in Figure 6, where Figure (a) shows a simple background and Figure (b) shows the complex background. The experiments were carried out on these two pictures to test the corner detection algorithm.

DSP obtained the image information from the camera. The image data is firstly converted from interlaced scanning data form to progressive scanning data form to obtain new image data. Then the image preprocessing procedure is performed, namely the gray value linear transformation and median filtering, after which the results shown in Figure 7 can be obtained. Compared to the input image, the output image's contrast expanded markedly. The output image becomes smoother with the reservations of edge features. The image quality has been greatly improved.

(a) Simple background   (b) complex background

Fig. 6. The input image

(a) Simple background   (b) complex background

Fig. 7. The output image after image preprocessing

The Harris corner detection algorithm is perform in the image after preprocessing, and the corner detected are labeled as a small black box, which are display to the monitor in real time , shown in Figure 8.

For checkerboard image under simple background, its corners are accurately detected. However, although the corner can be detected in the complex background, false corners have also been generated in the background , which is due to that the algorithm is based on the pixel points having the maximum gradient of the gray value in its neighborhood to determine whether it is a corner point, without considering the overall characteristics of the checkerboard. Therefore, to extract the checkerboard's corner under the complex background, further filter procedure needed to be performed.

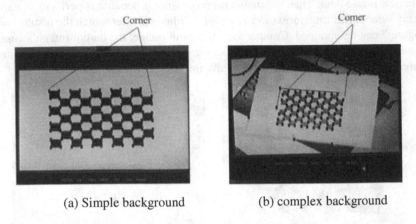

(a) Simple background   (b) complex background

Fig. 8. The output image after corner detection

4.2 Camera Calibration Experiment

The camera calibration method [10] needs to take at least three checkerboard images with different view. The pixel position of the corner points on the checkerboard can be extracted. Through the camera calibration, the camera parameters can be calculated.

Fig. 9. The checkerboard images in different Pose

The checkerboard used in the experiments is 10 × 7 squares array of alternating black and white, in which each square side length is 15mm. six different calibration plate images were obtained, as shown in Figure 9. Camera internal parameters matrix and the radial distortion coefficients is calculated by the calibration algorithm, given by:

$$A = \begin{bmatrix} 2416.8147 & 0 & 349.8417 \\ 0 & 2576.3327 & 323.9863 \\ 0 & 0 & 1 \end{bmatrix}$$

$$\begin{cases} k_1 = -0.2219 \\ k_2 = 3.7638 \end{cases}$$

Take the first checkerboard picture for example. The external parameter matrix is given by:

$$[R,t] = \begin{bmatrix} -0.9998 & -0.0132 & 67.4548 \\ 0.0137 & -0.9994 & 10.6665 \\ -0.0164 & -0.0306 & 891.9216 \end{bmatrix}$$

The main pixel coordinates measured is (350, 324), while the theoretical main pixel coordinates is (360, 288). The difference is mainly due to the lens mounting misalignment. Further, since the CCD sensor has the shape of rectangular, the size difference in the x, y direction can't be ignored. Thus the camera radial distortion coefficients k_1 and k_2 is not equal.

4.3   Monocular Positioning Experiment

After the calibration procedure of the camera, its internal parameter matrix A and the radial distortion coefficients k_1 and k_2 can be obtained. For a particular chessboard image, the external parameter matrix $[R\ t]$ can also be obtained. Thus a corner point on the image by corner detection algorithm is obtained in the form of pixel coordinates, and the actual coordinates can be calculated.

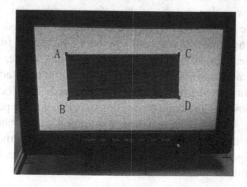

Fig. 10. The rectangular pattern image

A rectangular pattern with a size of 150 × 80, shown in Figure 10, is chosen to obtain the location of its four corner points A, B, C, D by the vision system, and thus obtain its actual size. By comparison with the theoretical value, positioning accuracy of the visual system can be obtained. The actual coordinates of each corner obtained are A (-66.6612, -45.8893), B (-68.5102, -125.5737), C (-216.2915, -42.9945), D (-217.9352, -122.4344). The measured and theoretical rectangular sizes are shown in Table 1.

Table 1. The comparison between the actual size and the theoretical size of the image

Line	theoretical size /mm	actual size /mm	Error/mm
$\overline{AB}$	80.00	80.0250	0.0250
$\overline{AC}$	150.00	149.0302	0.9698
$\overline{CD}$	80.00	79.0253	0.9747
$\overline{BD}$	150.00	149.0537	0.9463

From the table, the errors measured by the visual system are in 1mm, which is mainly caused by the following factors: the camera parameter calibration error, the calibration method which only considers the radial distortion of the camera, not considering the centrifugal thin lens distortion and aberration, and underexposed image quality problems caused by the experiment.

5   Conclusion

This paper has carried on the preliminary design of the glue dispensing robot vision positioning system. Binding the open robot control system, we proposed a visual positioning system based on embedded DSP architecture and introduces the hardware system components, with the analysis of the software architecture of the system. The key component of the system is DSP, which obtains the image signal from the CCD sensor, performs many image processing operation, including image preprocessing using Gray value linear transformation and median filter, corner detection applying Harris algorithm. Camera calibration method by Zhang is adopted. These algorithms are analyzed and a fair experiment results are achieved. The location method with single camera is proposed. The image processing experiments are performed, including image preprocessing, corner detection test, calibration experiments and monocular localization experiments. The experimental results show that the system has good visual positioning capability. The location results will be fed back to the robot controller with UDP communication via Ethernet.

References

1. Han, S.H., Seo, W.H., Yoon, K.S., et al.: Real-time control of an industrial robot using image-based visual servoing. In: IEEE/RSJ International Conference on Intelligent Robots and Systems, pp. 1762–1767 (1999)
2. Wang, Y., Jiang, X., Li, Z.: Research on the method of intelligent robot visual recognition and positioning. In: IEEE International Conference on Networking, Sensing and Control, pp. 916–919 (2008)
3. Liu, C.J., Zhu, J.G., Li, Y.B., et al.: Study on a real-time visual measuring and tracking system for industry robot welding. In: Advanced Laser Technologies 2005 (2006)
4. Ni, S.D., Zhang, M., Sheng, Z.G.: The research on visual industrial robot based on open platform. Manufacturing Automation (04), 44–48+62 (2014). (in Chinese)
5. Li, J.Y., Yang, C., Wang, J., et al.: A robot handling system based on visual Orientation. Manufacturing Automation 33(4), 40–42 (2011). (in Chinese)
6. Chen, H., Fuhlbrigge, T., Zhang, G., et al.: Integrated robotic system for high precision assembly in a semi-structured environment. Assembly Automation 27(3), 247–252 (2007)
7. Tokhi, O., Connolly, C.: A new integrated robot vision system from FANUC Robotics. Industrial Robot: An International Journal 34(2), 103–106 (2007)
8. Chen, W.H., Ma, Q.X., Chen, Y.J.: Vision Positioning System of Industrial Robot Based on The VisionPro. Modular Machine Tool & Automatic Manufacturing Technique (02), 81–83+87 (2012). (in Chinese)
9. Harris, C., Stephens, M.: A combined corner and edge detector. In: Alvey Vision Conference, Manchester, UK, p. 50 (1988)
10. Zhang, Z.Y.: Flexible camera calibration by viewing a plane from unknown orientations. In: The Seventh IEEE International Conference on Computer Vision, pp. 666–673(1999)

Efficient Collision Detection
for Industrial Robot Simulation System

Nianfeng Wang[✉], Chaochao Zheng, and Xianmin Zhang[✉]

School of Mechanical and Automobile Engineering,
South China University of Technology, Guangzhou, China
{menfwang,zhangxm}@scut.edu.cn

Abstract. As the result of an increasing number of industrial robots, the simulation of industrial robot has become a very active research field. Collision detection for industrial robot simulation system is an essential part. In this paper, we present two kinds of collision detection algorithms, namely, internal collision detection algorithm and external collision detection algorithm. These two kinds of algorithms are based mainly on the bounding volume technology. Internal collision detection algorithm is applicable to the collisions between robotic links and external collision detection algorithm is applicable to the collisions between robotic links and its surrounding obstacles. In addition, we also present two examples of simulation results. The examples prove that the collision detection module can satisfy the collision detection requirement of the industrial robot simulation system.

Keywords: Collision detection · Simulation system · Bounding volumes

1 Introduction

The industrial robots play an important role in the modern industry, especially in manufacturing industry. At present, the advantages of industrial robots have become increasingly prominent due to the growing demands for automatic productions. In addition, the simulations of industrial robots [1, 2] and their working environment could become a very good auxiliary tool. The purpose of the industrial robot simulation systems provides a virtual environment through computer graphics technology. The industrial robot simulation systems could be applied in many aspects such as painting robot simulation, welding robot simulation and stamping robot simulation [3] etc. In the fields of industrial robot simulation systems, collision detection is an essential part. A complete industrial robot simulation system must have the ability to detect collisions, including collisions between the industrial robot and its surrounding

This work is supported by National Natural Science Foundation of China (Grant Nos. 51205134, 91223201), Science and Technology Program of Guangzhou (Grant No. 2014Y2-00217), Research Project of State Key Laboratory of Mechanical System and Vibration (MSV201405), the Fundamental Research Funds for the Central University (Fund No. 2015ZZ007) and Natural Science Foundation of Guangdong Province (S2013030013355).

H. Liu et al. (Eds.): ICIRA 2015, Part III, LNAI 9246, pp. 552–562, 2015.
DOI: 10.1007/978-3-319-22873-0_49

obstacles. Collision detection is the foundation of carrying out a specific mission or doing path planning. If the industrial robot simulation system detects a collision when the industrial robot moves along a path, we can plan a new path.

To solve the problem of collision detection for industrial robot simulation system, we use the bounding volume technology [4~6] to achieve detection. Moreover, the collision detection module should have good accuracy and high efficiency. There is a tradeoff mode between good accuracy and high efficiency; i.e., the more accurate the detection, the less efficient in collision detection. So, we need to select the appropriate size of bounding volumes.

2   Previous Work

So far, many researchers have proposed different collision detection algorithms. In these algorithms, the most primitive method of collision detection algorithms is a brute-force method, namely, testing every basic geometrical element of industrial robot and its surrounding obstacles. Although this method can get correct results, the testing of every pair in the industrial robot simulation system is impractical when the number of geometrical elements is very large. So, we need to give some efficient collision detection algorithms for industrial robot simulation system. These efficient collision detection algorithms can be classified into two types: image based algorithms and graph based algorithms.

The main difference between image based algorithms and graph based algorithms is using depth information of image or using the geometric characteristics of objects in industrial robot simulation system. The graph based algorithms are classified into two types: spatial division techniques [7] and bounding volume hierarchies [8]. The purpose of spatial division techniques and bounding volume hierarchies is to reduce the amount of calculation. Spatial division techniques divide the computational space in uniform cells. On the other hand, the core idea of bounding volume hierarchies is using the bounding volume of which volume is slightly larger than object and geometric characteristics are simple to approximately describe the object in industrial robot simulation system. The typical bounding volume hierarchies mainly include bounding sphere [9], aligned axis bounding box (AABB), oriented bounding box (OBB) [10], discrete orientation polytope (k-DOP) and other kinds of bounding volumes [11, 12].

Comparing the advantages and disadvantages of these algorithms, we can know the method of the bounding volumes is very easy and efficient to detect collisions in industrial robot simulation system. In this paper, we develop two kinds of efficient collision detection algorithms based on the bounding volumes.

3   Algorithm Overview

In this section, we give an overview of our collision detection algorithms in industrial robot simulation system. These collisions are classified into two types: internal collisions of industrial robot and external collisions of industrial robot. Internal collisions of industrial robot mainly refer to the collisions between robotic links. The collisions between robotic joints and the collisions between robotic joints and robotic links are also internal collisions. External collisions of industrial robot mainly refer to the collisions between robotic links and their surrounding obstacles. According to the different situations of internal collisions and external collisions, we propose two kinds of collision detection algorithms which are called internal collision detection algorithm and external collision detection algorithm.

3.1   Internal Collision Detection Algorithm

The industrial robot has six degrees of freedom (DOF) characterized by rotation motions. In the industrial robot, the position of the end-effector is determined by the first three robotic links, while the orientation of the end-effector is determined by the three last robotic links. So the internal collisions in industrial robot simulation system mainly occur in the first three robotic links. Simplified models of internal collision detection algorithm are simplified by the bounding volumes of which volume is slightly larger than robotic links and geometric characteristics are simple to approximately describe robotic links.

The robotic link is approximated by a capsule composed by a cylinder and two hemispheres in the ends of the cylinder and the robotic joint is approximated by a sphere. The capsule model and sphere model are shown in Figure 1. The calculating of a collision between robotic links is turned into the calculating of the minimum distance between bounding volumes. The internal collision detection algorithm is not particularly accurate because the volume of the bounding volume is slightly larger than robotic link. However, this algorithm is suitable for the collisions between robotic links because robotic links are not too close to each other.

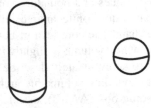

A) The capsule model B) The sphere model

Fig. 1. The models of bounding volumes.

Establishment of Bounding Volumes. In this section, taking the collision between the first robotic link and the third robotic link as an example illustrates internal collision detection algorithm. The first robotic link and the third robotic link are approximated by capsule c_1 and capsule c_2, as shown in Figure 2 and Figure 3. Furthermore, the first robotic joint, the second robotic joint, the third robotic joint and the fourth robotic joint are approximated by sphere s_1, s_2, s_3 and s_4. Point A, B, C and D are center of sphere s_1, s_2, s_3 and s_4, respectively.

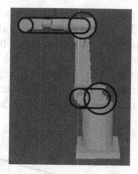

Fig. 2. The bounding volumes of robotic links and joints in the simulation system.

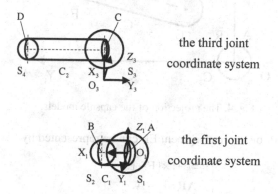

Fig. 3. Bounding volumes in joint coordinate system.

Distance Calculation. After robotic links are approximated by capsules and robotic joints are approximated by spheres, the problem of the collision detection is turned into the calculating of the minimum distance between bounding volumes. The minimum distance between bounding volumes are classified into two classes: point to point and point to line segment.

Point to Point. In order to detect collisions, we should firstly determine whether robotic joint and robotic joint collide. So we should calculate the minimum distance between spheres. According to link parameters, the coordinates of point A, B, C and D can be easy to get. Given point M (x_1, y_1, z_1) and N (x_2, y_2, z_2), the distance function d (M, N) between two points is defined as below.

$$d\,(M,\,N)=\sqrt{\left(x_2 - x_1\right)^2 + \left(y_2 - y_1\right)^2 + \left(z_2 - z_1\right)^2}\tag{1}$$

From equation (1), we can calculate the length of line segment AC, AD, BC and BD. Then the minimum distance can be easy to get. The radiuses of two corresponding spheres are R1 and R2. If the minimum distance is larger than the sum of R1 and R2, two robotic joints cannot collide. If the minimum distance is less than the sum of R1 and R2, two robotic joints will collide.

Point to Line segment. In order to detect collisions between robotic links, the bounding volumes of robotic links should be projected onto the same joint coordinate system. So the 3D collision problem becomes two-dimensional collision problem. The bounding volumes of the first robotic link and third robotic link are projected onto the plane $Y_1O_1Z_1$ of the first joint coordinate system, as shown in Figure 4. The radius of capsule is equal to the radius of hemisphere in the end of the capsule. Finally, capsule c_1 becomes a circle c_3 and capsule c_2 becomes c_4. The radius of c_3 is R_3 and the radius of c_4 is R_4.

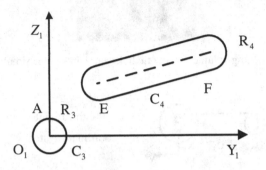

Fig. 4. The projection of the capsule models.

An arbitrary point P_1 on the line segment EF can be represented by

$$\begin{cases} P_1 = E + t \times (F - E) \\ \quad AP_1 = P_1 - A \\ \quad 0 \le t \le 1 \end{cases} \qquad (2)$$

where AP_1 is a column vector and $|AP_1|$ represents the distance between an arbitrary point on the line segment EF and point A

$$\begin{aligned} (AP_1)^2 &= (P_1 - A)^2 \\ \Rightarrow (AP_1)^2 &= F^2 t^2 + 2(E - A) \times F \times t + (E - A)^2 \\ \Rightarrow |AP_1|^2 &= at^2 + bt + c \end{aligned} \qquad (3)$$

where

$$A = \begin{pmatrix} y_1 & z_1 \end{pmatrix}^T$$
$$E = \begin{pmatrix} y_2 & z_2 \end{pmatrix}^T$$
$$F = \begin{pmatrix} y_3 & z_3 \end{pmatrix}^T$$
$$a = (y_3 - y_2)^2 + (z_3 - z_2)^2$$
$$b = 2((y_3 - y_2)(y_2 - y_1) + (z_3 - z_2)(z_2 - z_1))$$
$$c = (y_2 - y_1)^2 + (z_2 - z_1)^2$$

So the minimum distance between an arbitrary point on the line segment EF and point A is transformed into solving the minimum value of quadratic function. Solving the minimum value of quadratic function can be classified three classes. The results are as follows.

$$|AP_1|_{min} = \begin{cases} \sqrt{c}, -\dfrac{b}{2a} \le 0 \\ \sqrt{a+b+c}, -\dfrac{b}{2a} \ge 1 \\ \sqrt{a\left(-\dfrac{b}{2a}\right)^2 + b\left(-\dfrac{b}{2a}\right) + c}, 0 \le -\dfrac{b}{2a} \le 1 \end{cases} \tag{4}$$

If $|AP_1|_{min}$ ($R_3 + R_4$), there is no collision between robotic links. If $|AP_1|_{min} < (R_3 + R_4)$, there is a collision between robotic links.

Algorithm Steps. The steps of internal collision detection algorithm which is taking the collision detection between the first robotic link and the third robotic link as an example are as follows.

Step 1. Establish bounding volumes of robotic links and joints. Give coordinates of point A, B, C and D and give the radiuses of capsules and spheres.

Step 2. Obtain coordinates of point A, B, C and D in the first and third joint coordinate system by coordinate transformation.

Step 3. Solve the minimum distance between robotic joints. If the minimum distance is less than the sum of R_1 and R_2, then go to Step 9. Otherwise continue.

Step 4. Let c_1, s_3 and s_4 project onto the plane $Y_1O_1Z_1$ of the first joint coordinate system. Let c_2, s_1 and s_2 project onto the plane $X_1O_1Z_1$ of the third joint coordinate system. Solve the minimum distance d. The radius of corresponding capsule is R_5 and the radius of corresponding sphere is R_6. If $d < (R_5 + R_6)$, then go to Step 9. Otherwise continue.

Step 5. Let c_1 and c_2 project onto the plane $Y_1O_1Z_1$ of the first joint coordinate system. Solve the minimum distance d_{min1}. If d_{min1} ($R_3 + R_4$), then go to Step 8. Otherwise continue.

Step 6. Let c_1 and c_2 project onto the plane $X_1O_1Z_1$ of the first joint coordinate system. Point A_1, B_1, C_1 and D_1 are the projection of point A, B, C and D. d_{min2} is the minimum distance between point A_1 and line segment $C_1 D_1$. $|A_1B_1|$ is the length of line segment A_1B_1. If d_{min2} ($R_3 + R_4 + |A_1B_1|$), then go to Step 8. Otherwise continue.

Step 7. Let c_1 and c_2 project onto the plane $X_1O_1Z_1$ of the third joint coordinate system. Solve the minimum distance d_{min3}. If d_{min3} ($R_3 + R_4$), then go to Step 8. Otherwise continue.

Step 8. There is no collision.

Step 9. A collision occurs.

3.2   External Collision Detection Algorithm

External collision detection algorithm is to detect collisions between robotic links and their surrounding obstacles. The shapes of obstacles in industrial robot simulation system have no specific requirements. So the applicability of external collision detection algorithm is wider than that of internal collision detection algorithm. In this algorithm, we mainly use the bounding volume technology and the knowledge of STL (stereo lithography) file.

Introduction of STL File. STL file which is a 3D model file is an interface standard and is widely used in many fields. The surface profile of 3D model is composed of many triangles. The STL file stores a number of triangle data information. These data are mainly three vertex coordinates of triangles and normal vector of triangles. In addition, the formats of STL file are classified into two types: text file (ASCII) and binary file. In order to detect collisions quickly, we should select text file as the data source of collision detection.

A Line Segment and Triangles Intersection Judgment. In order to detect a collision between a robotic link and its surrounding obstacle, the robotic link should be approximated by capsule and the surface of the obstacle should move a certain distance which is equal to the radius of capsule along outer normal direction. So we can abstract vertex coordinates of triangles from the new STL file. Finally, the collision between a robotic link and its surrounding obstacle convert to the judgment of a line segment and triangles intersection. The model of a line segment and a triangle is shown in Figure 5.

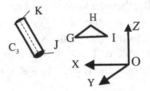

Fig. 5. The model of a line segment and a triangle.

An arbitrary point P_2 on the line segment JK can be represented by

$$\begin{cases} P_2 = J + s \times (K - J) \\ \quad 0 \le s \le 1 \end{cases} \tag{5}$$

$$
\left\{
\begin{aligned}
\mathbf{GP_2} &= \begin{pmatrix} s\times(x_5-x_4)+x_4-x_1 \\ s\times(y_5-y_4)+y_4-y_1 \\ s\times(z_5-z_4)+z_4-z_1 \end{pmatrix}^{\mathrm{T}} \\[2mm]
\mathbf{GH} &= \begin{pmatrix} x_2-x_1 \\ y_2-y_1 \\ z_2-z_1 \end{pmatrix}^{\mathrm{T}} \\[2mm]
\mathbf{GI} &= \begin{pmatrix} x_3-x_1 \\ y_3-y_1 \\ z_3-z_1 \end{pmatrix}^{\mathrm{T}}
\end{aligned}
\right.
\tag{6}
$$

where

$$
\mathbf{G} = \begin{pmatrix} x_1 & y_1 & z_1 \end{pmatrix}^{\mathrm{T}}
$$
$$
\mathbf{H} = \begin{pmatrix} x_2 & y_2 & z_2 \end{pmatrix}^{\mathrm{T}}
$$
$$
\mathbf{I} = \begin{pmatrix} x_3 & y_3 & z_3 \end{pmatrix}^{\mathrm{T}}
$$
$$
\mathbf{J} = \begin{pmatrix} x_4 & y_4 & z_4 \end{pmatrix}^{\mathrm{T}}
$$
$$
\mathbf{K} = \begin{pmatrix} x_5 & y_5 & z_5 \end{pmatrix}^{\mathrm{T}}
$$

$$
\mathbf{GP_2} \cdot (\mathbf{GH} \times \mathbf{GI}) = 0
\tag{7}
$$

Solve the above equation for s. If $0 \le s \le 1$, the line segment JK has a point on the plane GHI. Otherwise there is no point on the plane GHI. An arbitrary point P_3 on the triangle $\triangle$GHI can be represented by

$$
\left\{
\begin{aligned}
P_3 &= G + t_1 \times (H-G) + t_2 \times (I-G) \\
&\quad 0 \le t_1 \le 1 \\
&\quad 0 \le t_2 \le 1
\end{aligned}
\right.
\tag{8}
$$

$$
\begin{pmatrix} x_4 \\ y_4 \\ z_4 \end{pmatrix} + s\times \begin{pmatrix} x_5-x_4 \\ y_5-y_4 \\ z_5-z_4 \end{pmatrix} = \begin{pmatrix} x_1 \\ y_1 \\ z_1 \end{pmatrix} + t_1 \times \begin{pmatrix} x_2-x_1 \\ y_2-y_1 \\ z_2-z_1 \end{pmatrix} + t_2 \times \begin{pmatrix} x_3-x_1 \\ y_3-y_1 \\ z_3-z_1 \end{pmatrix}
\tag{9}
$$

Solve the above equation for s, t_1 and t_2. If $0 \le s \le 1$, $0 \le t_1 \le 1$, $0 \le t_2 \le 1$ and $0 \le (t_1+t_2) \le 1$, the line segment JK has a point on the triangle $\triangle$GHI. Otherwise there is no point on the triangle $\triangle$GHI.

Algorithm Steps. The steps of external collision detection algorithm which is taking the collision detection between a robotic link and its surrounding obstacle as an example are as follows.

Step 1. Establish a bounding volume of the robotic link. Let the surface of the obstacle move a certain distance which is equal to the radius of capsule along outer normal direction and abstract vertex coordinates of triangles from the new STL file.

Step 2. Let vertex coordinates of triangles and points of capsule transform to the same coordinate system.

Step 3. According to vertex coordinates of each triangle, solve the equation (7) for s. If $s \leq 0$ and s 1, then go to Step 5. Otherwise continue.

Step 4. Solve the equation (9) for s, t_1 and t_2. If $0 \leq s \leq 1$, $0 \leq t_1 \leq 1$, $0 \leq t_2 \leq 1$ and $0 \leq (t_1 + t_2) \leq 1$, then go to Step 6. Otherwise continue.

Step 5. There is no collision.

Step 6. A collision occurs.

4   Implementation and Results

The industrial robot simulation system is developed in visual C++ and includes a graphical representation based on OpenGL under Microsoft Windows (XP) operating system. In addition, we have implemented our collision detection algorithms in the industrial robot simulation system. We use our collision detection algorithms to study internal collisions and external collisions. In the industrial robot simulation system, our collision detection algorithms are able to detect all collisions.

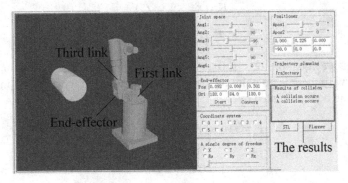

Fig. 6. The results of the internal collisions.

Taking the first robotic link and the third robotic link as an example illustrates internal collision detection algorithm. Once the articular variables change, bounding volumes of the robotic links will rebuild. Finally, the collision detection module will detect all collisions. The results of the internal collisions are shown in Figure 6. As we can see from the Figure 6, the first robotic link and the third robotic link collide. In addition, the first robotic link and the end-effector collide.

Taking the end-effector and the obstacle composed of several hundreds of triangles as an example illustrates external collision detection algorithm. The obstacle is placed at a fixed position in the industrial robot simulation system. When the articular variables change, the collision detection module will detect all collisions. The result of the external collision is shown in Figure 7. As we can see from the figure 7, the end-effector and the obstacle collide. Comparing external collision algorithm with internal collision algorithm, the applicability of external collision detection algorithm is wider than that of internal collision detection algorithm and the calculation of external collision detection algorithm is more complicated than that of internal collision detection algorithm.

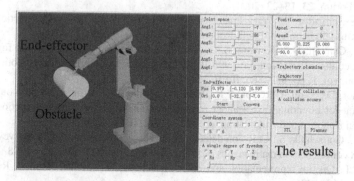

Fig. 7. The result of the external collision.

5   Conclusions

In this work, two kinds of collision detection algorithms are developed in order to detect collisions in the industrial robot simulation system. These algorithms are based mainly on the bounding volume technology and the knowledge of STL file. In detail, these algorithms are classified into two types: internal collision detection algorithm and external collision detection algorithm. Internal collision detection algorithm is suitable for the collisions between robotic links and external collision detection algorithm is suitable for the collisions between robotic links and their surrounding obstacles. Furthermore, we have implemented our collision detection algorithms in the industrial robot simulation system. Through two examples of simulation results, we can get that collision detection algorithms are accurate and reliable.

References

1. Stifter, S.: Collision detection in the robot simulation system SMART. Int. J. Adv. Manuf. Tech. **7**, 277–283 (1992)
2. Fawaz, K., Merzouki, R., Ould-Bouamama, B.: Model based real time monitoring for collision detection of an industrial robot. Mechatronics. **19**, 695–704 (2009)
3. Tesic, R., Banerjee, P.: Exact collision detection using virtual objects in virtual reality modeling of a manufacturing process. J. Manuf. Syst. **18**, 367–376 (1999)
4. Vemuri, B.C., Chen, L., Vu-Quoc, L., Zhang, X., Walton, O.: Efficient and accurate collision detection for granular flow simulation. graph. Models. **60**, 403–422 (1998)

5. Sato, Y., Hirata, M., Maruyama, T., Arita, Y.: Efficient collision detection using fast distance-calculation algorithms for convex and non-convex objects. In: 1996 IEEE International Conference on Robotics and Automation, vol. 1, pp. 771–778. IEEE, April 1996
6. Chang, C., Chung, M.J., Bien, Z.: Collision-free motion planning for two articulated robot arms using minimum distance functions. Robotica. **8**, 137–144 (1990)
7. Borro, D., García-Alonso, A., Matey, L.: Approximation of optimal voxel size for collision detection in maintainability simulations within massive virtual environments. Comput. Graph. Forum. **23**, 13–23 (2004)
8. Martínez-Salvador, B., Pérez-Francisco, M., Del Pobil, A.P.: Collision detection between robot arms and people. J. Intell. Robot. Syst. **38**, 105–119 (2003)
9. Ma, H., Cannon, D.J., Kumara, S.R.: A scheme integrating neural networks for real-time robotic collision detection. In: 1995 IEEE International Conference on Robotics and Automation, vol. 1, pp. 881–886. IEEE, May 1995
10. Albocher, D., Sarel, U., Choi, Y.K., Elber, G., Wang, W.: Efficient continuous collision detection for bounding boxes under rational motion. In: 2006 IEEE International Conference on Robotics and Automation, pp. 3017–3022. IEEE, May 2006
11. Lin, M., Gottschalk, S.: Collision detection between geometric models: A survey. In: Proc. of IMA conference on mathematics of surfaces, vol. 1, pp. 602–608, May 1998
12. Wu, C.J.: On the representation and collision detection of robots. J. Intell. Robot. Syst. **16**, 151–168 (1996)

Decoupling Control Based on Dynamic Model of 4-DOF Wafer Handling Robot

Lei Yang[1] and Nianfeng Wang[2](✉)

[1] R&D Department, Inovance Technology Co., Shenzhen, China
[2] School of Mechanical and Automobile Engineering,
South China University of Technology, Guangzhou, China
menfwang@scut.edu.cn

Abstract. To provide high velocity and performance, a dynamic model based on 4-DOF wafer handling robot in semiconductor industry is setup in the paper, formulas are derived, and also stability proof is given. In the final part of paper the experiment result of decoupling control is given, and it verified the correction of dynamic formulas.

Keywords: Decoupling control · Dynamic model · Wafer handling robot

1 Introduction

As fab construction costs escalated during the mid-1980s, it was recognized that there may be an advantage to miniaturizing clean rooms by producing products in a vacuum and transporting wafers using robots (see Figure 1). Wafer handling robots have been used in semiconductors manufacture for both 200mm and 300mm wafer widely, because the advantage is outstanding obviously, cleaner, more efficiently, safer and faster.

This work is supported by National Natural Science Foundation of China (Grant Nos. 51205134, 91223201), Science and Technology Program of Guangzhou (Grant No. 2014Y2-00217), Research Project of State Key Laboratory of Mechanical System and Vibration (MSV201405), the Fundamental Research Funds for the Central University (Fund No. 2015ZZ007) and Natural Science Foundation of Guangdong Province (S2013030013355).

H. Liu et al. (Eds.): ICIRA 2015, Part III, LNAI 9246, pp. 563–570, 2015.
DOI: 10.1007/978-3-319-22873-0_50

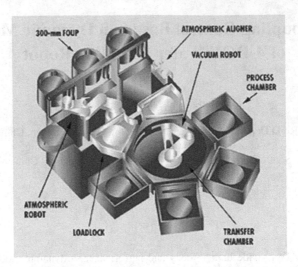

Fig. 1. State-of-the-art fabs use robots to handle wafers in the front end of process equipment [1]

2 Mechanical System

There are lots of different types in wafer handling robot, advances in direct-drive architecture versus gears and timing belts have led to a more than 10 times improvement in reliability. The wafer handling robot introduced in this paper(see figure 2) has been designed 3 DDR motors and 1 Z axis, is capable of greater than 10 million cycles between failures compared to conventional robot architectures which can complete fewer than 1 million cycles. Also provide greater positional repeatability and smoother movement, which ensures increased acceleration of wafer movement, and does not contribute to vibration. Direct-drive designs that attach motors directly to a driveshaft further minimize the number of moving parts that are subject to mechanical wear, and therefore provide extremely high reliability and optimize wafer throughput.

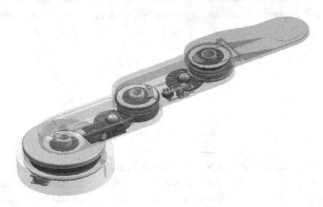

Fig. 2. Structure of wafer handling robot

3   Dynamic Model

Based on the real mechanical structure, buildup the mathematics model of the 3 arms in the wafer handling robot (See figure 3), for the 3 DDR arms have strong coupling in movement.

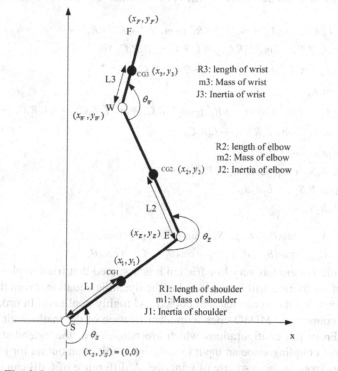

R3: length of wrist
m3: Mass of wrist
J3: Inertia of wrist

R2: length of elbow
m2: Mass of elbow
J2: Inertia of elbow

R1: length of shoulder
m1: Mass of shoulder
J1: Inertia of shoulder

Fig. 3. 3 arms mathematics model of the wafer handling robot

The dynamics of the wafer handing robot is derived from Lagrange's Equation:

$$\frac{d}{dt}(\frac{\partial T}{\partial \dot{q}_i}) - \frac{\partial T}{\partial q_i} + \frac{\partial u}{\partial q_i} = \tau_i \tag{1}$$

where T is kinetic energy and u is potential energy; q_i is angle (or displacement) and τ_i is moment (or force) of ith DOF.

$$
\begin{bmatrix} \tau_S \\ \tau_E \\ \tau_W \end{bmatrix} =
\begin{bmatrix} I_{11} & I_{12} & I_{13} \\ I_{12} & I_{22} & I_{23} \\ I_{13} & I_{23} & I_{33} \end{bmatrix}
\begin{bmatrix} \ddot{\theta}_S \\ \ddot{\theta}_E \\ \ddot{\theta}_W \end{bmatrix} +
\begin{bmatrix} 0 & V_{12} & V_{13} \\ -V_{12} & 0 & V_{23} \\ -V_{13} & -V_{23} & 0 \end{bmatrix}
\begin{bmatrix} \dot{\theta}_S^{\,2} \\ \dot{\theta}_E^{\,2} \\ \dot{\theta}_W^{\,2} \end{bmatrix} +
\begin{bmatrix} 2V_{12} & 2V_{13} & 2V_{13} \\ 0 & 2V_{23} & 2V_{23} \\ -2V_{23} & 0 & 0 \end{bmatrix}
\begin{bmatrix} \dot{\theta}_S\dot{\theta}_E \\ \dot{\theta}_S\dot{\theta}_W \\ \dot{\theta}_E\dot{\theta}_W \end{bmatrix} \tag{2}
$$

$$= D(\vec{q})\ddot{\vec{q}} + \vec{h}(\vec{q},\dot{\vec{q}}) = \tau + \tau_v(\vec{q},\dot{\vec{q}},\ddot{\vec{q}})$$

D is the inertia matrix and h is the vector due to centrifugal and Coriolis effects. τ_v is the collection of all perturbations from the rigid-body dynamics (e.g. friction, flexibilities and disturbances). The terms of the inertia matrix D and the matrices due to centrifugal and Coriolis effects are a function of the mass m_i, the length of the links R_i, the inertia Ii and the center of gravity of each link L_i and are given below.

$$
\begin{aligned}
I_{11} &= I_1 + I_2 + I_3 + m_1 L_1^2 + m_2(L_2^2 + R_1^2) + m_3(L_3^2 + R_1^2 + R_2^2) + \\
&\quad - 2m_2 L_2 R_1 C_E + 2m_3(L_3 R_1 C_{EW} - L_3 R_2 C_W - R_2 R_1 C_E) \\
I_{22} &= I_2 + I_3 + m_2 L_2^2 + m_3(L_3^2 + R_2^2) - 2m_3 L_3 R_2 C_W \\
I_{33} &= I_3 + m_3 L_3^2 \\
I_{12} &= I_2 + I_3 + m_2 L_2^2 + m_3(L_3^2 + R_2^2) - m_2 L_2 R_1 C_E + m_3(L_3 R_1 C_{EW} - 2L_3 R_2 C_W - R_2 R_1 C_E) \quad (3) \\
I_{13} &= I_3 + m_3 L_3^2 + m_3(L_3 R_1 C_{EW} - L_3 R_2 C_W) \\
I_{23} &= I_3 + m_3 L_3^2 - m_3 L_3 R_2 C_W \\
V_{12} &= m_2 L_2 R_1 S_E + m_3(-L_3 R_1 S_{EW} + R_2 R_1 S_E) \\
V_{13} &= m_3(-L_3 R_1 S_{EW} + L_3 R_2 S_W) \\
V_{23} &= m_3 L_3 R_2 S_W
\end{aligned}
$$

where,
$$
\begin{aligned}
S_{EW} &= \sin(\theta_E + \theta_W), \quad S_E = \sin(\theta_E), \quad S_W = \sin(\theta_W) \\
C_{EW} &= \cos(\theta_E + \theta_W), \quad C_E = \cos(\theta_E), \quad C_W = \cos(\theta_W)
\end{aligned}
$$

Since the system has very low friction it is assumed that friction plays a minor role and therefore friction will be neglected in the dynamic equation. From these equations it is obvious that the dynamics are coupled and highly nonlinear. In order to study the effect of coupling, a MIMO open loop identification of the mathematic model is performed. From the configurations which are considered, the extended configuration shows most coupling since an input to joint one result in an output for joint two which is twice as large as the response of joint one. A difference of 6 dB can be observed in spectrums where the velocity spectra from input to all the outputs are shown for the extended configuration. An input to the wrist shows least coupling since it doesn't affect the other two joints very much. The influence of several configurations is as expected and is the same as the spectra of the real system.

4   Stability Proof

To prove the coupled nonlinear system $\vec{\tau} = D(\vec{q})\ddot{\vec{q}} + \vec{h}(\vec{q}, \dot{\vec{q}})$ is stable with decoupled control law $\vec{\tau} = -k_p \vec{q} - k_v \dot{\vec{q}}$,

where $k_p = \begin{bmatrix} k_{p1} & 0 & 0 \\ 0 & k_{p2} & 0 \\ 0 & 0 & k_{p3} \end{bmatrix}$, $k_v = \begin{bmatrix} k_{v1} & 0 & 0 \\ 0 & k_{v2} & 0 \\ 0 & 0 & k_{v3} \end{bmatrix}$,

$k_{p1} > 0, k_{p2} > 0, k_{p3} > 0, \quad k_{v1} > 0, k_{v2} > 0, k_{v3} > 0$, and in general suppose

$$\vec{q}_d = \dot{\vec{q}}_d = \begin{bmatrix} 0 \\ 0 \\ 0 \end{bmatrix}.$$

Proof: construct a Lyapunov function $\upsilon = \frac{1}{2}\dot{\vec{q}}^T D(\vec{q})\dot{\vec{q}} + \frac{1}{2}\vec{q}^T k_p q$, which is larger than or equal to zero.

Then,

$$\dot{\upsilon} = \frac{1}{2}\dot{\vec{q}}^T \dot{D}(\vec{q})\dot{\vec{q}} + \dot{\vec{q}}^T D(\vec{q})\ddot{\vec{q}} + \vec{q}^T k_p \dot{\vec{q}}$$

$$= \dot{\vec{q}}^T \vec{h}(\vec{q},\dot{\vec{q}}) + [-\dot{\vec{q}}^T k_p \vec{q} - \dot{\vec{q}}^T k_v \dot{q} - \dot{\vec{q}}^T \vec{h}(\vec{q},\dot{\vec{q}})] + \vec{q}^T k_p \dot{\vec{q}} \qquad (4)$$

$$= -\dot{\vec{q}}^T k_v \dot{\vec{q}}$$

k_v is positive definite, $\dot{\upsilon} \le 0$. It's needed to prove that system will not sustain at somewhere $\vec{q} \ne \vec{q}_d$. Please note $\vec{q}_d = [0 \; 0 \; 0]^T$.

From above, if $\dot{\upsilon}$ keeps equal to zero, we get $\dot{\vec{q}} = \ddot{\vec{q}} = [0 \; 0 \; 0]^T$. Then the coupling system $\vec{\tau} = [0 \; 0 \; 0]^T$, from the feedback law and k_p is positive definite, it's deduced that $\vec{q} = [0 \; 0 \; 0]^T = \vec{q}_d$.

So it's proved that the control system is global asymptotic stable.

From above proof, the following decoupling control structure (see figure 4), which considers both servo control stability and highly coupled nonlinear variable model, can be used for robot.

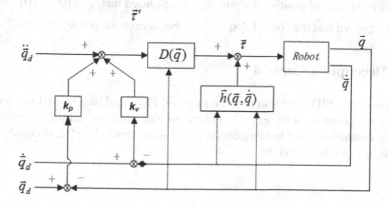

Fig. 4. Decoupling control structure

$$\vec{\tau} = \alpha\vec{\tau}' + \vec{\beta}, \ \alpha = D(\vec{q}), \ \vec{\beta} = \vec{h}(\vec{q}, \dot{\vec{q}})$$

$$\vec{\tau}' = k_p(\vec{q}_d - \vec{q}) + k_v(\dot{\vec{q}}_d - \dot{\vec{q}}) + \ddot{\vec{q}}_d$$

$$k_p = \begin{bmatrix} k_{p1} & 0 & 0 \\ 0 & k_{p2} & 0 \\ 0 & 0 & k_{p3} \end{bmatrix}, \ k_v = \begin{bmatrix} k_{v1} & 0 & 0 \\ 0 & k_{v2} & 0 \\ 0 & 0 & k_{v3} \end{bmatrix},$$

$$k_{p1} > 0, k_{p2} > 0, k_{p3} > 0, \ k_{v1} > 0, k_{v2} > 0, k_{v3} > 0,$$

$$\vec{\tau} = D(\vec{q})\ddot{\vec{q}} + \vec{h}(\vec{q}, \dot{\vec{q}}) = \alpha\vec{\tau}' + \vec{\beta} =$$

$$D(\vec{q})k_p(\vec{q}_d - \vec{q}) + D(\vec{q})k_v(\dot{\vec{q}}_d - \dot{\vec{q}}) + D(\vec{q})\ddot{\vec{q}}_d + \vec{h}(\vec{q}, \dot{\vec{q}})$$

So, $k_p(\vec{q}_d - \vec{q}) + k_v(\dot{\vec{q}}_d - \dot{\vec{q}}) + (\ddot{\vec{q}}_d - \ddot{\vec{q}}) = 0$, where k_p and k_v are diagonal matrix, ➜fully decoupled error equation.

Stability proof: construct a Lyapunov function $v = \frac{1}{2}\dot{\vec{q}}^T\dot{\vec{q}} + \frac{1}{2}\vec{q}^T k_p q$, which is larger than or equal to zero. In general, given $\vec{q}_d = \dot{\vec{q}}_d = [0 \ 0 \ 0]$.

Then,

$$\dot{v} = \dot{\vec{q}}^T\ddot{\vec{q}} + \vec{q}^T k_p \dot{\vec{q}}$$

$$= \dot{\vec{q}}^T(-k_p\vec{q} - k_v\dot{q}) + \vec{q}^T k_p \dot{\vec{q}} \tag{5}$$

$$= -\dot{\vec{q}}^T k_v \dot{\vec{q}}$$

k_v is positive definite, $\dot{v} \leq 0$. It's needed to prove that system will not sustain at somewhere $\vec{q} \neq \vec{q}_d$. Please note $\vec{q}_d = [0 \ 0 \ 0]^T$.

From above, if $\dot{v}$ keeps equal to zero, we get $\dot{\vec{q}} = \ddot{\vec{q}} = [0 \ 0 \ 0]^T$. Then from the error equation and positive definite k_p, it's deduced that $\vec{q} = [0 \ 0 \ 0]^T = \vec{q}_d$.

So it's proved that the control system is global asymptotic stable.

5   Decoupling Control

Based on simple PID control in position loop and PI control in velocity loop, and FFC (feed forward control) is increased to reduce position error. In real-time, the results of dynamic control are used to compensate into 3 arms controller. The decoupling control structure of robot is below:

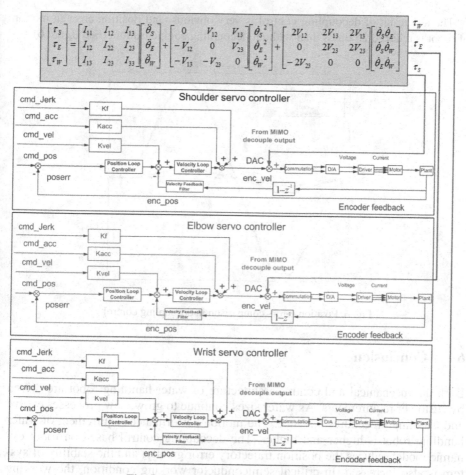

$$\begin{bmatrix} \tau_S \\ \tau_E \\ \tau_W \end{bmatrix} = \begin{bmatrix} I_{11} & I_{12} & I_{13} \\ I_{12} & I_{22} & I_{23} \\ I_{13} & I_{23} & I_{33} \end{bmatrix} \begin{bmatrix} \ddot{\theta}_S \\ \ddot{\theta}_E \\ \ddot{\theta}_W \end{bmatrix} + \begin{bmatrix} 0 & V_{12} & V_{13} \\ -V_{12} & 0 & V_{23} \\ -V_{13} & -V_{23} & 0 \end{bmatrix} \begin{bmatrix} \dot{\theta}_S^2 \\ \dot{\theta}_E^2 \\ \dot{\theta}_W^2 \end{bmatrix} + \begin{bmatrix} 2V_{12} & 2V_{13} & 2V_{13} \\ 0 & 2V_{23} & 2V_{23} \\ -2V_{23} & 0 & 0 \end{bmatrix} \begin{bmatrix} \dot{\theta}_S\dot{\theta}_E \\ \dot{\theta}_S\dot{\theta}_W \\ \dot{\theta}_E\dot{\theta}_W \end{bmatrix}$$

Fig. 5. Decoupling control structure of robot

We use the simulation parameters of robot in control, and the identification of these parameters is researched further in another paper.

Parameters of robot arms are listed in the following table.

m_1=5.14 kg	shoulder
m_2=3.87 kg,	elbow
m_3=2.0kg	wrist
I_1=0.05892343 Kg.m^2	
I_2=0.042337 Kg.m^2	
I_3=0.01477272 Kg.m^2	
R_1=0.21m	
R_2=0.21m	
R_3=0.21m	
L_1=0.091m	Shoulder CG to rotation centre
L_2=0.11m	Elbow CG to rotation centre
L_3=0.0542m	Wrist CG to rotation centre

The test result of decoupling control is very obvious, and position error in motion of robot is reduced 25 times than without decoupling compensation. (See figure 6)

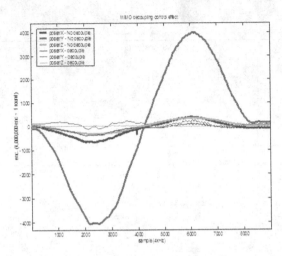

Fig. 6. Position errors comparison in decoupling control

6   Conclusion

Both the mechanical and control architecture of wafer handling robot are driven by customer requirements. As wafer sizes continue to grow, new processes evolve and handling procedures advance. The motion performance requirement of wafer handling robot is higher and higher. The decoupling control based on robot dynamic model reduces the position trajectory error greatly, and the stability of system is also increased. In critical semiconductor working condition, the working efficiency is enhanced obviously.

Reference

1. Armstrong, B., Khatib, O., Burdick, J.: The explicit dynamic model and inertial parameters of the PUMA 560 arm. In: IEEE International Conference on Robotics and Automation. IEEE (1986)

Motion Planning of Robot Arm with Rotating Table for a Multiple-Goal Task

Yanjiang Huang, Hao Ding, and Xianmin Zhang[✉]

Guangdong Provincial Key Laboratory of Precision Equipment and Manufacturing
Technology, South China University of Technology, Guangzhou, China
xmzhang@scut.edu.cn

Abstract. Motion planning of a manipulator system is important when using it
to complete a task. In this study, we consider a manipulator system with a robot
arm and a rotating table for a multiple-goal task. We first assign the goals
around the object into several clusters based on the face in the geometric shape
of the object. And then search for the shortest path for the goals in a cluster by
coordinating the motion of robot arm and the motion of rotating table based on
the particle swarm optimization (PSO). Finally connect the paths found in each
cluster. The collisions between the components of the manipulator system is
taken into account. The proposed method is compared to a method that only
uses nearest neighborhood algorithm (NNA) to coordinate the motion of the robot arm and the rotating table. The effectiveness of the proposed method is verified through a simulation with a set of tasks.

Keywords: Motion planning · Manipulator system · Multiple-goal task ·
Particle swarm optimization

1   Introduction

A manipulator system that consists of a robot arm and a rotating table has been commonly used to complete a multi-goal task, such as spot welding and inspection [1]. In
a multi-goal task, the manipulator end-effector has to reach several goals that are
defined by the position and orientation in the robot task space. In realizing a multi-
goal task by using the manipulator system with a robot arm and a rotating table, it is
important to coordinate the motion of the robot arm and the rotating table to increase
the productivity. However, motion planning of the robot arm with the rotating table is
complex since the manipulator system is redundant and the search space of motion
planning is large and nonlinear.

In previous studies, most researchers focused on the motion planning to find the
shortest path for the manipulator end-effector to minimize the task completion time.
Some researchers focused on the motion planning of a manipulator system for a multi-goal task [2, 3]. They divided the motion planning problem into two sub-problems:
the collision-free shortest path and the travelling-salesman problem. The probabilistic
roadmap method (PRM) and the sequential quadratic programming (SQP) were used
to solve the motion planning problem in [2] and [3], respectively. However, in these

© Springer International Publishing Switzerland 2015
H. Liu et al. (Eds.): ICIRA 2015, Part III, LNAI 9246, pp. 571–580, 2015.
DOI: 10.1007/978-3-319-22873-0_51

studies, the motion of manipulator was assumed to be collision-free, which is infeasible in real applications. PRM is considered as a state-of-the-art algorithm for solving motion planning because it can approximately represents the connectivity of the free configuration. Many researchers focused on the motion planning of robot based on PRM [4, 5]. In other studies, a phase-plane algorithm was used to minimize the trajectory path by considering the manipulator dynamics and constraints [6, 7]. In [8], a hybrid search algorithm that integrates the nearest neighbor algorithm (NNA) and the Dijkstra algorithm (DA) was used to deal with motion planning for a manipulator system with a 6-DOF manipulator and 1-DOF rotational positioning table. Recently, a multi-objective approach was proposed to solve the motion planning for redundant manipulators [9]. A constrained nonlinear programming incorporated into a sequential quadratic programming was proposed to realize the motion planning and control for redundant robots [10]. However, in these previous studies, some algorithms are computational time consuming, and some algorithms are complex.

In this study, we aim to realize the motion planning of a robot arm with a rotating table to minimize the task completion time for a multi-goal task. To realize the motion planning within a low computational time, we first classify the goals into several clusters, then search for the collision-free path for the goals in the same cluster, finally connect the collision-free paths in each cluster to obtain the minimal task completion time for the multi-goal task. The shortest collision-free path for the goals in the same cluster can be solved by many heuristics algorithms, which can be classified to single solution search algorithms (e.g., Simulated annealing, Tabu search), and population-based search algorithms (e.g., Particle swarm optimization, Genetic algorithm). The population-based algorithms concern a population of solutions at a time, which can overcome the drawback of single solution search algorithm by avoiding the local optimal. In this study, we propose the particle swarm optimization (PSO) to search for the shortest collision-free path for the goals in a cluster, because PSO can search for a good solution with fewer computational time [11, 12].

The problem is formulated in Section 2. The proposed method is described in Section 3. The simulation and discussion are presented in Section 4, and the conclusion is presented in Section 5.

2   Problem Formulation

This section describes the problem solved in this study, including the assumption, input parameters, objective function, constraints and design variables. As shown in Fig. 1, the robot manipulator system consists of a robot arm and a rotating table. The object is located on the rotating table. Serval goals are around the object. The end-effector of the manipulator has to reach the goals to perform some actions.

To solve the motion planning for the robot manipulator system with a robot arm and a rotating table, we maintain the following assumptions:

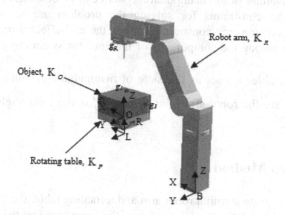

Fig. 1. A robot manipulator system with a robot arm and a rotating table. B is the base position of robot arm, L and O is base of positioning table and object that takes B as reference coordinate frame. R refers to rotational direction.

— The velocity of a robot joint and rotating table is constant when executing a task.
— The shape and size of an object, distribution of goals and goal order are known.
— The distance between the robot base and the rotating table base is known.
— The rotating table can rotate in $[-\pi, \pi]$.
— The specification of goals can be represented by position and orientation. Here, the orientation is the orientation of the end-effector required to perform some action.
— The object can be modelled as a rectangular box, and the goals around the object are outside the box.

The specifications of manipulator system (e.g., D-H parameters of robot arm, size of rotating table) and the specifications of the task (e.g., number of goals, position of goals) are set as input parameters.

The objective of the motion planning is to minimize the task completion time, which can be formulated as follows:

$$\text{Minimize} \quad T = \sum_{i=1}^{m} f(g_{i-1}, g_i) \tag{1}$$

$$f(g_{i-1}, g_i) = \max \left({}_{l \in \{1,...,6\}} \left\| \theta^l_{i-1} - \theta^l_i \right\| / V^{l,\max} , \left\| \theta^r_{i-1} - \theta^r_i \right\| / V^{r,\max} \right) \tag{2}$$

where, T is the task completion time that the end-effector of the manipulator from the initial position, passing through each goal. m is the number of goals that the end-effector has to pass through. $f(g_{i-1}, g_i)$ is the motion time of the robot joints and the rotating table from goal g_{i-1} to g_i. g_0 is the initial position. θ^l_i is the angle of robot joint l at goal g_i. $V^{l,\max}$ is the maximal speed of joint l. θ^r_i is the angle of the rotating table at goal g_i. $V^{r,\max}$ is the maximal speed of the rotating table.

The motion planning of the manipulator system can be considered as an optimization problem. The constraints for solving the problem are set as collision-free constraint. The collision-free constraint requires the end-effector reach to each goal without collision among the components of manipulator system (e.g., robot arm and rotating table).

The design variables are set as kinestate of manipulator system $q_i = \left(\theta_i^1 .... \theta_i^6, \theta_i^r\right)$, here, $\left(\theta_i^1 .... \theta_i^6\right)$ are the robot arm joint angles, θ_i^r is the joint angle of the rotating table.

3   Proposed Method

In the manipulator system with a robot arm and a rotating table, the travelling distance of the robot arm depends on the joint space of the robot arm and the rotating table. In this study, the configuration of manipulator system $q_i = \left(\theta_i^1 .... \theta_i^6, \theta_i^r\right)$ at i^{th} goal depends on the angle of rotating table θ_i^r. When the angle of rotating table is determined, the configuration of robot arm can be obtained based on the inverse kinematics (IK). Therefore, we can coordinate the motion of the rotating table to find the shortest collision-free path to pass through each goal. It is difficult to obtain the shortest collision-free path, because the angle of rotating table can be any value from $-\pi$ to π at each goal, which leads to a large and nonlinear joint space of robot arm. To realize the motion planning for the manipulator system within a reasonable computational time, we first classify the goals into several clusters, and then search for the collision-free path for the goals in the same cluster by coordinating the motion of robot arm and the motion of the rotating table, finally connect the path found in each cluster. The collision detection is taken into account in this study. Because the collision is easy to occur when the rotating table is in some angles, we limit the angle of the rotating table in some range. It is reasonable to do this by considering the computational time cost. To simplify the problem, the goal order and cluster order are known in advanced. The details of goal clustering, motion coordination, and collision detection are described later in this section.

1) Goal Clustering

We assign the goals into several clusters based on their topological location. Each cluster is determined by the face in the geometric shape of the object. Every goal is associated to a cluster $\{C_1 ... C_5\}$, as shown in Fig. 2. It is reasonable to assign the goals to several clusters and search for the shortest collision-free path in a cluster, because the number of dimension can be reduced by using this method. Because the goals in the same cluster are near, the motion time of robot arm and rotating table from a goal to another in the same cluster may be shorter than that moving from a goal to another in different clusters. Furthermore, the motion of robot arm between goals in the same cluster has a greater chance to avoid collision than goals from different clusters.

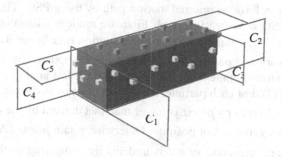

Fig. 2. Cluster of goals. The clusters $\{C_1...C_4\}$ correspond to the sides of the object while C_5 corresponds to the top of the object. Each goal can only be assigned to one cluster.

2) Motion Coordination

We apply the particle swarm optimization (PSO) to search for the shortest collision-free path in a cluster. PSO is a population-based heuristic algorithm, which is inspired by the movements of a block of birds or fishes when searching for food [13]. The PSO is an evolutionary computation algorithm and in each generation, the velocity and position of a particle are updated by using the following equations:

$$V_i^{h+1} = wV_i^h + c_1r_1(pbest_i^h - X_i^h) + c_2r_2(gbest^h - X_i^h) \tag{3}$$

$$X_i^{h+1} = X_i^h + V_i^{h+1} \tag{4}$$

Where, $pbest_i^h$ is the best-ever position of a particle; $gbest^h$ is the global best position in swarm; h is generation number; V_i^h is the velocity of the particle i at the hth generation, X_i^h is the position of the particle i at the hth generation, $i = 1,2,...,m$; w is an inertial weight; c_1 and c_2 are cognitive and social scaling parameters; r_1 and r_2 are uniform random numbers between 0 and 1. The PSO algorithm is terminated with a maximal number of generations is reached or the swarm's best position cannot be improved further after some continuous generations.

In the implementation of PSO to search for the configurations of robot arm and rotating table, we set the design parameter as the angle of rotating table at each goal in a cluster. The search space for the goals in a cluster is shown in Fig. 3. The number of possible paths to make the end-effector pass through each goal in a cluster is infinite. It is impractical to search for the shortest collision-free path by an exhaustive method. In this study, we set the number of dimension for a particle is the number of goals in a cluster. Therefore, a particle represents a motion path of the end-effector. The motion time of the robot arm and the rotating table from the first goal to the last goal in a cluster can be calculated based on Equation (2). We first search for a path based on nearest neighborhood algorithm, then limit the search range based on the obtained path. After that, search for an optimal motion path by using PSO. The process of implementation of PSO is shown in Fig. 4. First, we randomly initialize particles with different positions. The solution with minimal motion time is set as *gbest* . In one iteration, the velocity and position for each particle are updated by Equation (3) and (4). After that, the motion time from the first goal to the last goal in a cluster is calculated by Equation (2). For each particle, the current position is compared to its *pbest* . If the motion time derived by *pbest* is larger than that derived by the current position, then update *pbest* by the current position. Otherwise, retain *pbest* . After all particles are evaluated in one iteration, *gbest* is updated by comparing to the best *pbest* in one iteration. We tune the inertial weight and maximal velocity of the particle based on the Linearly Decreasing V_{max} method (LDVM) [14] when *gbest* is not improved in the continuous iterations. We terminate the process of motion planning for the goals in a cluster when the maximal number of iterations is reached.

3) Collision Detection

We conduct the collision detection based on oriented bounding boxes (OBBs). We assume that the manipulator system components can be approximately modelled as OBBs, which are defined by the size and orientation. The collision among the manipulator system components can be detected by checking the overlap of the OBBs. There is no collision among the manipulator system components if no collision occurs among the OBBs. Two OBBs are projected onto a potential separating axis. If the projections do not overlap, there is no collision between the two OBBs. If the projections do overlap, the OBBs will be projected onto other potential separating axis and the overlap among projections will be rechecked. 15 potential separating axes can be generated for two OBBs based on the separating axis theorem. A detailed discussion is presented in [15].

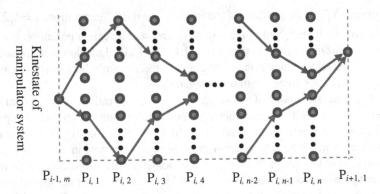

Fig. 3. Search space of motion planning for the goals in a cluster. *n* is the number of goals in a cluster. $P_{i,n}$ is the nth goal in the ith cluster. $P_{i-1,m}$ is the last goal in the $(i-1)^{th}$ cluster. The red arrow and blue arrow mean two different motion paths.

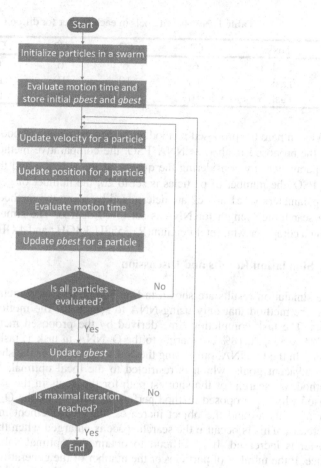

Fig. 4. Implementation of PSO for motion planning for the goals in a cluster

4   Simulation, Result and Discussion

1)   Simulation Parameter Setting

We conducted the simulation based on the model of a real robot named VS_6577, which is a robot produced by DENSO WAVE INCORPORATED. The specifications of VS_6577 (e.g., D-H parameters and joint velocity limit) can be found in [16]. The distance between the robot base and the rotating table base is set to 600mm. The table size (length, width, height) is set to (400mm, 400mm, 100mm). The initial position and orientation of the end-effector is set to (500, 0, 600, 0, 180, 0) referred to the robot base.

The object size (length, width, height) is set to (400mm, 400mm, 200mm). We tested three tasks with different number of goals in the simulation. The goals around the object can be assigned into different clusters. The number of goals in each cluster for different tasks is shown in Table 1.

Table 1. Number of goals in each cluster for different tasks

Tasks	C1	C2	C3	C4	C5
Task 1	6	6	6	6	6
Task 2	10	10	10	10	0
Task 3	10	10	10	10	10

We compare the proposed method to the nearest neighborhood algorithm (NNA). We call the proposed method as NNA_PSO, the comparative method as O_NNA. We set the parameters by considering the quality of the solution and the computational time. For PSO, the number of particles is set to 20, the number of generations is set to 100, the parameters w, c1 and c2 are determined by using the method described in [14, 17]. The search step length for NNA is set to 10 degree. The simulation is conducted by using a computer with Intel Pentium(R) 3558U 3.4GHz and 4 GB memory.

2) Simulation Results and Discussion

The simulation results are shown in Fig. 5. The proposed method can perform better than the method that only using NNA to search for the motion path in all the three tasks. The task completion time derived by the proposed method was shortened by 9.38%, 6.83%, 6.18% comparing to the O_NNA in task 1, task 2, and task 3, respectively. In the O_NNA, only using the NNA to search for the shortest path between the two adjacent goals, which is restricted to the local optimal. While in our proposed method, we search for the shortest path for the goals in the same cluster. This is the reason why the proposed method performed better than the O_NNA. When the number of goals around the object increases, the improvement in task completion time decreases. This is because the search space is enlarged when the number of goals in a cluster is increased. It is difficult to obtain the optimal solution in a large search space. If the number of particles or the number of the generations is increased, a better solution may be found. However, more computational time is required when the number of particles or the number of generations is increased. Therefore, the quality of the solution depends on the parameters of the PSO. In this study, we set the parameters for the PSO based on our previous studies by considering the quality of the solution and the computational time.

From these results, we can find that the relationship between the increment of task completion time and the increment of goals is nonlinear. When the goals are located in more faces in the geometric shape of the object, more task completion time is required. This is because more motion time is required for the robot arm moves from a face to another (i.e., the robot arm moves from a cluster to another in this study). Therefore, the optimization of the order for the clusters or the order of the goals in a cluster may be used to shorten the motion time for the robot arm moves from a cluster to another. This can also shorten the task completion time, as described in [1].

Considering the computational time, the proposed method requires longer computational time than that only using NNA, because the proposed method searches for the solution in the global search space. Although the O_NNA requires low computational time, it is easy to fall into local optimal. In a real application, an

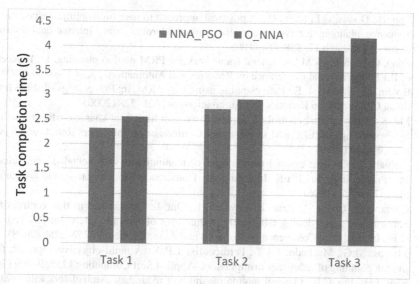

Fig. 5. Task completion time derived by the proposed method and comparative method for different tasks.

appropriate method can be chose based on the constraints in the quality of the solution and the computational time.

5   Conclusion

In this study, we proposed a method that integrates NNA and PSO to realize the motion planning for a robot manipulator system with a robot arm and a rotating table in a multi-goal task. The proposed method is verified to be effectiveness and practical by comparing to the method that only used NNA through a simulation. The task completion time derived by the proposed method is shorted. In the future, the real time motion planning for a robot manipulator system can be taken into account.

Acknowledgment. This work was partially supported by the Fundamental Research Funds for the Central University (Fund No. 20152M004).

References

1. Gueta, L.B., Chiba, R., Arai, T., Ueyama, T., Ota, J.: Practical point-to-point multiple-goal task realization in a robot arm with a rotating table. Advanced Robotics **25**, 717–738 (2011)
2. Saha, M., Roughgarden, T., Latmbe, J.-C.: Planning tours of robotic arms among partitioned goals. International Journal of Robotics Research **25**, 207–224 (2006)

3. Cao, B., Dodds, G.I., Irwin, G.: A practical approach to near time-optimal inspection-task-Sequence planning for two cooperative industrial robot arms. International Journal of Robotics Research **17**, 858–867 (1998)
4. Song, G., Amato, N.M.: A general framework for PRM motion planning. In: Proceedings of IEEE International Conference on Robotics and Automation, pp. 4445–4450 (2003)
5. Bohlin, R., Kavraki, L.E.: Path planning using lazy PRM. In: Proceedings of IEEE International Conference on Robotics and Automation, pp. 521–528 (2000)
6. Ma, S.: Time optimal control of manipulators with limit heat characteristics of actuators. In: Proceedings of IEEE/RSJ International Conference on Intelligent Robots and Systems, pp. 338–343 (1999)
7. Zlajpah, L.: On time optimal path control of manipulators with bounded joint velocities. In: Proceedings of IEEE International Conference on Robotics and Automation, pp. 1572–1577 (1996)
8. Gueta, L.B., Chiba, R., Arai, T., Ueyama, T., Ota, J.: Coordinated motion control of a robot arm and a positioning table with arrangement of multiple goals. In: Proceedings of IEEE International Conference on Robotics and Automation, pp. 2252–2258 (2008)
9. Marcos, M.G., Machado, J.A.T., Perdicoulis, T.P.A.: A multi-objective approach for the motion planning of redundant manipulators. Applied Soft Computing **12**, 589–599 (2012)
10. Kim, J.H., Joo, C.B.: Optimal motion planning of redundant manipulators with controlled task infeasibility. Mechanism and Machine Theory **64**, 155–174 (2013)
11. Kennedy, J., Eberhart, R.C.: Swarm intelligence. Morgan Kaufmann Publishers, San Francisco (2001)
12. Hassan, R., Cohanim, B., de Weck, O., Venter, G.: A comparison of particle swarm optimization and the genetic algorithm. In: Proceeding of the 1st AIAA Multidisciplinary Design Optimization Specialist Conference, pp. 1–13 (2005)
13. Fourie, P.C., Groenwold, A.A.: The particle swarm optimization algorithm in size and shape optimization. Structural and Multidisciplinary Optimization **23**, 259–267 (2002)
14. Gottschalk, S., Lin, M.C., Manocha, D.: OBB-tree: a hierarchical structure for rapid interference detection. In: Proceeding of ACM Annual Conference on Computer Graphics and Interactive Techniques, pp. 171–180 (1996)
15. Website of DENSO WAVE INCORPORATED. http://www.denso-wave.com/en/robot/product/five-six/vs__Spec.html
16. Huang, Y., Gueta, L.B., Chiba, R., Arai, T., Ueyama, T., Ota, J.: Selection of manipulator system for multiple-goal task by evaluating task completion time and cost with computational time constraints. Advanced Robotics. **27**, 233–245 (2013)

Correction to: Stiffness Control of Soft Robotic Manipulator for Minimally Invasive Surgery (MIS) Using Scale Jamming

S.M.Hadi Sadati, Yohan Noh, S. Elnaz Naghibi, Kaspar Althoefer,
and Thrishantha Nanayakkara

Correction to:
**Chapter "Stiffness Control of Soft Robotic Manipulator
for Minimally Invasive Surgery (MIS) Using Scale Jamming"
in: H. Liu et al. (Eds.): *Intelligent Robotics and Applications*,
LNAI 9246, https://doi.org/10.1007/978-3-319-22873-0_13**

The book was inadvertently published with an incorrect version of an author's name in Chapter 13 as "Althoefer Kaspar" whereas it has been updated as "Kaspar Althoefer".

The updated version of this chapter can be found at
https://doi.org/10.1007/978-3-319-22873-0_13

Author Index

Printed in the United States
By Bookmasters